FOUNDATIONS OF
DISCRETE MATHEMATICS

Second Edition

FOUNDATIONS OF DISCRETE MATHEMATICS

Albert D. Polimeni
H. Joseph Straight
State University of New York — Fredonia

Brooks/Cole Publishing Company
Pacific Grove, California

Consulting Editor: Robert L. Wisner

Brooks/Cole Publishing Company
A Division of Wadsworth, Inc.

Printed in the United States of America
10 9 8 7 6 5 4 3 2 1

Library of Congress Cataloging in Publication Data
Polimeni, Albert D., [date]
    Foundations of discrete mathematics.

    1. Mathematics.   I. Straight, H. Joseph
    II. Title.
QA39.2.P67   1989   510   89-22147
ISBN 0-534-12402-X

Sponsoring Editor: Sue Ewing/Jeremy Hayhurst
Editorial Assistant: Nancy Champlin
Production Editor: Linda Loba
Manuscript Editor: Carol Reitz
Interior and Cover Design: Flora Pomeroy
Art Coordinator: Sue C. Howard/Cloyce Wall
Typesetting: Science Typographers, Inc.
Cover Printing: The Lehigh Press Co.
Printing and Binding: Arcata Graphics—Fairfield

Preface

This book is intended as an introduction to discrete mathematics for sophomore-level mathematics and computer science majors. It is meant to be a first encounter with abstract mathematics and, at the same time, an introduction to those topics in mathematics basic to the study of computer science. It can be used for a course that is devoted exclusively to mathematics; indeed, the text material and exercises on computer science can easily be omitted. To facilitate this, the exercises pertaining to computer science concepts have been placed at the end of the exercise sets. If the book is used for such a course, then we recommend a minimum prerequisite of one year of college-level mathematics, including at least one semester of calculus. If one prefers to make use of programming skills and other topics from computer science, then a programming course in a high-level language, such as Pascal, should also be required. Examples in this book that involve programming are most often presented in pseudocode or Pascal.

We use the text for a one-semester (four-credit) course in mathematics and feel that it is important to establish a solid mathematical foundation from which one can smoothly proceed to more advanced courses in mathematics and computer science. For mathematics majors, we find that having had a properly paced treatment of the rudiments of naive set theory, residue class arithmetic, elementary properties of divisibility, mathematical induction, relations, functions, elementary counting techniques, and binary operations provides a much needed edge for courses in linear algebra, abstract algebra, combinatorics, geometry, and analysis. Many of the same ideas are relevant in computer science, for courses in data structures, database systems, algorithms, the theory of computation, and artificial intelligence. Hence, aside from any specific algorithms or applications, it is our hope that a computer science major who completes our course will have developed a good understanding of both the language of mathematics and the fundamental ideas promoted in the course. Such students will have many opportunities, in subsequent course work, to analyze and code specific algorithms.

In a one-semester, four credit course, we find that we can reasonably cover Chapters 1–3, with the exception of the material on program verification (in 1.6) and proofs of program correctness (in 3.5 and 3.6). In Chapters 4 and 5, we generally omit Sections 4.5 and 5.5. Finally, in Chapters 6 and 8, we cover Sections 6.1–6.4, 6.6, 8.1, and as much of 8.6–8.8 as time permits.

We now feel that this text contains enough material for a one-year sequence. Here we would naturally include inclusion-exclusion, recursive algorithms, a good selection of topics from the chapter on graph theory, and further ideas from algebraic structures. Some possible outlines for such a course are presented in the following table:

Chapter	Sections
1	all
2	all
3	all
4	all
5	5.1–5.4
6	6.1–6.7
7	7.1–7.5, 7.9 or 7.1, 7.2, 7.6–7.9
8	8.1, 8.2, 8.5–8.8 or 8.1–8.6

SECOND-EDITION CHANGES

The decision to revise the first edition was based primarily on comments and suggestions solicited from users and selected reviewers. We raised specific questions regarding several aspects of the text and, in most cases, responses were consistent with our own views. We also made use of the report of the *Committee on Discrete Mathematics in the First Two Years*, published by the Mathematical Association of America. The major issues dealt with expanding the coverage on combinatorics, restructuring the chapters on algebraic structures and graph theory, and adding, throughout the text, more exercises of a routine nature. Although most of our efforts were devoted to these areas, other changes have also been effected.

CHAPTER REVISIONS

Chapters 1 through 4 show no significant change. In Chapter 1, Logic, for example, we introduce the notion of a set, along with the standard notations for well-known sets, and define the set relations. This makes it much easier to converse throughout the chapter and also allows one to deal more effectively with restricted quantifiers. In Chapter 2, Set Theory, the student is given the opportunity to use the ideas of proof that he or she has learned in Chapter 1. Since this is the first encounter with problems that use the wording "Prove that...," we have deliberately tried to be careful and somewhat formal in our approach to proofs. This idea is relaxed as one progresses in the text. In Chapter 3, Number Theory and Mathematical Induction, we have made a special effort to include a good number of elementary and instructive exercises. In this chapter students are given the opportunity to develop number theoretic algorithms and, if desired, to implement them in pseudocode or Pascal. In Chapter 4, Relations, we make use of matrix properties of relations. An appendix on matrices has been added at the end of the text.

In Chapter 5, Functions, we have added sections on the big-O notation and set equivalence. The big-O notation was written into the text to furnish the computer science major with an elementary treatment of how one estimates the running time of an algorithm. Set equivalence is more mathematical and is intended to extend the idea of cardinal equivalence to infinite sets.

Chapter 6, Combinatorics, is a new chapter and includes a selection of standard topics. Section 6.3 on permutations and combinations appeared in the second chapter of the first edition and the material on recursion and iteration was formerly in Chapter 5. Among other things, we have added material on the pigeonhole principle, the binomial theorem, and the principle of inclusion-exclusion. Also, the treatment of recurrence relations and recursive algorithms has been expanded.

Chapter 7, Graph Theory, has been changed significantly in that there is generally more emphasis on algorithms and directed graphs. For instance, Warshall's algorithm for determining the reachability matrix of a digraph, as well as the topological sort and depth-first-search algorithms, are included. We have also added sections on eulerian graphs and hamiltonian graphs, and a section on finite state automata. One can provide a basic introduction to graphs by covering Sections 7.1, 7.2 (Eulerian graphs), and 7.6 (Trees). A unit on directed graphs can be given by starting with Section 7.3 (an introduction), followed by 7.4 (Subdigraphs, Reachability, and Distance) and 7.5 (Acyclic Digraphs and Rooted Trees). Sections 7.7 (Hamiltonian Graphs), 7.8 (Vertex Coloring and Planar Graphs), and 7.9 (Finite-State Automata) are more advanced in nature.

In Chapter 8, Algebraic Structures, the discussion of groups has been softened and spread over three sections, as compared to one compressed section in the first edition. Some rather fundamental ideas, which were formerly in the exercise sets, have now become text material. A treatment of normal forms for Boolean switching functions has been inserted after the section on Boolean algebras, and Section 8.8 deals with switching circuits and logic gates. The layout of the chapter allows one to easily pick a small selection of topics to cover. For instance, in our case we try to cover the section on binary operations and the sections on Boolean algebras.

MATHEMATICAL STYLE

Just as with the first edition, we have made a serious effort to maintain a level of exposition that is suitable for a sophomore-level student in mathematics or computer science. At the same time, we have made a special attempt to respect rigor in the mathematics. It is essential at this level that the student learn to read and do mathematical proofs. This is initially dealt with in Chapter 1, where the formal language of mathematics and various proof techniques are covered. In the remaining chapters, theorems are clearly displayed and stated. If the proof of a theorem is given and the method used is other than direct, then the particular proof technique to be used is specified. In all cases we strived to make the presentation both precise and readable.

EXAMPLES

Users and reviewers of the first edition considered the examples to be a strong feature of the text. In this edition we have included most of the examples given in the first edition, with some corrections and improvements; in addition, we have expanded the list of examples in areas where there was a clear need. Our belief has been that there should be examples to illustrate each new concept and to show how theorems and algorithms are applied. Just as in the first edition, we continue to phrase most examples in problem form and encourage students to participate in the development of a solution, to attempt to solve these problems independently and check their work with the given solution.

EXERCISES AND CHAPTER PROBLEMS

Although most users and reviewers felt that there is a good selection of exercises and chapter problems in the first edition, many of them stated that there was a strong need for more routine and elementary exercises that illustrate concepts. This was especially noted in Chapter 3 on number theory and induction. We made this a priority for the current edition and have added many new exercises that we hope will allow the student to move progressively to the more difficult problems in the text. As with the first edition, there are exercises at the end of each section and problems at the end of each chapter; the exercises focus on the material covered in a section, while chapter problems are intended not only for review but to deepen the student's exposure to key ideas. Some problems integrate concepts from several previous chapters or invite the student to explore some aspect of a topic not covered in detail in the text. Exercises and problems that are especially challenging are prefixed with a star (★).

ANSWERS TO EXERCISES AND PROBLEMS

Answers to selected exercises and problems are given in Appendix B, with answers to most of the remaining exercises and problems included in the Instructor's Manual. The rational used in selecting exercises and problems for which to provide answers was twofold: We tried to achieve a balance between the needs of the students, who want to check whether they have done problems correctly, and those of the instructor, who may wish to assign some problems (for which no solution is given) to be collected and graded. With the current emphasis on problem solving and writing in mathematics, we felt that these latter problems should be of a type requiring a more intricate analysis or the writing of a proof.

We would like to thank the following reviewers for their helpful comments and suggestions: Professor Jonathan Barron, Eastern College, St. Davids, Pennsylvania; Professor David Berman, University of New Orleans; Professor Joseph Buoni, Youngstown State University, Youngstown, Ohio; Professor Elwyn Davis, Pittsburgh State University, Pittsburgh, Kansas; Professor Joseph Gallian, University of Minnesota, Duluth; Professor Charles Parry, Virginia Polytechnic Institute and State University; Professor George Peck, Arizona State University; Professor Wayne Powell, Oklahoma State University; Professor Tina Straley, Kennesaw State College, Marietta, Georgia; and Professor Thomas Upson, Rochester Institute of Technology, Rochester, New York.

We wish to express our sincere thanks to our colleagues Nancy Boynton, Fred Byham, and Richard Dowds who used the text and made many valuable suggestions for this revision. In addition, we would like to express our sincere appreciation to our editors, Jeremy Hayhurst and Sue Ewing, for their valuable assistance and patience throughout the preparation of this text.

Albert D. Polimeni
H. Joseph Straight

For the Student

What is discrete mathematics? When addressing this question, one usually draws a contrast between discrete mathematics and *continuous mathematics*. We shall, by illustration and intuition, attempt to convey some feeling for both discrete and continuous mathematics. For instance, the following are examples of problems from discrete mathematics:

1. In how many ways can 12 different problems be distributed among 20 students if no student gets more than one problem and each problem is assigned to at most one student.

2. How many social security numbers have no repeated digits?

3. Given n points in the plane, no three collinear, how many triangles do they determine?

4. Find all integers n such that $n^2 + 1$ is a multiple of 11.

5. Prove that $1^2 + 2^2 + 3^2 + \cdots + n^2 = n(n + 1)(2n + 1)/6$ for all positive integers n.

6. (Travelling Salesman Problem) A salesman must go on a sales trip, beginning at his hometown H, visiting each of the cities A, B, C, D, E, F, and G. Under what conditions is it possible for him to do so in such a way that he visits each of the cities A, B, C, E, F, and G exactly once and then returns home.

7. If possible, design an algorithm (step-by-step process) that provides a solution to the Travelling Salesman Problem.

Notice that each problem deals with a finite collection of objects, a finite process, the positive integers $1, 2, 3, \ldots$, or all the integers $\ldots -3, -2, -1, 0, 1, 2, 3, \ldots$. Any problem that involves a finite collection of objects or a finite process is a problem in discrete mathematics. More generally, any problem involving any collection of integers is a problem from discrete mathematics. Formally, one should think of discrete mathematics as the mathematics of *discrete sets*. If we think of numbers as points of a coordinate line, then one can think of a discrete set of numbers as a set S of points on the line having the following property

Each point x of S is contained in some interval containing no other points of S.

For instance, the set of integers is a discrete set but the set of rational numbers is not. An interval can be thought of as a "continuous set" of points of the line, and any problem or process that deals with continuous sets can be said to fall in the area of "continuous mathematics." For example, the problems

1. Find the absolute maximum and minimum values of the function

$$f(x) = x^3 - x \text{ on the given closed interval } [-2, 2].$$

2. Evaluate the definite integral

$$\int_{-i}^{3} x^4 + x^3 + 3x + 2 \, dx$$

belong to the realm of continuous mathematics. However, the student may recall that there are methods for approximating the value of a definite integral by a Riemann sum (Simpson's Rule provides one such method). A Riemann sum is a finite sum of numbers and, hence, falls within the scope of discrete mathematics. One can use the power of a computer to rapidly compute the value of such a sum and thus obtain a good approximation to the exact value of the definite integral. In many and varied cases one can use discrete mathematics to approximate solutions to problems from continuous mathematics. In this text you will encounter ideas from areas of mathematics that are basic to both discrete and continuous mathematics; however, in the larger share of cases, we will deal with problems and processes involving discrete sets.

Contents

FOUR   Relations   134

FIVE   Functions   183

SIX   Combinatorial Mathematics   222

O N E

Logic

INTRODUCTION
TO LOGIC AND SETS

Like other scientists, mathematicians engaged in research are primarily interested in investigating assertions and determining whether a given assertion is true or false. Research problems in mathematics may arise in various ways. For example, it often happens that interesting problems arise as a result of seminar discussions with colleagues, poring over research articles, or attempts to solve other problems. Once an assertion is clearly stated, a mathematician uses formal training and acquired knowledge and experience to try to resolve it. Valuable insight and understanding are gained by building examples that illustrate the ideas involved. Of course, if an example denies the assertion, then one need go no further; the assertion is false. If all the examples considered seem to support the assertion, along with other related information (such as computer-generated data and previously established results), there may be good reason to suspect that the assertion is true. Then, to be called a mathematical fact, or *theorem*, the result must be logically deduced from known facts; that is, it must be proved.

The logic of mathematics furnishes a set of ground rules that govern the development of a mathematical proof. These rules allow us to determine whether a proposed proof is valid. They are one aspect of proof that can be learned easily; the rest is intuition, creativity, imagination, and instinct.

The deductive method is important in computer science as well. Here, there is great interest in the process of designing, coding, and testing programs, particularly large, complicated programs. Ideally, the rules of logic are used in this process, with the aim of ensuring program correctness. Indeed, one often is required to give or read a proof that a given algorithm "works." There is also a relatively new programming methodology, called logic programming, that explicitly embodies the ideas of mathematical logic. A popular language for logic programming is PROLOG. It is widely used to program expert systems—programs that simulate the deductive analysis of the human expert in some narrow domain, such as medical diagnosis.

Emphasis in this chapter is placed on certain common forms of exposition and reasoning, apart from any particular applications. A student who learns these well finds it easier to follow the line of reasoning in advanced mathematics and computer science courses. This allows the student to focus on the content of a course and not be distracted by the logical forms being used.

Hereafter the term *argument* is used in its mathematical sense, as a logical discussion that establishes the validity of some mathematical fact. We ask, then, What is the logic of an argument? Crudely put, the logic of an argument is what is left over when the particular meaning of the argument is ignored. In other words, the logic of an argument is its form or syntax.

A similar distinction is made concerning the statements in a programming language like Pascal. These statements must obey certain syntactic rules, apart from their actual semantic interpretation. For example, consider the assignment statement

```
SUM := SUM+NEXT
```

We can say that this statement is syntactically correct but, when taken out of the context of a particular program, its meaning is not clear.

Perhaps an example will help to clarify the preceding comments.

Example 1.1   Consider the following argument about Randy, who is a student in the discrete mathematics course at a particular college.

> If Randy has the ability and works hard, then Randy is successful in this course. Therefore, if Randy is not successful in this course, then Randy does not have the ability or Randy does not work hard.

This argument is symbolized by letting p, q, and r represent the following statements:

> p: Randy has the ability.
>
> q: Randy works hard.
>
> r: Randy is successful in this course.

The logical form of the argument is then represented as follows:

> If p and q, then r. Therefore, if not r, then not p or not q.

It is easy to think of other arguments with the same logical or abstract form, such as:

> If the Red Sox hit well and avoid injuries, then they will win the pennant. Therefore, if the Red Sox do not win the pennant, then they did not hit well or they did not avoid injuries.   □

Note that the abstracted form of an argument such as the one in the preceding example is not bound to any particular content. Logic provides a means for validating such an abstract form in a way that is independent of the truth or falsity of the constituent statements represented. This is done through the use of "symbolic logic" and the construction of "truth tables." We discuss how this is done later in the chapter.

It is important to understand the meaning of the term *statement* as it is used in symbolic logic. Sentences such as

> Everyone in this class gets an A grade.

or

> The number 72 is positive.

are "declarative sentences"—each makes an assertion. On the other hand, the question

> What time is it?

and the exclamation

> Holy cow!

are not declarative sentences. For purposes of mathematical logic, our interest is in declarative sentences that are either true or false; these are called *statements* or *propositions*. If a statement is true, then its truth-value is denoted T, whereas the truth-value of a false statement is denoted F.

For instance, each of the sentences

> The quotient obtained when 7 is divided by 3 is an odd integer.

and

> The integer 6 is a factor of the product of 117 and 118.

is a statement. In fact, the first is false and the second is true.

Next consider the sentences

> The number x is positive.

and

> He is a baseball pitcher.

These are not statements because, as they are presented, we cannot determine the truth-value of either one. If we replace x by -3 in the first sentence, then we obtain the (false) statement

> The number -3 is positive.

Similarly, we may replace "He" in the second sentence by "Fernando Valenzuela" to obtain the (true) statement

> Fernando Valenzuela is a baseball pitcher.

Symbols like the x in the example above are called *variables*; such symbols are used to represent any one of a number of possible values. We say more later about sentences with variables that become statements when the variables are given particular values.

Lest the student be misled, it should be pointed out that some sentences with variables are statements. For instance, the sentence

> For all x, if $x > 2$, then $x^3 > 8$.

is a (true) statement. We discuss statements like this in Section 1.4.

We also accept as statements sentences like

> The billionth digit in the decimal expansion of $\sqrt{2}$ is 7.

or

> Every even integer greater than 2 can be expressed
> as the sum of two primes.

We take the attitude that such sentences are either true or false, although we may not know the truth-value at the present time. The truth-value of the first sentence can, at least in theory, be computed. The second sentence is an example of a mathematical *conjecture*; there exists a great deal of empirical evidence for its truth, but as yet it has not been proven.

We have said that an understanding of logic is central to the process of doing mathematics. Another central idea that we shall use throughout is that of a set. A set can be thought of as a collection of objects. The Greek alphabet, a baseball team, and the Euclidean plane are all examples of sets. The Greek alphabet consists of letters, a baseball team is composed of its players, and the Euclidean plane is made up of points. In general, the objects that make up a set are referred to as its *elements* or *members*. It is common to think of the terms *set* and *member* as undefined or primitive notions. Most of us have an intuitive feeling (based on examples like those just given) about what these terms mean. We use the term *collection* interchangeably with *set*.

Sets are the building blocks for most mathematical structures. For example, Euclidean plane geometry is based on the interpretation of the Euclidean plane as a set of points. Other important examples arise in the field of abstract algebra, which studies the properties of sets on which one or more "binary operations" are defined. For example, ordinary addition is a binary operation on the set of positive integers. Because the associative law

$$(x + y) + z = x + (y + z)$$

holds for all positive integers x, y, and z, we say that the set of positive integers is a "semigroup." Another example of a semigroup comes from the area of computer science. Consider the set of alphanumeric character strings under the operation & of catenation. It is not difficult to see that this operation satisfies the associative law; for instance,

```
(`MATH`&`E`) &`MATICS`=`MATHE`&`MATICS`
                    =`MATHEMATICS`
                    =`MATH`&`EMATICS`
                    =`MATH`& (`E`&`MATICS`)
```

Certain special sets of numbers are used frequently in this book, so we adopt some special notations for them. The numbers $1, 2, 3, \ldots$ are called the *positive integers* (or *natural numbers*), and the notation $\mathbb{N}$ is used to denote the set of positive integers. The positive integers together with 0 and the negative integers $-1, -2, -3, \ldots$ form the set of *integers*; this set is denoted by $\mathbb{Z}$. A *rational number* is any number expressible in the form m/n, where m and n are integers and $n \neq 0$. For example, $2/3, -5/11$, $17 = 17/1$, and $0.222 \ldots = 2/9$ are rational numbers. The set of rational numbers is denoted by $\mathbb{Q}$. Numbers like $\sqrt{2}, \sqrt[3]{5}$, and π are not rational; in general, such numbers are called *irrational numbers*. The rational numbers and the irrational numbers together make up the set of *real numbers*, which is denoted $\mathbb{R}$. To repeat, then, we adopt the following notational conventions:

$$\mathbb{N} = \text{the set of positive integers}$$
$$\mathbb{Z} = \text{the set of integers}$$
$$\mathbb{Q} = \text{the set of rational numbers}$$
$$\mathbb{R} = \text{the set of real numbers}$$

The basic relationship between an element x and a set A is that of membership. If x is an element of A, then we write $x \in A$, and if x is not an element of A, we write $x \notin A$. For instance, $-3 \in \mathbb{Z}$, $-3 \notin \mathbb{N}$, $\sqrt{2} \in \mathbb{R}$, and $\sqrt{2} \notin \mathbb{Q}$.

Except for certain special sets like $\mathbb{R}$, sets are denoted in this book by uppercase italic letters like A, B, C, and elements by lowercase italic letters like a, b, c.

If a set A consists of a "small" number of elements, then we can exhibit A by explicitly listing its elements between braces. For example, if A is the set of odd positive integers less than 16, then we write

$$A = \{1, 3, 5, 7, 9, 11, 13, 15\}$$

However, some sets contain too many elements to be listed in this way. In many such cases, the "three-dot notation" is used to mean "and so on" or "and so on up to," depending on the context. For instance, the set $\mathbb{N}$ can be exhibited as

$$\mathbb{N} = \{1, 2, 3, \ldots\}$$

the set $\mathbb{Z}$ as

$$\mathbb{Z} = \{\ldots, -3, -2, -1, 0, 1, 2, 3, \ldots\}$$

and the set B of integers between 17 and 93 as

$$B = \{17, 18, 19, \ldots, 93\}$$

Often a set A is described as consisting of those elements x in some set B that satisfy a specified property. As an example, let E be the set of even integers. Then we may write

$$E = \{\ldots, -4, -2, 0, 2, 4, \ldots\}$$

or

$$E = \{x \in \mathbb{Z} \mid x \text{ is even}\}$$

Here the symbol $\mid$ translates as "such that" or "for which."

Note that every positive integer is an integer, every integer is a rational number, and every rational number is a real number. In general, for two sets A and B, it is possible that each element of A is also an element of B.

DEFINITION 1.1

A set A is called a *subset* of a set B, denoted $A \subseteq B$, provided every element of A is also an element of B.

Special types of subsets of $\mathbb{R}$, called *intervals*, are used frequently in this book; this is a good point at which to review the standard notations for them. Let a and b be real numbers with $a < b$. Then

$$(a, b) = \{x \in \mathbb{R} \mid a < x < b\}$$
$$[a, b] = \{x \in \mathbb{R} \mid a \le x \le b\}$$
$$[a, b) = \{x \in \mathbb{R} \mid a \le x < b\}$$
$$(a, b] = \{x \in \mathbb{R} \mid a < x \le b\}$$

Undoubtedly, this notation has been encountered in previous courses. The interval (a, b) is called an *open interval* and $[a, b]$ is called a *closed interval*, while both $[a, b)$ and $(a, b]$ are called *half-open intervals*. In each case, a and b are called the *endpoints* of the interval. Some other intervals are

$$(a, \infty) = \{x \in \mathbb{R} \mid x > a\}$$
$$[a, \infty) = \{x \in \mathbb{R} \mid x \ge a\}$$
$$(-\infty, b) = \{x \in \mathbb{R} \mid x < b\}$$
$$(-\infty, b] = \{x \in \mathbb{R} \mid x \le b\}$$
$$(-\infty, \infty) = \mathbb{R}$$

Exercises 1.1

1. Indicate which of the following are declarative sentences and which are statements.
   (a) The integer 24 is even.
   (b) Is the integer $3^{15} - 1$ even?
   (c) The product of 2 and 3 is 7.
   (d) The sum of x and y is 3.
   (e) If the integer x is odd, is x^2 odd?
   (f) It is not possible for $3^{15} - 1$ to be both even and odd.
   (g) The product of x^2 and x^3 is x^6.
   (h) The integer $2^{524287} - 1$ is prime. (Recall that a positive integer greater than 2 is prime provided its only positive factors are itself and 1.)

2. Write each of the sets by listing its elements.
   (a) $A = \{ n \in \mathbb{Z} \mid -4 < n < 5 \}$
   (b) $B = \{ x \in \mathbb{R} \mid x^3 - x^2 - 2x = 0 \}$

3. Write each of the sets in the form $\{ n \in \mathbb{Z} \mid p(n) \}$, where $p(n)$ is some property of n; for example, $\{ \ldots, -4, -2, 0, 2, 4, \ldots \} = \{ n \in \mathbb{Z} \mid n$ is even$\}$.
   (a) $\{ -1, -2, -3, \ldots \}$
   (b) $\{ 0, 1, 4, 9, 16, \ldots \}$
   (c) $\{ \ldots, -27, -8, -1, 0, 1, 8, 27, \ldots \}$
   (d) $\{ \ldots, -8, -4, 0, 4, 8, \ldots \}$
   (e) $\{ \ldots, -15, -9, -3, 3, 9, 15, \ldots \}$

1.2   LOGICAL CONNECTIVES

In research, in the classroom, or at a roadside pub, mathematicians (and perhaps others) are frequently interested in determining the truth-value of a given mathematical statement. Most mathematical statements are combinations of simpler statements formed through some choice of the words *or*, *and*, *not*, or the phrases *if*—and *then*—and, and *if and only if*. These are called the *logical connectives* (or simply *connectives*) and are defined in this section.

In this chapter we use lowercase italic letters, such as p, q, r, or subscripted versions of these letters, such as p_1, q_2, r_3, to denote statements; these are called *propositional variables*. Many mathematical statements are symbolized by a single propositional variable or by some meaningful combination of propositional variables and logical connectives. A statement represented by a single propositional variable is called a *primitive* (or *simple*) *statement*, whereas a statement represented by some combination of propositional variables and logical connectives is called a *compound statement*. The symbolic representation of a primitive or compound statement is called a *formula*.

DEFINITION 1.2

Let p and q represent statements.

1. The *disjunction* of p and q is the compound statement

$$p \text{ or } q$$

 It is true provided at least one of p or q is true; otherwise, it is false. We denote the disjunction of p and q by $p \vee q$.
2. The *conjunction* of p and q is the compound statement

$$p \text{ and } q$$

 It is true provided both p and q are true; otherwise, it is false. We denote the conjunction of p and q by $p \wedge q$.
3. The *negation* of p is the compound statement

$$\text{not } p$$

 It has truth-value opposite that of p and is denoted by $\sim p$.

The truth-value of a given statement is determined by the truth-values of the primitive statements of which it is composed. We will demonstrate how a *truth table* is used to examine a given statement. Such a table contains a column for each propositional variable in the statement and a column for the whole statement. The table has a row corresponding to each possible combination of truth-values for the propositional variables involved. If the statement is particularly complex, additional propositional variables may be introduced to represent parts of it, and then the table contains a column for each of these as well.

In Figure 1.1 are shown the truth tables for disjunction, conjunction, and negation. Note that there are four possible combinations for the truth-values of two propositional variables p and q.

p	q	$p \vee q$
T	T	T
T	F	T
F	T	T
F	F	F

p	q	$p \wedge q$
T	T	T
T	F	F
F	T	F
F	F	F

p	$\sim p$
T	F
F	T

Figure 1.1 Truth tables for $p \vee q$, $p \wedge q$, and $\sim p$

Example 1.2 For which integers n is the condition

$$n > -2 \quad \text{and} \quad n < 3$$

satisfied?

Solution   The condition $n > -2$ is satisfied by the integers $-1, 0, 1, 2, 3, 4, \ldots$, while the condition $n < 3$ is satisfied by $2, 1, 0, -1, -2, -3, \ldots$. We want those integers that satisfy both conditions—namely, those that are both greater than -2 and less than 3. There are four such integers n: -1, 0, 1, and 2.   □

Example 1.3   The Boston Red Sox, Cleveland Indians, Milwaukee Brewers, and New York Yankees are professional baseball teams. Suppose on a given night the Red Sox play the Yankees and the Indians play the Brewers. Both games are completed, and the next morning someone makes the statement:

> The Brewers or the Red Sox won last night.

Find the negation of this compound statement.

Solution   The given compound statement is of the form $p \vee q$, where p represents the statement

> The Brewers won last night.

and q represents the statement

> The Red Sox won last night.

By definition, $p \vee q$ is false provided both p and q are false. Thus, $\sim (p \vee q)$ is true provided both p and q are false. It follows that the negation of the given compound statement is

> Both the Brewers and the Red Sox lost last night.

In the next section we show that the negation of the formula $p \vee q$ is the formula $\sim p \wedge \sim q$.   □

Example 1.4   Determine the output of the following Pascal program segment:

```
X := 1;
WHILE NOT (X > 50) DO
 BEGIN
  WRITELN(X);
  X := 2*X
 END;
```

Solution   The value of the variable X is initialized to 1. Then, as long as the value of X is not greater than 50, that value is output and doubled. Thus the values 1, 2, 4, 8, 16, and 32 will be output. When the value of X reaches 64, the condition

```
NOT (X > 50)
```

is false, which completes the program segment.   □

The preceding examples, though somewhat elementary, demonstrate the use of the logical connectives in both mathematical and programming settings. Indeed, these connectives are used in a notable way—as "Boolean" operators—in many programming languages; their use in Pascal is typical.

Pascal provides a built-in BOOLEAN data type. A variable that is declared to be of type BOOLEAN has exactly one of the two possible values: FALSE, TRUE. For instance, suppose a program is to be written to keep score in a volleyball game. It might be desirable to use variables A and B, of type INTEGER, for the scores of the two teams, and a BOOLEAN variable AWINS that is TRUE if team A wins the game and FALSE if team B wins. For team A to win the game, they must score at least 15 points and be at least 2 points ahead of team B. It thus makes sense and is syntactically correct for the program to contain the assignment statement

```
AWINS := (A > 14) AND (A - B > 1)
```

The expression "(A > 14) AND (A − B > 1)" is called a "Boolean expression" because its value is either TRUE or FALSE, rather than an integer, or a real number, or a value of some other type. Hence it is perfectly valid to assign the value of this expression to the BOOLEAN variable AWINS. (Note that each of the conditions "A > 14" and "A − B > 1" is a Boolean expression as well.)

Pascal provides the Boolean operators NOT, AND, and OR, whose operands are Boolean expressions. Specifically, the syntax is

$$\text{NOT } \langle \text{Boolean expression} \rangle$$

$$\langle \text{Boolean expression} \rangle \text{ AND } \langle \text{Boolean expression} \rangle$$

$$\langle \text{Boolean expression} \rangle \text{ OR } \langle \text{Boolean expression} \rangle$$

In addition, Pascal uses Boolean expressions in conditional statements and in statements used to program loops. Consider, for example, the syntax for the two-branch conditional

$$\text{IF } \langle \text{Boolean expression} \rangle \text{ THEN } \langle \text{statement} \rangle \text{ ELSE } \langle \text{statement} \rangle$$

or that for the WHILE loop

$$\text{WHILE } \langle \text{Boolean expression} \rangle \text{ DO } \langle \text{statement} \rangle$$

Example 1.5   A data structure composed of a fixed number of elements of the same type, organized as a simple linear sequence, is called a *vector*. The elements of a vector are *subscripted*, normally using an integer subrange or some enumeration. A vector is also called a *one-dimensional* or *linear array*.

A standard special case is that of a vector subscripted by the positive integers $1, 2, \ldots, $ MAX, where MAX is a constant. The Pascal declaration of a type VECTORTYPE, having 50 elements of type REAL, is given as follows:

```
CONST MAX = 50;
TYPE SUBSCRIPTTYPE = 1..MAX;
     ELEMENTTYPE = REAL;
     VECTORTYPE = ARRAY [SUBSCRIPTTYPE] OF ELEMENTTYPE;
```

(Note that the type VECTORTYPE is determined by the types SUB-SCRIPTTYPE and ELEMENTTYPE.)

An important problem concerning vectors is that of searching a given vector V for a given value KEY. Suppose we wish to know whether one of the elements V[1], V[2], \ldots, V[MAX] is equal to KEY, and if so, then we would also like to know a subscript POSITION such that V[POSITION] = KEY. A simple and general method for solving this problem is called *linear searching*; it simply compares KEY with V[1], V[2], and so on, in turn. How should such a search be controlled? Let the variable CURRENT, of type INTEGER, be the subscript of the element of V currently being compared with KEY, and let the variable FOUND, of type BOOLEAN, indicate whether an element equal to KEY has been found. Initially, then, CURRENT = 1 and FOUND = FALSE. The search is to continue as long as FOUND = FALSE (KEY has not been found) and CURRENT ≤ MAX (the whole vector V has not been searched). This suggests using a WHILE loop to control the search, as seen in the following program segment:

```
(*begin search*)
CURRENT := 1;
FOUND := FALSE;
WHILE (NOT FOUND) AND (CURRENT <= MAX) DO
 IF V[CURRENT] = KEY
  THEN BEGIN
    FOUND := TRUE;
    POSITION := CURRENT
    END;
   ELSE CURRENT := CURRENT + 1;
(*end search*)                                              □
```

Consider next the compound statement

$$\text{If } p, \text{ then } q.$$

How is the truth-value of such a statement determined by the truth-values of the primitive statements p and q? An example will help to clarify the situation.

Example 1.6   Consider the statement

> If you score 90 or more on the final exam, then you will get an A grade for the course.

which an instructor might make to a student. We let p represent the statement

> You score 90 or more on the final exam.

and q represent the statement

> You will get an A grade for the course.

Then the instructor's statement is represented by the formula

> If p, then q.

Let us analyze this statement for each of the four possible truth-value combinations of p and q. (Assume the semester is over and the student's final exam score and course grade are known.)

Case 1 Both p and q are true. In this case the student scored 90 or more on the final exam and got an A, just as the instructor promised. The instructor's statement is true.

Case 2 p is true and q is false. The student scored 90 or more on the final exam but for some reason did not get an A. Perhaps the instructor made an error in recording the grade; at any rate, based on the evidence we must conclude that the instructor's statement is false.

Case 3 p is false and q is true. The student scored less than 90 on the final exam but got an A anyway. Perhaps the student got an 89 and the instructor, being in a good mood, decided to give the student the A. The point is that the facts do not contradict the instructor's statement; in a sense, the statement has not been tested. Thus we take the statement to be true.

Case 4 Both p and q are false. The student scored less than 90 on the final and did not get an A. This is much like the previous case, so we take the instructor's statement to be true. (Note that the instructor's statement does not say what the student's grade will be if the student scores less than 90 on the final exam.)

We see in this example that the only case in which the formula

> If p, then q.

is false is when p is true and q is false. Examples like this one serve to motivate the next definition.   □

DEFINITION 1.3

Let p and q be propositional variables. The compound statement

If p, then q.

is called the *implication of q by p* and is denoted by $p \rightarrow q$. It is false when p is true and q is false, and it is true otherwise. It is common to refer to $p \rightarrow q$ as a *conditional statement* or simply as an *implication*. We often read $p \rightarrow q$ as "p implies q" and call p the *hypothesis* and q the *conclusion*.

The truth table for the implication is shown in Figure 1.2.

p	q	$p \rightarrow q$
T	T	T
T	F	F
F	T	T
F	F	T

Figure 1.2 Truth table for $p \rightarrow q$

Example 1.7   Consider the following program segment in Pascal:

```
A := 1;
IF (X < -1) OR (Y > 2) THEN A := 2;
WRITELN(A);
```

What is output as the value of A if:

(a) $X = -2$ and $Y = 1$?
(b) $X = 1$ and $Y = 4$?
(c) $X = -2$ and $Y = 4$?
(d) $X = 1$ and $Y = 1$?

Solution   (a) 2   (b) 2   (c) 2   (d) 1

(Note that A is assigned the value 2 unless both the conditions $X < -1$ and $Y > 2$ are false.)   □

Next consider the statements

(a)   If $x = 1$, then $x^2 = 1$.

(b)   If $x^2 = 1$, then $x = 1$.

about a real number x. It is clear that these statements are different, yet related. In fact, (a) is true and (b) is false. On the other hand, the statements

(a)   If $x = 1$ or $x = -1$, then $x^2 = 1$.

(b)   If $x^2 = 1$, then $x = 1$ or $x = -1$.

though related in the same way, are both true. In each case the statements, taken together, are of the type

(a)   If p, then q.

(b)   If q, then p.

DEFINITION 1.4

Let p and q be propositional variables. The conditional statement

If q, then p.

is called the *converse* of the conditional statement

If p, then q.

In addition to the converse of an implication $p \to q$, there are two other related conditional statements that are used frequently in mathematical logic. They are formed using the negation and are given by

(c)   If $\sim q$, then $\sim p$.

(d)   If $\sim p$, then $\sim q$.

We call (c) the *contrapositive* of $p \to q$ and we call (d) the *inverse* of $p \to q$. In the next section we see a close connection between an implication and its contrapositive, and also between the converse and the inverse.

It happens frequently in mathematics that we need to examine a statement with the logical form

$$(p \to q) \wedge (q \to p)$$

that is, we need to examine the conjunction of an implication and its converse. Hence there is good reason to make the next definition.

DEFINITION 1.5

Let p and q be propositional variables. The compound statement

$$(p \to q) \wedge (q \to p)$$

is called the *biconditional*, and it is denoted by $p \leftrightarrow q$. This statement is read "p if and only if q." A common shorthand form of "p if and only if q" is "p iff q."

p	q	$p \to q$	$q \to p$	$p \leftrightarrow q$
T	T	T	T	T
T	F	F	T	F
F	T	T	F	F
F	F	T	T	T

Figure 1.3 Truth table for $p \leftrightarrow q$

The truth table for the biconditional, shown in Figure 1.3, is not difficult to complete from the truth tables for the conjunction and the implication. Note that $p \leftrightarrow q$ is true only when p and q have the same truth-value. It is very important to keep in mind, once again, that in order to show that $p \leftrightarrow q$ is true, one must show that both $p \to q$ and $q \to p$ are true.

This is a good spot to discuss the use of parentheses in a formula. For example, consider the formula $\sim p \lor q$. Does it mean $(\sim p) \lor q$ or does it mean $\sim(p \lor q)$? In fact, convention has it that $(\sim p) \lor q$ is the correct interpretation. Questions of this sort can always be avoided by using parentheses, but their overuse can be quite tedious and can, at times, make formulas difficult to interpret. Hence we adopt basic rules that allow certain parentheses to be omitted from formulas. These rules are based on the idea of "precedence" of the logical connectives.

Precedence Rules for the Logical Connectives

In a parenthesis-free formula, the logical connectives are applied in the following order:

Connective	*Precedence*
$\sim$	First
$\land$	Second
$\lor$	Third
$\to$	Fourth
$\leftrightarrow$	Fourth

We say that negation has higher precedence than conjunction, conjunction has higher precedence than disjunction, and so on. (These precedence rules agree with the precedence rules for the Boolean operators NOT, AND, and OR in Pascal.) Note that the implication and the biconditional have the same precedence level. As a consequence, if a formula involves both of these connectives, then we shall agree to use parentheses to make it clear how they are to be applied.

Example 1.8 Here are several formulas and their correct interpretations.

Formula	Interpretation
(a) $\sim p \lor q$	$(\sim p) \lor q$
(b) $p \lor \sim q \land r$	$p \lor ((\sim q) \land r)$
(c) $\sim q \to \sim p$	$(\sim q) \to (\sim p)$
(d) $(p \to q) \leftrightarrow \sim p \lor q$	$(p \to q) \leftrightarrow ((\sim p) \lor q)$
(e) $p \land q \to (p \leftrightarrow q)$	$(p \land q) \to (p \leftrightarrow q)$

$\square$

Exercises 1.2

1. Let p, q, and r represent the following statements:

p: Ralph reads the *New York Times*.

q: Ralph watches the *MacNeil, Lehrer Newshour*.

r: Ralph jogs 3 miles.

Give a formula for each of these statements.
(a) Ralph reads the *New York Times* and watches the *MacNeil, Lehrer Newshour*.
(b) Ralph reads the *New York Times* or jogs 3 miles.
(c) If Ralph reads the *New York Times*, then he doesn't watch the *MacNeil, Lehrer Newshour*.
(d) Ralph reads the *New York Times* if and only if he jogs 3 miles.
(e) It is not the case that if Ralph jogs 3 miles then he reads the *New York Times*.
(f) Ralph watches the *MacNeil, Lehrer Newshour* or jogs 3 miles, but not both.

2. Let the propositional variables p, q, and r be defined as in Exercise 1. Express each of these formulas in words.
(a) $p \land r$ **(b)** $q \lor r$
(c) $(p \land q) \lor r$ **(d)** $\sim p \lor \sim q$
(e) $p \to q$ **(f)** $q \leftrightarrow r$

3. Consider the following statements p and q:

p: Dodger pitcher Fernando Valenzuela had a sore arm in 1981.

q: The Los Angeles Dodgers won the 1981 World Series.

The statement p is false; q is true. Give a formula for each of these statements and determine its truth-value.
(a) Dodger pitcher Fernando Valenzuela had a sore arm in 1981 or the Dodgers won the 1981 World Series.
(b) Dodger pitcher Fernando Valenzuela had a sore arm in 1981 and the Dodgers won the 1981 World Series.

(c) If Dodger pitcher Fernando Valenzuela had a sore arm in 1981, then the Dodgers won the 1981 World Series.

(d) If the Dodgers did not win the 1981 World Series, then Dodger pitcher Fernando Valenzuela had a sore arm that year.

(e) Dodger pitcher Fernando Valenzuela had a sore arm in 1981 if and only if the Dodgers won the World Series that year.

(f) The Dodgers won the 1981 World Series if and only if pitcher Fernando Valenzuela did not have a sore arm that year.

4. For each of these compound statements, first identify the primitive statements p, q, r, and so on, of which it is composed. Then give a formula for the statement.

(a) If $\sqrt{2709}$ is an integer, then either $\sqrt{2709}$ is even or $\sqrt{2709}$ is odd.

(b) If 53 is prime and 53 is greater than 2, then 53 is odd.

(c) If $\sqrt[3]{7}$ is not negative and its square is less than 4, then either $\sqrt[3]{7} = 0$ or $\sqrt[3]{7}$ is positive and less than 2.

(d) If 26 is even and 26 is greater than 2, then 26 is not prime.

5. Express each of these compound statements symbolically.

(a) If triangle ABC is equilateral, then it is isosceles.

(b) The number $x = 3$ if and only if $2x - 3 = 3$.

(c) If π^π is a real number, then either π^π is rational or π^π is irrational.

(d) The product $xy = 0$ if and only if either $x = 0$ or $y = 0$.

(e) If $\sqrt{47089}$ is greater than 200, then, if $\sqrt{47089}$ is prime, it is greater than 210.

(f) If line k is perpendicular to line m and line m is parallel to line n, then line k is perpendicular to line n.

(g) If n is an integer, then either n is positive or n is negative or $n = 0$.

(h) If a and b are integers and $b \neq 0$, then a/b is a rational number.

6. In mathematics the connective *or* is used inclusively, meaning "one or the other or both." However, in everyday language *or* is often used in the exclusive sense, as in the sentence

> With your order you may have french fries *or* potato salad.

Used in this way, the *or* is interpreted as "one or the other but not both." Using the symbol $\veebar$ to represent the connective "exclusive or," construct the truth table for $p \veebar q$.

7. Consider the Pascal program segment for the linear search in Example 1.5.

(a) Rewrite the program segment using a REPEAT–UNTIL loop instead of a WHILE loop.

(b) How is the Boolean expression that controls the REPEAT–UNTIL loop related to the Boolean expression that controls the WHILE loop?

8. We mentioned the precedence rules in Pascal in relation to the Boolean operators NOT, AND, and OR; namely, NOT precedes AND and AND precedes OR. Pascal also has arithmetic operators, such as + and *, and relational operators, such as = and < .

(a) How do the Boolean operators fit into the overall precedence scheme?

(b) Is the expression "A > 14 AND A − B > 1" valid? Explain.

(c) Is the expression "NOT FOUND AND (CURRENT <= MAX)" valid? Explain.

(d) Answer the questions in parts (a)–(c) for the programming language Ada.

★**9.** A *list* is a linear data structure composed of a variable number of elements. If all the elements of a list have the same type, then the list is said to be *homogeneous*. A Pascal data type LIST for homogeneous lists may be declared using a record with two fields—a field ELEMENT to hold the elements of the list and a field LENGTH to hold the current length of the list—as follows:

```
CONST MAX = 50; (*the maximum length*)
TYPE SUBSCRIPTTYPE = 1..MAX;
     ELEMENTTYPE = REAL(*type for the elements*)
     LIST = RECORD
     ELEMENT: ARRAY [SUBSCRIPTTYPE] OF ELEMENTTYPE;
     LENGTH:0..MAX
   END;
```

If L is a variable of type LIST, then L.LENGTH is the current length of L and, if L.LENGTH > 0, then L.ELEMENT[J] is the Jth element of L, where $1 \le J \le$ L.LENGTH.

(a) Write a Pascal procedure SEARCH to perform a linear search of a list L for a value KEY. Return FOUND and POSITION, as defined in Example 1.5.

Assume that the elements of L are sorted so that

```
L.ELEMENT[1] ≤ L.ELEMENT[2] ≤...≤ L.ELEMENT[L.LENGTH]
```

(b) Modify the procedure SEARCH of part (a) to take advantage of the fact that L is sorted.

(c) Write a procedure COMPACT that removes any duplicate values from a sorted list L; that is, COMPACT returns a list of distinct values.

10. In volleyball it is important to know which team is serving because a team scores a point only if that team is serving and wins a volley. If the serving team loses the volley, then the other team gets to serve. Thus in a program to keep score in a volleyball game (between teams A and B), it may be useful to have variables ASERVES and AWINSVOLLEY of type BOOLEAN. ASERVES is true if team A is serving and false if team B is serving; AWINSVOLLEY is true if team A wins the current volley and false if team B wins it.

(a) Write a Boolean expression that is true if team A scores a point and is false otherwise.

(b) Write a Boolean expression that is true if team B scores a point and is false otherwise.

(c) Write a Boolean expression that is true if the serving team loses the current volley and is false otherwise.

(d) Write an assignment statement that changes the serving team.

1.3   LOGICAL EQUIVALENCE

In mathematics, as in all subjects, there may be several different ways to say the same thing. In this section we formally define what this means for logical statements.

DEFINITION 1.6

Let u and v be formulas. We say that u is *logically equivalent* to v, denoted $u \equiv v$, provided u and v have the same truth-value for every possible choice of truth-values for the propositional variables involved in u and v.

The two examples of logical equivalence that follow are important in that both are used often in mathematics.

Example 1.9   Show that $\sim(p \lor q) \equiv \sim p \land \sim q$.

Solution   We do this by constructing a truth table and comparing the columns labeled $\sim(p \lor q)$ and $\sim p \land \sim q$. The truth table is shown in Figure 1.4. Since the columns headed by $\sim(p \lor q)$ and $\sim p \land \sim q$ agree in every row, these formulas are logically equivalent.

p	q	$\sim p$	$\sim q$	$p \lor q$	$\sim(p \lor q)$	$\sim p \land \sim q$
T	T	F	F	T	F	F
T	F	F	T	T	F	F
F	T	T	F	T	F	F
F	F	T	T	F	T	T

Figure 1.4 Truth table showing $\sim(p \lor q) \equiv \sim p \land \sim q$

□

p	q	$\sim p$	$p \rightarrow q$	$\sim p \vee q$
T	T	F	T	T
T	F	F	F	F
F	T	T	T	T
F	F	T	T	T

Figure 1.5 Truth table showing $p \rightarrow q \equiv \sim p \vee q$

Example 1.10   Use a truth table to verify that $p \rightarrow q \equiv \sim p \vee q$.

Solution   See Figure 1.5.   □

Example 1.11   Recall that the contrapositive of the implication $p \rightarrow q$ is the implication $\sim q \rightarrow \sim p$. In words, the contrapositive states

If not q, then not p.

We wish to show that an implication and its contrapositive are logically equivalent. A truth table can be used for this, but we show how to do it without a truth table.

We need several facts. First, if u, v, and w are formulas with $u \equiv v$ and $v \equiv w$, then clearly $u \equiv w$. Second, from Example 1.10 we have that

$$s \rightarrow t \equiv \sim s \vee t \tag{1}$$

It is also not difficult to see that

$$s \vee t \equiv t \vee s \tag{2}$$

(We use the propositional variables s and t here to avoid confusion with the variables p and q, which appear in the logical equivalence we are trying to establish.) Now, by (1),

$$p \rightarrow q \equiv \sim p \vee q$$

Next, by (2),

$$\sim p \vee q \equiv q \vee \sim p$$

Finally, the hard step: by (1), with s replaced by $\sim q$ and t replaced by $\sim p$, we have

$$q \vee \sim p \equiv \sim q \rightarrow \sim p$$

Thus it follows that

$$p \rightarrow q \equiv \sim q \rightarrow \sim p \qquad \qquad □$$

Some formulas have the seemingly dull property of always being true. However, they are given an interesting name.

DEFINITION 1.7

A formula that is true for all possible truth-values of its constituent propositional variables is called a *tautology*. A formula that is false for all possible truth-values of its constituent propositional variables is called a *contradiction*.

---

p	q	$p \rightarrow q$	$\sim p \vee q$	$(p \rightarrow q) \leftrightarrow (\sim p \vee q)$
T	T	T	T	T
T	F	F	F	T
F	T	T	T	T
F	F	T	T	T

Figure 1.6 Truth table showing $(p \rightarrow q) \leftrightarrow (\sim p \vee q)$ is a tautology

Example 1.12 In Figure 1.6 is the truth table for the formula

$$(p \rightarrow q) \leftrightarrow (\sim p \vee q)$$

Since this formula is always true, it is a tautology. It is interesting to compare the result of this example with the result of Example 1.10.   □

Example 1.13 Verify that the following formula is a tautology:

$$[(p \wedge q) \rightarrow r] \rightarrow [\sim r \rightarrow (\sim p \vee \sim q)]$$

Solution Here we have our first instance of a formula that involves three propositional variables. As shown in the truth table in Figure 1.7, there are eight possible truth-value combinations that must be considered. For convenience, we let u and v denote the formulas $(p \wedge q) \rightarrow r$ and $\sim r \rightarrow (\sim p \vee \sim q)$, respectively.

p	q	r	$\sim p$	$\sim q$	$\sim r$	$p \wedge q$	$\sim p \vee \sim q$	u	v	$u \rightarrow v$
T	T	T	F	F	F	T	F	T	T	T
T	T	F	F	F	T	T	F	F	F	T
T	F	T	F	T	F	F	T	T	T	T
T	F	F	F	T	T	F	T	T	T	T
F	T	T	T	F	F	F	T	T	T	T
F	T	F	T	F	T	F	T	T	T	T
F	F	T	T	T	F	F	T	T	T	T
F	F	F	T	T	T	F	T	T	T	T

Figure 1.7 Truth table showing $[(p \wedge q) \rightarrow r] \rightarrow [\sim r \rightarrow (\sim p \vee \sim q)]$ is a tautology

Alternative
Solution

It is possible to verify that $u \to v$ is a tautology without using a truth table. We know that the only case in which $u \to v$ is false is where u is true and v is false. Suppose that v is false; then it follows that $\sim r$ is true and $\sim p \vee \sim q$ is false. Thus r is false and p and q are both true; that is, r is false and $p \wedge q$ is true. Therefore, $(p \wedge q) \to r$—namely, u—is false. Thus whenever v is false, u is also false. This shows that $u \to v$ is a tautology.   □

Example 1.14   Express the formula $\sim (p \to q)$ as a conjunction.

Solution   From Example 1.10,

$$p \to q \equiv \sim p \vee q$$

Thus we have

$$\sim (p \to q) \equiv \sim (\sim p \vee q)$$
$$\equiv \sim (\sim p) \wedge \sim q \qquad (\text{by Example 1.9})$$
$$\equiv p \wedge \sim q$$

Here we again use the equivalence $\sim (\sim p) \equiv p$.   □

Note that the formula of Example 1.12 is of the form $u \leftrightarrow v$, and, as established in Example 1.10, $u \equiv v$. This is not just a coincidence, for it can be shown that formulas u and v are logically equivalent if and only if the compound statement $u \leftrightarrow v$ is a tautology.

A note about logical equivalence. In a later section we describe what it means to prove a mathematical statement or theorem. The notion of logical equivalence will turn out to be of frequent use in this regard. Suppose u and v are formulas representing mathematical statements and we wish to prove u. If $u \equiv v$, then it suffices to prove v. For example, suppose we want to prove a mathematical statement of the form $p \to q$. Since $p \to q \equiv \sim q \to \sim p$, we may instead prove $\sim q \to \sim p$. This technique, called "proof by contrapositive," is a standard proof technique in mathematics, and we explore it further in Section 1.5.

Here is a list of some of the important logical equivalences that are frequently encountered and used in mathematics.

Logical Equivalences

1. The commutative properties
   (a) $p \vee q \equiv q \vee p$ (b) $p \wedge q \equiv q \wedge p$
2. The associative properties
   (a) $(p \vee q) \vee r \equiv p \vee (q \vee r)$
   (b) $(p \wedge q) \wedge r \equiv p \wedge (q \wedge r)$

3. The distributive properties
   (a) $p \vee (q \wedge r) \equiv (p \vee q) \wedge (p \vee r)$
   (b) $p \wedge (q \vee r) \equiv (p \wedge q) \vee (p \wedge r)$

4. The idempotent laws
   (a) $p \vee p \equiv p$ (b) $p \wedge p \equiv p$

5. DeMorgan's laws
   (a) $\sim(p \vee q) \equiv \sim p \wedge \sim q$ (b) $\sim(p \wedge q) \equiv \sim p \vee \sim q$

6. Law of the excluded middle
   (a) $p \vee \sim p$ is a tautology
   (b) $p \wedge \sim p$ is a contradiction

7. An implication and its contrapositive are logically equivalent:

$$p \rightarrow q \equiv \sim q \rightarrow \sim p$$

8. The converse and inverse of the implication $p \rightarrow q$ are logically equivalent:

$$q \rightarrow p \equiv \sim p \rightarrow \sim q$$

9. Let T denote a tautology and F denote a contradiction. Then
   (a) $p \vee T \equiv T$ (b) $p \wedge T \equiv p$
   (c) $p \vee F \equiv p$ (d) $p \wedge F \equiv F$

A brief comment is in order concerning formulas such as $(p \wedge q) \wedge r$ or $((p \wedge q) \wedge r) \wedge s$. In view of the fact that $(p \wedge q) \wedge r \equiv p \wedge (q \wedge r)$, it makes good sense to define $p \wedge q \wedge r$ to be $(p \wedge q) \wedge r$. It follows that $p \wedge q \wedge r \equiv p \wedge (q \wedge r)$, so we can insert parentheses in $p \wedge q \wedge r$ in both ways without loss of meaning. Similarly, we can insert parentheses in $p \wedge q \wedge r \wedge s$ in five meaningful ways, and since all the resulting formulas are logically equivalent, we agree to let $p \wedge q \wedge r \wedge s$ denote any one of them. Analogous remarks apply for disjunction.

Example 1.15 The formula

$$(p \wedge (p \rightarrow q)) \rightarrow q$$

is called *modus ponens*. It represents any statement of the form

If both p and p implies q, then q.

For example, the statement

If it is raining and if rain implies that I study discrete mathematics, then I study discrete mathematics.

is of this form. Use properties of logical equivalence to show that modus ponens is a tautology.

Solution   We apply the properties as follows:

$$(p \wedge (p \rightarrow q)) \rightarrow q$$

$$\equiv \sim(p \wedge (p \rightarrow q)) \vee q \qquad \text{(by Example 1.10)}$$

$$\equiv (\sim p \vee \sim(p \rightarrow q)) \vee q \qquad \text{(by 5(b))}$$

$$\equiv (\sim p \vee (p \wedge \sim q)) \vee q \qquad \text{(by Example 1.14)}$$

$$\equiv ((\sim p \vee p) \wedge (\sim p \vee \sim q)) \vee q \qquad \text{(by 3(a))}$$

$$\equiv (T \wedge (\sim p \vee \sim q)) \vee q \qquad \text{(by 1(a) and 6(a))}$$

$$\equiv (\sim p \vee \sim q) \vee q \qquad \text{(by 1(b) and 9(b))}$$

$$\equiv \sim p \vee (\sim q \vee q) \qquad \text{(by 2(a))}$$

$$\equiv \sim p \vee T \qquad \text{(by 1(a) and 6(a))}$$

$$\equiv T \qquad \text{(by 9(a))} \qquad \square$$

Example 1.16   Use properties of logical equivalence to show that

$$\sim(p \leftrightarrow q) \equiv (\sim q) \leftrightarrow p$$

Solution   We apply the properties as follows:

$$\sim(p \leftrightarrow q) \equiv \sim[(p \rightarrow q) \wedge (q \rightarrow p)] \qquad \text{(by Definition 1.4)}$$

$$\equiv \sim(p \rightarrow q) \vee \sim(q \rightarrow p) \qquad \text{(by 5(b))}$$

$$\equiv (p \wedge \sim q) \vee (q \wedge \sim p) \qquad \text{(by Example 1.14)}$$

$$\equiv [(p \wedge \sim q) \vee q] \wedge [(p \wedge \sim q) \vee \sim p] \qquad \text{(by 3(a))}$$

$$\equiv [(p \vee q) \wedge (\sim q \vee q)]$$

$$\quad \wedge [(p \vee \sim p) \wedge (\sim q \vee \sim p)] \qquad \text{(by 1(a) and 3(a))}$$

$$\equiv [(p \vee q) \wedge T] \wedge [T \wedge (\sim q \vee \sim p)] \qquad \text{(by 1(a) and 6(a))}$$

$$\equiv (p \vee q) \wedge (\sim q \vee \sim p) \qquad \text{(by 1(b) and 9(b))}$$

$$\equiv (q \vee p) \wedge (\sim p \vee \sim q) \qquad \text{(by 1(a))}$$

$$\equiv (\sim q \rightarrow p) \wedge (p \rightarrow \sim q) \qquad \text{(by Example 1.10)}$$

$$\equiv (\sim q) \leftrightarrow p \qquad \text{(by Definition 1.4)}$$

$$\qquad \square$$

At this point it is convenient to discuss some alternate forms for the implication and biconditional that are frequently used in mathematical discussions.

The formula $p \rightarrow q$ has two interpretations in words thus far:

If p, then q.

and

p implies q.

We know that $p \to q$ is logically equivalent to its contrapositive $\sim q \to \sim p$, which is read

If not q, then not p.

Assume $p \to q$ is true. Then $\sim q \to \sim p$ is also true, which means that if q is false, then p must also be false. Thus p is true only under the condition that q is true. This last statement is written

p only if q.

and it is another way of saying p implies q. Another common phrase in mathematics is

p is sufficient for q.

It states that the condition that p holds is enough to guarantee that q holds. Thus, it is another way of saying p implies q. Finally, one last phrase equivalent to $p \to q$ is

q is necessary for p.

This says that in order for p to be true, q must be true, so that q being false implies that p is false, again giving us the contrapositive $\sim q \to \sim p$. To summarize, then, the following statements are equivalent:

If p, then q.

p implies q.

p only if q.

p is sufficient for q.

q is necessary for p.

Example 1.17   Consider the statement

If $x = 1$, then $x^2 - 1 = 0$.

concerning a real number x. Rewrite this statement in four equivalent ways.

Solution   The given statement is equivalent to each of the following:

$x = 1$ implies $x^2 - 1 = 0$.

$x = 1$ only if $x^2 - 1 = 0$.

$x = 1$ is sufficient for $x^2 - 1 = 0$.

$x^2 - 1 = 0$ is necessary for $x = 1$. $\square$

Now consider the biconditional $p \leftrightarrow q$, which reads

p if and only if q.

Recall that $p \leftrightarrow q$ is by definition a shorthand for $(p \to q) \wedge (q \to p)$. One way to read this is

p is sufficient for q and p is necessary for q.

This last statement is usually shortened to

p is necessary and sufficient for q.

and is an alternative way of saying p if and only if q. For instance, the statements

$$x = 2 \text{ if and only if } x^3 - 8 = 0.$$

and

$$x = 2 \text{ is necessary and sufficient for } x^3 - 8 = 0.$$

are equivalent.

Exercises 1.3

1. Use a truth table to verify DeMorgan's law 5(b).
2. Use truth tables to verify the associative properties 2(a) and 2(b).
3. Verify property 8 that the inverse and converse of the implication $p \rightarrow q$ are logically equivalent.
4. Each implication given concerns integers x and y. Find (i) the inverse, (ii) the converse, and (iii) the contrapositive.
   (a) If $x = 2$, then $x^4 = 16$.
   (b) If $y > 0$, then $y \neq -3$.
   (c) If x is odd and y is odd, then xy is odd.
   (d) If $x^2 = x$, then either $x = 0$ or $x = 1$.
   (e) If $x = 17$ or $x^3 = 8$, then x is prime.
   (f) If $xy \neq 0$, then both $x \neq 0$ and $y \neq 0$.
5. Find (i) the inverse, (ii) the converse, and (iii) the contrapositive of each of the following implications.
   (a) If quadrilateral $ABCD$ is a rectangle, then $ABCD$ is a parallelogram.
   (b) If triangle ABC is isosceles and contains an angle of 45 degrees, then ABC is a right triangle.
   (c) If quadrilateral $ABCD$ is a square, then it is both a rectangle and a rhombus.
   (d) If quadrilateral $ABCD$ has two sides of equal length, then it is a rectangle or a rhombus.
   ★(e) If polygon P has the property that P is equiangular if and only if P is equilateral, then P is a triangle.
6. The formula

$$[\sim p \wedge (p \vee q)] \rightarrow q$$

   is known as the *disjunctive syllogism*.
   (a) Express the formula in words.
   (b) Show that the disjunctive syllogism is a tautology. (A truth table may be used, but it is instructive to use the properties of logical equivalence.)

7. The formula

$$[(p \to q) \wedge \sim q] \to \sim p$$

is called *modus tollens*.
(a) Express the formula in words.
(b) Show that modus tollens is a tautology.

8. Give a truth table for each formula. Which are tautologies? Which are contradictions?
(a) $(\sim p \wedge q) \to p$ (b) $p \leftrightarrow \sim(\sim p)$
(c) $(p \to q) \wedge (p \to \sim q)$ (d) $(p \to q) \vee (q \to p)$
(e) $(p \to q) \to r$ (f) $p \wedge q \wedge (p \to \sim q)$

9. Use truth tables to verify the distributive properties 3(a) and 3(b).

10. Verify these logical equivalences. (Again, a truth table may be used, but try to use the properties of logical equivalence.)
(a) $(p \to \sim q) \wedge (p \to \sim r) \equiv \sim[p \wedge (q \vee r)]$
(b) $(p \wedge q) \leftrightarrow p \equiv p \to q$
(c) $(p \wedge q) \to r \equiv p \to (\sim q \vee r)$
(d) $p \to (q \vee r) \equiv \sim q \to (\sim p \vee r)$

11. Determine whether formulas u and v are logically equivalent.
(a) u: $(p \to q) \wedge (p \to \sim q)$; v: $\sim p$
(b) u: $p \to q$; v: $q \to p$
(c) u: $p \leftrightarrow q$; v: $q \leftrightarrow p$
(d) u: $(p \to q) \to r$; v: $p \to (q \to r)$
(e) u: $(p \leftrightarrow q) \leftrightarrow r$; v: $p \leftrightarrow (q \leftrightarrow r)$
(f) u: $p \to (q \to r)$; v: $(p \to q) \to (p \to r)$
(g) u: $p \to (q \vee r)$; v: $(p \to q) \vee (p \to r)$
(h) u: $p \vee (q \to r)$; v: $(p \vee q) \to (p \vee r)$

12. Given that the formula $(q \vee r) \to \sim p$ is false and q is false, determine the truth-values of r and p.

13. Identify the form of each implication as one of the following:
  (i) If p, then q.
  (ii) p implies q.
  (iii) p only if q.
  (iv) p is sufficient for q.
  (v) q is necessary for p.

  Then rewrite the implication in each of the other four forms.
  (a) If $x = -2$, then $x^3 = -8$.
  (b) Being intelligent is necessary for Randy to pass this course.
  (c) Working hard is sufficient for Susan to pass this course.
  (d) 11111 is prime only if 11111 is not a multiple of 4.
  (e) In order for triangle ABC to be a right triangle, it is necessary that the side lengths of ABC satisfy the Pythagorean theorem.
  (f) The fact that Susan is a good student implies that Susan studies hard.

      (g) In order for $\sqrt{3}$ to be rational, it is sufficient that its decimal expansion terminates.

      (h) Randy passes this course if Randy passes the final exam.

      (i) Randy passes this course only if Randy passes the final exam.

14. Rewrite each statement in an equivalent way.

      (a) $x^3 - x^2 + x - 1 = 0$ is necessary and sufficient for $x = 1$.

      (b) Randy passes this course if and only if Susan helps him study.

1.4   LOGICAL QUANTIFIERS

Consider the sentence

$$\text{The integer } x \text{ is prime and } x > 17.$$

(Recall that a prime number is an integer $n > 1$ whose only positive factors are itself and 1. We discuss prime numbers in Chapter 3.) This sentence is not a statement; if $x = 19$, it is true, whereas if $x = 24$, it is false. Until the variable x is given a specific value, the truth-value of the sentence cannot be determined. This sentence is an example of a "propositional function."

DEFINITION 1.8

A *propositional function* in the variable x is a sentence $p(x)$ about x that becomes a statement when x is given a particular value (or meaning). Propositional functions in one variable x are denoted $p(x), q(x), r(x)$, and so on.

Consider the sentence

$$\text{If } x \text{ is prime, then } x \text{ is not a multiple of 4.}$$

This sentence has the logical form $p(x) \to q(x)$. On the other hand, sentences like

$$\text{There exists an } x \text{ such that } x \text{ is prime and } x + 10 \text{ is prime.}$$

or

$$\text{For all } x, \text{ if } x \text{ is prime, then } x^2 + 5 \text{ is not prime.}$$

cannot be symbolized using the logical connectives presented thus far. The reason for this is the presence of the phrases "there exists an x" and "for all x." They are used frequently enough in mathematics to warrant symbolic representation.

DEFINITION 1.9

The statement

There exists an x such that $p(x)$.

is symbolized by the formula

$$\exists x \, p(x)$$

The symbol $\exists$ is called the *existential quantifier* and translates as "there exists." Other common phrases for $\exists$ are "for some" and "there is some." Also, the words "such that" are often replaced by "for which" or "satisfying." The statement

$$\exists x \, p(x)$$

is true provided there is at least one value of x for which $p(x)$ is true.

DEFINITION 1.10

The statement

For all x, $p(x)$.

is symbolized by the formula

$$\forall x \, p(x)$$

The symbol $\forall$ is called the *universal quantifier* and translates as "for all." Other common phrases for $\forall$ are "for each," "for every," and "given any." The statement

$$\forall x \, p(x)$$

is true provided $p(x)$ is true for every value of x.

The quantifiers $\exists$ and $\forall$, together with the logical connectives, are collectively referred to as the *logical symbols*.

Example 1.18   Consider the following statements about a real number x:

(a) $\exists x(x^2 = 2)$ (b) $\exists x(x^2 < 0)$
(c) $\forall x(x + 1 > x)$ (d) $\forall x(\sqrt{x^2} = x)$

Statement (a) is true because it asserts the existence of the number $\sqrt{2}$.
Statement (b) is false, since every real number x has the property that $x^2 \geq 0$.
Statement (c) is clearly true.
Statement (d) is false; see what happens if x is a negative number.   □

Notice that the statement

$$\text{There is some } x \text{ such that } x^2 = 2.$$

is false if we are considering only rational numbers x. If the possible values of the variable x in such a statement are not assumed in advance, then such assumptions are often included as part of the statement. For example, the statement

$$\text{There is some real number } x \text{ such that } x^2 = 2.$$

is written symbolically as

$$\exists x (x \in \mathbb{R} \wedge x^2 = 2)$$

In this case, we have used a *restricted form* of the quantifier $\exists$.

As another example, consider the statement

$$\text{For every real number } x, x^2 + 1 > 0.$$

This statement uses the restricted form of the quantifier $\forall$. It says that, given any x, if x happens to be a real number, then $x^2 + 1 > 0$; it is represented in symbolic form as

$$\forall x (x \in \mathbb{R} \rightarrow x^2 + 1 > 0)$$

In general, the statement

$$\text{For every } x \text{ with property } r(x), \text{ property } p(x) \text{ holds.}$$

translates as

$$\text{For every } x, \text{ if } r(x) \text{ holds, then } p(x) \text{ holds.}$$

and is represented by the formula

$$\forall x (r(x) \rightarrow p(x))$$

Next consider the statement

$$\text{There is some real number } x \text{ such that } x^3 - 3x + 1 = 0.$$

It states that there is some x having the two properties that x is a real number and x is a solution of the equation $x^3 - 3x + 1 = 0$. It provides an example of the restricted use of the existential quantifier $\exists$ and is represented by the formula

$$\exists x (x \in \mathbb{R} \wedge x^3 - 3x + 1 = 0)$$

Any statement having the general form

$$\text{There is some } x \text{ with property } r(x) \text{ such that property } p(x) \text{ holds.}$$

translates as

There is some x such that both $r(x)$ and $p(x)$ hold.

and is represented by the formula

$$\exists x (r(x) \wedge p(x))$$

Example 1.19   Express each of the following statements in symbolic form.

(a) For every positive integer x, either x is prime or $x^2 + 1$ is prime.
(b) There is some positive integer x such that both $x^2 + 1$ is prime and x is a perfect square.

Solution   Statement (a) is represented by the formula

$$\forall x \left[x \in \mathbb{N} \rightarrow (x \text{ is prime} \vee x^2 + 1 \text{ is prime}) \right]$$

Notice how the restricted use of the quantifier $\forall$ in "For every positive integer $x \ldots$" becomes an implication:

$$\forall x [x \in \mathbb{N} \rightarrow \ldots]$$

Statement (b) is written

$$\exists x \left[x \in \mathbb{N} \wedge (x^2 + 1 \text{ is prime} \wedge x \text{ is a perfect square}) \right]$$

Here it is important to understand how the restricted use of the quantifier $\exists$ in "There is some positive integer $x \ldots$" becomes a conjunction:

$$\exists x [x \in \mathbb{N} \wedge \ldots] \qquad \qquad \square$$

Consider again the formula

$$\forall x (x \in \mathbb{R} \rightarrow x^2 + 1 > 0)$$

This formula may also be written as

$$\forall x \in \mathbb{R} (x^2 + 1 > 0)$$

or

$$x^2 + 1 > 0 \, \forall x \in \mathbb{R}$$

In general, if A is a set and $p(x)$ is a propositional function, then all of the statements

$$\forall x (x \in A \rightarrow p(x))$$
$$\forall x \in A \, p(x)$$
$$p(x) \, \forall x \in A$$

say the same thing—namely, that $p(x)$ holds for every element x of A. Similarly, the formula

$$\exists x (x \in \mathbb{R} \wedge x^3 - 3x + 1 = 0)$$

may also be written as

$$\exists x \in \mathbb{R}(x^3 - 3x + 1 = 0)$$

In general, both of the statements

$$\exists x(x \in A \wedge p(x))$$
$$\exists x \in A\, p(x)$$

signify that there is some element x of the set A for which $p(x)$ is true.

It is also possible to have statements that involve several quantifiers. Consider the following two statements:

(a) There exists an integer x such that for every integer y, $x + y = 4$.

(b) For every integer y there exists an integer x such that $x + y = 4$.

These statements are represented by the following formulas:

(a) $\exists x \in \mathbb{Z}[\forall y \in \mathbb{Z}(x + y = 4)]$

(b) $\forall y \in \mathbb{Z}[\exists x \in \mathbb{Z}(x + y = 4)]$

(Note that here we have a propositional function, "$x + y = 4$," that involves two variables.) What can we say about the truth-values of statements (a) and (b)? Statement (a) asserts the existence of an integer x such that, no matter what integer y is chosen, $x + y = 4$. This statement is clearly false, for once y is chosen, $x = 4 - y$ is uniquely determined. Statement (b), on the other hand, states that for any integer y, there is some integer x such that $x + y = 4$. This statement is certainly true.

Example 1.20   For each statement, give a formula that represents it.

(a) For every positive integer n there exists a prime p such that $p > n$.
(b) There exist two primes p and q whose sum is also prime.
(c) For all rational numbers x and y, the sum $x + y$ is also rational.
(d) Every even integer $n \geq 4$ is the sum of two primes p and q.

Solution   Let P denote the set of primes.

(a) $\forall n \in \mathbb{N}\ [\exists p(p \text{ is prime} \wedge p > n)]$ or $\forall n \in \mathbb{N}\ [\exists p \in P(p > n)]$
(b) $\exists p\, \exists q(p \text{ is prime} \wedge q \text{ is prime} \wedge p + q \text{ is prime})$ or
$\exists p \in P\, \exists q \in P(p + q \in P)$
(c) $\forall x \in \mathbb{Q}\, \forall y \in \mathbb{Q}(x + y \in \mathbb{Q})$
(d) $\forall n \in \mathbb{N}\ [(n \text{ even} \wedge n \geq 4) \rightarrow \exists p \in P\, \exists q \in P(n = p + q)]$ □

What is the negation of the statement $\exists x p(x)$? Of course, one can readily say, "It is not the case that there is some x such that $p(x)$." It seems clear, however, that this is equivalent to saying "For all x, not $p(x)$." Hence we see that

$$\sim [\exists x p(x)] \equiv \forall x [\sim p(x)]$$

Similarly, the negation of the statement $\forall x\, p(x)$ is "It is not the case that, for every x, $p(x)$." Equivalently we can say, "There is some x such that not $p(x)$." Thus we arrive at

$$\sim [\forall x\, p(x)] \equiv \exists x\, [\sim p(x)]$$

What about negating the restricted forms of the quantifiers? Letting A be a set, it is not difficult to see that

$$\sim [\exists x \in A\, p(x)] \equiv \forall x \in A\, [\sim p(x)]$$

and

$$\sim [\forall x \in A\, p(x)] \equiv \exists x \in A\, [\sim p(x)]$$

Example 1.21 Find the negation of each formula.

(a) $\forall x \in \mathbb{N}$ (x is prime $\to x^2 + 1$ is even)
(b) $\exists x \in \mathbb{Q}$ ($x > 0 \wedge x^3 = 2$)
(c) $\exists x\, [\forall y\, (xy = y)]$
(d) $\forall x\, \forall y\, [x < y \to \exists z(x < z \wedge z < y)]$

Solution Formula (a) is of the type

$$\forall x \in \mathbb{N}(\, p(x) \to q(x))$$

We determine the negation of such a formula as follows:

$$\sim [\forall x \in \mathbb{N}(\, p(x) \to q(x))] \equiv \exists x \in \mathbb{N}[\sim (\, p(x) \to q(x))]$$
$$\equiv \exists x \in \mathbb{N}[\, p(x) \wedge \sim q(x)]$$

So the negation of formula (a) is

$$\exists x \in \mathbb{N}(x \text{ is prime } \wedge x^2 + 1 \text{ is odd})$$

Formula (b) is of the type

$$\exists x \in \mathbb{Q}(\, p(x) \wedge q(x))$$

Hence the negation is

$$\sim [\exists x \in \mathbb{Q}(\, p(x) \wedge q(x))] \equiv \forall x \in \mathbb{Q}[\sim (\, p(x) \wedge q(x))]$$
$$\equiv \forall x \in \mathbb{Q}[\sim p(x) \vee \sim q(x)]$$

Thus the negation of formula (b) is

$$\forall x \in \mathbb{Q}(x \le 0 \vee x^3 \ne 2)$$

Recalling that $p \to q \equiv \sim p \vee q$, it is preferable to restate the negation as

$$\forall x \in \mathbb{Q}(x > 0 \to x^3 \ne 2)$$

Formula (c) is of the type

$$\exists x\, [\forall y\, p(x, y)]$$

where $p(x, y)$ represents the propositional function $xy = y$. We determine the negation as follows:

$$\sim[\exists x[\forall y p(x, y)]] \equiv \forall x[\sim[\forall y p(x, y)]] \equiv \forall x[\exists y[\sim p(x, y)]]$$

So the negation of formula (c) is

$$\forall x[\exists y(xy \neq y)]$$

Formula (d) has the shape

$$\forall x \,\forall y[p(x, y) \rightarrow \exists z[q(x, z) \wedge r(y, z)]]$$

It is left to the reader to verify that the negation has the form

$$\exists x \,\exists y[p(x, y) \wedge \forall z[\sim q(x, z) \vee \sim r(y, z)]]$$

so that the negation of formula (d) is

$$\exists x \,\exists y[(x < y) \wedge \forall z(x \geq z \vee z \geq y)] \qquad \square$$

Exercises 1.4

1. Let x be a positive integer and define the following propositional functions:

$$p(x): \quad x \text{ is prime}$$
$$q(x): \quad x \text{ is even}$$
$$r(x): \quad x > 2$$

Write out these statements.
(a) $\exists x\, p(x)$ (b) $\forall x\, r(x)$
(c) $\exists x[p(x) \wedge q(x)]$ (d) $\forall x[r(x) \rightarrow (p(x) \vee q(x))]$
(e) $\exists x[p(x) \wedge (q(x) \vee r(x))]$ (f) $\forall x[(p(x) \wedge q(x)) \rightarrow \sim r(x)]$

2. Give a formula for each statement.
(a) For every even integer n there exists an integer m such that $n = 2m$.
(b) Every integer is a rational number.
(c) There exists a right triangle T that is an isosceles triangle.
(d) Given any quadrilateral Q, if Q is a parallelogram and Q has two adjacent sides that are perpendicular, then Q is a rectangle.
(e) There exists an even prime integer.
(f) There exist integers s and t such that $1 < s < t < 187$ and $st = 187$.
(g) There is an integer x such that $x/2$ is an integer and, for every integer y, $x/(2y)$ is not an integer.
(h) Given any real numbers x and y, $2x^2 - xy + 5 > 0$.

3. Find the negation (in simplest form) of each formula.
(a) $\forall x[p(x) \vee q(x)]$ (b) $\forall x \,\forall y[p(x, y) \rightarrow q(x, y)]$
(c) $\forall x[\exists y(p(x, y) \rightarrow q(x, y))]$
(d) $\exists x[(\forall y p(x, y) \rightarrow q(x, y)) \wedge \exists z\, r(x, z)]$

4. For each statement, (i) give a formula for it, (ii) give a formula for its negation, and (iii) express the negation in words.

   (a) For every x and for every y, $x + y = y + x$.

   (b) For every x there exists y such that $y^2 = x$.

   (c) There exists y such that for every x, $2x^2 + 1 > x^2 y$.

   (d) There exist x and y such that $x < y$ and $x^3 - x > y^3 - y$.

   (e) For all x and y there exists z such that $2z = x + y$.

   (f) For every x and y, if $x^3 + x - 2 = y^3 + y - 2$, then $x = y$.

5. Let x represent an integer. Use the propositional functions

$$p(x): \quad x \text{ is even}$$
$$q(x): \quad x \text{ is odd}$$
$$r(x): \quad x^2 < 0$$

to show that formulas u and v are not logically equivalent.

   (a) u: $\forall x[\, p(x) \lor q(x)]$; v: $[\forall x\, p(x)] \lor [\forall x\, q(x)]$

   (b) u: $\exists x[\, p(x) \land q(x)]$; v: $[\exists x\, p(x)] \land [\exists x\, q(x)]$

   (c) u: $\forall x[\, p(x) \to q(x)]$; v: $[\forall x\, p(x)] \to [\forall x\, q(x)]$

   (d) u: $\exists x[\, p(x) \to r(x)]$; v: $[\exists x\, p(x)] \to [\exists x\, r(x)]$

6. Let A be a set. Verify that

   (a) $\exists x \in A[\, p(x) \lor q(x)] \equiv [\exists x \in A\, p(x)] \lor [\exists x \in A\, q(x)]$

   (b) $\forall x \in A[\, p(x) \land q(x)] \equiv [\forall x \in A\, p(x)] \land [\forall x \in A\, q(x)]$

   (c) $\exists x \in A[\, p(x) \to q(x)] \equiv [\forall x \in A\, p(x)] \to [\exists x \in A\, q(x)]$

7. Let the Pascal vector V be declared as follows:

```
CONST MAX = 30;
TYPE  SUBSCRIPTTYPE = 1..MAX;
      ELEMENTTYPE = REAL;
      VECTORTYPE = ARRAY [SUBSCRIPTTYPE] OF ELEMENTTYPE;
VAR   V:VECTORTYPE;
      I,J:SUBSCRIPTTYPE;
```

where the variables I and J represent subscripts for V. Give a formula for each statement about V. (For example, a formula for "Every element of V is positive" is $\forall I\ V[I] > 0$.)

   (a) There is some element of V that is equal to 0.

   (b) Every element of V is equal to 0.

   (c) The elements of V are distinct.

   (d) There exist two elements of V that sum to 0.

   (e) For each element of V there is another element that differs from it by exactly 3.

   (f) The vector V has a unique largest element.

   (g) Each element of V (from the second element on) is greater than its predecessor.

   (h) Any element with an even subscript is greater than every element with an odd subscript.

| 1.5 | ## METHODS OF PROOF |

It may be said that mathematics is uniquely characterized among human endeavors by the practice known as proof. To quote from *The Mathematical Experience* by Phillip Davis and Reuben Hersh:

> Mathematics, then, is the subject in which there are proofs. Traditionally, proof was first met in Euclid, and millions of hours have been spent in class after class, in country after country, in generation after generation, proving and reproving the theorems in Euclid. After the introduction of the "new math" in the mid-nineteen fifties, proof spread to other high school mathematics such as algebra, and subjects such as set theory were deliberately introduced so as to be a vehicle for the axiomatic method and proof. In college, a typical lecture in advanced mathematics, especially a lecture given by an instructor with "pure" interests, consists entirely of definition, theorem, proof, definition, theorem, proof, in solemn and unrelieved concatenation. Why is this? If, as claimed, proof is validation and certification, then one might think that once a proof has been accepted by a competent group of scholars, the rest of the scholarly world would be glad to take their word for it and to go on. Why do mathematicians and their students find it worthwhile to prove again and yet again the Pythagorean theorem or the theorems of Lebesgue or Wiener or Kolmogoroff?
>
> Proof serves many purposes simultaneously. In being exposed to the scrutiny and judgement of a new audience, the proof is subject to a constant process of criticism and revalidation. Errors, ambiguities, and misunderstandings are cleared up by constant exposure. Proof is respectability. Proof is the seal of authority.
>
> Proof, in its best instances, increases understanding by revealing the heart of the matter. Proof suggests new mathematics. The novice who studies proofs gets closer to the creation of new mathematics. Proof is mathematical power, the electric voltage of the subject which vitalizes the static assertions of the theorems.
>
> Finally, proof is ritual, and a celebration of the power of pure reason. Such an exercise in reassurance may be very necessary in view of all the messes that clear thinking clearly gets us into.

A *theorem* is a mathematical statement that is true, and a *proof* is a logical argument that verifies the truth of the theorem. Writing clear and correct proofs is an art, and to become good at it takes much practice. One of the purposes of this textbook is to help the student become proficient at writing proofs. We hope to do this by providing examples of proofs, and also by providing the student many opportunities to test and improve his or her skills.

Initially, however, it is important to understand the proofs given in this and other courses. And to understand a proof, it is necessary to understand the method of proof being used. In this section we discuss the two basic strategies of proof; these are the *direct method* and the *indirect method*. We also briefly introduce the problem of program verification. Later, in Chapter 3, an important method of proof known as mathematical induction is discussed.

Since a proof is a logical argument, a proof is valid provided the logical form of the argument is a tautology. Certain statements that appear in a proof are known to be true; these include axioms as well as theorems that have already been established. Other statements that appear in a proof are hypotheses of the theorem; these are assumed to be true. Still other statements are derived from previous ones using the rules of logical inference.

Example 1.22   Consider the following example of a valid argument:

Fred passed the final exam in CS 260. If Fred passed the final exam in CS 260, then Fred passed the CS 260 course. Therefore, Fred passed the CS 260 course.

Let p represent the statement

Fred passed the final exam in CS 260.

and q represent the statement

Fred passed the CS 260 course.

Then the logical form of the argument is

$$[p \wedge (p \to q)] \to q$$

This is the role of modus ponens, which was shown to be a tautology in Example 1.15. Hence we may conclude that the argument is valid. On the other hand, the following argument is not valid:

If Fred got a grade of B or better in the CS 260 course, then Fred passed the CS 261 course. Fred passed the CS 261 course. Therefore, Fred got a grade of B or better in CS 260.

To see that this is not a valid argument, let s represent the statement

Fred got a grade of B or better in the CS 260 course.

and let t represent the statement

Fred passed the CS 261 course.

Then the logical form of the argument is

$$[(s \to t) \wedge t] \to s$$

But this is not a tautology because it fails to hold when s is false and t is true. (Perhaps Fred got a C+ in CS 260, but by working hard he was able to pass CS 261.)   □

Suppose that we are confronted with the problem of proving a theorem of the form $p \to q$. We must then show that $p \to q$ is true. This will always be the case if p is false. Also, we know that if p is true and q is false, then $p \to q$ is false. Thus, our problem is reduced to showing that if p is true, then q must also be true. This method of proof is called the *direct method*, and its basic outline is as follows:

Problem: Prove $p \to q$.

Direct method: (a)   Assume p is true.

(b)   Show that q is true.

For the direct method of proof, the most common form of argument is known as the "syllogism." We illustrate this form with an example.

Example 1.23   Suppose the following statements are known to be true:

If you get to bed by 3 A.M., then you get up for your 9 A.M. class.

If you win at poker, then you get to bed by 3 A.M.

We may then conclude that the next statement is true:

If you win at poker, then you get up for your 9 A.M. class.   □

The type of reasoning displayed in Example 1.23 is called *syllogistic reasoning*, the logical form of the argument being

$$[(p \to r) \wedge (r \to q)] \to (p \to q)$$

It is easy to verify that the above formula is a tautology (see Exercise 2). This formula reads

If p implies r and r implies q, then p implies q.

Now suppose we wish to prove a theorem of the form $p \to q$. By itself this may be quite difficult. However, suppose we can come up with a statement s such that we can prove $(p \to s) \wedge (s \to q)$. Then the truth of $p \to q$ follows and our proof is complete. Often we may need to string several implications together, such as

$$(p \to s_1) \wedge (s_1 \to s_2) \wedge \cdots \wedge (s_n \to s_{n+1}) \wedge (s_{n+1} \to q)$$

Example 1.24   We give a direct proof of the statement

If $x^2 - 4 = 0$, then either $x = -2$ or $x = 2$.

where x is assumed to be a real number.

Proof   It is given that $x^2 - 4 = 0$, and since $x^2 - 4 = (x + 2)(x - 2)$, we have that $(x + 2)(x - 2) = 0$. If $(x + 2)(x - 2) = 0$, then either $x + 2 = 0$ or $x - 2 = 0$. Therefore, either $x = -2$ or $x = 2$.

Discussion   Let p, r, s, and q represent the following statements (actually propositional functions):

$$p: \quad x^2 - 4 = 0$$
$$r: \quad (x + 2)(x - 2) = 0$$
$$s: \quad x + 2 = 0 \text{ or } x - 2 = 0$$
$$q: \quad x = -2 \text{ or } x = 2$$

Our aim was to prove $p \rightarrow q$. What we actually proved was

$$(p \rightarrow r) \wedge (r \rightarrow s) \wedge (s \rightarrow q)$$

and $p \rightarrow q$ followed. So our proof is valid; being a direct proof that uses syllogistic reasoning. Of course, several other facts were used along the way: $x^2 - 4$ was factored as $(x + 2)(x - 2)$, and a very important property of the real numbers was used: if a and b are real numbers and $ab = 0$, then either $a = 0$ or $b = 0$.   □

Example 1.25   Consider a triangle ABC, such as the one shown in Figure 1.8. We give a direct proof of the theorem:

> If the perpendicular from vertex A to side BC bisects BC, then ABC is isosceles.

Proof   Let D be the point where the perpendicular from A to BC meets BC, as shown in the figure. Since D is the midpoint of BC, segments DB and DC have the same length. Now compare triangles ADB and ADC. These triangles have side AD in common, angles ADB and ADC are both right angles, and sides DB and DC have the same length. Hence, by side–angle–side, triangles ADB and ADC are congruent. It follows that the corresponding sides AB and AC have the same length. Therefore, triangle ABC is isosceles.

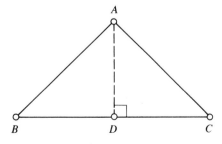

Figure 1.8

Discussion   Let p, u, v, w, and q represent the following statements:

> p: The perpendicular AD from A to side BC bisects BC.
>
> u: Comparing triangles ADB and ADC, they have side AD in common, angles ADB and ADC are both right angles, and sides DB and DC have the same length.
>
> v: Triangles ADB and ADC are congruent.
>
> w: Sides AB and AC have the same length.
>
> q: Triangle ABC is an isosceles triangle.

The proof is a direct proof of $p \rightarrow q$ with the logical form

$$[(p \rightarrow u) \wedge (u \rightarrow v) \wedge (v \rightarrow w) \wedge (w \rightarrow q)] \rightarrow (p \rightarrow q) \qquad \square$$

There are two methods of proof that we refer to as *indirect methods*. The first is known (in Latin) as "reductio ad absurdum," meaning "reduction to an absurdity." It is more commonly called *proof by contradiction*. The second indirect method is called *proof by contrapositive*.

Proof by contradiction is probably the hardest method to understand. It is based on the law of the excluded middle. As mentioned in Section 1.3, this law states that $r \vee \sim r$ is a tautology and $r \wedge \sim r$ is a contradiction for any statement r. Suppose we wish to prove some statement t—that is, to show t is true. To do this we attempt to find a statement r for which

$$\sim t \rightarrow (r \wedge \sim r)$$

is true. In this case the statement $\sim t$ implies a statement that is always false. The only way that such an implication can be true is for the hypothesis $\sim t$ to be false also. Hence it must be that t is true. In other words, the formula

$$[\sim t \rightarrow (r \wedge \sim r)] \rightarrow t$$

is a tautology. We summarize the method of proof by contradiction as follows:

> *Problem*:   Prove t.
>
> *Method of proof by contradiction*:
>   (a)   Assume t is false.
>   (b)   Argue to a contradiction $r \wedge \sim r$ for some statement r.

Before giving an example of a proof by contradiction, we take a moment to discuss some of the history of irrational numbers. While still a young man, the Greek mathematician Pythagoras (circa 550 B.C.) ventured to the seaport of Cratona, in southern Italy, and there founded his famous school of philosophy, mathematics, and natural science. The Pythagoreans were primarily interested in the study of numbers; it is to them that we credit the discovery of irrational numbers. Actually, this discovery came as quite a shock and surprise because much of Pythagorean mathematics had been based on the assumption that any number is rational—that it can be expressed as the ratio of two integers. In fact, many regard the discovery of irrational numbers as an extremely important event in the history of mathematics, for it showed the danger of relying solely on empirical evidence, and it showed the power of the deductive method to set matters straight.

In the next example we give a proof that $\sqrt{2}$ is irrational. Note that $\sqrt{2}$ is the length of the diagonal of a unit square, so it occurs quite naturally in the study of plane geometry.

Example 1.26   Let p denote the statement

$$\sqrt{2} \text{ is irrational.}$$

We use the method of proof by contradiction to prove p—that is, to show p is true.

Proof   Suppose $\sqrt{2}$ is rational. Then there exist positive integers a and b such that $\sqrt{2} = a/b$. Furthermore, we may assume without loss of generality that the fraction a/b is in lowest terms. Then

$$\sqrt{2}\,b = a \rightarrow 2b^2 = a^2 \rightarrow a^2 \text{ is even} \rightarrow a \text{ is even.}$$

Let $a = 2n$. Then

$$2b^2 = a^2 \rightarrow 2b^2 = (2n)^2 \rightarrow 2b^2 = 4n^2 \rightarrow$$
$$b^2 = 2n^2 \rightarrow b^2 \text{ is even} \rightarrow b \text{ is even}$$

But now we have that both a and b are even, contradicting the assumption that a/b is in lowest terms. Therefore, we conclude that $\sqrt{2}$ is irrational.

Discussion   Let r be the statement

Any rational number can be expressed as a fraction in lowest terms.

Here we are accepting r as a known fact. Assuming $\sim p$, we showed $\sim r$, so we proved that

$$\sim p \rightarrow (r \wedge \sim r)$$

is true. Therefore our conclusion is that p is true.   □

The method of proof by contradiction may be used to prove an implication $p \rightarrow q$. Simply replace t by $p \rightarrow q$ in the discussion preceding Example 1.26. Recalling that $\sim(p \rightarrow q)$ is logically equivalent to $p \wedge \sim q$, we may summarize this case of proof by contradiction as follows:

> *Problem*:   Prove $p \rightarrow q$.
>
> *Method of proof by contradiction*:
>
> (a)   Assume p is true and q is false.
>
> (b)   Argue to a contradiction $r \wedge \sim r$ for some statement r.

Example 1.27   From plane geometry, recall the axiom

> r: Given two distinct points, there is exactly one line containing them.

(Remember, an axiom is true by assumption.) Making use of this axiom, prove the following:

> If two distinct lines m and n intersect, then their intersection consists of exactly one point.

Proof   Suppose that two distinct lines m and n intersect in more than one point. Let A and B denote two distinct points where m and n intersect. Then both lines m and n contain points A and B, contradicting the axiom r. Therefore, if m and n intersect, then their intersection must consist of exactly one point.

Discussion   Let p and q represent the statements

> p: Distinct lines m and n intersect.
>
> q: The intersection of m and n consists of exactly one point.

We wanted to prove $p \rightarrow q$. We assumed $p \wedge \sim q$, which led to $\sim r$. Thus the contradiction $r \wedge \sim r$ was obtained. Therefore, $p \rightarrow q$ must be true.   □

The last method of proof we discuss in this section is called proof by contrapositive. Recall that an implication $p \rightarrow q$ and its contrapositive $\sim q \rightarrow \sim p$ are logically equivalent. Thus to prove $p \rightarrow q$ it suffices to prove its contrapositive. Note that once the statements $\sim q$ and $\sim p$ have been identified, the direct method is used to prove $\sim q \rightarrow \sim p$. We summarize the method of proof by contrapositive as follows:

> *Problem*:   Prove $p \rightarrow q$.
>
> *Method of proof by contrapositive*:
>
> (a)   Assume q is false.
>
> (b)   Show that p is false.

Example 1.28   Let x be a real number. Use proof by contrapositive to prove

> If $x^3 - x^2 + x - 1 = 0$, then $x = 1$.

Then prove the same result directly and by contradiction. Compare the three methods.

Proof by contrapositive

Assume $x \neq 1$. Then $x - 1 \neq 0$. Also, since x is real, $x^2 + 1 \neq 0$. It follows that $(x - 1)(x^2 + 1) \neq 0$. Upon multiplication we obtain the result $x^3 - x^2 + x - 1 \neq 0$.

Direct proof

Assume $x^3 - x^2 + x - 1 = 0$. Then $(x - 1)(x^2 + 1) = 0$, which implies that either $x - 1 = 0$ or $x^2 + 1 = 0$. But we know that $x^2 + 1 \neq 0$, and so it must be the case that $x - 1 = 0$. Hence $x = 1$.

Proof by contradiction

Assume $x^3 - x^2 + x - 1 = 0$ and $x \neq 1$. Then $x - 1 \neq 0$. Since $x^3 - x^2 + x - 1 = (x - 1)(x^2 + 1)$ and $x - 1 \neq 0$, it must be that $x^2 + 1 = 0$. But this is a contradiction, given that x is a real number. $\square$

In this section we have discussed three basic strategies for proving a mathematical statement of the form $p \rightarrow q$: the direct method, proof by contradiction, and proof by contrapositive. As illustrated by the last example, it is often the case that more than one method can be applied. As a general rule, however, proofs by contradiction are more involved and harder to follow; they should be done with extreme care. It is perhaps a good practice to see whether a direct proof or a proof by contrapositive can be found first.

Now consider the problem of showing that a statement of the form $p \rightarrow q$ is false. Since this occurs only when p is true and q is false, we must find an instance where the hypothesis p holds but the conclusion q does not. Most often the statement will be of the form:

$$\forall x \in A(r(x))$$

Since

$$\sim [\forall x \in A(r(x))] \equiv \exists x \in A(\sim r(x))$$

we must find an $x_0 \in A$ for which the statement $r(x_0)$ is false. In this case x_0 is a *counterexample* to the statement $\forall x \in A(r(x))$. For instance, to disprove the statement

If n is a positive integer, then $n^2 + n + 41$ is prime.

it suffices to find one positive integer n for which $n^2 + n + 41$ is not prime. Letting $n = 41$, we see that $n^2 + n + 41 = 41^2 + 41 + 41 = 41(41 + 1 + 1) = 41 \cdot 43$, which is clearly not prime. So $n = 41$ is a counterexample to the statement.

Many times the statement $r(x)$ will have the form of an implication $q(x) \rightarrow s(x)$, so that the statement we are trying to disprove becomes

$$\forall x \in A[q(x) \rightarrow s(x)]$$

Since

$$\sim [q(x) \rightarrow s(x)] \equiv q(x) \land \sim s(x)$$

we seek a value $x_0 \in A$ for which the hypothesis $q(x_0)$ is true and the conclusion $s(x_0)$ is false. For example, consider the statement

$$\forall n \in \mathbb{N} \text{ (if } n \text{ is even and } n > 2, \text{ then } n \text{ is expressible}$$
$$\text{as a sum of two primes)}$$

(For instance, notice that $24 = 5 + 19$ and $32 = 13 + 19$.) This very famous conjecture, known as the *Goldbach conjecture*, was made in 1742 and still remains unresolved. To disprove it one would have to find a positive integer n such that n is even, $n > 2$, and n is not expressible as a sum of two primes.

What about the problem of proving a mathematical statement whose form is the biconditional $p \leftrightarrow q$? Recall that the biconditional is defined as the conjunction of two implications:

$$(p \leftrightarrow q) \equiv [(p \rightarrow q) \land (q \rightarrow p)]$$

Therefore, the problem of proving $p \leftrightarrow q$ reduces to that of proving both the implications $p \rightarrow q$ and $q \rightarrow p$. Alternatively, it can be shown (see Exercise 4) that

$$[(p \leftrightarrow r) \land (r \leftrightarrow q)] \rightarrow (p \leftrightarrow q)$$

is a tautology. Thus it is common to prove $p \leftrightarrow q$ by finding a statement r for which $p \leftrightarrow r$ and $r \leftrightarrow q$ both hold.

Before ending this section, we consider a special topic in computer science in which proofs play an important role. This pertains to the area of *algorithm* or *program verification*, in which the problem is to prove that a given algorithm or program performs its task as specified.

Think about the problem of writing, testing, and debugging a reasonably complicated program. Let's suppose, for the sake of discussion, that we have a program written in Pascal. The first thing we might try is to compile the program, but chances are the program contains several syntactic errors and does not compile right off. *Syntactic errors* are those caused by constructs in the program that violate the rules of the Pascal language; for example, using $=$ instead of $:=$ in an assignment statement. With practice and the aid of good error diagnostics from the compiler, such errors are not difficult to fix. So eventually we get the program running. Now we are interested in whether the program gives the correct results for various sets of test data. If, for some set of valid data, the program gives an incorrect result, then the program is said to contain a *logical error*. Even if the program works as expected for each set of test data, this does not guarantee that all logical errors have been eliminated. It is usually impossible to test every possible data set, and although by choosing good test data we can make the probability of error quite small, we still aren't positive that the program is correct. (This is evidenced by the numerous examples of commercial software in which subtle, and sometimes gross, errors are found by the users.) In general, program verification requires the use of formal proof techniques not unlike those discussed in this section.

Example 1.29 The Pascal program segment for the linear search (see Example 1.5) is repeated below. We wish to prove it correct.

```
            (*INITIAL STATE: We assume the value of MAX is a
             positive integer, and that V[1],...,V[MAX] and KEY
             have values of type ELEMENTTYPE.*)
(*begin search*)
CURRENT := 1;
FOUND := FALSE;
WHILE (NOT FOUND) AND (CURRENT <= MAX) DO
  IF V[CURRENT] = KEY
   THEN BEGIN
     FOUND := TRUE;
     POSITION := CURRENT
     END;
   ELSE CURRENT := CURRENT + 1;
(*end search*)

            (*FINAL STATE: The Boolean expression
            (FOUND AND V[POSITION] = KEY) OR (NOT FOUND AND
            V[J] <> KEY)
            is true for all J, 1 <= J <= MAX.*)
```

In order to prove such a program segment is correct, we must have a precise statement of its intended purpose. This involves specifying the *initial state* of computation before the segment is executed and the desired *final state* of computation after the segment has been executed. Specifying the state of computation involves, among other things, specifying the values of certain relevant variables and constants. We have given the initial and final states of computation as comments preceding and following the program segment, respectively.

The "action" in this program segment centers about the execution of the WHILE loop. In fact, the segment terminates when the WHILE loop is exited, which happens when the Boolean expression

```
(NOT FOUND) AND (CURRENT <= MAX)
```

evaluates to FALSE. Whenever execution of the segment is about to evaluate this condition, we say that execution has reached the *top of the loop*.

For most segments that contain loops, the key to proving the segment is correct is finding a statement about the state of computation that holds whenever execution is at the top of the loop and that remains true when the loop is exited. Such a statement is known as a *loop invariant condition*. Often the most difficult part of program verification is coming up with the right loop invariant condition.

What is the correct loop invariant condition in this case? A hint comes from considering the final state, since the final state should be true when the loop is exited. The final state is

```
(FOUND AND V[POSITION] = KEY) OR (NOT FOUND AND V[J]≠KEY,
1 ≤ J ≤ MAX)
```

At some intermediate stage, if we reach the top of the loop and KEY has not been found, then the most we can say is that V[J] ≠ KEY, $1 \le J <$ CURRENT. Thus the loop invariant condition (LIC) is as follows:

```
(FOUND AND V[POSITION] = KEY) OR (NOT FOUND AND V[J]≠KEY,
1 ≤ J < CURRENT)
```

Proving that the segment is correct now amounts to showing that LIC holds whenever execution reaches the top of the loop. This reduces to showing two things:

1. When execution first reaches the top of the loop, LIC holds.
2. If execution reaches the top of the loop and LIC holds, then either the program segment terminates or, the next time execution reaches the top of the loop, LIC still holds.

It follows that LIC holds when the segment terminates, giving us the desired final state of computation. (The segment does terminate because at each pass through the loop, either KEY is found or the value of CURRENT is incremented by 1. Hence the condition (NOT FOUND) AND (CURRENT ≤ MAX) is false eventually.)

Proof of 1   When execution first reaches the top of the loop, FOUND = FALSE and CURRENT = 1, so LIC is satisfied (vacuously).

Proof of 2   Assume execution reaches the top of the loop and LIC holds. We consider three cases, depending on the values of FOUND and CURRENT.

Case 1 FOUND = TRUE. Then the program segment terminates and, since LIC holds, V[POSITION] = KEY.

Case 2 FOUND = FALSE and CURRENT > MAX. Then the program segment terminates and, since LIC holds, V[J] ≠ KEY, 1 ≤ J ≤ MAX.

Case 3 FOUND = FALSE and CURRENT ≤ MAX. In this case execution proceeds to the body of the WHILE loop. IF V[CURRENT] = KEY, then FOUND is set to TRUE and POSITION is assigned the value of CURRENT. Thus, when execution returns to the top of the loop, LIC still holds. On the other hand, if V[CURRENT] ≠ KEY, then the value of CURRENT is incremented by 1. Hence, when execution returns to the top of the loop, LIC (in particular, (NOT FOUND) AND V[J] ≠ KEY, 1 ≤ J < CURRENT) still holds. □

As is evidenced by the preceding example, program verification is often a difficult and costly process, although perhaps not as costly as the consequences of an incorrect program. Computer scientists continue to explore formal techniques for program verification. It is hoped that this work will result in computer programs to aid in the task of program verification, thus reducing the cost and effort involved.

Exercises 1.5

1. Determine the validity of each argument by giving a formula that represents the argument and then deciding whether the formula is a tautology.
    (a) Randy does not work hard. In order for Randy to pass this course, it is necessary that he work hard. Therefore, Randy will not pass this course.
    (b) If the President wins in California, then he will be reelected. But the President will lose in California. Therefore, the President will not be reelected.
    (c) Either the Redskins or the Raiders will win the Super Bowl. The Raiders will not win the Super Bowl. Therefore, the Redskins will win the Super Bowl.
    (d) If Dr. Boynton is good, then her MA 351 course is interesting. Either Dr. Boynton's assignments are challenging or the MA 351 course is not interesting. The assignments are not challenging. Therefore, Dr. Boynton is not good.
    (e) For the President to be reelected it is sufficient that he negotiate an arms reduction treaty with the Soviet Union. He will negotiate such a treaty only if the Soviets pull their ground troops out of eastern Europe. But the Soviets will not pull their ground troops out of eastern Europe. Therefore, the President will not be reelected.
2. Verify that the syllogism $[(p \rightarrow r) \wedge (r \rightarrow q)] \rightarrow (p \rightarrow q)$ is a tautology.
3. Let x and y be real numbers. Give (a) a direct proof and (b) an indirect proof of the statement $(x = y) \rightarrow (x^2 = y^2)$.
4. Verify that the formula $[(p \leftrightarrow r) \wedge (r \leftrightarrow q)] \rightarrow (p \leftrightarrow q)$ is a tautology.
5. Given triangle ABC, prove that if sides AB and AC have the same length, then angles ABC and ACB have the same measure.

6. Recall the fifth postulate (axiom) of Euclid:

> Given a line m and a point P not on m (in a plane), then there is exactly one line through P and parallel to m.

Let u, v, and w be distinct lines in a plane. Use the axiom and the method of proof by contradiction to prove the following result: If u is parallel to v and v is parallel to w, then u is parallel to w.

7. Let m and t be integers. Recall that t is a factor of m provided $m = t \cdot k$ for some integer k; m is even provided 2 is a factor of m.
   (a) Prove directly: If m is even, then m^2 is even.
   (b) Prove directly: If 3 is a factor of m, then 3 is a factor of m^2.
   (c) Prove indirectly: If m^2 is even, then m is even.
   (d) Prove indirectly: If m^3 is even, then m is even.
   (e) Prove indirectly: If 3 is a factor of m^2, then 3 is a factor of m.

8. Use the method of proof by contradiction to prove:
   (a) $\sqrt{3}$ is irrational.   (b) $\sqrt[3]{2}$ is irrational.

9. Let x be a real number. Prove by contradiction: If $x^3 + 4x = 0$, then $x = 0$.

10. Use a counterexample to disprove the statement

> If p is an odd prime, then $p^2 + 4$ is prime.

11. Prove that each program segment is correct.

(a) ```
(*INITIAL STATE: A = a, B = b, where a and b are
   integers and a >= 0.*)

PRODUCT := 0;
WHILE A > 0 DO
  BEGIN
    PRODUCT := PRODUCT + B;
    A := A - 1
  END;

(*FINAL STATE: PRODUCT = ab.*)
(*Hint: Use the loop invariant condition PRODUCT =
   ab - Ab.*)
```

(b) ```
(*INITIAL STATE: The values of A, B, and C are of
   type REAL.*)

MAXI := A;
IF B > MAXI THEN MAXI := B;
IF C > MAXI THEN MAXI := C;
(*FINAL STATE: ((MAXI = A) OR (MAXI = B) OR
     (MAXI = C)) AND (MAXI >= A) AND (MAXI >= B)
     AND (MAXI >= C);
   that is, MAXI is the maximum of A, B, and C.*)
```

CHAPTER PROBLEMS

1. The statements

> p: Ralph reads the *New York Times*.
>
> q: Ralph watches the *MacNeil, Lehrer Newshour*.
>
> r: Ralph jogs 3 miles.

are those of Exercise 1 in Exercises 1.2. Express each formula in words.

(a) $(p \wedge q) \to r$ (b) $\sim(p \leftrightarrow q)$

(c) $\sim(p \wedge q) \to r$ (d) $\sim p \vee \sim q \to r$

(e) $p \to (q \vee r)$ (f) $r \to [\sim(p \to q)]$

(g) $p \wedge q \wedge \sim r$ (h) $(p \to q) \to r$

2. Determine the truth-value of each statement.

(a) $1 + 1 = 2$ or 10 is even.   (b) $1 + 1 = 2$ and 10 is even.

(c) $1 + 1 = 2$ or 10 is odd.   (d) $1 + 1 = 2$ and 10 is odd.

(e) $1 + 1 = 3$ or 10 is odd.   (f) $1 + 1 = 3$ and 10 is odd.

(g) $1 + 1 = 2$ implies 10 is even.   (h) $1 + 1 = 2$ implies 10 is odd.

(i) $1 + 1 = 3$ implies 10 is even.   (j) $1 + 1 = 3$ implies 10 is odd.

3. Let p, q, and r denote the following statements:

> p: Fred knows BASIC.
>
> q: Fred knows Pascal.
>
> r: Fred has taken CS 260.

Express, unambiguously, each formula in words.

(a) $r \leftrightarrow (p \vee q)$ (b) $r \to q$

(c) $r \wedge \sim p$ (d) $q \to (r \wedge \sim p)$

(e) $(p \wedge q) \vee \sim r$ (f) $p \wedge (r \to q)$

4. For each statement, (i) give a formula for it, (ii) find the negation of the formula, and (iii) express the negation in words.

(a) The function $f(x) = x^3 - x$ is both one-to-one and onto.

(b) $2^{1234} - 1$ is prime or even.

(c) Knowing that the graph G is hamiltonian is sufficient to say that G is connected.

(d) If 6 is related to 18 and 18 is related to 72, then 6 is related to 72.

(e) The function $f(x) = 2^x$ is one-to-one if and only if it is onto.

5. Find and simplify the negation of each formula.

(a) $p \wedge q \wedge r$ (b) $p \leftrightarrow q$

(c) $p \to (q \to r)$ (d) $p \wedge (q \vee r)$

(e) $p \wedge (p \to q) \wedge (q \to r)$ (f) $\sim p \wedge (q \to p)$

(g) $[p \wedge (q \to r)] \vee (\sim q \wedge p)$ (h) $p \to (q \vee r)$

6. Verify each of these logical equivalences and indicate how it provides a strategy for proving an implication involving a compound hypothesis or conclusion.
   (a) $[(p \wedge q) \to r] \equiv [p \to (q \to r)]$
   (b) $[(p \vee q) \to r] \equiv [(p \to r) \wedge (q \to r)]$
   (c) $[p \to (q \wedge r)] \equiv [(p \to q) \wedge (p \to r)]$
   (d) $[p \to (q \vee r)] \equiv [(p \wedge \sim r) \to q]$

7. Often in mathematics we wish to prove that a statement of the form

$$(p \leftrightarrow q) \wedge (q \leftrightarrow r) \wedge (r \leftrightarrow p)$$

   is true; in this case we say that the statements p, q, and r are *equivalent*. To show that p, q, and r are equivalent, show that it suffices to prove

$$(p \to q) \wedge (q \to r) \wedge (r \to p)$$

8. For each formula, find a logically equivalent formula that uses only the logical connectives $\wedge$ and $\sim$.
   (a) $p \vee q$ (b) $p \to q$
   (c) $p \leftrightarrow q$ (d) $p \vee (q \to \sim r)$

9. The *nand* operator, denoted by $|$, is defined by $p \mid q \equiv \sim(p \wedge q)$ (hence the acronym "nand" for "not and").
   (a) Give the truth table for $p \mid q$.
   (b) Show that $\sim p \equiv p \mid p$.
   For each formula, find a logically equivalent formula that uses only nand.
   (c) $p \wedge q$ (d) $p \vee q$
   (This problem shows that any formula involving the logical connectives may be expressed using nand only.)

10. Given a statement p, let p' denote the statement

$$p \text{ is true.}$$

   Show that $p \equiv p'$. (Sometimes the phrase "p implies q" is used to mean "p implies q is true.")

11. Find (i) the inverse, (ii) the converse, and (iii) the contrapositive of each implication.
   (a) If x is odd, then x^2 is odd.
   (b) x^2 is even only if x is even.
   (c) If f is differentiable, then f is continuous.
   (d) $x^3 = y^3$ implies $x = y$.
   (e) $F(x) > F(y)$ only if $x > y$.
   (f) The conditions x is prime and $x > 2$ are sufficient for x to be odd.
   (g) For the function f to be defined at 0 it is necessary that f be continuous at 0.
   (h) Being connected and either having no cycles or having its order be one greater than its size are together sufficient conditions for a graph G to be a tree.

12. Consider the connective $\veebar$ (exclusive or), as defined in Exercise 6 of Exercises 1.2.
    (a) Show that $p \veebar q \equiv (p \wedge \sim q) \vee (\sim p \wedge q)$.
    (b) Show that $p \veebar q \equiv \sim (p \leftrightarrow q)$.
13. The implications $(p \wedge \sim r) \to \sim q$ and $(q \wedge \sim r) \to \sim p$ are called *partial contrapositives* of the implication $(p \wedge q) \to r$.
    (a) Show that $(p \wedge q) \to r \equiv (p \wedge \sim r) \to \sim q$.
    (b) Show that $(p \wedge q) \to r \equiv (q \wedge \sim r) \to \sim p$.
    Find both partial contrapositives of each statement.
    (c) If n is prime and $n > 2$, then n is odd.
    (d) If $f'(x) = 2x + 1$ and $f(0) = 3$, then $f(x) = x^2 + x + 3$.
14. Write each statement using the logical symbols.
    (a) Given any real numbers x and y, $2^x = 2^y$ if and only if $x = y$.
    (b) If 7 is the greatest common divisor of 119 and 154, then there exist integers s and t such that $7 = 119s + 154t$.
    (c) For every positive integer n, either $n = 1$ or n is prime or there exist integers s and t such that $1 < s \le t < n$ and $n = st$.
    (d) There is a function f such that both f is one-to-one and f has the property $f(x) \ne x$ for every real number x.
15. For each truth table in Figure 1.9, find a formula u that has that truth table.

a.

p	q	u
T	T	F
T	F	F
F	T	F
F	F	T

b.

p	q	u
T	T	F
T	F	T
F	T	T
F	F	F

c.

p	q	u
T	T	F
T	F	T
F	T	F
F	F	T

d.

p	q	r	u
T	T	T	F
T	T	F	T
T	F	T	F
T	F	F	T
F	T	T	F
F	T	F	F
F	F	T	F
F	F	F	F

e.

p	q	r	u
T	T	T	T
T	T	F	F
T	F	T	T
T	F	F	F
F	T	T	T
F	T	F	T
F	F	T	T
F	F	F	F

f.

p	q	r	u
T	T	T	F
T	T	F	T
T	F	T	T
T	F	F	T
F	T	T	T
F	T	F	T
F	F	T	T
F	F	F	F

Figure 1.9

16. For each pair of statements about positive real numbers x and y, give formulas for both statements and note the difference between them. Also determine each statement's truth-value.
    (a) For every x there exists y such that $y < x$.
    There exists y such that $y < x$ for every x.
    (b) For every x there exists y such that $xy = 1$.
    There exists y such that $xy = 1$ for every x.

17. Let x, y, and z be real numbers and let n be a positive integer. Find and simplify the negation of each of the following statements.
    (a) $\exists y[\forall x(xy \geq 3)]$ (b) $\forall x[\exists y(\forall z(x + y = z))]$
    (c) $\forall x(x > 0 \rightarrow x^2 \geq x)$ (d) $\forall x[\exists y(x < y \wedge 2^x \geq 2^y)]$
    (e) $\exists x[x \in \mathbb{Q} \wedge \forall y(y \notin \mathbb{Q} \rightarrow x + y \notin \mathbb{Q})]$
    (f) $\forall x \, \forall y[(x > 0 \wedge y > 0) \rightarrow \exists n(nx > y)]$
    (g) $\forall y[y > 0 \rightarrow \exists x(\log x > y)]$

18. For each statement, (i) give a formula for it, (ii) find a formula for its negation, and (iii) write out the negation in words.
    (a) There exist real numbers x and y such that $x^2 + y^2 = -1$.
    (b) There exists a positive integer N such that for every real number $x \neq 1$, $N > 1/(x - 1)$.
    (c) Given any real numbers x and y, $2x^2 - xy + 5 > 0$.
    (d) For every real number x there is some positive integer n such that x^n is rational.
    (e) For every integer n, if n is a multiple of 6, then there exist integers s and t such that $n = 12s + 18t$.
    (f) For every positive real number ε there is some positive real number δ such that if $|x - 2| < \delta$, then $|x^2 - 4| < \varepsilon$.

19. Find the contrapositive of each of the following implications.
    (a) If, for all x and y in the set A, the fact that x is related to y implies that y is related to x, then the relation R is symmetric.
    (b) If 7 is the greatest common divisor of 119 and 154, then there exist integers s and t such that $7 = 119s + 154t$.
    (c) In order for the function $\log x$ to be onto, it is necessary that, given any real number b, there is some positive real number a such that $\log a = b$.
    (d) If the relation R on the set A is transitive, then, for all a, b, and c in A, the conditions a is related to b and b is related to c are sufficient to imply that a is related to c.

20. Let x and y represent integers.
    (a) Prove: If both x and y are odd, then $x + y$ is even.
    (b) Prove the result of part (a) by contradiction. (Hint: Derive that 1 is even, a contradiction.)
    (c) Prove by contrapositive: If xy is odd, then both x and y are odd.

21. Determine whether each of the following arguments is valid.
    (a) Knowing Pascal is sufficient for passing CS 260. Nancy knows Pascal; therefore, Nancy passed CS 260.

(b) Fred is required to take MA 350 or MA 337. Fred will not take MA 350; therefore, he will take MA 337.

(c) Al is taking both MA 350 and MA 337. If he doesn't take MA 350, then he must take MA 337. Therefore, Al is not taking MA 337.

(d) If one learns Pascal, then one passes CS 260. If one doesn't pass CS 260, then one can't take CS 261. Therefore, if one is taking CS 261, then one learned Pascal.

(e) Janet will pass this course if and only if she works hard. Either she'll work hard or she won't graduate. But Janet will graduate; therefore, she will pass this course.

22. Use the method of proof by contradiction to prove these statements about real numbers x and y.

(a) If x is rational and y is irrational, then $x + y$ is irrational.

(b) If x is rational, $x \neq 0$, and y is irrational, then xy is irrational.

(c) If x and y are both positive, then $\sqrt{x + y} \neq \sqrt{x} + \sqrt{y}$.

23. Prove the statement:

$$\text{If } x \text{ is a positive real number, then } x + (1/x) \geq 2.$$

(a) by contrapositive and **(b)** by contradiction.

24. Given lines m_1, m_2, and m_3 and angles α and β, as shown in Figure 1.10, prove that if $\alpha = \beta$, then lines m_1 and m_2 are parallel. (Hint: Prove the contrapositive. If m_1 and m_2 intersect at some point C, consider triangle ABC.)

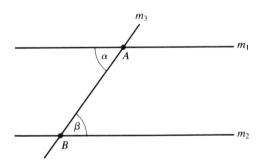

Figure 1.10

25. Supply a conclusion, in the form of an implication, to the following argument so as to make it valid.

If Joe goes to a party, he does not fail to brush his hair. To look fascinating it is necessary that Joe be tidy. If Joe is an opium eater, then he has no self-command. If Joe brushes his hair, he looks fascinating. Joe wears white kid gloves only if he goes to a party. Having no self-command is sufficient to make Joe look untidy. Therefore, ...

T W O

Set Theory

SETS AND SUBSETS

It is the purpose of this chapter to define relations and operations on sets and to derive several of their basic properties. This material provides some very important background for the later chapters of this book and also for subsequent courses in mathematics and computer science.

The following notational conventions were introduced in Chapter 1:

$$\mathbb{N} = \text{the set of positive integers}$$
$$\mathbb{Z} = \text{the set of integers}$$
$$\mathbb{Q} = \text{the set of rational numbers}$$
$$\mathbb{R} = \text{the set of real numbers}$$

Except for these and some other special sets, uppercase italic letters, such as A, B, C are used to denote sets. Lowercase italic letters, such as a, b, c, are used to denote elements.

In any particular discussion of sets, we are almost always concerned with subsets of some specified set U. The set U is called a *universal set*. For instance, in a calculus lecture it may be that $U = \mathbb{R}$, while in a Pascal class, U might be the set of values of the type INTEGER for some particular implementation.

There is a special set that is used often in mathematics; this is the set that has no elements. It is aptly called the *empty set* and is denoted $\emptyset$. Unlike U, the set $\emptyset$ is unique (there is exactly one set that contains no elements), although there are numerous examples of it. For example, consider the set of former female Presidents of the United States or the set of integers m such that m^2 is negative. An alternative notation for the empty set is { }.

A set A is called *finite* if A consists of n elements for some nonnegative integer n. Sets that are not finite are called *infinite*; in particular, the sets $\mathbb{N}$, $\mathbb{Z}$, $\mathbb{Q}$, and $\mathbb{R}$ are infinite.

As mentioned in the first chapter, if a set A is finite with n elements and n is not too large, then A can be exhibited by explicitly listing its elements between braces. For example, if A is the set of prime numbers less than 20, then we write

$$A = \{2, 3, 5, 7, 11, 13, 17, 19\}$$

Recall that the three-dot (or ellipsis) notation, $\ldots$, was also introduced, as well as the notation

$$A = \{x \in S \mid q(x)\}$$

Recall that this is read "A is the set of those elements x belonging to the set S such that (the propositional function) $q(x)$ holds."

Example 2.1   The set A of positive even integers less than 100 is written as

$$A = \{2, 4, 6, \ldots, 98\}$$

The set H of "harmonic fractions" is defined by

$$H = \left\{ x \in \mathbb{Q} \mid x = \frac{1}{n} \text{ for some } n \in \mathbb{N} \right\}$$

Alternatively one may write

$$H = \left\{ 1, \tfrac{1}{2}, \tfrac{1}{3}, \tfrac{1}{4}, \ldots \right\}$$

The set P of primes can be exhibited as either

$$P = \{2, 3, 5, 7, 11, \ldots\}$$

or

$$P = \{ p \in \mathbb{N} \mid p \text{ is prime} \}$$

The latter method of writing P may be preferable as it is less likely to be misunderstood.   ■

For the remainder of this section it is assumed that all sets are subsets of some universal set U.

Perhaps the most fundamental relationship that can exist between two sets is equality.

DEFINITION 2.1

Two sets A and B are called *equal*, denoted $A = B$, provided they consist of the same elements. If A and B are not equal, we write $A \neq B$.

Set equality may be restated as

$$A = B \leftrightarrow \forall x (x \in A \leftrightarrow x \in B)$$

Under what condition is $A \neq B$? Interpreted as $\sim(A = B)$, the logical negation is derived as follows:

$$
\begin{aligned}
A \neq B &\equiv \sim [\forall x (x \in A \leftrightarrow x \in B)] \\
&\equiv \exists x [\sim (x \in A \leftrightarrow x \in B)] \\
&\equiv \exists x [\sim ((x \in A \rightarrow x \in B) \wedge (x \in B \rightarrow x \in A))] \\
&\equiv \exists x [\sim (x \in A \rightarrow x \in B) \vee \sim (x \in B \rightarrow x \in A)] \\
&\equiv \exists x [(x \in A \wedge x \notin B) \vee (x \in B \wedge x \notin A)]
\end{aligned}
$$

In words, $A \neq B$ if and only if there is some element x such that either $x \in A$ and $x \notin B$ or $x \in B$ and $x \notin A$. More briefly, $A \neq B$ if and only if there is an element x that belongs to one of the sets but not to the other.

Example 2.2   Consider the following sets:

$$
\begin{aligned}
B &= \{0, 1\} \\
D &= \{0, 1, 0\} \\
A &= \{0, 1, \{0, 1\}\} \\
G &= \{\emptyset, \{0\}, \{1\}, \{0, 1\}\} \\
C &= \{y \mid y = \{0, 1\} \vee y \in \{0, 1\}\}
\end{aligned}
$$

Which of these sets, if any, are equal?

Solution   First note that sets B and D are equal because both contain exactly the elements 0 and 1. We do not normally write a set as we have written D; in fact, we adopt the convention that in listing the elements of a set, each element is to be listed exactly once. The set A contains three elements: $0, 1$, and the set $\{0, 1\}$. The set G contains four elements: the empty set, and the sets $\{0\}, \{1\}$, and $\{0, 1\}$. Thus A and G are not equal and neither is equal to B. What about C? Note that the set $\{0, 1\}$ belongs to C, as do the elements (0 and 1) of this set. Hence, $C = \{0, 1, \{0, 1\}\} = A$. ∎

The definition of *subset* is given in Chapter 1 and is repeated here for convenience.

DEFINITION 2.2

A set A is called a *subset* of a set B, denoted $A \subseteq B$, provided every element of A is also an element of B. If A is not a subset of B, then we write $A \nsubseteq B$.

It should be noted that for sets A and B,

$$A \subseteq B \leftrightarrow \forall x (x \in A \rightarrow x \in B)$$

It is evident from this condition that the relations

$$\varnothing \subseteq A$$

and

$$A \subseteq A$$

hold for any set A. It is also not difficult to see that

$$A \nsubseteq B \leftrightarrow \exists x (x \in A \wedge x \notin B)$$

Thus to show that $A \nsubseteq B$ it suffices to show there is some $x \in A$ such that $x \notin B$.

Example 2.3   Consider the following sets:

$$A = \{ x \in \mathbb{Z} \mid -20 \leq x \leq 20 \}$$
$$B = \{ n \in \mathbb{N} \mid n^2 \leq 49 \}$$
$$C = \{ A, B, -3, 4, 6 \}$$
$$D = \{ -3, 4, 6, 7 \}$$

We have the following facts about these sets:

$B \subseteq A$

$C \nsubseteq A$, since $A \in C$ but $A \notin A$

$D \subseteq A$

$C \nsubseteq B$

$D \nsubseteq B$, since $-3 \in D$ but $-3 \notin B$

$D \nsubseteq C$, since $7 \in D$ but $7 \notin C$ ■

Frequently in mathematics we are given two sets A and B and asked to prove that $A \subseteq B$. In keeping with Definition 2.2, it is common practice to start such a proof with the sentence

Let x be an arbitrary element of A.

Then we proceed to show that x is an element of B. This scheme is illustrated in the proof of the next theorem.

THEOREM 2.1   Let A, B, and C be subsets of a universal set U. If $A \subseteq B$ and $B \subseteq C$, then $A \subseteq C$.

Proof   Let x be an arbitrary element of A. We must show that $x \in C$. Since $A \subseteq B$ and $x \in A$, it follows that $x \in B$. Then, since $B \subseteq C$ and $x \in B$, we may conclude that $x \in C$. Thus we have shown that for every x, if $x \in A$, then $x \in C$. Therefore, $A \subseteq C$. □

The student should observe that syllogistic reasoning is used in the preceding proof; in particular, the logical form of the argument is

$$[(x \in A \to x \in B) \wedge (x \in B \to x \in C)] \to (x \in A \to x \in C)$$

It has been pointed out that to prove a mathematical statement whose form is the biconditional $p \leftrightarrow q$, it is necessary to prove both $p \to q$ and $q \to p$. It is also possible to prove $p \leftrightarrow q$ by proving both $p \leftrightarrow r$ and $r \leftrightarrow q$ for some proposition r. This technique is used to prove the next theorem.

THEOREM 2.2 For any two subsets A and B of a universal set U,

$$A = B \leftrightarrow (A \subseteq B \wedge B \subseteq A)$$

Proof We proceed as follows:

$$A = B \leftrightarrow \forall x (x \in A \leftrightarrow x \in B)$$
$$\leftrightarrow \forall x [(x \in A \to x \in B) \wedge (x \in B \to x \in A)]$$
$$\leftrightarrow [\forall x (x \in A \to x \in B)] \wedge [\forall x (x \in B \to x \in A)]$$
$$\leftrightarrow (A \subseteq B) \wedge (B \subseteq A)$$ □

THEOREM 2.3 For any two subsets A and B of a universal set U,

$$A \subseteq B \leftrightarrow \forall C (C \subseteq A \to C \subseteq B)$$

Proof First we show that

$$A \subseteq B \to \forall C (C \subseteq A \to C \subseteq B)$$

Assume $A \subseteq B$ and let C be any set such that $C \subseteq A$. Then we have $C \subseteq A$ and $A \subseteq B$, and we may conclude from Theorem 2.1 that $C \subseteq B$. Next it must be shown that

$$\forall C (C \subseteq A \to C \subseteq B) \to A \subseteq B$$

Here our hypothesis is that *for any set* C, if $C \subseteq A$, then $C \subseteq B$. In particular, let $C = A$. Then $A \subseteq A$ implies $A \subseteq B$, which is precisely what we wanted to prove. □

THEOREM 2.4 Let A be any subset of a universal set U. Then, for all $x \in U$,

$$x \in A \leftrightarrow \{x\} \subseteq A$$

The proof of Theorem 2.4 is left to Exercise 4.

> **DEFINITION 2.3**
>
> A set A is called a *proper subset* of a set B provided $A \subseteq B$ and $A \neq B$. We write $A \subset B$ to denote that A is a proper subset of B.

THEOREM 2.5   For any two subsets A and B of a universal set U,

$$A \subset B \leftrightarrow [A \subseteq B \wedge \exists x(x \in B \wedge x \notin A)]$$

The proof of Theorem 2.5 is left to Exercise 6.

Given a finite collection of sets, it is sometimes useful to have a "picture" of the subset relationships that exist among them. We construct such a picture as follows. Corresponding to each set in the collection is a point, called a *vertex*, that is labeled with the name of the set. Next suppose X and Y are two sets in the collection and $X \subset Y$. Further assume that there is no set Z in the collection such that $X \subset Z$ and $Z \subset Y$. Then a directed line segment, called an *arc*, is drawn joining the vertex X to the vertex Y, as shown in Figure 2.1(a). (It is also customary to place the vertex Y somewhere above the vertex X. In addition, some people prefer to join X and Y with a simple line segment, called an "edge," with the implicit assumption that all edges are "directed upward.") Such a picture is called a *subset diagram* for the given collection of sets.

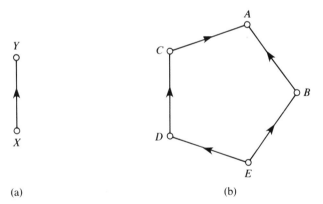

(a)   (b)

Figure 2.1 Subset diagrams

Example 2.4   Consider the collection of sets $\{A, B, C, D, E\}$, where $A = \{1, 2, 3, 4\}$, $B = \{1, 3\}$, $C = \{2, 3, 4\}$, $D = \{2, 3\}$, and $E = \{3\}$. For this collection we have the subset diagram shown in Figure 2.1(b). Notice that no arc is drawn, for example, from D to A. This is because $D \subset C$ and $C \subset A$. From the subset diagram and Theorem 2.1, it can be deduced that $D \subset A$. ∎

Given a set A, we are often interested in the collection of all subsets of A.

DEFINITION 2.4

The set consisting of all subsets of a given set A is called the *power set* of A and is denoted $\mathscr{P}(A)$.

Example 2.5   Describe $\mathscr{P}(A)$ if the set A has:
      (a) 0 elements   (b) 1 element   (c) 2 elements

Solution   (a) If the set A has 0 elements, then $A = \emptyset$ and the only subset of A is itself. So $\mathscr{P}(\emptyset) = \{\emptyset\}$. (Note that $\{\emptyset\}$ is not empty; it is a set with exactly one element, that element being $\emptyset$.)
(b) If A has exactly one element, then the subsets of A are $\emptyset$ and A, and so $\mathscr{P}(A) = \{\emptyset, A\}$.
(c) Suppose $A = \{a, b\}$. Then, besides $\emptyset$ and A, $\mathscr{P}(A)$ includes two nonempty proper subsets of A—namely, $\{a\}$ and $\{b\}$. Thus $\mathscr{P}(A) = \{\emptyset, \{a\}, \{b\}, A\}$. ∎

Two important properties of $\mathscr{P}(A)$ worth remembering are that $\emptyset \in \mathscr{P}(A)$ and $A \in \mathscr{P}(A)$.

Exercises 2.1

1. Determine which of the following assertions are true and which are false.
   (a) $\emptyset \in \emptyset$ (b) $1 \in \{1\}$
   (c) $\{1, 2\} = \{2, 1\}$ (d) $\emptyset = \{\emptyset\}$
   (e) $\emptyset \subset \{\emptyset\}$ (f) $1 \subseteq \{1\}$
   (g) $\{1\} \subset \{1, 2\}$ (h) $\emptyset \in \{\emptyset\}$

2. Identify which of these sets are equal.

$$A = \{m \in \mathbb{Z} \mid |m| < 2\}$$

$$B = \{m \in \mathbb{Z} \mid m^3 = m\}$$

$$C = \{m \in \mathbb{Z} \mid m^2 \leq 2m\}$$

$$D = \{m \in \mathbb{Z} \mid m^2 \leq 1\}$$

$$E = \{0, 1, 2\}$$

3. Find the power set of each set.
   (a) $\emptyset$ (b) $\{1\}$
   (c) $\{1, 2\}$ (d) $\{1, 2, 3\}$
   (e) $\{\emptyset, \{1\}\}$ (f) $\mathscr{P}(\mathscr{P}(\emptyset))$

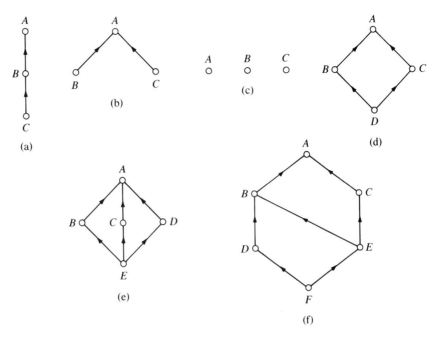

Figure 2.2

4. Prove Theorem 2.4.
5. In view of Exercise 3, make a guess as to the number of subsets of a finite set having n elements.
6. Prove Theorem 2.5.
7. Draw the subset diagram for the collection $\{A, B, C, D, E, F, G\}$, where $A = \{1, 2, 3, 4, 5\}$, $B = \{1, 2, 3\}$, $C = \{2, 3, 4\}$, $D = \{2, 4\}$, $E = \{3, 4\}$, $F = \{4, 5\}$, and $G = \{4\}$.
8. Draw the subset diagram for the collection $\{\mathbb{R}, \mathbb{Q}', \mathbb{Q}, \mathbb{Z}, \mathbb{N}\}$, where $\mathbb{Q}'$ denotes the set of irrational numbers.
9. For each part of Figure 2.2, give a collection of sets having that subset diagram.
10. Give examples of sets A, B, and C such that:
    (a) $A \in B$ and $B \in C$ and $A \notin C$
    (b) $A \in B$ and $B \in C$ and $A \in C$
    (c) $A \in B$ and $A \subset B$
11. Draw the subset diagram for the power set of the set $\{1, 2, 3\}$.

2.2 SET OPERATIONS

The most frequently used operations in set theory are intersection, union, difference, and product. These operations and some of their properties are discussed in this section.

DEFINITION 2.5

Let A and B be subsets of some universal set U. The set operations of intersection, union, and difference are defined as follows:

1. The *intersection* of A and B is the set $A \cap B$ defined by

$$A \cap B = \{x \mid x \in A \wedge x \in B\}$$

2. The *union* of A and B is the set $A \cup B$ defined by

$$A \cup B = \{x \mid x \in A \vee x \in B\}$$

3. The *difference set*, $A - B$, is defined by

$$A - B = \{x \mid x \in A \wedge x \notin B\}$$

We also call $A - B$ the *relative complement* of B in A. It is important to realize that, in general, $A - B \neq B - A$. The relative complement of A in U is called the *complement* of A and is denoted by A'. Thus, $A' = U - A$.

It is often useful to picture sets using what are called *Venn diagrams*. Venn diagrams for $A \cap B$, $A \cup B$, $A - B$, and A' are shown in Figure 2.3.

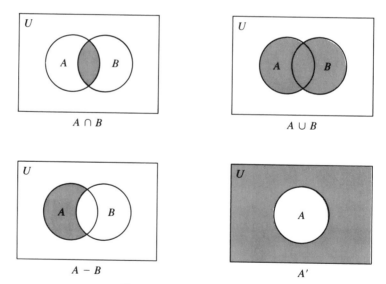

Figure 2.3 Venn diagrams

Example 2.6   Given $U = \{0, 1, 2, 3, 4, 5, 6\}$, $A = \{0, 2, 4, 6\}$, and $B = \{2, 3, 5\}$, determine $A \cap B$, $A \cup B$, $A - B$, $B - A$, A', and B'.

Solution

$$A \cap B = \{2\}$$

$$A \cup B = \{0, 2, 3, 4, 5, 6\}$$

$$A - B = \{0, 4, 6\}$$

$$B - A = \{3, 5\}$$

$$A' = \{1, 3, 5\}$$

$$B' = \{0, 1, 4, 6\}$$ ∎

Example 2.7 Given $U = \mathbb{N}$, $A = \{1, 3, 5, 7, \dots\}$, $P = \{p \in \mathbb{N} \mid p \text{ is prime}\}$, and $T = \{3, 6, 9, 12, \dots\}$, find A', P', $A \cap P$, $A - P$, $P - A$, $A' \cap T$, and $A \cup T'$.

Solution The set A is the set of odd positive integers, so A' is the set of even positive integers: $A' = \{2, 4, 6, 8, \dots\}$. The set P is the set of primes, so $P' = \{n \in \mathbb{N} \mid n \text{ is not prime}\} = \{1, 4, 6, 8, 9, \dots\}$. The set $A \cap P$ is the set of odd primes: $A \cap P = \{3, 5, 7, 11, 13, 17, \dots\} = P - \{2\}$. The set of odd positive integers that are not prime is $A - P = \{1, 9, 15, 21, 25, 27, \dots\}$, whereas $P - A$ is the set of primes that are not odd, namely, $P - A = \{2\}$. The set T is the set of positive multiples of 3, so $A' \cap T$ consists of those elements of T that are even: $A' \cap T = \{6, 12, 18, 24, \dots\}$. Finally, $A \cup T'$ includes any positive integer that is either odd or not a multiple of 3, so $A \cup T' = \{1, 2, 3, 4, 5, 7, 8, 9, 10, 11, 13, \dots\}$. Note that $A \cup T' = \{6, 12, 18, 24, \dots\}' = (A' \cap T)'$. ∎

Example 2.8 Let U, A, and B be the same as in Example 2.6, and let $C = \{0, 1, 5, 6\}$. Find and compare each pair of sets:

(a) $A \cap (B \cup C)$ and $(A \cap B) \cup (A \cap C)$
(b) $A \cup (B \cap C)$ and $(A \cup B) \cap (A \cup C)$
(c) $(A \cup B)'$ and $A' \cap B'$
(d) $(A \cap B)'$ and $A' \cup B'$

Solution For (a) we find that

$$A \cap (B \cup C) = \{0, 2, 4, 6\} \cap \{0, 1, 2, 3, 5, 6\} = \{0, 2, 6\}$$

whereas

$$(A \cap B) \cup (A \cap C) = \{2\} \cup \{0, 6\} = \{0, 2, 6\}$$

Note that $A \cap (B \cup C) = (A \cap B) \cup (A \cap C)$. In (b) the student should verify that

$$A \cup (B \cap C) = (A \cup B) \cap (A \cup C) = \{0, 2, 4, 5, 6\}$$

For (c) we have

$$(A \cup B)' = \{0, 2, 3, 4, 5, 6\}' = \{1\}$$

whereas

$$A' \cap B' = \{1, 3, 5\} \cap \{0, 1, 4, 6\} = \{1\}$$

so again the two sets are equal. And in (d) notice that

$$(A \cap B)' = A' \cup B' = \{0, 1, 3, 4, 5, 6\}$$ ■

Example 2.8 illustrates some general properties of the set operations, which are among those presented in the next two theorems. The student should compare the properties given in Theorem 2.6 with the properties of logical equivalence given in Section 1.3.

THEOREM 2.6   The following properties hold for any subsets A, B, and C of a universal set U:

1. Commutative properties
   (a) $A \cup B = B \cup A$ (b) $A \cap B = B \cap A$
2. Associative properties
   (a) $(A \cup B) \cup C = A \cup (B \cup C)$
   (b) $(A \cap B) \cap C = A \cap (B \cap C)$
3. Distributive properties
   (a) $A \cup (B \cap C) = (A \cup B) \cap (A \cup C)$
   (b) $A \cap (B \cup C) = (A \cap B) \cup (A \cap C)$
4. Idempotent laws
   (a) $A \cup A = A$ (b) $A \cap A = A$
5. DeMorgan's laws
   (a) $(A \cup B)' = A' \cap B'$ (b) $(A \cap B)' = A' \cup B'$
6. Law of the excluded middle
   (a) $A \cup A' = U$ (b) $A \cap A' = \emptyset$
7. $A \subseteq B$ if and only if $B' \subseteq A'$
8. (a) $A \cup U = U$ (b) $A \cup \emptyset = A$
   (c) $A \cap U = A$ (d) $A \cap \emptyset = \emptyset$

Proof   We prove properties 2(b), 3(a), and 5(b), leaving the remaining properties to Exercise 2.

Proof of 2(b):   Let x be an arbitrary element of $(A \cap B) \cap C$. Then

$$x \in (A \cap B) \cap C$$
$$\rightarrow [x \in (A \cap B) \wedge x \in C] \qquad \text{(by Definition 2.5(a))}$$
$$\rightarrow [(x \in A \wedge x \in B) \wedge x \in C] \qquad \text{(by Definition 2.5(a))}$$
$$\rightarrow [x \in A \wedge (x \in B \wedge x \in C)] \qquad \text{(by associativity of } \wedge)$$
$$\rightarrow [x \in A \wedge x \in (B \cap C)] \qquad \text{(by Definition 2.5(a))}$$
$$\rightarrow x \in A \cap (B \cap C) \qquad \text{(by Definition 2.5(a))}$$

Thus, $x \in (A \cap B) \cap C \rightarrow x \in A \cap (B \cap C)$ and consequently

$$(A \cap B) \cap C \subseteq A \cap (B \cap C)$$

In a similar fashion (simply reverse the implications) one may prove that $A \cap (B \cap C) \subseteq (A \cap B) \cap C$. Therefore, $(A \cap B) \cap C = A \cap (B \cap C)$.

Proof of 3(a):   In this case we prove, for all x, that

$$x \in A \cup (B \cap C) \leftrightarrow x \in (A \cup B) \cap (A \cup C)$$

using a string of biconditionals.

$x \in A \cup (B \cap C)$

$\leftrightarrow [x \in A \vee x \in (B \cap C)]$	(by Definition 2.5(2))
$\leftrightarrow [x \in A \vee (x \in B \wedge x \in C)]$	(by Definition 2.5(1))
$\leftrightarrow [(x \in A \vee x \in B) \wedge (x \in A \vee x \in C)]$	($\vee$ distributes over $\wedge$)
$\leftrightarrow [(x \in A \cup B) \wedge (x \in A \cup C)]$	(by Definition 2.5(2))
$\leftrightarrow x \in (A \cup B) \cap (A \cup C)$	(by Definition 2.5(1))

Therefore, $A \cup (B \cap C) = (A \cup B) \cap (A \cup C)$.

Proof of 5(b):   Here we have, for any x,

$$\begin{aligned} x \in (A \cap B)' &\leftrightarrow x \notin A \cap B \\ &\leftrightarrow [\sim(x \in A \cap B)] \\ &\leftrightarrow [\sim(x \in A \wedge x \in B)] \\ &\leftrightarrow [\sim(x \in A) \vee \sim(x \in B)] \\ &\leftrightarrow [x \notin A \vee x \notin B] \\ &\leftrightarrow x \in A' \vee x \in B' \\ &\leftrightarrow x \in A' \cup B' \end{aligned}$$

Hence $(A \cap B)' = A' \cup B'$. □

DeMorgan's law $(B \cap C)' = B' \cup C'$ pertains directly to those elements of the universal set U that are not in $B \cap C$. Another way to express this law is

$$U - (B \cap C) = (U - B) \cup (U - C)$$

Given subsets A, B, and C of U, the above expression can be generalized by considering those elements of A that are not in $B \cap C$, namely, $A - (B \cap C)$. A similar expression arises from considering $A - (B \cup C)$. The resulting properties are called the "generalized DeMorgan laws."

THEOREM 2.7   **(Generalized DeMorgan laws)**

The following two properties hold for any subsets A, B, and C of a universal set U.

1. $A - (B \cup C) = (A - B) \cap (A - C)$
2. $A - (B \cap C) = (A - B) \cup (A - C)$

Proof   We prove part 1 and leave part 2 to Exercise 4. For any x,

$$x \in A - (B \cup C) \leftrightarrow [x \in A \wedge x \notin (B \cup C)]$$
$$\leftrightarrow [x \in A \wedge (x \notin B \wedge x \notin C)]$$
$$\leftrightarrow [(x \in A \wedge x \notin B) \wedge (x \in A \wedge x \notin C)]$$
$$\leftrightarrow (x \in A - B) \wedge (x \in A - C)$$
$$\leftrightarrow x \in (A - B) \cap (A - C)$$

So we have $A - (B \cup C) = (A - B) \cap (A - C)$.   □

In the preceding definitions and theorems we have exercised care and patience in the use of qualifying phrases like "Let A, B, and C be subsets of a universal set U" and "For subsets $A_1, A_2, \ldots, A_n$ of a universal set $U \ldots$." Henceforth, unless the situation demands it, we omit any explicit reference to U and tacitly assume that all sets in a given discussion are subsets of some universal set. Thus, for example, in place of the above phrases, we could write "Let A, B, and C be any sets" and "For any sets $A_1, A_2, \ldots, A_n \ldots$."

Often it helps to complete a Venn diagram to feel convinced that a given result about sets is true. In the case of Theorem 2.7, part 2, the associated Venn diagrams are shown in Figure 2.4. In the diagram for $(A - B) \cup (A - C)$, the set $A - B$ is shaded with horizontal line segments and $A - C$ is shaded with vertical line segments. It should be emphasized that such a representation does not constitute a proof; it is merely an indication that the result holds.

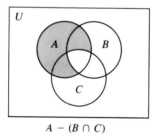
$A - (B \cap C)$

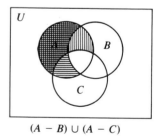
$(A - B) \cup (A - C)$

Figure 2.4

The order in which the elements of a set are listed is of no importance in the definition of a set; for example, it is clear that $\{1, 2\} = \{2, 1\}$. Thus it really makes no sense to speak of the "first element" of $\{1, 2\}$. In many cases, however, it turns out to be important to distinguish the order of appearance of two elements. This leads to the notion of an *ordered pair* of two elements a and b, denoted (a, b), where a is the *first element* and b is the *second element*. In formal set theory, the ordered pair (a, b) is defined as

$$\{\{a\}, \{a, b\}\}$$

so that (a, b) is in fact a set in which a and b play different roles. It then turns out that $(a, b) = (c, d)$ if and only if $a = c$ and $b = d$. We choose not to dwell on the formal definition of ordered pairs; instead we rely on the student's intuition and past experience in using this concept.

DEFINITION 2.6

Given any two sets A and B, the *Cartesian product* of A and B is the set $A \times B$ defined by

$$A \times B = \{(a, b) \mid a \in A \wedge b \in B\}$$

Example 2.9 Given $A = \{0, 1, 2\}$ and $B = \{1, 3\}$, find $A \times B$ and $B \times A$.

Solution
$$A \times B = \{(0, 1), (0, 3), (1, 1), (1, 3), (2, 1), (2, 3)\}$$
$$B \times A = \{(1, 0), (1, 1), (1, 2), (3, 0), (3, 1), (3, 2)\}$$

Note that $A \times B \neq B \times A$, although both sets have six elements. ∎

We see in Example 2.9 that the Cartesian product operation does not obey a commutative law. The next theorem presents several distributive laws that are satisfied when the Cartesian product, as a set operation, is mixed with each of the operations union, intersection, and difference.

THEOREM 2.8 For any sets A, B, and C, the following properties are satisfied.

1. $A \times (B \cup C) = (A \times B) \cup (A \times C)$
2. $A \times (B \cap C) = (A \times B) \cap (A \times C)$
3. $A \times (B - C) = (A \times B) - (A \times C)$

Proof We prove property 1 and leave 2 and 3 to Exercise 6. For any x and y,

$$(x, y) \in A \times (B \cup C) \leftrightarrow [x \in A \wedge y \in (B \cup C)]$$
$$\leftrightarrow [x \in A \wedge (y \in B \vee y \in C)]$$
$$\leftrightarrow [(x \in A \wedge y \in B) \vee (x \in A \wedge y \in B)]$$
$$\leftrightarrow [(x, y) \in A \times B \vee (x, y) \in A \times C]$$
$$\leftrightarrow (x, y) \in (A \times B) \cup (A \times C)$$

Therefore, $A \times (B \cup C) = (A \times B) \cup (A \times C)$. ∎

The idea of an ordered pair can be extended to more than two elements. In other words, given k elements $a_1, a_2, \ldots, a_k$, where $k \geq 3$, we can define the *ordered k-tuple* $(a_1, a_2, \ldots, a_k)$, in which a_1 is the first element (or first *coordinate*), a_2 is the second element, and so on, and a_k is the kth element. It is then possible to generalize the definition of Cartesian product.

DEFINITION 2.7

Given any k sets $A_1, A_2, \ldots, A_k$, the (k-fold) *Cartesian product* of $A_1, A_2, \ldots, A_k$ is the set defined by

$$A_1 \times A_2 \times \cdots \times A_k = \{(a_1, a_2, \ldots, a_k) \mid a_i \in A_i \text{ for each } i, 1 \leq i \leq k\}$$

We remark that $(a_1, a_2, \ldots, a_k) = (b_1, b_2, \ldots, b_k)$ if and only if $a_i = b_i$ for each i, $1 \leq i \leq k$. Also, notice that $A_1 \times A_2 \times \cdots \times A_k = \emptyset$ if and only if $A_i = \emptyset$ for some i, $1 \leq i \leq k$. In working with Cartesian products of sets, it is normally the case that each of the sets is nonempty.

Example 2.10   Let $A_1 = A_2 = A_3 = \{0, 1\}$. Find $A_1 \times A_2 \times A_3$.

Solution $A_1 \times A_2 \times A_3 = \{(0, 0, 0), (0, 0, 1), (0, 1, 0), (0, 1, 1), (1, 0, 0), (1, 0, 1), (1, 1, 0), (1, 1, 1)\}$ ∎

The language Pascal is rather unique among high-level languages in that it provides a built-in SET data type. The elements of a Pascal set must all have the same type; this type is referred to as the *base type* and is restricted to being an enumeration or integer subrange. Pascal sets are also limited in that they may contain only up to a specified number of elements. The maximum number of elements a set is allowed to contain is implementation-dependent; it ordinarily corresponds to the number of bits in one or a few words of memory.

The Pascal declarations

```
TYPE BASETYPE = 1..30;
     SETTYPE = SET OF BASETYPE;
VAR  A,B,C: SETTYPE;
```

declare sets A, B, and C whose elements are integers in the range 1 to 30. Similarly,

```
TYPE BASETYPE = CHAR;
     SETTYPE = SET OF BASETYPE;
VAR  X,Y,Z: SETTYPE;
```

declare sets X, Y, and Z whose elements are characters. (Caution: Such a base type is not possible in some implementations because it would exceed the maximum number of elements allowed.)

Given A, B, and C as declared above, the assignment statements

```
A := [2, 3, 5, 7, 11];
C := [10..20];
```

assign to A the set $\{2, 3, 5, 7, 11\}$ and to C the set $\{10, 11, 12, \ldots, 20\}$. Notice that Pascal uses square brackets, [and], rather than set braces to enclose the elements of a set. Also notice the two methods of specifying the elements to be included in a set; these two methods may be combined, as in the assignment statement

```
B := [1, 3, 7..11, 19];
```

which assigns to B the set $\{1, 3, 7, 8, 9, 10, 11, 19\}$.

The operations that Pascal provides for sets include union (syntax $+$), intersection ($*$), and difference ($-$). For instance, with A and B as assigned before, the statement

```
C := A * B;
```

assigns to C the set $A \cap B = \{3, 7, 11\}$, whereas

```
C := A + B;
```

assigns to C the set $A \cup B = \{1, 2, 3, 5, 7, 8, 9, 10, 11, 19\}$. The relational operators $=$, $<>$, $<=$, and $<$ are also provided. They are interpreted for sets A and B as follows:

$$A = B \text{ means A equals B}$$

$$A <> B \text{ means A does not equal B}$$

$$A <= B \text{ means A is a subset of B}$$

$$A < B \text{ means A is a proper subset of B}$$

Thus, a program could contain the statement

```
IF ['A', 'E', 'I', 'O', 'U'] <= X THEN
   WRITELN('SET X CONTAINS ALL OF THE VOWELS.');
```

Perhaps the most useful set operation is the test for membership of a given element in a given set. Here the operator IN is used. Suppose in a particular program it is important to know whether a given input character NEXTCHR is equal to one of the operator symbols $+$, $-$, $*$, or $/$. Rather than using the cumbersome

```
IF (NEXTCHR='+') OR (NEXTCHR='-') OR (NEXTCHR='*') OR
(NEXTCHR = '/') THEN...
```

it is much simpler and nicer to write

```
IF NEXTCHR IN ['+', '-', '*', '/'] THEN...
```

We have already mentioned a limitation of the Pascal SET data type— namely, that the base type must be an enumeration or integer subrange containing no more than some implementation-specified number of values. This excludes, for example, sets whose elements are of type REAL or sets whose elements are themselves sets. In addition, it should be pointed out that Pascal provides no direct facilities for the input/output of sets. Thus, although it does simplify some aspects of programming, the uses of the Pascal SET data type are limited. Nevertheless, the set as a general data structure is important in the study of computer science, and various methods for implementing sets are further explored in the exercises and in later portions of this text.

Exercises 2.2

1. Given $A = \{1, 2, 3\}$ and $B = \{2, 4, 5\}$, construct a subset diagram for the collection $\{\emptyset, A, B, A \cup B, A \cap B, A - B, B - A\}$.

2. Prove these properties of Theorem 2.6:
   (a) parts 1(a) and 1(b)   (b) part 2(a)
   (c) part 3(b)   (d) part 4
   (e) part 5(a)   (f) parts 6(a) and 6(b)
   (g) part 7   (h) parts 8(a), 8(b), 8(c), and 8(d)

3. Given $U = \mathbb{Z}, A = \{\ldots, -4, -2, 0, 2, 4, \ldots\}, B = \{\ldots, -6, -3, 0, 3, 6, \ldots\}$, and $C = \{\ldots, -8, -4, 0, 4, 8, \ldots\}$, find these sets.
   (a) $A \cap B$ (b) $B - A$
   (c) $A - C$ (d) $A \cap C'$
   (e) $C - A$ (f) $B \cup C$
   (g) $(A \cup B) \cap C$ (h) $(A \cup B) - C$

4. Prove Theorem 2.7, part 2.

5. Let A and B be arbitrary sets.
   (a) Prove that $A \cap B \subseteq A \cup B$.
   (b) Under what condition does $A \cap B = A \cup B$?
   (c) Prove that $A - B = A \cap B'$.
   (d) Show that $(A \cap B) \cap (A - B) = \emptyset$.
   (e) Show that $A = (A \cap B) \cup (A - B)$.

6. Prove Theorem 2.8, (a) part 2 and (b) part 3.

7. Each statement concerns arbitrary sets A and B. Complete the statement so as to make it true.
   (a) $A \subseteq B \leftrightarrow A \cap B =$
   (b) $A \subseteq B \leftrightarrow A \cup B =$
   (c) $A \subseteq B \leftrightarrow A - B =$
   (d) $A \subset B \leftrightarrow (A - B = \qquad \wedge B - A \neq \qquad)$
   (e) $A \subset B \leftrightarrow (A \cap B = \qquad \wedge A \cap B \neq \qquad)$
   (f) $A - B = B - A \leftrightarrow$

8. Let A and B denote arbitrary sets. Prove or disprove:
   (a) $(A \times B)' = A' \times B'$ (b) $A \times A = A$
   (c) $A \times U = U$ (d) $A \times \emptyset = \emptyset$

9. Let $A = \{x, y\}$, $B = \{0, 1\}$, and $C = \{-1, 0, 1\}$. Find these sets.
   (a) $A \times B$ (b) $B \times C$
   (c) $A \times B \times C$ (d) $(A \times B) \times C$
   (e) $A \times (B \times C)$ (f) $B \times B \times B \times B$
   (g) $\mathscr{P}(B \times B)$ (h) $\mathscr{P}(B) \times \mathscr{P}(B)$

10. Let A, B, and C be arbitrary sets. Prove that if $B \subseteq C$, then $A \times B \subseteq A \times C$.

11. Given that $\mathbb{R}$ is the universal set, describe the set $\mathbb{Q}'$.

12. Let A and B be arbitrary nonempty sets.
    (a) Under what condition does $A \times B = B \times A$?
    (b) Under what condition is $(A \times B) \cap (B \times A)$ empty?

13. A telephone number like 673-3459 can be thought of as a 7-tuple of digits: $(6, 7, 3, 3, 4, 5, 9)$. Thus, with $D = \{0, 1, 2, \ldots, 9\}$, we can think of a telephone number as an element of the Cartesian product $D \times D \times D \times D \times D \times D \times D$. Describe how each of the following can be thought of as an element of some appropriate Cartesian product of sets.
    (a) a social security number
    (b) a typical automobile license plate identification
    (c) a valid Pascal identifier having exactly three characters—for example, SUM

14. Each part of this exercise refers to the following declarations:

    ```
    CONST MIN = 1; MAX = 25;
    TYPE BASETYPE = MIN..MAX;
         SETTYPE = SET OF BASETYPE;
    VAR  A, B, C: SETTYPE;
    ```

    Write Pascal assignment statements that assign to C the following:
    (a) the set of numbers (in the base type) that are prime
    (b) the set of numbers between 3 and 17, inclusive
    (c) the set of numbers (in the base type) that are not in $\{7, 14, 21\}$
    (d) the set of those numbers that are in exactly one of the sets A or B (assume that A and B have been defined)
    (e) the set of numbers that are in C but in neither A nor B

15. Assume that MIN and MAX have been declared as (integer or character) constants, and we have the declarations

    ```
    TYPE BASETYPE = MIN..MAX;
         SETTYPE = SET OF BASETYPE;
    ```

    (a) Write a procedure SETIN that inputs the elements of a set A. The procedure should first prompt the user to enter the number of elements in the set A, and then prompt the user to enter the elements. (Hint: If X is a variable of type BASETYPE, then the statement

```
A := A + [X]
```

assigns to A the union of A and the set containing the value of X.)

(b) Write a procedure SETOUT that outputs the elements of a set A.

16. A type STRING, for character strings having between 0 and MAXLEN characters, may be declared in Pascal as follows:

```
CONST MAXLEN = 80;
TYPE  STRING = RECORD
        VALUE: PACKED ARRAY [1..MAXLEN] OF CHAR;
        LENGTH: 0..MAXLEN
      END;
```

The idea is that if WORD is a variable of type STRING, then WORD.LENGTH is the number of characters in WORD, and WORD.VALUE[J] is the Jth character in WORD, for $1 <= J <= $ WORD.LENGTH. Assume that SETTYPE is declared as in Exercise 14, with MIN = 'A' and MAX = 'Z' (assuming the ASCII character set).

(a) Write a procedure FIND that takes a character string WORD and returns the set VOWS of vowels occurring in WORD and the set CONS of consonants occurring in WORD. For example if WORD = "KALAMAZOO," then VOWS = {'A', 'O'} and CONS = {'K', 'L', 'M', 'Z'}.

(b) Assuming that V1, V2, C1, C2, L1, L2, and L3 are variables of type SETTYPE, and WORD1 and WORD2 are character strings, complete the following Pascal program segment so that it does what the comments indicate.

```
FIND(WORD1,V1,C1);
FIND(WORD2,V2,C2);
L1 :=              ;(*Assign to L1 the set of let-
                     ters in WORD1.*)
L2 :=              ;(*Assign to L2 the set of
                     consonants that are in both
                     WORD1 and WORD2.*)
L3 :=              ;(*Assign to L3 the set of vowels
                     that are in WORD1 but not in
                     WORD2.*)
```

17. If X is a variable of type BASETYPE and A is a set of type SETTYPE, what is the negation of the condition "X IN A"?

18. Suppose we wish to work with sets whose base type is like that of Exercise 14. Such sets may be implemented as Boolean arrays; in Pascal we may declare sets A, B, and C as follows:

```
CONST MIN = -99; MAX = 99;
TYPE BASETYPE = MIN..MAX;
     SETTYPE = ARRAY [BASETYPE] OF BOOLEAN;
VAR  A, B, C: SETTYPE;
```

The idea here is that A[J] is true if and only if J is an element of A, for MIN ≤ J ≤ MAX. Assuming that such declarations have been made, and that the sets A and B have been defined, write a FOR statement of the form

```
FOR J := MIN TO MAX DO
  C[J] :=                ;
```

that assigns to C the set
(a) $A \cup B$ (b) $A \cap B$ (c) A'

2.3   ARBITRARY UNIONS AND INTERSECTIONS

In Section 2.2 the union $A \cup B$ and the intersection $A \cap B$ were defined for two sets A and B. In this section these notions are extended to cover unions and intersections of any number of sets.

If A_1 and A_2 are two sets, then

$$A_1 \cup A_2 = \{ x \mid x \in A_1 \vee x \in A_2 \}$$

$$A_1 \cap A_2 = \{ x \mid x \in A_1 \wedge x \in A_2 \}$$

If A_1, A_2, and A_3 are any three sets, then by the associative laws

$$(A_1 \cup A_2) \cup A_3 = A_1 \cup (A_2 \cup A_3)$$

$$(A_1 \cap A_2) \cap A_3 = A_1 \cap (A_2 \cap A_3)$$

Hence there is no confusion if one simply writes $A_1 \cup A_2 \cup A_3$ or $A_1 \cap A_2 \cap A_3$. We then have

$$A_1 \cup A_2 \cup A_3 = \{ x \mid x \in A_1 \vee x \in A_2 \vee x \in A_3 \}$$

$$= \{ x \mid x \in A_i \text{ for some } i, 1 \leq i \leq 3 \}$$

and

$$A_1 \cap A_2 \cap A_3 = \{x \mid x \in A_1 \wedge x \in A_2 \wedge x \in A_3\}$$
$$= \{x \mid x \in A_i \text{ for each } i, 1 \leq i \leq 3\}$$

DEFINITION 2.8

Given any n sets $A_1, A_2, \ldots, A_n$, we define

1. their *union* to be the set

$$A_1 \cup A_2 \cup \cdots \cup A_n = \{x \mid x \in A_i \text{ for some } i, 1 \leq i \leq n\}$$

2. their *intersection* to be the set

$$A_1 \cap A_2 \cap \cdots \cap A_n = \{x \mid x \in A_i \text{ for each } i, 1 \leq i \leq n\}$$

These sets are expressed as

$$\bigcup_{i=1}^{n} A_i \quad \text{and} \quad \bigcap_{i=1}^{n} A_i$$

respectively.

Example 2.11 For $i \in \{1, 2, \ldots, 10\}$, define

$$A_i = [-i, 10 - i]$$

Then $A_1 = [-1, 9]$, $A_2 = [-2, 8], \ldots, A_{10} = [-10, 0]$. Hence

$$\bigcup_{i=1}^{10} A_i = [-10, 9] \quad \text{and} \quad \bigcap_{i=1}^{10} A_i = [-1, 0] \qquad \blacksquare$$

Example 2.12 For $k \in \{1, 2, \ldots, 100\}$, define $B_k = \{r \in \mathbb{Q} \mid -1 \leq k \cdot r \leq 1\}$. Then

$$B_1 = \{r \in \mathbb{Q} \mid -1 \leq r \leq 1\}$$
$$B_2 = \{r \in \mathbb{Q} \mid -1 \leq 2r \leq 1\} = \{r \in \mathbb{Q} \mid -\tfrac{1}{2} \leq r \leq \tfrac{1}{2}\}$$
$$\vdots$$
$$B_{100} = \{r \in \mathbb{Q} \mid -1 \leq 100r \leq 1\} = \{r \in \mathbb{Q} \mid -\tfrac{1}{100} \leq r \leq \tfrac{1}{100}\}$$

Notice that

$$B_{100} \subset B_{99} \subset \cdots \subset B_2 \subset B_1$$

It follows that

$$\bigcup_{k=1}^{100} B_k = B_1 \quad \text{and} \quad \bigcap_{k=1}^{100} B_k = B_{100} \qquad \blacksquare$$

DEFINITION 2.9

Given sets $A_1, A_2, \ldots, A_n, \ldots$, we define

1. their *union* to be the set

$$\bigcup_{n=1}^{\infty} A_n = \{x \mid x \in A_n \text{ for some } n \in \mathbb{N}\}$$

2. their *intersection* to be the set

$$\bigcap_{n=1}^{\infty} A_n = \{x \mid x \in A_n \text{ for each } n \in \mathbb{N}\}$$

Example 2.13   Given that $A_n = \left[0, \dfrac{1}{n}\right]$ for $n \in \mathbb{N}$, find

$$S = \bigcup_{n=1}^{\infty} A_n \quad \text{and} \quad T = \bigcap_{n=1}^{\infty} A_n$$

Solution   We first note that $A_2 \subset A_1$, $A_3 \subset A_2$, and so on. Thus $S = A_1 = [0, 1]$. To find the intersection, T, first notice that $0 \in A_n$ for every $n \in \mathbb{N}$, and so $0 \in T$. We claim that $T = \{0\}$. To verify this, we must show that if x is any positive real number, then $x \notin T$. Let $x > 0$. Consider the harmonic fractions $1, 1/2, 1/3, \ldots$; there is a term $1/m$ in this sequence such that $1/m < x$. Since $A_m = [0, 1/m]$, we see that $x \notin A_m$. Therefore $x \notin T$.   ■

It should be pointed out that a very important principle of the real numbers was used in the preceding example. This is the Archimedean principle, and it states that if x and y are any two positive real numbers, then there exists a positive integer m such that $m \cdot x > y$. It was applied in the example to conclude that there is a positive integer m such that $1/m < x$ (here $y = 1$).

Example 2.14   For $n \in \mathbb{N}$, define

$$A_n = \left(\frac{-1}{n}, \frac{2n - 1}{n}\right)$$

Then $A_1 = (-1, 1)$, $A_2 = (-1/2, 3/2)$, $A_3 = (-1/3, 5/3)$, and so on. Notice that the numbers $-1/n$ are increasing and approaching 0 (from the negative side), while the numbers $(2n - 1)/n$ are increasing and approaching 2 (through numbers that are less than 2). Using these facts, we can argue that

$$\bigcup_{n=1}^{\infty} A_n = (-1, 2) \quad \text{and} \quad \bigcap_{n=1}^{\infty} A_n = [0, 1)$$ ■

If A and B are sets and $A \subseteq B$, then we also write $B \supseteq A$. Also, if $A_2 \subseteq A_1$, $A_3 \subseteq A_2$, and so on, as in Example 2.13, then we write

$$A_1 \supseteq A_2 \supseteq A_3 \supseteq \cdots$$

THEOREM 2.9   Given sets $A_1, A_2, \ldots, A_n, \ldots$, the following two properties hold:

1. If $A_1 \subseteq A_2 \subseteq A_3 \subseteq \cdots$, then

$$\bigcap_{n=1}^{\infty} A_n = A_1$$

2. If $A_1 \supseteq A_2 \supseteq A_3 \supseteq \cdots$, then

$$\bigcup_{n=1}^{\infty} A_n = A_1$$

Proof   The proof of part 1 is given here; part 2 is similar and is left for Exercise 4. Let

$$T = \bigcap_{n=1}^{\infty} A_n$$

and let x be an arbitrary element of T. Then $x \in A_k$ for each $k \in \mathbb{N}$, and this certainly implies that $x \in A_1$. So $T \subseteq A_1$. Next assume $x \in A_1$. Since $A_1 \subseteq A_2 \subseteq A_3 \subseteq \cdots$, we have that $A_1 \subseteq A_k$ for every $k \in \mathbb{N}$. Thus $x \in A_k$ for every $k \in \mathbb{N}$, and hence $x \in T$. This shows that $A_1 \subseteq T$; hence we conclude that $T = A_1$.   □

Example 2.15   For $n \in \mathbb{N}$ let $A_n = \{m \in \mathbb{Z} \mid -n \le m \wedge 2^m \le n\}$. Then $A_1 = \{-1, 0\}$, $A_2 = \{-2, -1, 0, 1\}$, $A_3 = \{-3, -2, -1, 0, 1\}$, $A_4 = \{-4, -3, -2, -1, 0, 1, 2\}$, and so on. Note that $A_1 \subseteq A_2 \subseteq A_3 \subseteq \cdots$. Hence, by Theorem 2.9,

$$\bigcap_{n=1}^{\infty} A_n = A_1$$

We leave it to the student to verify that

$$\bigcup_{n=1}^{\infty} A_n = \mathbb{Z} \qquad \blacksquare$$

We now define the operations of union and intersection in the most general setting, for any collection of sets.

DEFINITION 2.10

Let I be a nonempty set, and suppose that for each element $i \in I$ there is associated a set A_i. We then call I an *index set* for the collection of sets

$$\mathscr{A} = \{ A_i \mid i \in I \}$$

1. The *union* of the collection $\mathscr{A}$ is defined to be the set

$$\bigcup_{i \in I} A_i = \{ x \mid x \in A_i \text{ for some } i \in I \}$$

2. The *intersection* of the collection $\mathscr{A}$ is defined to be the set

$$\bigcap_{i \in I} A_i = \{ x \mid x \in A_i \text{ for each } i \in I \}$$

Note that Definitions 2.8 and 2.9 are just special cases of Definition 2.10. In Definition 2.8 the index set $I = \{1, 2, \ldots, n\}$, whereas in Definition 2.9 the index set $I = \mathbb{N}$.

Example 2.16   Let $I = (0, 1)$, and for $i \in I$, define $A_i = (-i, i)$. For example,

$$A_{1/2} = \left(\frac{-1}{2}, \frac{1}{2} \right) \quad \text{and} \quad A_{\sqrt{2}/3} = \left(\frac{-\sqrt{2}}{3}, \frac{\sqrt{2}}{3} \right)$$

Notice that if $0 < i < j < 1$, then $\{0\} \subset A_i \subset A_j \subset (-1, 1)$. It follows that

$$\bigcup_{i \in I} A_i = (-1, 1) \quad \text{and} \quad \bigcap_{i \in I} A_i = \{0\} \qquad \blacksquare$$

Example 2.17   Let $I = \mathbb{R}$. For $r \in \mathbb{R}$, define $A_r = \{(x, y) \mid y = rx\}$. Geometrically, A_r is a line in the coordinate plane that passes through the origin and has slope r. The only point that is on every such line is the origin; hence

$$\bigcap_{r \in \mathbb{R}} A_r = \{(0, 0)\}$$

On the other hand, if $a \neq 0$, then the point (a, b) lies on the line $y = \dfrac{b}{a} x$, so that $(a, b) \in A_r$ with $r = b/a$. Also those points on the y-axis other than the origin belong to no set A_r. Hence

$$\bigcup_{r \in \mathbb{R}} A_r = \{(a, b) \mid a \neq 0 \lor a = b = 0\}$$

In other words, this set includes the origin plus all other points of the plane except those on the y-axis. $\qquad \blacksquare$

Example 2.18   Let S denote the set of ASCII character strings, and for $s \in S$, let C_s be the set of ASCII character strings that contain the string s as a substring. For instance, if $s =$ "MA," then the strings "MATRIX" and "KALAMAZOO" both belong to C_s. The intersection

$$\bigcap_{s \in S} C_s$$

is empty, since no string of finite length contains every possible substring. On the other hand,

$$\bigcup_{s \in S} C_s = S$$

since $s \in C_s$ (every character string contains itself as a substring).   ∎

THEOREM 2.10   **(Extended DeMorgan laws)**

Let $\mathscr{A} = \{ A_i \mid i \in I \}$ be a collection of sets indexed by the nonempty set I. Then

1. $\left(\bigcup_{i \in I} A_i \right)' = \bigcap_{i \in I} A_i'$

2. $\left(\bigcap_{i \in I} A_i \right)' = \bigcup_{i \in I} A_i'$

Proof   We prove part 1 and leave part 2 to Exercise 6. For each x,

$$x \in \left(\bigcup_{i \in I} A_i \right)' \leftrightarrow x \notin \bigcup_{i \in I} A_i$$

$$\leftrightarrow x \notin A_i \text{ for all } i \in I$$

$$\leftrightarrow x \in A_i' \text{ for all } i \in I$$

$$\leftrightarrow x \in \bigcap_{i \in I} A_i'$$

Therefore

$$\left(\bigcup_{i \in I} A_i \right)' = \bigcap_{i \in I} A_i'$$ □

Exercises 2.3

1. For $n \in \{1, 2, \ldots, 50\}$, define $C_n = (-1/n, 2n)$. Find $\bigcup_{n=1}^{50} C_n$ and $\bigcap_{n=1}^{50} C_n$.

2. Prove: If $A_1 \subseteq A_2 \subseteq \cdots \subseteq A_k$, then $\bigcap_{n=1}^{k} A_n = A_1$ and $\bigcup_{n=1}^{k} A_n = A_k$.

3. For each collection $\{ A_n \mid n \in \mathbb{N} \}$, find $\bigcup_{n=1}^{\infty} A_n$ and $\bigcap_{n=1}^{\infty} A_n$.

   (a) $A_n = \{ k \in \mathbb{Z} \mid -n \le k \le 2n \}$
   (b) $A_n = (-1/n, (n+1)/n)$ if n is odd; $A_n = ((1-n)/n, 1/n)$ if n is even
   (c) $A_1 = \mathbb{N}$ and, for $n \ge 2$, $A_n = \{ m \in \mathbb{N} \mid m \ge n \wedge m/n \notin \mathbb{N} \}$
   (d) $A_n = \{ kn \mid k \in \mathbb{Z} \}$

4. Prove Theorem 2.9, part 2.
5. Given the universal set U, the index set I, and the collection $\{A_i \mid i \in I\}$, find $\bigcup_{i \in I} A_i$ and $\bigcap_{i \in I} A_i$.

   (a) $U = \mathbb{R}$, $I = (0, 1)$, $A_i = [1 - i, 1/i]$
   (b) $U = \mathbb{R} \times \mathbb{R}$, $I = [0, \infty)$, $A_i = \{(x, y) \mid x^2 + y^2 = i^2\}$
   (c) $U = \{(a_0, a_1, \ldots, a_n) \mid \text{each } a_i = 0 \text{ or } 1\} = \{\text{binary words of length } n + 1\}$, $I = \{0, 1, \ldots, n\}$, $A_i = \{(a_0, a_1, \ldots, a_n) \mid a_i = 1\}$
   (d) $U = \{\text{identifiers in a given Pascal program } P\}$,
       $I = \{\text{procedures in } P \text{ (including the main procedure)}\}$,
       $A_i = \{\text{identifiers that may be referenced in procedure } i\}$
   (e) $U = \mathscr{P}(B)$, where B is a given nonempty set, $I = B$,
       $A_i = \{C \in \mathscr{P}(B) \mid i \in C\}$
   (f) $U = \mathbb{R} \times \mathbb{R}$, $I = (0, \infty)$, $A_i = \{(x, y) \mid (x - i)^2 + y^2 = i^2\}$
6. Prove Theorem 2.10, part 2.

2.4   CARDINALITY

Recall that a set S is finite if S consists of m elements for some nonnegative integer m.

DEFINITION 2.11

Let S be a finite set. The *cardinality* of S, denoted by $|S|$, is defined to be the number of elements in S. We also say that S has *cardinal number* $|S|$. An alternate notation for $|S|$ is $n(S)$.

In Chapter 5 a more rigorous definition of cardinality is given, and the definition is extended to the case when S is infinite. Note that the empty set, $\emptyset$, is the only set with cardinality 0.

Example 2.19   For each set A, determine the cardinality of the power set $\mathscr{P}(A)$ of A.
(a) $A = \emptyset$ (b) $A = \{1\}$ (c) $A = \{1, 2\}$ (d) $A = \{1, 2, 3\}$

Solution
(a) If $A = \emptyset$, then $\mathscr{P}(A) = \{\emptyset\}$, so that $|\mathscr{P}(A)| = 1$.
(b) If $A = \{1\}$, then $\mathscr{P}(A) = \{\emptyset, \{1\}\}$, so that $|\mathscr{P}(A)| = 2$.
(c) If $A = \{1, 2\}$, then $\mathscr{P}(A) = \{\emptyset, \{1\}, \{2\}, \{1, 2\}\}$, so that $|\mathscr{P}(A)| = 4$.
(d) If $A = \{1, 2, 3\}$, then $\mathscr{P}(A) = \{\emptyset, \{1\}, \{2\}, \{3\}, \{1, 2\}, \{1, 3\}, \{2, 3\}, \{1, 2, 3\}\}$, so that $|\mathscr{P}(A)| = 8$. ∎

In view of Example 2.19, it seems reasonable to conjecture that for a finite set A,

$$|A| = n \to |\mathscr{P}(A)| = 2^n$$

This result is proved in Chapter 6, along with other results that generally fall under the heading of combinatorial mathematics. In this section, however, we present two of the most basic principles of cardinality: the addition principle and the multiplication principle. The addition principle requires the notion of a "pairwise disjoint" collection of sets.

DEFINITION 2.12

Two sets A and B are called *disjoint* if $A \cap B = \emptyset$. More generally, if $A_1, A_2, \ldots, A_n$ are sets and $A_i \cap A_j = \emptyset$ for all i and j such that $1 \leq i < j \leq n$, then we say that these sets are *pairwise disjoint*, and the collection $\{A_1, A_2, \ldots, A_n\}$ is called a *pairwise disjoint collection* of sets.

Example 2.20   Let $A = \{-1, 2, 3, 5, 8\}$, $B = \{1, 3, 5, 7\}$, $C = \{-1, 2, 4, 9\}$, $D = \{-2, 0, 6\}$, and $E = \{-3, 8\}$. Then A and D are disjoint and $\{B, C, D, E\}$ is a pairwise disjoint collection.   ∎

Note that the sets A and B in Example 2.20 are not disjoint; however, it is the case that $A - B = \{-1, 2, 8\}$ and $A \cap B = \{3, 5\}$ are disjoint and that $A = (A - B) \cup (A \cap B)$. This relation holds in general and turns out to be quite useful; you were asked to prove it in Exercise 5(e) in Exercises 2.2. Here we state it as a theorem.

THEOREM 2.11   For any sets A and B, the sets $A - B$ and $A \cap B$ are disjoint and

$$A = (A - B) \cup (A \cap B)$$

If A and B are disjoint finite sets, say $A = \{a_1, a_2, \ldots, a_r\}$ and $B = \{b_1, b_2, \ldots, b_s\}$, then the elements $a_1, a_2, \ldots, a_r, b_1, b_2, \ldots, b_s$ are distinct and we see that

$$|A \cup B| = |\{a_1, a_2, \ldots, a_r, b_1, b_2, \ldots, b_s\}| = r + s$$

Hence

$$|A \cup B| = |A| + |B|$$

There is a natural generalization of this result to any k pairwise disjoint finite sets, where $k \geq 2$. It is referred to as the *addition principle*.

THEOREM 2.12   **(Addition principle)**

If $A_1, A_2, \ldots, A_k$ are pairwise disjoint finite sets, then

$$|A_1 \cup A_2 \cup \cdots \cup A_k| = |A_1| + |A_2| + \cdots + |A_k|$$

What if the sets $A_1, A_2, \ldots, A_k$ are not pairwise disjoint? In this case the formula for $|A_1 \cup A_2 \cup \cdots \cup A_k|$ is rather complicated and its proof requires a detailed combinatorial argument. The formula is known as the "principle of inclusion–exclusion" (or inclusion–exclusion formula) and is given in Chapter 6. It is instructive at this point, however, to present the case $k = 2$. To facilitate our discussion, recall that $A - B = A \cap B'$ (see Exercise 5(c) in Exercises 2.2). Also recall the relations $A \cap \emptyset = \emptyset$ and $A \cap U = A$, which hold for any subset A of the universal set U.

LEMMA 2.13 Given any two sets A and B, the sets $A - B$, $A \cap B$, and $B - A$ are pairwise disjoint and

$$A \cup B = (A - B) \cup (A \cap B) \cup (B - A)$$

(A "lemma" is a result that is used to prove some other, usually more important result.)

Proof We prove that $A - B$ and $B - A$ are disjoint. It was shown $(A - B)$ and $(A \cap B)$ are disjoint in Exercise 5(d) of Exercises 2.2; by interchanging the roles of A and B we obtain $(A \cap B) \cap (B - A) = \emptyset$. Now,

$$\begin{aligned}
(A - B) \cap (B - A) &= (A \cap B') \cap (B \cap A') \\
&= [A \cap (B' \cap B)] \cap A' \\
&= (A \cap \emptyset) \cap A' \\
&= \emptyset \cap A' \\
&= \emptyset
\end{aligned}$$

Therefore, $A - B$ and $B - A$ are disjoint.

The proof that $A \cup B = (A - B) \cup (A \cap B) \cup (B - A)$ then proceeds as follows:

$$\begin{aligned}
(A - B) \cup (A \cap B) \cup (B - A) &= [(A \cap B') \cup (A \cap B)] \cup (B \cap A') \\
&= [A \cap (B' \cup B)] \cup (B \cap A') \\
&= (A \cap U) \cup (B \cap A') \\
&= A \cup (B \cap A') \\
&= (A \cup B) \cap (A \cup A') \\
&= (A \cup B) \cap U \\
&= A \cup B \qquad \square
\end{aligned}$$

The following result is a special case of the principle of inclusion–exclusion.

THEOREM 2.14 If A and B are finite sets, then

$$|A \cup B| = |A| + |B| - |A \cap B| \qquad (1)$$

Proof It follows from Theorem 2.12 and the addition principle that $|A| = |A - B| + |A \cap B|$. Thus,

$$|A - B| = |A| - |A \cap B|$$

Similarly, with the roles of A and B interchanged, we have that

$$|B - A| = |B| - |A \cap B|$$

Using these two facts together with Lemma 2.13, we have that

$$
\begin{aligned}
|A \cup B| &= |A - B| + |A \cap B| + |B - A| \\
&= (|A| - |A \cap B|) + |A \cap B| + (|B| - |A \cap B|) \\
&= |A| + |B| - |A \cap B| \qquad\qquad \square
\end{aligned}
$$

The next result is a direct consequence of Theorem 2.14; in fact, it is nothing more than a disguised version of (1) in which one makes use of the relation $U = (A \cup B)' \cup (A \cup B)$. (Hence we also refer to it as a special case of inclusion–exclusion.) The term *corollary*, as seen below, is used in mathematics to signify a result that is a direct consequence of a preceding theorem.

COROLLARY 2.15 Let A and B be subsets of a finite universal set U. Then

$$|A' \cap B'| = |U| - |A| - |B| + |A \cap B|$$

Proof   Using Theorem 2.14 and the fact that $U = (A \cup B)' \cup (A \cup B)$ is a disjoint union, we obtain

$$
\begin{aligned}
|A' \cap B'| &= |(A \cup B)'| \\
&= |U| - |A \cup B| \\
&= |U| - (|A| + |B| - |A \cap B|) \\
&= |U| - |A| - |B| + |A \cap B| \qquad\qquad \square
\end{aligned}
$$

The rationale behind the use of the terminology *inclusion–exclusion* can be nicely illustrated with an alternative proof of Corollary 2.15 and we present it here.

Alternative proof of Corollary 2.15

Let x be any element of U and consider the contribution, corresponding to x, to each side of the formula in Corollary 2.15. Four cases are clearly distinguished:

>**Case 1.** $x \in A'$ and $x \in B'$
>
>**Case 2.** $x \in A'$ and $x \notin B'$
>
>**Case 3.** $x \notin A'$ and $x \in B'$
>
>**Case 4.** $x \notin A'$ and $x \notin B'$

In Case 1, x is included once in $A' \cap B'$ and hence contributes 1 to the left-hand side of the formula. On the right-hand side, x is included in U once, is excluded from A, is excluded from B, and is excluded from $A \cap B$.

Hence, x contributes 1 to the right side of the formula. So the contribution to each side is 1.

In Case 2, x is excluded from $A' \cap B'$ and hence contributes 0 to the left-hand side of the formula. On the right-hand side, x is included in U once, is excluded from A, is included in B once, and is excluded from $A \cap B$. Thus, x contributes 0 to the right-hand side. So the contribution of x to each side of the formula is 0.

The same type of reasoning can be used to handle the remaining two cases. □

The preceding argument is referred to as an "inclusion–exclusion" type proof of the stated result.

Example 2.21   A certain manufactured item may have one type of defect, another type of defect, or both types of defects. Call the defect types 1 and 2. Of 100 of these items that were inspected, 15 had defect 1, ten had defect 2, and seven had both types of defects. (a) How many of these 100 items were defective in some way? (b) How many were not defective?

Solution   Let U be the set of items that were inspected, let A be the set of items with defect 1, and let B be the set of items with defect 2. We are given that $|U| = 100$, $|A| = 15$, $|B| = 10$, and $|A \cap B| = 7$. In (a) we are asked to find $|A \cup B|$, the number of items having either type of defect. Using Theorem 2.14, we obtain

$$|A \cup B| = |A| + |B| - |A \cap B| = 15 + 10 - 7 = 18$$

For (b) we are asked to determine $|A' \cap B'|$, the number of items with neither type of defect. Using Corollary 2.15, we obtain

$$|A' \cap B'| = |U| - |A| - |B| + |A \cap B| = 100 - 15 - 10 + 7 = 82 \quad \blacksquare$$

Example 2.22   An insurance company classifies its policyholders according to age and marital status. Of 500 policyholders surveyed, 350 are married, 110 are married and under 25 years of age, and 60 are not married and 25 years of age or older. How many of the 500 policyholders are under 25 years of age?

Solution   Let A and B denote the sets of married policyholders and policyholders under 25, respectively, where U is the set of 500 policyholders included in the survey. We are given that $|A| = 350$, $|A \cap B| = 110$, and $|A' \cap B'| = 60$, and asked to find $|B|$. First of all, note that

$$|A \cup B| = |U| - |A' \cap B'| = 500 - 60 = 440$$

Then, using Theorem 2.14, we have

$$|B| = |A \cup B| - |A| + |A \cap B| = 440 - 350 + 110 = 200$$

Hence, 200 of the 500 policyholders surveyed are under twenty-five. (The

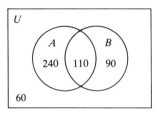

Figure 2.5

Venn diagram in Figure 2.5 shows the cardinalities of the various disjoint subsets of U.)   ■

If A and B are finite sets, how is the cardinality of $A \times B$ related to the cardinalities of A and B? As an example, let $A = \{1, 2\}$ and $B = \{1, 2, 3\}$. Then

$$A \times B = \{(1,1), (1,2), (1,3), (2,1), (2,2), (2,3)\}$$

Note that $|A| = 2$, $|B| = 3$, and $|A \times B| = 6$, so that

$$|A \times B| = |A| \cdot |B|$$

This relation holds in general for two finite sets A and B and is commonly called the *multiplication principle*.

THEOREM 2.16   **(Multiplication principle)**

Let A and B be finite subsets of some universal set U. Then $A \times B$ is a finite set and

$$|A \times B| = |A| \cdot |B|$$

Proof   If either $A = \emptyset$ or $B = \emptyset$, then the result is immediate. So assume both A and B are nonempty, say, $A = \{a_1, a_2, \ldots, a_r\}$ and $B = \{b_1, b_2, \ldots, b_s\}$. Then

$$\begin{aligned}
A \times B &= \{(a_i, b_j) \mid 1 \le i \le r, 1 \le j \le s\} \\
&= \{(a_1, b_1), \ldots, (a_1, b_s), \ldots, (a_r, b_1), \ldots, (a_r, b_s)\} \\
&= \{(a_1, b_1), \ldots, (a_1, b_s)\} \cup \{(a_2, b_1), \ldots, (a_2, b_s)\} \cup \cdots \\
&\quad \cup \{(a_r, b_1), \ldots, (a_r, b_s)\} \\
&= (\{a_1\} \times B) \cup (\{a_2\} \times B) \cup \cdots \cup (\{a_r\} \times B)
\end{aligned}$$

It is evident that each of the r sets $\{a_i\} \times B$, $1 \le i \le r$, consists of s elements. Furthermore, the sets $\{a_i\} \times B$, $1 \le i \le r$, form a pairwise disjoint collection. Hence, by the addition principle,

$$\begin{aligned}
|A \times B| &= |\{a_1\} \times B| + |\{a_2\} \times B| + \cdots + |\{a_r\} \times B| \\
&= s + s + \cdots + s \\
&= rs
\end{aligned}$$

Thus, $|A \times B| = |A| \cdot |B|$ and the proof is complete.   □

An alternative formulation of the multiplication principle that is quite often useful in applications is the following:

> **Multiplication principle.**   If a first object can be selected in r ways and, once this choice has been made, a second object can be selected in s ways, then the two selections can be made together in $r \cdot s$ ways.

For example, suppose there are 13 highways that connect Kalamazoo, Michigan, to Ottumwa, Iowa, and 17 highways that connect Ottumwa to San Francisco. Using only these highways, in how many ways can one travel (by car) from Kalamazoo to San Francisco? Using the multiplication principle, we see that the answer is $(13)(17) = 221$ ways.

A convenient and helpful way to illustrate the multiplication principle is with the aid of a *tree diagram*. Consider the case where $A = \{a_1, a_2, a_3\}$ and $B = \{b_1, b_2, b_3, b_4\}$. The associated tree diagram for $A \times B$ is shown in Figure 2.6. Notice that there are three initial branches corresponding to the elements a_1, a_2, and a_3 of A. Attached to each of these initial branches are four branches corresponding to the elements b_1, b_2, b_3, and b_4 of B. To obtain an element of $A \times B$, we start at the "root" of the tree and choose an initial branch to follow. The choice of an initial branch determines the first coordinate of the ordered pair. Then, to determine the second element, we choose a secondary branch to follow. For example, the ordered pair (a_2, b_3) is obtained by choosing the initial branch labeled a_2 followed by the secondary branch labeled b_3. Note that each path from the root, along an initial branch, and then along a secondary branch to a "leaf" (or

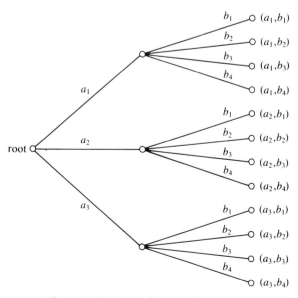

Figure 2.6 A tree diagram for $A \times B$

"terminus") of the tree determines a different ordered pair; thus the leaves are labeled with the elements of $A \times B$, as shown.

As with the addition principle, the multiplication principle can be extended to an arbitrary finite number of finite sets.

COROLLARY 2.17   If $A_1, A_2, \ldots, A_k$ are finite subsets of some universal set U, then

$$|A_1 \times A_2 \times \cdots \times A_k| = |A_1| \cdot |A_2| \cdot \cdots \cdot |A_k|$$

The proof of Corollary 2.17, which depends on mathematical induction, is considered at a later point. If we let $A_1 = A_2 = \cdots = A_k$, we obtain the following special case of the multiplication principle.

COROLLARY 2.18   If A is a finite set and $P = A \times A \times \cdots \times A$ is the k-fold product of A with itself, then

$$|P| = |A|^k$$

Example 2.23   An experiment in probability consists of tossing a fair die three times in succession. Letting $A = \{1, 2, 3, 4, 5, 6\}$, note that an outcome for this experiment can be considered as an element of the set $A \times A \times A$. For instance, the ordered triple $(3, 1, 4)$ indicates that 3 is obtained on the first toss of the die, 1 on the second toss, and 4 on the third toss. Hence there are $|A|^3 = 6^3 = 216$ possible outcomes for this experiment. How many outcomes have no number repeated? Well, there are six possible outcomes for the first toss; call it x. Then the second toss must result in an element of $A - \{x\}$, so there are five possible results for the second toss. If the second toss is y, then the third toss must result in an element of $A - \{x, y\}$, so there are four possible results for the third toss. Therefore, by the multiplication principle, there are $(6)(5)(4) = 120$ outcomes for the experiment that have no number repeated.   ∎

Example 2.24   An identifier in FORTRAN consists of from one to six alphanumeric characters (letters and digits), where the first character must be a letter. (Also assume all letters are uppercase.) How many such identifiers are there? How many end with a digit?

Solution   Let $L = \{A, B, \ldots, Z\}$ be the set of 26 uppercase letters and let $D = \{0, 1, \ldots, 9\}$ be the set of ten digits. Then $A = L \cup D$ is the set of alphanumeric characters from which FORTRAN identifiers are formed. There are $|L| = 26$ single-character FORTRAN identifiers. An identifier with m characters, $2 \leq m \leq 6$, can be thought of as an element of the set $L \times A \times \cdots \times A$, where A is used as a factor $m - 1$ times. Hence there are $(26)(36^{m-1})$ identifiers with m characters. Then, by the addition principle, the number of FORTRAN identifiers is

$$26 + (26)(36) + (26)(36^2) + \cdots + (26)(36^5) = 1{,}617{,}038{,}306$$

Using a similar analysis, the number of FORTRAN identifiers that end with a digit is

$$(26)(10) + (26)(36)(10) + (26)(36^2)(10) + (26)(36^3)(10)$$

$$+ (26)(36^4)(10) = 449{,}177{,}300 \qquad \blacksquare$$

Exercises 2.4

1. Of the 100 manufactured items discussed in Example 2.21, suppose that ten have defect 1, eight have defect 2, and 85 are nondefective. How many of these items have both types of defects?
2. There are 30 students in a certain discrete mathematics class. Ten students failed the midterm exam, five students failed the final exam, and three students failed both the midterm and the final.
   (a) How many students passed both the midterm and the final?
   (b) How many students passed the midterm or the final?
3. A social security number is a sequence of nine digits—for example, 080-55-1618.
   (a) How many possible social security numbers are there?
   (b) How many possible social security numbers do not contain the digit 0?
   (c) How many possible social security numbers neither begin nor end with the digit 0?
   (d) How many possible social security numbers have no digit repeated?
4. Given any three sets A, B, and C, show that the sets $(A - B) - C$, $(A \cap B) - C$, $(A \cap C) - B$, and $A \cap B \cap C$ are pairwise disjoint and
$$A = [(A - B) - C] \cup [(A \cap B) - C]$$
$$\cup [(A \cap C) - B] \cup [A \cap B \cap C]$$
5. An identifier in Ada consists of letters, digits, and underline characters. An identifier must begin with a letter, may not end with an underline character, and may not contain two consecutive underline characters. Also no distinction is made between lowercase and uppercase letters.
   (a) How many three-character Ada identifiers are there?
   (b) How many four-character Ada identifiers are there?
   (c) How many five-character Ada identifiers contain at most one underline character?
   (d) How many five-character Ada identifiers contain exactly two underline characters?
   (e) How many five-character Ada identifiers are there?
6. Determine an inclusion–exclusion formula for $|A \cup B \cup C|$. (Hint: Write $D = B \cup C$ and apply Theorem 2.14 to find $|A \cup D|$.)
7. Consider the experiment of choosing, at random, a sequence of four cards from a standard deck of 52 playing cards. The cards are chosen

one at a time with replacement, meaning that after a card is chosen, it is replaced and the deck is reshuffled. We are interested in the sequence of cards that is obtained—for example, (2 of clubs, ace of hearts, king of spades, 2 of clubs). Determine each of the following:

(a) the number of possible sequences.

(b) the number of sequences having no spades.

(c) the number of sequences such that all four cards are spades.

(d) the number of sequences such that all four cards are the same suit.

(e) the number of sequences such that the first card is a spade and the third card is not an ace.

(f) the number of sequences such that the first card is a spade or the third card is not an ace.

8. There are 225 computer science majors at a certain university. This semester 100 of these students are taking the data structures course, 75 are taking the database management course, 50 are taking the artificial intelligence course, 7 are taking both data structures and database management, 4 are taking both data structures and artificial intelligence, 31 are taking both database management and artificial intelligence, and 1 student is taking all three of these courses.

(a) Use the result of Exercise 6 to determine the number of computer science majors who are taking at least one of these three courses.

(b) How many computer science majors are taking none of these three courses?

(c) Draw a Venn diagram like that of Figure 2.5, where U is the set of 225 computer science majors.

9. A box contains three balls colored red, blue, and green. An experiment consists of choosing at random a sequence of three balls, one at a time with replacement. An outcome for this experiment is written as an ordered triple; for example, (B, R, G) denotes that the first ball chosen is blue, the second is red, and the third is green.

(a) How many possible outcomes are there for this experiment?

(b) Draw a tree diagram that helps answer the question, How many outcomes for the experiment have the property that no two balls chosen consecutively have the same color?

(c) Draw a tree diagram that helps answer the question, How many outcomes for the experiment have the property that at least two of the balls chosen are blue?

10. In 1986 the Boston Red Sox and the California Angels played in the playoff series to determine the American League (baseball) champion. The winner of such a series is the first team to win four games (best four out of seven). An outcome for such a series can be represented as an ordered n-tuple, where $4 \leq n \leq 7$; for example, (B, B, B, B) indicates that Boston won the series in four games, and (C, B, C, C, C) indicates that California won the series in five games, with Boston winning the second game. Use tree diagrams to help answer these questions.

(a) How many ways are there for Boston to win the series in five games?

(b) How many ways are there for California to win the series in six games?

(c) How many ways are there for Boston to win the series?

(d) How many possible outcomes for the series are there?

11. Three distinct mathematics and two distinct computer science final examinations must be scheduled during a five-day period. How many ways are there to schedule the examinations if

(a) there are no restrictions?

(b) no two exams can be scheduled for the same day?

(c) no two mathematics exams, nor two computer science exams, can be scheduled for the same day?

(d) each mathematics exam must be the only exam scheduled for a given day?

12. A box contains 12 distinct colored balls, numbered 1 through 12. Balls 1 through 3 are red, balls 4 through 7 are blue, and balls 8 through 12 are green. An experiment consists of choosing three balls at random from the box, one at a time with replacement; an outcome is the sequence of balls chosen, $(7, 4, 9)$ or $(3, 8, 3)$, for example.

(a) How many outcomes are there?

How many outcomes are there such that

(b) the first ball is red, the second is blue, and the third is green?

(c) exactly two balls are green?

(d) all three balls are red?

(e) all three balls are the same color?

(f) the first ball is red or the third ball is green?

(g) at least one of the three balls is red?

13. Consider four-letter sequences, such as (A, B, C, D) and (B, E, E, P). There are 26^4 such sequences; how many of these have

(a) exactly one A?

(b) at most one A?

(c) at least one A?

(d) exactly one vowel (A, E, I, O, or U)?

(e) at least one vowel?

14. In the game of Mastermind, one player, the "codemaker," selects as the "code" a sequence of four colors, each chosen from a set of six possible colors. For example, (green, red, blue, red) and (white, black, yellow, red) are possible codes.

(a) How many possible codes are there?

(b) How many possible codes use four colors?

15. Ron has n friends who enjoy playing bridge. Every Wednesday evening Ron invites three of these friends over to his home for a bridge game. Ron always sits in the south position at the bridge table; he decides which friends are to sit in the west, north, and east positions. How large

(at least) must n be if Ron is able to do this for 210 weeks before repeating a seating arrangement?

CHAPTER PROBLEMS

1. In separate Venn diagrams like the one shown in Figure 2.7, shade the region corresponding to each set.

(a) $A \cap B$
(b) $B \cap C' \cap D$
(c) $(A \cup B) \cap D'$
(d) $(A \cap B') \cup (C \cap D)$
(e) $(A \cup B \cup C \cup D)'$
(f) $((A' \cup B) \cap C)'$

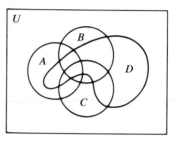

Figure 2.7

2. Let U be the set of those people who voted in the 1984 U.S. presidential election. Define the following subsets of U:

$$D = \{ x \in U \mid x \text{ registered as a Democrat} \}$$

$$R = \{ x \in U \mid x \text{ voted for Ronald Reagan} \}$$

$$W = \{ x \in U \mid x \text{ belonged to a union} \}$$

Describe each subset of U in terms of D, R, and W and draw an appropriate Venn diagram for each.

(a) people who did not vote for Reagan
(b) union members who voted for Reagan
(c) registered Democrats who voted for Reagan but did not belong to a union
(d) union members who either were not registered as Democrats or voted for Reagan
(e) people who voted for Reagan but were not registered as Democrats and were not union members
(f) people who were either registered as Democrats, were union members, or did not vote for Reagan

3. An insurance company classifies its set U of policyholders by using the following sets:

$$A = \{\, x \in U \mid x \text{ drives a subcompact car} \,\}$$

$$B = \{\, x \in U \mid x \text{ drives a car that is more than 5 years old} \,\}$$

$$C = \{\, x \in U \mid x \text{ is married} \,\}$$

$$D = \{\, x \in U \mid x \text{ is over 20 years of age} \,\}$$

$$E = \{\, x \in U \mid x \text{ is male} \,\}$$

Express each subset of U in terms of A, B, C, D, and E.
(a) female policyholders over 20 years of age
(b) policyholders who are male or drive cars more than 5 years old
(c) female policyholders over 20 years of age who drive subcompact cars
(d) male policyholders who are either married or over 20 years of age and do not drive subcompact cars

4. For sets A, B, and C, consider the following conditions:
(i) $A \cap B = A \cap C$, (ii) $A \cup B = A \cup C$, and (iii) $B - A = C - A$.
For each pair of these conditions,
(a) (i) and (ii)   **(b)** (i) and (iii)   **(c)** (ii) and (iii)
either prove that the two conditions together imply that $B = C$, or give an example in which the two conditions hold but $B \neq C$.

5. For arbitrary sets A, B, C, and D, prove that

$$(A \subseteq C \land B \subseteq D) \leftrightarrow (A \times B \subseteq C \times D)$$

6. Let A, B, C, and D be sets such that $A \subseteq C$ and $B \subseteq D$.
Prove that:
(a) $A \cap B \subseteq C \cap D$ **(b)** $A \cup B \subseteq C \cup D$ **(c)** $A - D \subseteq C - B$

7. For any sets A, B, and C, prove or disprove:
(a) $(A - B) - C = A - (B - C)$
(b) $(A - B) - C = (A - C) - B$
(c) $(A - B) - C = (A - C) - (B - C)$

8. Let A, B, and C be arbitrary sets. Prove:
(a) $A - (B - C) = A \cap (B' \cup C)$
(b) $(A - B) - C = A - (B \cup C)$

9. For any sets A and B, prove or disprove:
(a) $\mathcal{P}(A \cap B) = \mathcal{P}(A) \cap \mathcal{P}(B)$
(b) $\mathcal{P}(A \cup B) = \mathcal{P}(A) \cup \mathcal{P}(B)$
(c) $\mathcal{P}(A - B) = \mathcal{P}(A) - \mathcal{P}(B)$

10. For any sets A and B, the *symmetric difference* of A and B is the set $A * B$ defined by

$$A * B = (A - B) \cup (B - A)$$

Prove:
(a) $A * B = B * A$ ★(b) $(A * B) * C = A * (B * C)$
(c) $A * B = (A \cup B) - (A \cap B)$
(d) $A \cap (B * C) = (A \cap B) * (A \cap C)$
(e) $A \cap B = \emptyset \leftrightarrow A * B = A \cup B$ (f) $A * A = \emptyset$

11. Let A, B, and C be any sets and prove each of the following:
    (a) If $C \subseteq A$ and $C \subseteq B$, then $C \subseteq A \cap B$.
    (b) If $A \subseteq C$ and $B \subseteq C$, then $A \cup B \subseteq C$.

12. Let A, B, and C be finite subsets of a universal set U, as shown in Figure 2.8.
    (a) Express each of regions 1 through 7 in terms of A, B, and C.
    (b) Give a formula for $|A \cup B \cup C|$ in terms of the cardinalities of regions 1 through 7.

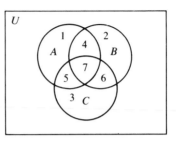

Figure 2.8

13. In a survey of 100 students, the following information was obtained:

    12 are taking courses in English, mathematics, and biology.

    22 are taking courses in English and mathematics, but not biology.

    3 are taking courses in English and biology, but not mathematics.

    7 are taking courses in mathematics and biology, but not English.

    20 are taking courses in English, but not biology or mathematics.

    17 are taking courses in biology, but not English or mathematics.

    Also, each of the 100 students is taking a course in at least one of these three areas. How many students are taking courses in mathematics, but not English or biology? (Hint: Use the result of Problem 12, part (b).)

In Problems 14–21, the universal set is $\mathbb{R}$.

14. For $i \in \{1, 2, \ldots, 100\}$, define $A_i = \{-i, -i + 1, \ldots, i\}$. Find $\bigcup_{i=1}^{100} A_i$ and $\bigcap_{i=1}^{100} A_i$.

15. For $n \in \mathbb{N}$, define $A_n = \{n, n + 1, \ldots, 2n + 1\}$. Find $\bigcup_{n=1}^{\infty} A_n$ and $\bigcap_{n=1}^{\infty} A_n$.

16. For $n \in \mathbb{N}$, define $A_n = \{-n, -n + 1, \ldots, n^2\}$. Find $\bigcup_{n=1}^{\infty} A_n$ and $\bigcap_{n=1}^{\infty} A_n$.

17. Let $I = (0, 1)$ and for $i \in I$, define $A_i = (i, 1/i)$. Find $\bigcup_{i \in I} A_i$ and $\bigcap_{i \in I} A_i$.

18. For $n \in \mathbb{N}$, define $A_n = \{m/n \mid m \in \mathbb{Z}\}$; that is, A_n is the set of fractions with denominator n. Find $\bigcup_{n=1}^{\infty} A_n$ and $\bigcap_{n=1}^{\infty} A_n$.

19. Let $I = (-1, 1)$ and for $i \in I$, define $A_i = (2i - 1, i^2)$. Find $\bigcup_{i \in I} A_i$ and $\bigcap_{i \in I} A_i$. (Hint: Graph the equations $y = 2x - 1$ and $y = x^2$ for $x \in (-1, 1)$.)

20. Let $I = [1, \infty)$ and for $i \in I$, define $A_i = [\sqrt{i} - 1, \sqrt{i} + 1]$. Find $\bigcup_{i \in I} A_i$ and $\bigcap_{i \in I} A_i$.

21. For $n \in \mathbb{N}$, define $A_n = \left(\dfrac{n + 1}{2n}, \dfrac{3n - 1}{n} \right)$. Find $\bigcup_{n=1}^{\infty} A_n$ and $\bigcap_{n=1}^{\infty} A_n$.

22. A chessboard has 64 squares arranged into eight rows and eight columns.
    (a) In how many ways can eight identical pawns be placed on a chessboard so that no two are in the same row or column?
    (b) In how many ways can two white rooks and two black rooks be placed on a chessboard so that no two rooks of the opposite color are in the same row or column? (Consider two rooks of the same color to be identical.)

23. Let $I = (0, \infty)$ and for $i \in I$, define $A_i = \{(x, y) \mid y = i \cdot x^2\}$; that is, A_i is the set of points on the parabola $y = i \cdot x^2$. Find $\bigcup_{i \in I} A_i$ and $\bigcap_{i \in I} A_i$. (The universal set in this problem is $\mathbb{R} \times \mathbb{R}$.)

24. Consider a quiz with ten questions. How many ways are there to complete the quiz if
    (a) each question is answered "true" or "false?"
    (b) each question has four possible answers: a, b, c, or d?

25. How many possible telephone numbers are there if
    (a) a number consists of seven digits and the first digit is not allowed to be zero?
    (b) a number consists of seven digits with first digit 6 and second digit 7?

(c) a number consists of ten digits and neither the first digit nor the fourth digit is allowed to be 0.

26. Let A, B, and C be (finite) sets and consider the sets $A \times B \times C$ and $(A \times B) \times C$.
   (a) Explain why, according to the definition of Cartesian product, these two sets are not equal.
   (b) Show that $|A \times B \times C| = |(A \times B) \times C|$.

27. Let I be a nonempty set and let $\{A_i \mid i \in I\}$ be a collection of sets indexed by I. For any set B, prove:
   (a) $[\forall i \in I(B \subseteq A_i)] \leftrightarrow \left(B \subseteq \bigcap_{i \in I} A_i \right)$
   (b) $[\forall i \in I(A_i \subseteq B)] \leftrightarrow \left(\bigcup_{i \in I} A_i \subseteq B \right)$

28. Let I be a nonempty set and let $\{A_i \mid i \in I\}$ be a collection of sets indexed by the set I. Let $B = \bigcup_{i \in I} A_i$ and $C = \bigcap_{i \in I} A_i$. Prove that

$$C = B - \bigcup_{i \in I} (B - A_i)$$

29. Assuming the declarations

```
TYPE BASETYPE = MIN..MAX;
     SETTYPE = SET OF BASETYPE;
```

(where MIN and MAX are constants), write a Pascal function CARD that returns the cardinality of a set A of type SETTYPE.

30. This problem illustrates a very general method for implementing sets in Pascal. Consider the declarations

```
CONST MAX = 50;
TYPE BASETYPE = REAL;
     SETTYPE = RECORD
       ELEMENT: ARRAY[1..MAX] OF BASETYPE;
       CARD: 0..MAX
     END;
```

This declares a type SETTYPE for sets of up to 50 elements of type REAL. If A is a variable of type SETTYPE, then A.CARD is the cardinality of A and, if A.CARD > 0, then A.ELEMENT[1], A.ELEMENT[2], ..., A.ELEMENT[A.CARD] are the elements of A. Of course, by using a different value for MAX or a different type for BASETYPE, other types of sets may be implemented in this way.
   (a) Write a function MEMBER that tests whether an element X belongs to a set A. (MEMBER returns a result of type BOOLEAN.)
   Use MEMBER to write the following procedures
   (b) INTERSECTION
   (c) UNION
   (d) DIFFERENCE
   to find $A \cap B$, $A \cup B$, and $A - B$, respectively, for two sets A and B.

THREE

Number Theory and Mathematical Induction

INTRODUCTION

Number theory is that area of mathematics dealing with the properties of the integers under the ordinary operations of addition, subtraction, multiplication, and division. It is one of the oldest and, without dispute, one of the most beautiful branches of mathematics. Its problems have been studied by mathematicians, scientists, and others for well over 2000 years.

In a large measure, the subject is characterized by the simplicity with which difficult problems can be stated and the ease with which they can be understood and appreciated by persons without much mathematical background. Thus it should come as no surprise that such problems have attracted the attention of professional mathematicians and amateurs alike.

Several of the most basic and interesting problems in number theory involve prime numbers. Here is an example of one such problem:

Are there infinitely many primes of the form $n^2 + 1$, where $n \in \mathbb{N}$?

This question is certainly easy to understand and yet, to this day, no one has determined the answer to it.

Another outstanding conjecture, due to the famous French mathematician Pierre de Fermat (1601–1665), goes like this:

For every integer $n \geq 3$, the equation $x^n + y^n = z^n$ has no solution in nonzero integers x, y, and z.

Fermat had the practice of making notes in his copy of the works of the Greek mathematician Diophantus (circa A.D. 300) and would quite often state, without proof, a result he had discovered. The preceding conjecture is one of these discoveries; in fact, it is the only one for which the mathematical community has been unable to supply a proof. Fermat himself wrote, "For this I have discovered a truly wonderful proof, but the margin is too small." Because of this claim the conjecture is called "Fermat's last theorem." It should be mentioned that attempts to prove it have given rise to a wealth of important mathematics, including a good portion of abstract algebra.

One of the aims of this chapter is to provide the student with some basic information from elementary number theory. This includes a treatment of several number theoretic algorithms. In addition, we introduce the "principle of mathematical induction" and show how it is applied in proving statements about the positive integers. Many of the ideas and results presented are used in succeeding chapters of this book. This material is also encountered in subsequent course work in various areas of mathematics and computer science.

One of the most basic principles used in discrete mathematics, especially in number theory, is the "principle of well-ordering."

> **Principle of well-ordering.** Every nonempty set of positive integers has a smallest element.

It is not possible to prove this statement using the familiar properties satisfied by the integers under addition and multiplication. However, after a little thought, the principle should seem truly self-evident. Hence we shall take it as an axiom. At a later point in this chapter, when the principle of mathematical induction is introduced, we show that it is a logical consequence of the principle of well-ordering (in fact, it turns out that these two principles are logically equivalent).

In general, a subset T of $\mathbb{R}$ is said to be *well-ordered* provided any nonempty subset S of T has a smallest element. Thus, the principle of well-ordering states that $\mathbb{N}$ is well-ordered.

Example 3.1   Find the smallest element in each subset of $\mathbb{N}$.

(a) $S_1 = \{n \in \mathbb{N} \mid n \text{ is prime}\}$
(b) $S_2 = \{n \in \mathbb{N} \mid n \text{ is a multiple of } 7\}$
(c) $S_3 = \{n \in \mathbb{N} \mid n = 110 - 17m \text{ for some } m \in \mathbb{Z}\}$
(d) $S_4 = \{n \in \mathbb{N} \mid n = 12s + 18t \text{ for some } s, t \in \mathbb{Z}\}$

Solution   (a) The smallest prime number is 2.

(b) The smallest positive multiple of 7 is 7.

(c) Here we must find the smallest positive number n of the form $110 - 17m$, where m is an integer. The number $110 = 110 - (17)(0)$ is of this form, and as m increases, n decreases. As m takes on the values

$1, 2, 3, \ldots,$ the values of n form the sequence

$$93, 76, 59, \ldots, 8, -9, \ldots$$

Hence the smallest element of S_3 is 8. The number 8 just happens to be the remainder when 110 is divided by 17. This is more than just a coincidence, as we see in the next section when we discuss the "division algorithm."

(d) We are looking for the smallest positive number n of the form $12s + 18t$, where s and t are integers. Note that $12s + 18t = 6(2s + 3t)$, so that an element of S_4 must be a multiple of 6. Also, $6 = (12)(-1) + (18)(1)$, so that 6 is an element of S_4. This shows that 6 is the smallest element of S_4. The number 6 happens to be the "greatest common divisor" of 12 and 18, an idea that is explored in Section 3.3.   ■

Example 3.2   We give two examples of subsets of $\mathbb{R}$ that are not well-ordered. First consider the set $\mathbb{Z}$ of integers; $\mathbb{Z}$ is not well-ordered because $\mathbb{Z}$ itself does not have a smallest element. In particular, given any integer m, the integer $m - 1$ is less than m. Second, consider the set $T = [0, 1]$ of real numbers between 0 and 1, inclusive. The set T has a smallest element, namely 0, but this does not mean that T is well-ordered. In fact, the set $S = (0, 1)$ is a subset of T that does not have a smallest element. To see this, note that if x is a real number such that $0 < x < 1$, then $0 < x/2 < x$.   ■

We often make use of the following slight extension of the principle of well-ordering.

THEOREM 3.1   If S is a nonempty set of nonnegative integers, then S has a smallest element.

Proof   We consider two cases depending on whether $0 \in S$. If $0 \in S$, then clearly 0 is the smallest element of S. On the other hand, if $0 \notin S$, then S is a nonempty set of positive integers and the principle of well-ordering implies that S has a smallest element.   □

Exercises 3.1

1. Determine whether these subsets of $\mathbb{R}$ are well-ordered.
   (a) T_1 is a finite subset of $\mathbb{R}$. (In particular, what if $T = \emptyset$?)
   (b) $T_2 = \{ t \in \mathbb{Q} \mid t \geq 0 \}$
   (c) T_3 is the set of even integers.
   (d) T_4 is a subset of $\mathbb{Z}$ and T_4 itself has a smallest element t_0.
2. Let T_1 and T_2 be subsets of $\mathbb{R}$ such that $T_1 \subseteq T_2$. Prove: If T_2 is well-ordered, then T_1 is well-ordered.
3. Find the smallest element of each subset of $\mathbb{N}$.
   (a) $A = \{ n \in \mathbb{N} \mid n = m^2 - 10m + 28 \text{ for some integer } m \}$
   (b) $B = \{ n \in \mathbb{N} \mid n = 5q + 2 \text{ for some integer } q \}$
   (c) $C = \{ n \in \mathbb{N} \mid n = -150 - 19m \text{ for some integer } m \}$
   (d) $D = \{ n \in \mathbb{N} \mid n = 5s + 8t \text{ for some integers } s \text{ and } t \}$.

3.2   THE DIVISION ALGORITHM

One of the fundamental concepts included in any introduction to number theory is that of factoring integers. In particular, given an integer $n > 1$, we are interested in expressing n as a product of primes. For example, if $n = 132$, then $n = 2 \cdot 2 \cdot 3 \cdot 11$. Is it always possible to do this? Can it be done in several possible ways?

Before these questions can be answered, it is necessary to define and work with certain fundamental terms, like *divisor* and *prime*.

DEFINITION 3.1

Let a and b be integers, with $a \neq 0$.

1. We say that a *divides* b, denoted $a \mid b$, provided there is an integer q such that $aq = b$. We say that a is a *divisor* (or factor) of b, and we call b a *multiple* of a.

2. A positive integer d is called a *proper divisor* of b if $d \mid b$ and $d < |b|$.

3. An integer $p > 1$ is called a prime *number* (or simply a prime) provided 1 is the only proper divisor of p. An integer $c > 1$ that is not prime is called *composite*.

In view of the definition of *prime*, it is clear that an integer $p > 1$ is prime if and only if it is impossible to express p as $p = ab$, where a and b are integers and both $1 < a < p$ and $1 < b < p$. Thus an integer $c > 1$ is composite if and only if there exist integers a and b such that $c = ab$, with both $1 < a < c$ and $1 < b < c$.

Example 3.3   (a) $2 \mid 6$, since $(2)(3) = 6$.
(b) $-3 \mid 27$, since $(-3)(-9) = 27$.
(c) $12 \mid -72$, since $(12)(-6) = -72$.
(d) 4 does not divide 7, since there is no integer q such that $4q = 7$.
(e) -8 does not divide 28, since there is no $q \in \mathbb{Z}$ such that $-8q = 28$.
(f) If $b \neq 0$, then $b \mid 0$, since $b \cdot 0 = 0$.   ∎

Example 3.4   Find the divisors of 126 and factor 126 as a product of primes.

Solution   The positive divisors of 126 are 1, 2, 3, 6, 7, 9, 14, 18, 21, 42, 63, and 126. Note that if m is a divisor of n, then so is $-m$. So the set of divisors of

126 is

$$\{-126, -63, -42, -21, -18, -14, -9, -7, -6, -3, -2, -1\}$$

$$\cup\,\{1, 2, 3, 6, 7, 9, 14, 18, 21, 42, 63, 126\}$$

The prime divisors of 126 are 2, 3, and 7, and

$$126 = 2 \cdot 3 \cdot 3 \cdot 7 \qquad\blacksquare$$

A number of basic properties of the relation divides are used in this and subsequent chapters. The next theorem lists a few of these properties.

THEOREM 3.2 For integers a, b, and c, with $a \neq 0$, the following hold:

1. If $a \mid b$, then $a \mid bx$ for any integer x.
2. If $a \mid b$ and $b \mid c$ (where $b \neq 0$), then $a \mid c$.
3. If $a \mid b$ and $a \mid c$, then $a \mid (bx + cy)$ for any integers x and y.
4. If $b \neq 0$ and both $a \mid b$ and $b \mid a$, then $a = b$ or $a = -b$.
5. If $a \mid b$ and both a and b are positive, then $a \leq b$.

Proof The proofs of parts 2 and 3 are given here and provide a good illustration of how the remaining parts should be proved; these are left to Exercise 2.

For the proof of part 2, we are given that $a \mid b$ and $b \mid c$, so by definition there exist integers q_1 and q_2 such that $aq_1 = b$ and $bq_2 = c$. Substitution for b then yields

$$c = bq_2 = (aq_1)q_2 = a(q_1 q_2)$$

It follows that $a \mid c$.

The hypothesis of part 3 is that $a \mid b$ and $a \mid c$, so there exist integers q_1 and q_2 such that $aq_1 = b$ and $aq_2 = c$. Hence for any integers x and y,

$$bx + cy = (aq_1)x + (aq_2)y = a(q_1 x) + a(q_2 y) = a(q_1 x + q_2 y)$$

Therefore, $a \mid (bx + cy)$. $\qquad\square$

Once again the meaning of the statement "a divides b" should be emphasized—namely, that there is some integer q such that $aq = b$. No doubt the reader recalls the process of "long division." Here one integer b is divided by another integer a, obtaining a "quotient" and a "remainder." For example, if $a = 7$ and $b = 23$, then the relation $23 = (7)(3) + 2$ is obtained. In this case, 3 is the quotient and 2 is the remainder. In general, if a and b are integers with $a > 0$, then there exist unique integers q and r such that $b = aq + r$ and $0 \leq r < a$. This property is called the *division algorithm*. Here the numbers a, b, q, and r are called the *divisor, dividend, quotient* and *remainder*, respectively.

THEOREM 3.3   **(Division algorithm)**

Given integers a and b with $a > 0$, there exist integers q and r such that $b = aq + r$ and $0 \leq r < a$. Moreover, q and r are uniquely determined by a and b.

Proof   We first show that there exist integers q and r such that $b = aq + r$ and $0 \leq r < a$. In order to do this, we apply the principle of well-ordering (the extended version, Theorem 3.1) to the set

$$S = \{b - ax \mid x \in \mathbb{Z} \wedge b - ax \geq 0\}$$

So S is a set of nonnegative integers. In order to apply Theorem 3.1, we must first show that S is nonempty. If $b \geq 0$, then choosing $x = 0$, we obtain $b = b - a \cdot 0$, so that $b \in S$. Suppose, on the other hand, that $b < 0$. Since $a \geq 1$, choosing $x = b$ yields $b - ab = b(1 - a) \geq 0$. Hence $b - ab \in S$. Thus we have $S \neq \emptyset$. Therefore, by the principle of well-ordering, S has a smallest element, say, r. Thus there is some $x \in \mathbb{Z}$, say $x = q$, such that $r = b - aq$, or $b = aq + r$.

Since $r \in S$, it follows that $r \geq 0$, so it remains to show that $r < a$. We proceed by contradiction and suppose that $r \geq a$. Then $t = r - a \geq 0$ and, since $a > 0$, $t < r$. Moreover,

$$t = r - a = (b - aq) - a = b - (aq + a) = b - a(q + 1)$$

Since $t \geq 0$ and $t = b - a(q + 1)$, we have that $t \in S$. But then $t \in S$ and $t < r$, contradicting the fact that r is the smallest element of S. Thus $r < a$, and we have shown that $b = aq + r$, where $0 \leq r < a$.

We next show that the quotient q and the remainder r are uniquely determined by a and b. Suppose that $b = aq + r$ and $b = ap + s$, where q, r, p, and s are integers and both $0 \leq r < a$ and $0 \leq s < a$. We must show that $p = q$ and $s = r$. Assume, without loss of generality, that $s \leq r$; hence $r - s \geq 0$. Since $ap + s = b = aq + r$, we obtain

$$a(p - q) = r - s$$

Thus $a \mid (r - s)$. Since $0 \leq r - s < a$, it must be the case that $r - s = 0$, so that $r = s$. Then to see that $p = q$, observe that $a(p - q) = r - s = 0$ and $a \neq 0$, so $p - q = 0$. ☐

The proof of Theorem 3.3 is an "existence proof," in that it concentrates on showing the existence of integers q and r satisfying the properties stated in the theorem, rather than on giving a method for finding q and r. However, the proof implicitly suggests an algorithm for finding q and r from a and b, using only the operations of addition and subtraction. We give an informal description of this algorithm, with the implementation details left to Exercise 15.

If $b \geq 0$, then consider the sequence

$$b = b - a \cdot 0, \; b - a \cdot 1, \; b - a \cdot 2, \ldots$$

Since $a > 0$, this sequence is eventually negative; q is the largest nonnegative integer for which $b - aq \geq 0$. On the other hand, if $b < 0$, then consider the sequence

$$b = b - a \cdot 0, b - a \cdot (-1), b - a \cdot (-2), \ldots$$

This sequence is eventually nonnegative; if k is the smallest positive integer for which $b - a \cdot (-k) \geq 0$, then $q = -k$. In either case $r = b - aq$.

The division algorithm can be extended to handle any nonzero divisor.

COROLLARY 3.4 Given integers a and b with $a \neq 0$, there exist integers q and r such that $b = aq + r$, where $0 \leq r < |a|$. Moreover, q and r are uniquely determined by a and b.

The proof of Corollary 3.4 is left to Exercise 10.

The division algorithm turns out to be very useful in determining further divisibility properties of the integers. Corollary 3.4 is used, somewhat implicitly, in many programming languages. For example, the operators "DIV" and "MOD" in Pascal have the effect of yielding the quotient and remainder, respectively, when applied to positive integers b and a. The general definition of DIV and MOD given next concurs with some versions of Pascal.

DEFINITION 3.2

Let a and b be integers, with $a \neq 0$, and suppose $b = aq + r$, where $0 \leq r < |a|$. We define the operators DIV and MOD by

$$b \operatorname{DIV} a = q$$

$$b \operatorname{MOD} a = r$$

The application of DIV to two integers is called *integer division*.

Example 3.5 Find $b \operatorname{DIV} a$ and $b \operatorname{MOD} a$.

(a) $a = 17, b = 110$ (b) $a = 7, b = -59$
(c) $a = -11, b = 41$ (d) $a = -5, b = -27$

Solution   (a) As was seen in Example 3.1(c),

$$110 = 17 \cdot 6 + 8$$

Hence $110 \operatorname{DIV} 17 = 6$ and $110 \operatorname{MOD} 17 = 8$.

(b) Here it is easy to make the mistake of saying $-59 \text{ DIV } 7 = -8$ and $-59 \text{ MOD } 7 = -3$. But remember, $b \text{ MOD } a$ must always satisfy

$$0 \leq b \text{ MOD } a < |a|$$

The correct answers are $-59 \text{ DIV } 7 = -9$ and $-59 \text{ MOD } 7 = 4$.

(c) Since $41 = (-11)(-3) + 8$, it follows that $41 \text{ DIV } -11 = -3$ and $41 \text{ MOD } -11 = 8$.

(d) Similarly, $-27 = (-5) \cdot 6 + 3$, so $-27 \text{ DIV } -5 = 6$ and $-27 \text{ MOD } -5 = 3$. ■

Given any integer x, the number $x \text{ MOD } 3$ belongs to the set $\{0, 1, 2\}$. Define the sets S_0, S_1, and S_2 by

$$S_0 = \{x \in \mathbb{Z} \mid x \text{ MOD } 3 = 0\}$$
$$S_1 = \{x \in \mathbb{Z} \mid x \text{ MOD } 3 = 1\}$$
$$S_2 = \{x \in \mathbb{Z} \mid x \text{ MOD } 3 = 2\}$$

Then, for $r = 0, 1, 2$, S_r is the set of all those integers x that yield a remainder of r when divided by 3. For instance, $11 \text{ MOD } 3 = 2$, so $11 \in S_2$, whereas $-11 \text{ MOD } 3 = 1$, so $-11 \in S_1$. By the division algorithm, each integer x belongs to exactly one of the sets S_0, S_1, or S_2. It follows that

1. $\mathbb{Z} = S_0 \cup S_1 \cup S_2$
2. $\{S_0, S_1, S_2\}$ is a pairwise disjoint collection of nonempty sets.

Because of these two properties, we say that $\{S_0, S_1, S_2\}$ is a "partition" of the set $\mathbb{Z}$; the important concept of partition is one we study further in Chapter 4.

Explicitly, what are the sets S_0, S_1, and S_2? First consider S_0. If $x \in S_0$, then $x \text{ MOD } 3 = 0$, so there is some integer q such that $x = 3q$. This is the same thing as saying that x is a multiple of 3. Conversely, if $x = 3q$ for some integer q, then $x \text{ MOD } 3 = 0$. Thus

$$S_0 = \{3q \mid q \in \mathbb{Z}\}$$

More generally, for $r = 0, 1, 2$, $x \in S_r$ if and only if there is some integer q such that $x = 3q + r$. Thus

$$S_1 = \{3q + 1 \mid q \in \mathbb{Z}\}$$
$$S_2 = \{3q + 2 \mid q \in \mathbb{Z}\}$$

Explicitly,

$$S_0 = \{\ldots, -9, -6, -3, 0, 3, 6, 9, \ldots\}$$
$$S_1 = \{\ldots, -8, -5, -2, 1, 4, 7, 10, \ldots\}$$
$$S_2 = \{\ldots, -7, -4, -1, 2, 5, 8, 11, \ldots\}$$

Note that two integers are in the same set S_r if and only if they differ by a multiple of 3. This idea is generalized in Exercise 8.

Example 3.6   Let x and y be integers such that

$$x \text{ DIV } 5 = q_1 \qquad x \text{ MOD } 5 = 2$$
$$y \text{ DIV } 5 = q_2 \qquad y \text{ MOD } 5 = 3$$

Find:

(a) $(x + y) \text{ DIV } 5$ and $(x + y) \text{ MOD } 5$
(b) $(xy) \text{ DIV } 5$ and $(xy) \text{ MOD } 5$

Solution   Since $x \text{ DIV } 5 = q_1$, $x \text{ MOD } 5 = 2$, $y \text{ DIV } 5 = q_2$, and $y \text{ MOD } 5 = 3$, we have that,

$$x = 5q_1 + 2 \quad \text{and} \quad y = 5q_2 + 3$$

So for (a) we obtain

$$x + y = (5q_1 + 2) + (5q_2 + 3)$$
$$= 5q_1 + 5q_2 + 5$$
$$= 5(q_1 + q_2 + 1)$$

which shows that $x + y$ is a multiple of 5. Therefore $(x + y) \text{ MOD } 5 = 0$ and $(x + y) \text{ DIV } 5 = q_1 + q_2 + 1$. For (b) we find

$$xy = (5q_1 + 2)(5q_2 + 3)$$
$$= 25q_1q_2 + 15q_1 + 10q_2 + 6$$
$$= 25q_1q_2 + 15q_1 + 10q_2 + 5 + 1$$
$$= 5(5q_1q_2 + 3q_1 + 2q_2 + 1) + 1$$

Hence, $(xy) \text{ MOD } 5 = 1$ and $(xy) \text{ DIV } 5 = 5q_1q_2 + 3q_1 + 2q_2 + 1$.   ∎

Exercises 3.2

1. Find $b \text{ DIV } a$ and $b \text{ MOD } a$.
   (a) $a = 11$, $b = 297$ (b) $a = 9$, $b = -63$
   (c) $a = 8$, $b = 77$ (d) $a = 6$, $b = -71$
   (e) $a = -5$, $b = 35$ (f) $a = 6$, $b = 39$
2. Prove Theorem 3.2, parts 1, 4, and 5.
3. Prove: If a, b, c, and d are integers such that a and c are nonzero and both $a \mid b$ and $c \mid d$, then $ac \mid bd$.
4. Prove: If a, b, and c are integers such that a and c are nonzero and $ac \mid bc$, then $a \mid b$.
5. Prove or disprove: If a, b, and c are integers such that $a \neq 0$ and $a \mid bc$, then either $a \mid b$ or $a \mid c$.
6. Recall that an integer m is *even* if $m = 2k$ for some $k \in \mathbb{Z}$.
   (a) Complete: An integer m is *odd* if $m = \underline{\qquad}$ for some $k \in \mathbb{Z}$.
   Prove each of the following:
   (b) The sum of two even integers or of two odd integers is even.
   (c) The sum of an even integer and an odd integer is odd.

(d) The product of two even integers is even and of two odd integers is odd.

(e) The product of an even integer and an odd integer is even.

7. Prove: If a is any integer, then one of the integers a, $a + 2$, and $a + 4$ is divisible by 3. (Hint: Consider three cases depending on the value of a MOD 3.)

8. Let m, x, and y be integers, with $m \neq 0$. Prove: x MOD $m = y$ MOD m if and only if $x - y$ is a multiple of m.

9. Given that x MOD 7 = 2 and y MOD 7 = 6, find:
   (a) $(x + 5)$ MOD 7   **(b)** $(2x)$ MOD 7
   (c) $(-y)$ MOD 7   **(d)** $(x + y)$ MOD 7
   (e) $(2x + 3y)$ MOD 7   **(f)** (xy) MOD 7

10. Prove Corollary 3.4; in fact, show for $a < 0$ that

$$b \text{ DIV } a = -(b \text{ DIV}(-a)) \quad \text{and} \quad b \text{ MOD } a = b \text{ MOD}(-a)$$

   (Hint: If $a < 0$, then $-a > 0$; so by Theorem 3.3 there exist integers q and r such that $b = (-a)q + r$, where $0 \leq r < -a$.)

11. Prove that, given any three consecutive integers, one of them is a multiple of 3. (Hint: Let n, $n + 1$, and $n + 2$ be the three integers and read the hint for Exercise 7.)

12. Let a, b, c, d, and m be integers with $m \neq 0$. Prove: If a MOD $m = c$ MOD m and b MOD $m = d$ MOD m, then:
   (a) $(a + b)$ MOD $m = (c + d)$ MOD m
   (b) (ab) MOD $m = (cd)$ MOD m
   As a consequence of part (b), it can be shown that

$$a \text{ MOD } m = c \text{ MOD } m \rightarrow a^n \text{ MOD } m = c^n \text{ MOD } m$$

   for any positive integer n. Use part (a) and this result to find:
   (c) 3^{40} MOD 7
   (d) $(1^3 + 2^3 + 3^3 + 4^3 + \cdots + 100^3)$ MOD 4

13. Let x be an integer. Prove: x^2 MOD 5 $\in \{0, 1, 4\}$. (Hint: Apply Exercise 12 to show that x^2 MOD 5 = $(x$ MOD 5$)^2$ MOD 5.)

14. Let n be an integer.
   (a) Show that no integer of the form $n^2 + 1$ is a multiple of 7.
   (b) Find the two possible values for n MOD 13 given that $n^2 + 1$ is a multiple of 13.

15. Implement the division algorithm as a Pascal procedure that takes integers a and b, with $a \neq 0$, and finds $q = b$ DIV a and $r = b$ MOD a. (Hint: Consider the case where both a and b are positive. Initialize q to 0 and r to b. While $r \geq a$, increment q by 1 and subtract a from r.)

16. For some (if not all) Pascal compilers, the operators DIV and MOD give results that sometimes do not agree with Definition 3.2. Write a Pascal program to compute b DIV a and b MOD a (using the built-in operators) for various values of a and b. Compare the results with

those given by Definition 3.2. (The problem seems to occur when the dividend is negative. For $b > 0$, Pascal defines $b \,\text{DIV}\, a$ to be the integer part of b/a and $-b \,\text{DIV}\, a$ to be $-(b \,\text{DIV}\, a)$.)

3.3

THE EUCLIDEAN ALGORITHM

In this section we define the greatest common divisor of two integers (not both 0) and exhibit a method for finding it, given the integers.

DEFINITION 3.3

An integer $c \neq 0$ is called a *common divisor* of the integers a and b provided $c \mid a$ and $c \mid b$. If a and b are not both 0, then we define the *greatest common divisor* of a and b to be the largest common divisor of a and b. The greatest common divisor of a and b is denoted by $\gcd(a, b)$.

Example 3.7 Find:

(a) $\gcd(12, 18)$ (b) $\gcd(-5, 10)$ (c) $\gcd(-60, -24)$

Solution

(a) $\gcd(12, 18) = 6$
(b) $\gcd(-5, 10) = 5$
(c) $\gcd(-60, -24) = 12$ ■

Given integers a and b, not both 0, we can make some observations about $\gcd(a, b)$ that are easily verified. First, since $1 \mid a$ and $1 \mid b$, we see that $1 \leq \gcd(a, b)$. Also notice that $\gcd(a, b) = \gcd(b, a)$ and $\gcd(-a, b) = \gcd(a, -b)$. If $b \neq 0$, then it is clear that $\gcd(0, b) = |b|$. Thus to determine $\gcd(a, b)$ it suffices, without loss of generality, to consider only the case $1 \leq a \leq b$. In this case notice that

$$1 \leq \gcd(a, b) \leq a$$

If $1 \leq a \leq b$, then a simple method for finding $d = \gcd(a, b)$ is to search the integers $a, a - 1, a - 2, \ldots, 1$, looking for the first of these that is a common divisor of a and b. This method is implemented in the following Pascal segment.

ALGORITHM 3.1

```
(* Initial state: Variables A, B, and D are of type INTEGER; the values
    a and b of A and B, respectively, satisfy 1 <= a <= b. *)
D := A;
WHILE (A MOD D <> 0) OR (B MOD D <> 0) DO
    D := D - 1;
(* Final state: The value d of D is such that d = gcd(a, b). *)
```

When Algorithm 3.1 is executed to find $d = \gcd(a, b)$, the body of the WHILE loop is repeated $a - d$ times. If a is quite large relative to d, then Algorithm 3.1 is rather inefficient.

A faster method, which goes back to Euclid, is based on repeated application of the division algorithm. Given a and b, with $1 \le a \le b$, there exist integers q_1 and r_1 such that $b = aq_1 + r_1$, where $0 \le r_1 < a$. Suppose $d = \gcd(a, b)$ and $e = \gcd(r_1, a)$. Since $r_1 = b - aq_1$ and d is a common divisor of a and b, we see, by Theorem 3.2, part 3, that $d \mid r_1$. Thus d is a common divisor of r_1 and a, so that $d \le e$. Likewise, since $b = aq_1 + r_1$ and e is a common divisor of a and r_1, we see that $e \mid b$. Thus e is a common divisor of a and b, so that $e \le d$. Hence $d = e$, and we have $\gcd(a, b) = \gcd(r_1, a)$. If $r_1 = 0$, then $\gcd(a, b) = \gcd(0, a) = a$; otherwise, the division algorithm may be applied again to r_1 and a to obtain integers q_2 and r_2 such that $a = r_1q_2 + r_2$, where $0 \le r_2 < r_1$. Then

$$\gcd(a, b) = \gcd(r_1, a) = \gcd(r_2, r_1)$$

so that if $r_2 = 0$, then $\gcd(a, b) = r_1$. If $r_2 \ne 0$, then the division algorithm may be applied again. Setting $r_0 = a$, notice that the sequence of remainders $r_0, r_1, r_2, \ldots$ obtained by this process satisfies

$$r_0 > r_1 > r_2 > \cdots \ge 0$$

Hence the process must eventually terminate with a 0 remainder. Indeed, if r_{k+1} is the first 0 remainder ($r_{k+1} = 0$ and $r_k \ne 0$), then $\gcd(a, b) = r_k$. This method is known as the *Euclidean algorithm* and it is considered by some computer scientists to be the oldest known algorithm.

The following equations are obtained by applying the division algorithm:

$$b = aq_1 + r_1$$
$$a = r_1q_2 + r_2$$
$$r_1 = r_2q_3 + r_3$$
$$\vdots$$
$$r_{k-2} = r_{k-1}q_k + r_k$$
$$r_{k-1} = r_kq_{k+1}$$

Notice that the number of equations exhibited is $k + 1$; this represents the number of divisions required to determine $\gcd(a, b)$ using the Euclidean algorithm. Much is known about this algorithm, and it turns out to be significantly faster than Algorithm 3.1. Indeed, a result of G. Lame (1845) states that

$$10^k < a^5$$

According to Algorithm 3.1, the number of steps required to compute $\gcd(a, b)$ is roughly a. For example, if $a = 1,000,000 = 10^6$, then roughly 10^6 steps are needed for Algorithm 3.1. Lame's inequality yields

$$10^k < \left(10^6\right)^5 = 10^{30}$$

from which it follows that $k + 1 \leq 30$. Hence at most 30 steps are required by the Euclidean algorithm. Certainly 30 steps are far better than a million!

An implementation of the Euclidean algorithm as a Pascal program segment is presented next.

ALGORITHM 3.2   **Euclidean algorithm**

> (∗Initial state: Variables A, B, OLDR, R, and RNEW are of type INTEGER;
> The values a and b of A and B, respectively, satisfy 1 <= a <= b.∗)
> R := A;
> RNEW := B MOD A;
> WHILE RNEW > 0 DO
>   BEGIN
>   OLDR := R;
>   R := RNEW;
>   RNEW := OLDR MOD R;
>   END;
> (∗Final state: The value r of R is such that r = gcd(a, b).∗)

Example 3.8   Use Algorithm 3.2 to find gcd(228, 528).

Solution

$$528 \text{ MOD } 228 = 72$$
$$228 \text{ MOD } 72 = 12$$
$$72 \text{ MOD } 12 = 0$$

Hence gcd(228, 528) = 12.   ∎

Given integers a and b, a *linear combination* of a and b (over $\mathbb{Z}$) is any expression of the form

$$as + bt$$

where $s, t \in \mathbb{Z}$. The next result provides an important characterization of $gcd(a, b)$, showing that it is the smallest positive integer that can be expressed as a linear combination of a and b. The proof of the theorem applies the division algorithm in a strong way.

THEOREM 3.5   Let a and b be integers, not both 0. Then $gcd(a, b)$ is the smallest positive integer expressible in the form $as + bt$ for some $s, t \in \mathbb{Z}$.

Proof   Let d be the smallest positive integer expressible in the form $d = as + bt$, where $s, t \in \mathbb{Z}$. To show that $d = gcd(a, b)$, we must verify the following:

1. $d \mid a$
2. $d \mid b$
3. If $e \mid a$ and $e \mid b$, then $e \leq d$.

To show 1, apply the division algorithm to a and d to obtain integers q and r such that $a = dq + r$, where $0 \leq r < d$. We wish to show that $r = 0$.

Since $d = as + bt$,

$$r = a - dq = a - (as + bt)q = a(1 - sq) + b(-tq)$$

and both $1 - sq$ and $-tq$ are integers. So r is expressible as $r = ax + by$, where x and y are integers. Since d is the smallest positive integer that is expressible in this way (and since $r < d$), it must be that $r = 0$. Hence $a = dq$ and we have $d \mid a$.

The proof of 2 is exactly analogous to the proof of 1.

To show 3, let e be an integer such that $e \mid a$ and $e \mid b$. If $e < 0$, then it is clear that $e \le d$, so assume that $e > 0$. Since e is a common divisor of a and b, it follows from Theorem 3.2, part 3, that $e \mid d$. Then, since both d and e are positive, it follows from Theorem 3.2, part 5, that $e \le d$. □

Example 3.9   Use the Euclidean algorithm to find $\gcd(119, 154)$ and express it as a linear combination of 119 and 154.

Solution   We wish to express $d = \gcd(119, 154)$ as

$$d = 119s + 154t$$

for some integers s and t. Recall that d is the last nonzero remainder obtained by applying the Euclidean algorithm. If we take care to express, at each stage of the process, the current remainder as a linear combination of a and b, then d is so expressed when the process terminates.

Since $154 \operatorname{MOD} 119 = 35$ and $154 \operatorname{DIV} 119 = 1$, we have that

$$35 = (119)(-1) + (154)(1)$$

Next, $119 \operatorname{MOD} 35 = 14$ and $119 \operatorname{DIV} 35 = 3$, so that

$$
\begin{aligned}
14 &= 119 - (35)(3) \\
&= 119 - (154 - 119)(3) \\
&= (119)(4) + (154)(-3)
\end{aligned}
$$

Next, $35 \operatorname{MOD} 14 = 7$ and $35 \operatorname{DIV} 14 = 2$, so that

$$
\begin{aligned}
7 &= 35 - (14)(2) \\
&= [(119)(-1) + (154)(1)] - [(119)(4) + (154)(-3)](2) \\
&= (119)(-9) + (154)(7)
\end{aligned}
$$

Finally, $14 \operatorname{MOD} 7 = 0$, so that $7 = \gcd(119, 154)$. Furthermore, the equation

$$7 = (119)(-9) + (154)(7)$$

expresses 7 as a linear combination of 119 and 154. ■

As stated in Theorem 3.5, if $d = \gcd(a, b)$, then d is the smallest positive integer expressible as a linear combination of a and b, namely,

$d = as + bt$, where $s, t \in \mathbb{Z}$. It is important to note, however, that if a positive integer e is expressible as a linear combination of a and b, then it does not necessarily follow that $e = \gcd(a, b)$. For example, $10 = (2)(11) + (3)(-4)$, but $10 \neq \gcd(2, 3)$.

There is an exceptional case that deserves special attention, however. Given $a, b \in \mathbb{Z}$, suppose that there exist $x, y \in \mathbb{Z}$ such that $ax + by = 1$. Then 1 must be the smallest positive integer expressible as a linear combination of a and b, so $\gcd(a, b) = 1$. Hence we can conclude the following:

COROLLARY 3.6 If a and b are integers, not both 0, then $\gcd(a, b) = 1$ if and only if $as + bt = 1$ for some integers s and t.

DEFINITION 3.4

Two integers a and b are called *relatively prime* provided $\gcd(a, b) = 1$.

Note that a and b are relatively prime if and only if 1 is the only common positive divisor of a and b. For instance, 10 and 21 are relatively prime, as are 63 and 88.

Example 3.10 Show that $5m + 3$ and $7m + 4$ are relatively prime for any $m \in \mathbb{Z}$.

Solution The trick here is to apply Corollary 3.6. Notice that

$$1 = (5m + 3)(7) + (7m + 4)(-5)$$

Hence, by Corollary 3.6, the two integers $5m + 3$ and $7m + 4$ are relatively prime. ∎

There are many interesting, intriguing, and useful results concerning two integers that are relatively prime. We present two of these here, with further applications presented in the exercises and chapter problems.

THEOREM 3.7 **Euclid's lemma**

Let a, b, and c be integers. If c and a are relatively prime and $c \mid ab$, then $c \mid b$.

Proof Since a and c are relatively prime, there exist integers s and t such that $1 = as + ct$. It follows that

$$b = b(as + ct) = b(as) + b(ct) = (ab)s + c(bt)$$

By hypothesis, $c \mid ab$; also, $c \mid c$. Thus, part 3 of Theorem 3.2 implies that $c \mid b$. □

In the following corollary we use the notation $p \nmid a$ to mean that p does not divide a.

COROLLARY 3.8 Let a and b be integers and let p be a prime. If $p \mid ab$, then $p \mid a$ or $p \mid b$.

Proof Either $p \mid a$ or $p \nmid a$. If $p \mid a$, then the conclusion follows. Hence assume $p \nmid a$. In this case, $\gcd(a, p) = 1$ and Theorem 3.7 applies (with $c = p$) to yield $p \mid b$. □

The preceding corollary can be extended to allow any finite number of factors; in particular, if $a_1, a_2, \ldots, a_n$ are integers and p is a prime such that $p \mid a_1 a_2 \cdots a_n$, then $p \mid a_i$ for some i, where $1 \leq i \leq n$. The proof of this fact is given as Problem 30 at the end of the chapter.

Theorem 3.5 characterizes the greatest common divisor of two integers a and b as a special linear combination of a and b. Another very important and useful characterization of $\gcd(a, b)$ is presented in the next theorem. The condition stated is quite often taken as the definition of $\gcd(a, b)$.

THEOREM 3.9 Let a and b be integers, not both 0, and let d be a positive integer. Then $d = \gcd(a, b)$ if and only if d satisfies the following:

1. $d \mid a$ and $d \mid b$
2. If $e \mid a$ and $e \mid b$, then $e \mid d$.

Proof We prove the "only if" portion of the theorem and leave the remaining part to Exercise 4. Thus we assume $d = \gcd(a, b)$ and we must show that d satisfies conditions 1 and 2. Condition 1 follows directly from the definition of $\gcd(a, b)$. To obtain condition 2, suppose that $e \mid a$ and $e \mid b$. By Theorem 3.5, d is expressible in the form $d = as + bt$ for some integers s and t. And since e is a common divisor of a and b, Theorem 3.2, part 3, implies that $e \mid (as + bt)$, that is, $e \mid d$. Hence condition 2 holds and the result is established. □

Exercises 3.3

1. Use Algorithm 3.2 to find $d = \gcd(a, b)$ and integers s and t such that $d = as + bt$.
   (a) $a = 412$, $b = 936$ (b) $a = 378$, $b = 532$
2. Let a and b be integers, not both 0. Prove: An integer e is a linear combination of a and b if and only if e is a multiple of $\gcd(a, b)$.
3. Suppose the Euclidean algorithm is applied to find $\gcd(a, b)$, and at some stage a remainder r_{i+1} is obtained that is exactly 1 less than the previous remainder r_i. What does this imply?
4. Let a and b be integers, not both 0, and let d be a positive integer. Prove: If d satisfies conditions 1 and 2 of Theorem 3.9, then $d = \gcd(a, b)$.
5. Prove or disprove:
   (a) Any two consecutive integers are relatively prime.
   (b) $\forall m \in \mathbb{Z}$ ($2m$ and $4m + 3$ are relatively prime)
   (c) $\forall m \in \mathbb{Z}$ ($2m + 1$ and $3m + 2$ are relatively prime)

6. Let m be an integer, let n be a positive integer, and let p be a prime such that p does not divide m. Show that m and p^n are relatively prime.

7. Suppose the Euclidean algorithm is applied to find $\gcd(a, b)$, and at some stage we obtain a remainder r_i that we recognize is prime.
   (a) Show that $\gcd(a, b) = r_i$ or $\gcd(a, b) = 1$.
   (b) How can we tell which of the alternatives in part (a) is the case? Apply the observations made in parts (a) and (b) to find:
   (c) $\gcd(40, 371)$ (d) $\gcd(52, 325)$

8. Let a, b, and m be integers with $m \neq 0$. Prove or disprove:
   (a) $a \operatorname{MOD} m = b \operatorname{MOD} m \rightarrow \gcd(a, m) = \gcd(b, m)$
   (b) $\gcd(a, m) = \gcd(b, m) \rightarrow a \operatorname{MOD} m = b \operatorname{MOD} m$

9. Let a, b, c, and d be integers, with a and b different from 0. Prove:
   (a) If $a \mid c$, $b \mid c$, and $d = \gcd(a, b)$, then $ab \mid cd$.
   (b) If $a \mid c$, $b \mid c$, and a and b are relatively prime, then $ab \mid c$.

10. Let a, b, and d be integers with $0 < d \leq a \leq b$. Prove: If d is a common divisor of a and b and d can be expressed as a linear combination of a and b, then $d = \gcd(a, b)$.

11. Let a, b, n, n_1, and n_2 be integers, with $n_1 \neq 0$, $n_2 \neq 0$, and $n = \gcd(n_1, n_2)$. Prove: If $a \operatorname{MOD} n_1 = b \operatorname{MOD} n_1$ and $a \operatorname{MOD} n_2 = b \operatorname{MOD} n_2$, then $a \operatorname{MOD} n = b \operatorname{MOD} n$.

12. Algorithm 3.2 may be extended to output, in addition to $d = \gcd(a, b)$, values of s and t such that $d = as + bt$. The resulting algorithm is called the "extended Euclidean algorithm." Complete the statements in the following Pascal segment for this algorithm.

```
(*Initial state: The variables A, B, OLDR, Q, R, RNEW,
   OLDS, S, SNEW, OLDT, T, and TNEW are of type INTEGER
   and the values a and b of A and B, respectively, are
   such that 1 <= a <= b.*)
(*Initialize OLDS, OLDT, R, S, T, RNEW and Q.*)
OLDS := ----------;
OLDT := ----------;
R := A;
S := ----------;
T := ----------;
RNEW := B MOD A;
Q := B DIV A;


(*Let r, s, and t denote the values of R, S, T, respec-
   tively. The loop invariant condition is that
   r = a * s + b * t. * )
WHILE RNEW > 0 DO
  BEGIN
    SNEW := --------------------;
    TNEW := --------------------;
```

```
(*Hint: Let oldr, olds, oldt, rnew, snew, and tnew de-
   note the values of OLDR, OLDS, OLDT, RNEW, SNEW, and
   TNEW, respectively. In general, oldr=a*olds+b*oldt
   and r=a*s+b*t. The values of snew and tnew must sat-
   isfy rnew = a*snew+b*tnew. These may be obtained us-
   ing the fact that rnew = oldr - r * q.*)
OLDR := R;
OLDS := S;
OLDT := T;
R := RNEW;
S := SNEW;
T := TNEW;
RNEW := ---------------;
Q := ---------------;
END;
(*Final state: r = gcd(a, b) = a * s + b * t.*)
```

13. Apply the algorithm of Exercise 12 to find $d = \gcd(189, 520)$ and integers s and t such that $d = 189s + 520t$.

3.4   THE FUNDAMENTAL THEOREM OF ARITHMETIC

As stated in Section 3.2, one of the basic notions in number theory is that an integer $a > 1$ may be factored as a product of primes. We prove this result in this section, along with the fact that such a factorization is, in a certain sense, unique.

THEOREM 3.10   **Fundamental theorem of arithmetic**

Given any integer $a > 1$, there exist primes $p_1, p_2, \ldots, p_n$ such that

$$a = p_1 p_2 \cdots p_n \qquad \text{where} \quad p_1 \leq p_2 \leq \cdots \leq p_n \qquad (*)$$

Furthermore, this factorization of a is unique in the sense that, if $q_1, q_2, \ldots, q_k$ are primes with $q_1 \leq q_2 \leq \cdots \leq q_k$ and $a = q_1 q_2 \cdots q_k$, then $n = k$ and $q_i = p_i$ for $i = 1, 2, \ldots, n$.

Proof   We first prove the existence of such a factorization and then show uniqueness. We proceed by contradiction and suppose that the existence part of the theorem is false. This means that there is some integer $b > 1$ that is not expressible in the form $(*)$. Let S be the set of integers $c > 1$ that are not expressible in the form $(*)$. Since $b \in S$, S is a nonempty set of positive integers. Hence, by the principle of well-ordering, S has a smallest element, say d. It must be that d is not prime since otherwise, $d = d$ is an expression of the form $(*)$. Thus d is composite and therefore $d = s \cdot t$, where s and t are integers such that $1 < s < d$ and $1 < t < d$. But then neither

s nor t belongs to S, and consequently both s and t are expressible in the form (∗); that is, there exist primes $q_1, q_2, \ldots, q_i, r_1, r_2, \ldots, r_j$, with $q_1 \leq q_2 \leq \cdots \leq q_i$ and $r_1 \leq r_2 \leq \cdots \leq r_j$, such that

$$s = q_1 q_2 \cdots q_i \quad \text{and} \quad t = r_1 r_2 \cdots r_j$$

This yields

$$d = (q_1 q_2 \cdots q_i)(r_1 r_2 \cdots r_j)$$

which, after suitable rearrangement of the prime factors, yields an expression of the form (∗). However, this contradicts the fact that $d \in S$. Therefore we are forced to conclude that S is empty and hence that every integer $a > 1$ can be factored as a product of primes as stated in the theorem.

Next we prove uniqueness. Again we proceed by contradiction and assume that there is some integer $b > 1$ for which

$$b = p_1 p_2 \cdots p_n = q_1 q_2 \cdots q_k$$

where $p_1, p_2, \ldots, p_n, q_1, q_2, \ldots, q_k$ are primes such that $p_1 \leq p_2 \leq \cdots \leq p_n$ and $q_1 \leq q_2 \leq \cdots \leq q_k$, and it is not the case that both $n = k$ and $p_i = q_i$ for $i = 1, 2, \ldots, n$. Without loss of generality, assume $n \leq k$. Now compare p_i with q_i for $i = 1, 2, \ldots, n$, looking for the smallest value of i for which $p_i \neq q_i$. First of all, it might happen that there is no such i, that is, that $p_i = q_i$ for each $i = 1, 2, \ldots, n$. Then, since $p_1 p_2 \cdots p_n = q_1 q_2 \cdots q_k$, it must be that $n = k$. Thus it may be assumed that there is a smallest value of i, for $1 \leq i \leq n$, such that $p_i \neq q_i$. Now observe that, since $p_1 p_2 \cdots p_n = q_1 q_2 \cdots q_k$, we have $p_i \mid q_1 q_2 \cdots q_k$ and $q_i \mid p_1 p_2 \cdots p_n$. Assume without loss of generality that $p_i < q_i$. Then by the extended form of Euclid's lemma (Corollary 3.8), $p_i \mid q_j$ for some j, where $1 \leq j \leq k$. But then, since q_j is prime, $p_i = q_j$. However, $q_j \geq q_i$, which contradicts our assumption that $p_i < q_i$. It follows that indeed $n = k$ and $p_i = q_i$ for each $i = 1, 2, \ldots, n$.   □

Suppose now that an integer $a > 1$ is expressed as a product of primes, say, $a = q_1 q_2 \cdots q_n$. The primes $q_1, q_2, \ldots, q_n$ need not be distinct; however, we can collect all equal prime factors and express a in the form

$$a = p_1^{\alpha_1} \cdot p_2^{\alpha_2} \cdot \cdots \cdot p_k^{\alpha_k}$$

where $p_1, p_2, \ldots, p_k$ are primes, $\alpha_1, \alpha_2, \ldots, \alpha_k$ are positive integers, and $p_1 < p_2 < \cdots < p_k$. We call this the *canonical factorization* of a.

Example 3.11   Find the canonical factorization of

(a) 14553   (b) 53508   (c) 149

Solution   For part (a) we proceed as follows:

$$14553 \bmod 2 \neq 0$$
$$14553 \bmod 3 = 0, \quad 14553 \operatorname{DIV} 3 = 4851$$
$$4851 \bmod 3 = 0, \quad 4851 \operatorname{DIV} 3 = 1617$$
$$1617 \bmod 3 = 0, \quad 1617 \operatorname{DIV} 3 = 539$$
$$539 \bmod 3 \neq 0$$
$$539 \bmod 5 \neq 0$$
$$539 \bmod 7 = 0, \quad 539 \operatorname{DIV} 7 = 77$$
$$77 \bmod 7 = 0, \quad 77 \operatorname{DIV} 7 = 11$$

and 11 is prime. Therefore the canonical factorization of 14553 is

$$14553 = 3^3 \cdot 7^2 \cdot 11^1$$

Similarly, for part (b)

$$53508 = 2^2 \cdot 3^1 \cdot 7^3 \cdot 13^1$$

For part (c) note that $149 \bmod 2 \neq 0$, $149 \bmod 3 \neq 0$, $149 \bmod 5 \neq 0$, $149 \bmod 7 \neq 0$, and $149 \bmod 11 \neq 0$. At this point we may conclude that 149 is prime, since $13^2 > 149$. Note that, if an integer $a > 1$ is composite, then a has a prime factor p such that $p^2 \leq a$. ∎

As an interesting sidelight to the preceding example, consider the problem of finding $\gcd(14553, 53508)$. With this purpose in mind, it is convenient to express these numbers as follows:

$$14553 = 2^0 \cdot 3^3 \cdot 7^2 \cdot 11^1 \cdot 13^0$$
$$53508 = 2^2 \cdot 3^1 \cdot 7^3 \cdot 11^0 \cdot 13^1$$

so that each factorization includes the same primes. Then we have

$$\gcd(14553, 53508) = \gcd(2^0 \cdot 3^3 \cdot 7^2 \cdot 11^1 \cdot 13^0, 2^2 \cdot 3^1 \cdot 7^3 \cdot 11^0 \cdot 13^1)$$
$$= 2^0 \cdot 3^1 \cdot 7^2 \cdot 11^0 \cdot 13^0$$
$$= 3^1 \cdot 7^2$$
$$= 147$$

Thus, for each of the primes involved, we choose the smaller of the two exponents to determine its contribution to $\gcd(14553, 53508)$. This procedure can be formulated in general terms without much difficulty (see Exercise 2). It should be mentioned that there are additional applications of the above notation.

Before ending this section we prove, for the sake of completeness, that the number of primes is infinite. The student is no doubt aware of this fact but perhaps has never seen a proof.

THEOREM 3.11   The number of primes is infinite.

Proof   We proceed by contradiction and assume that the number of primes is finite. Suppose that $P = \{ p_1, p_2, \ldots, p_n \}$ is the set of all primes. Consider the integer $m = p_1 p_2 \cdots p_n + 1$. Clearly $m \geq 2$. We leave it to the student to verify that m MOD $p_i = 1$ for each $i = 1, 2, \ldots, n$; thus none of the p_i's is a factor of m. However, by the fundamental theorem of arithmetic, m can be factored as a product of primes; let p be a prime factor of m. Then $p \neq p_i$ for each $i = 1, 2, \ldots, n$, so $p \notin P$. This contradicts the supposition that P is the set of all primes and thus proves the result. $\square$

Exercises 3.4

1. Find the canonical factorizations of these integers.
    (a) 4725   (b) 9702   (c) 25625
2. Given integers a and b with $1 < a \leq b$, let

$$P_{a,b} = \{ p \mid p \text{ is prime} \wedge p \text{ divides } ab \}$$
$$= \{ p_1, p_2, \ldots, p_n \}$$

   where $p_1 < p_2 < \cdots < p_n$; further, suppose that

$$a = p_1^{\alpha_1} \cdot p_2^{\alpha_2} \cdot \cdots \cdot p_n^{\alpha_n}$$
$$b = p_1^{\beta_1} \cdot p_2^{\beta_2} \cdot \cdots \cdot p_n^{\beta_n}$$

   where α_i and β_i are nonnegative integers for each $i \in \{1, 2, \ldots, n\}$. (For example, $P_{200, 504} = \{2, 3, 5, 7\}$, since $300 = 2^2 \cdot 3^1 \cdot 5^2 \cdot 7^0$ and $504 = 2^3 \cdot 3^2 \cdot 5^0 \cdot 7^1$.)
    (a) Using the above expressions for a and b, give a formula for $\gcd(a, b)$ in terms of $p_1, p_2, \ldots, p_n$.
    (b) Use the result of part (a) to find $\gcd(4725, 9702)$.
3. Let p be a prime and b a positive integer. If $p^m \mid b$ and $p^{m+1} \nmid b$, then we say that p^m *exactly divides* b and we write $p^m \| b$. For example, $2^3 \| 24$, since $2^3 \mid 24$ and $2^4 \nmid 24$. Prove each of the following:
    (a) If $p^m \| b$ and $p^n \| c$, then $p^{m+n} \| bc$.
    (b) If $p^m \| b$, then $p^{km} \| b^k$ for any $k \in \mathbb{N}$.
4. If p is a prime such that $p > 3$, then show that p is of the form:
    (a) $4n + 1$ or $4n + 3$ for some $n \in \mathbb{N}$
    (b) $6n + 1$ or $6n + 5$ for some nonnegative integer n
5. Show that no integer $x = 4n + 3$, where $n \in \mathbb{Z}$, can be written in the form $x = a^2 + b^2$, where a and b are integers.
6. If p is a prime, prove that there do not exist positive integers a and b such that $a^2 = pb^2$. (Hence $\sqrt{p}$ is irrational.)
7. Write a Pascal procedure that accepts a positive integer N and returns an array PRIME containing the first N primes (PRIME[1] = 2, PRIME[2] = 3, etc.).
8. Write a Pascal procedure that takes a positive integer N $>=$ 2 and returns an array PRIMEFACTOR containing the prime factors of N. That is, if K is the number of prime factors of N, then

```
N = PRIMEFACTOR[1] * PRIMEFACTOR[2] *. . .* PRIMEFACTOR[K]
```

with PRIMEFACTOR[1] < = PRIMEFACTOR[2] < = $\cdots$ < = PRIMEFACTOR[K].

9. Write a Pascal procedure that accepts a positive integer N > = 2 and returns two arrays PFACTOR and EXPONENT such that PFACTOR[1] < PFACTOR[2] < $\cdots$ < PFACTOR[K] are the distinct prime factors of N and

N = PFACTOR[1] ＊＊ EXPONENT[1] ＊. . .＊ PFACTOR[K] ＊＊ EXPONENT[K]

is the canonical factorization of N.

3.5

THE PRINCIPLE OF MATHEMATICAL INDUCTION

In mathematics many problems have the following general form:

1. Let $P(n)$ be a statement about the positive integer n.
2. Prove that $P(n)$ is true for every $n \in \mathbb{N}$.

Several examples taken from various areas of mathematics give an idea of the frequency with which such problems occur.

Example 3.12   The following are statements about an arbitrary positive integer n:

(a) $1 + 2 + 3 + \cdots + n = \dfrac{n(n+1)}{2}$

(b) For any real numbers x and y,

$$(x + y)^n = \sum_{k=0}^{n} \frac{n(n-1)\cdots(n-k+1)}{k(k-1)\cdots(1)} x^{n-k} y^k$$

(This statement is known as the "binomial theorem" and is presented in Chapter 6.)

(c) The nth derivative of $y = \dfrac{1}{1 - 2x}$ is $\dfrac{2^n(1 \cdot 2 \cdot 3 \cdot \cdots \cdot n)}{(1 - 2x)^{n+1}}$.

(d) $1 + \dfrac{1}{4} + \dfrac{1}{9} + \cdots + \dfrac{1}{n^2} \leq 2 - \dfrac{1}{n}$

(e) Let A be a finite set. If $|A| = n$, then $|\mathscr{P}(A)| = 2^n$.

(f) $6 \mid (n^3 + 5n)$ ∎

How are problems of this general form solved? Suppose we can prove that the following two conditions hold:

1. $P(1)$ is true.
2. For an arbitrary $k \in \mathbb{N}$, if $P(k)$ is true, then $P(k + 1)$ is true.

The second condition, stated another way, says that, for an arbitrary $k \in \mathbb{N}$, the truth of $P(k)$ implies the truth of $P(k + 1)$. How can these conditions aid in proving that $P(n)$ is true for every $n \in \mathbb{N}$? Condition 1 tells us that $P(1)$ is true. Hence condition 2, applied with $k = 1$, states that $P(2)$ is true. And since $P(2)$ is true, we may apply condition 2 again, this time with $k = 2$, to give us that $P(3)$ is true. Since $P(3)$ is true, $P(4)$ is true, and so on. It definitely seems inviting now to state that $P(n)$ is true for every $n \in \mathbb{N}$. This is exactly what the "principle of mathematical induction" allows us to do. It can be shown that this principle is equivalent to the principle of well-ordering, which we restate here.

Principle of well-ordering.   Every nonempty set of positive integers has a smallest element.

THEOREM 3.12   **Principle of mathematical induction**

Let S be a set of positive integers such that the following are true:

1. $1 \in S$
2. For an arbitrary $k \geq 1$, if $k \in S$, then $k + 1 \in S$.

Then $S = \mathbb{N}$.

Proof   We use the principle of well-ordering to establish the principle of mathematical induction. We proceed by contradiction and assume that S is a set of positive integers satisfying conditions 1 and 2 of the theorem, but $S \neq \mathbb{N}$. Let $A = \mathbb{N} - S$. Then $A \neq \emptyset$ and $\mathbb{N} = A \cup S$. Since $A \neq \emptyset$, the principle of well-ordering implies that A has a smallest element, say a. Since it is given that $1 \in S$, we see that $a \geq 2$; hence $a - 1 \geq 1$. And since a is the smallest element of A, it follows that $a - 1 \notin A$. But $a - 1 \in \mathbb{N}$ and $\mathbb{N} = A \cup S$, so it must be that $a - 1 \in S$. Now condition 2 of the theorem, applied with $k = a - 1$, implies that $k + 1 = a \in S$. But this is a contradiction. It follows that $S = \mathbb{N}$ and the theorem is proved.   □

In Exercise 2 the student is asked to show that the principle of well-ordering is implied by the principle of mathematical induction, thereby showing that these two principles are indeed logically equivalent.

With the principle of mathematical induction (PMI) established, we now describe how it can be used to prove statements like those presented in Example 3.12. It must first be determined that a given problem has the form exhibited at the beginning of this section. If this is the case and a decision is made to use the PMI to prove the statement, then the following outline for the proof is strongly recommended.

Outline for a Proof by Mathematical Induction

Proceed by induction on n, and let S be the set of positive integers for which the statement $P(n)$ is true.

Step 1. Show $1 \in S$.
Step 2. Assume $k \in S$ for an arbitrary $k \geq 1$.
Step 3. Show $k + 1 \in S$.

It follows by the principle of mathematical induction that $S = \mathbb{N}$; hence $P(n)$ is true for every $n \in \mathbb{N}$.

Notice the use of the phrase "Proceed by induction on $n \ldots$." It is standard practice to use this wording to begin a proof by mathematical induction. Step 1 is commonly called the *anchor step*; once it has been shown that $1 \in S$, we quite often say that the induction is "anchored." Steps 2 and 3 make up the implication

$$k \in S \to k + 1 \in S$$

For emphasis and clarity we state the premise and conclusion separately. We call Step 2 the *induction hypothesis* (IHOP for short), and we call Step 3 the *inductive step*. When doing a proof by mathematical induction, it is most important to correctly identify the induction hypothesis, since the main idea is to apply the induction hypothesis in the inductive step. Several examples will serve to illustrate the usefulness and convenience of this outline when doing a proof by mathematical induction.

Example 3.13   Use mathematical induction to prove the statement $P(n)$,

$$1 + 2 + \cdots + n = \frac{n(n + 1)}{2}$$

holds for every positive integer n.

Proof   Proceed by induction on n and let S be the set of positive integers for which $P(n)$ is true.

1. Show $1 \in S$. Since $1 = 1(1 + 1)/2$, it follows that $1 \in S$.
2. Assume $k \in S$ for an arbitrary $k \geq 1$. Hence we assume

$$\text{IHOP:} \quad 1 + 2 + \cdots + k = \frac{k(k + 1)}{2}$$

3. Show $k + 1 \in S$. We must show

$$1 + 2 + \cdots + k + (k + 1) = \frac{(k + 1)(k + 2)}{2}$$

Adding $k + 1$ to both sides of the IHOP equation yields

$$1 + 2 + \cdots + k + (k + 1) = \frac{k(k + 1)}{2} + (k + 1)$$

$$= (k + 1)\left(\frac{k}{2} + 1\right)$$

$$= (k + 1)\left(\frac{k + 2}{2}\right)$$

$$= \frac{(k + 1)(k + 2)}{2}$$

showing that $k + 1 \in S$. It follows by the PMI that $S = \mathbb{N}$ and hence that $P(n)$ is true for every $n \in \mathbb{N}$.   □

The implication

$$k \in S \rightarrow k + 1 \in S$$

in Steps 2 and 3 of the outline may be established using an indirect proof. For example, if a proof by contradiction is used, then we assume $k \in S$ and $k + 1 \notin S$ and attempt to derive a contradiction. This technique is illustrated by the next example.

Example 3.14   Use mathematical induction to prove that the statement $P(n)$,

$$1 + \frac{1}{4} + \frac{1}{9} + \cdots + \frac{1}{n^2} \leq 2 - \frac{1}{n}$$

holds for every positive integer n.

Proof   Proceed by induction on n and let S be the set of positive integers for which $P(n)$ is true.

1. Show $1 \in S$. Since $1 \leq 1 = 2 - 1$, it follows that $1 \in S$.
2. Assume $k \in S$ for an arbitrary $k \geq 1$. Hence we have

$$\text{IHOP:}\quad 1 + \frac{1}{4} + \cdots + \frac{1}{k^2} \leq 2 - \frac{1}{k}$$

3. Show $k + 1 \in S$. We proceed by contradiction and assume $k + 1 \notin S$; that is,

$$1 + \frac{1}{4} + \cdots + \frac{1}{k^2} + \frac{1}{(k+1)^2} > 2 - \frac{1}{k+1}$$

Then, making use of the IHOP inequality, we have

$$\left(1 + \frac{1}{4} + \cdots + \frac{1}{(k+1)^2} \right) - \left(1 + \frac{1}{4} + \cdots + \frac{1}{k^2} \right)$$

$$> \left(2 - \frac{1}{k+1} \right) - \left(2 - \frac{1}{k} \right)$$

Hence it follows that

$$\frac{1}{(k+1)^2} > \frac{1}{k} - \frac{1}{k+1}$$

or

$$k > (k+1)^2 - k(k+1)$$

or
$$k > k + 1$$
which is a contradiction. Therefore $k + 1 \in S$. Hence by the PMI $S = \mathbb{N}$, so $P(n)$ is true for all $n \in \mathbb{N}$.   □

In some cases we are asked to prove that a statement $P(n)$ is true for all integers $n \geq n_0$, where n_0 is some given fixed integer. For instance, we may be asked to prove $n(n - 1) \cdots (1) > 2^n$ for every integer $n \geq 4$. (Note that this inequality fails for $1 \leq n \leq 3$.) In cases such as this, the following corollary to Theorem 3.12 provides the necessary proof technique.

COROLLARY 3.13   Let $n_0 \in \mathbb{Z}$ and let $M = \{n \in \mathbb{Z} \mid n \geq n_0\}$. Let S be a subset of M such that the following are true:

1. $n_0 \in S$
2. For an arbitrary $k \geq n_0$, if $k \in S$, then $k + 1 \in S$.

Then $S = M$.

The proof of Corollary 3.13 is left to Exercise 4. To apply this corollary to do a proof by mathematical induction, it is necessary only to replace 1 by n_0 and $\mathbb{N}$ by M in Steps 1–3 of the suggested outline.

Example 3.15   Use mathematical induction (Corollary 3.13) to prove that
$$n(n - 1) \cdots (1) > 2^n$$
holds for every integer $n \geq 4$.

Proof   Proceed by induction on n; let $M = \{n \in \mathbb{N} \mid n \geq 4\}$ and let $S = \{n \in M \mid n(n - 1) \cdots (1) > 2^n\}$.

1. Show $4 \in S$. Since $4 \cdot 3 \cdot 2 \cdot 1 = 24 > 16 = 2^4$, it follows that $4 \in S$.
2. Assume $k \in S$ for an arbitrary $k \geq 4$. Hence we assume

IHOP: $k(k - 1) \cdots (1) > 2^k$

3. Show $k + 1 \in S$. We must show that
$$(k + 1)k(k - 1) \cdots 1 > 2^{k+1}$$

Multiplying both sides of the IHOP inequality by $k + 1$, we obtain
$$(k + 1)k(k - 1) \cdots 1 > 2^k(k + 1)$$
$$> 2^k \cdot 2 \quad (\text{since } k + 1 > 2)$$
$$= 2^{k+1}$$

which shows that $k + 1 \in S$. It follows by Corollary 3.13 that $S = M$ and hence that $n(n - 1) \cdots 1 > 2^n$ holds for every integer $n \geq 4$.   □

Example 3.16   Use mathematical induction (Corollary 3.13) to prove that, for each nonnegative integer n, there are precisely 2^n subsets of a set with n elements. Let M denote the set of nonnegative integers and note that the statement to be proved, $P(n)$, can be conveniently rephrased as follows:

$$P(n): \quad \text{For any set } A, \text{ if } |A| = n, \text{ then } |\mathscr{P}(A)| = 2^n.$$

Proof   Proceed by induction on n and let $S = \{n \in M \mid P(n) \text{ is true}\}$.

1. Show $0 \in S$. If $|A| = 0$, then $A = \emptyset$ and $\mathscr{P}(A) = \{\emptyset\}$, so that $|\mathscr{P}(A)| = 1 = 2^0$. This shows that $0 \in S$.
2. Assume $k \in S$ for an arbitrary $k \geq 0$. Hence we assume

$$\text{IHOP:} \quad \text{For any set } A, \text{ if } |A| = k, \text{ then } |\mathscr{P}(A)| = 2^k.$$

3. Show $k + 1 \in S$. We must show that:

$$\text{For any set } A, \text{ if } |A| = k + 1, \text{ then } |\mathscr{P}(A)| = 2^{k+1}.$$

Let A be a set such that $|A| = k + 1$, and let a be some element of A. Consider the set $B = A - \{a\}$. Note that $|B| = k$; hence by the induction hypothesis, $|\mathscr{P}(B)| = 2^k$. Furthermore, to each subset C of B there correspond two subsets of A, namely, C and $C \cup \{a\}$. In other words, A has twice as many subsets as B has. Thus

$$|\mathscr{P}(A)| = 2 \cdot |\mathscr{P}(B)| = 2 \cdot 2^k = 2^{k+1}$$

showing that $k + 1 \in S$. It follows that $S = M$ and hence that $P(n)$ holds for every nonnegative integer n.   □

We now consider an application of mathematical induction to the problem of proving program correctness. (See the end of Section 1.5 for a discussion of this topic.)

Consider the following Pascal program segment to compute the nth power of a real number x.

```
(*Initial state: The variables X and Y are of type REAL, J
  and N are type INTEGER, the value of X is x and the value
  of N is n >= 0.*)
Y := 1;
J := 0;
WHILE J < N DO
  BEGIN
    Y := X * Y;
    J := J + 1
  END;
  (*Final state: The value y of Y is such that y = x ** n,
    where ** denotes exponentiation.*)
```

The key feature of this program segment is that it contains a loop, in this case a WHILE loop. (In fact, the computation is more naturally represented using a FOR loop. However, using a FOR loop hides the incrementing and exit-testing steps that are performed with the loop control variable.) The loop control variable is J, with final value n. Since n can, in theory, be any nonnegative integer, the body of the loop may be executed 0, 1, 2, or, in general, any nonnegative-integer number of times. Thus there is a natural connection between mathematical induction and verifying the correctness of programs that contain loops.

We wish to prove that this program segment is correct; that is, if the conditions specified in the initial state hold, then the condition specified in the final state holds. The key to proving this is to focus attention on the loop in the program segment. Suppose execution reaches the top of the loop (where the condition J < N is about to be evaluated), and the variables X, Y, and J have values x, y, and j, respectively. Our claim is that $y = x ** j$. (This is our loop invariant condition.) If we can establish this claim, then the program segment is correct, for when the loop is exited the value of J is n, and so $y = x ** n$ holds.

To prove this claim, we proceed by induction on the value j of J.

1. (Anchor step.) If execution reaches the top of the loop and $j = 0$, then execution has reached the top of the loop for the very first time (since each iteration of the loop increments j). Thus $y = 1 = x ** 0 = x ** j$, and our induction is anchored.

2. Let k be an arbitrary nonnegative integer, and assume that if execution reaches the top of the loop and the value of J is k, then $y = x ** k$. (Note that the values x and y are fixed but arbitrary; they can be any values of type REAL.)

3. Suppose execution reaches the top of the loop and the value of J is $k + 1$. We must show that $y = x ** (k + 1)$. Since $k + 1 > 0$, the body of the loop has been executed at least once. Let us "back up" execution to the previous time that execution was at the top of the loop. Then the value of J was k; suppose the value of Y was *oldy*. Then by the induction hypothesis, *oldy* $= x ** k$. The body of the loop was executed, giving Y the value $y = x * oldy$ and incrementing the value of J to $k + 1$. Thus

$$y = x * oldy = x * (x ** k) = x ** (k + 1)$$

as was to be shown. This establishes our claim and so proves the correctness of the program segment.

Exercises 3.5

1. Use mathematical induction to prove that

$$1 + 3 + \cdots + (2n - 1) = n^2$$

holds for every $n \in \mathbb{N}$.

2. Show that the principle of well-ordering is implied by the principle of mathematical induction.

3. Use mathematical induction to prove that

$$1^2 + 2^2 + \cdots + n^2 = \frac{n(n + 1)(2n + 1)}{6}$$

holds for every $n \in \mathbb{N}$.

4. Prove Corollary 3.13.
5. Use mathematical induction to prove that
$$6 \mid (n^3 + 5n)$$
holds for every $n \in \mathbb{N}$.
6. For $n \in \mathbb{N}$, let $P_1(n)$ be the statement that $n^2 + n + 11$ is prime and let $P_2(n)$ be the statement that $3 \mid (3n + 2)$.
   (a) Note that $P_1(1)$, $P_1(2)$, ..., $P_1(9)$ are all true. Is $P_1(n)$ true for every $n \in \mathbb{N}$?
   (b) Note that the implication $P_2(k) \to P_2(k + 1)$ holds for every $k \in \mathbb{N}$. Can we conclude that $P_2(n)$ is true for every $n \in \mathbb{N}$?
7. Use mathematical induction (Corollary 3.13) to prove
   (a) $2n + 1 < n^2$ for all $n \geq 3$
   (b) $n^2 < 2^n$ for all $n \geq 5$ (Hint: Use the result of part (a).)
8. Use mathematical induction (Corollary 3.13) to prove these formulas hold for any nonnegative integer n.
   (a) $1 + \dfrac{1}{3} + \cdots + \dfrac{1}{3^n} = \dfrac{3}{2}\left(1 - \dfrac{1}{3^{n+1}}\right)$
   (b) $a + ar + \cdots + ar^n = \dfrac{a(r^{n+1} - 1)}{r - 1}$

   where a and r are real numbers, $r \neq 1$ (Note that since a is a factor of both sides of the formula, it suffices to prove it for the case $a = 1$.)
9. Use mathematical induction to prove that
$$2 + \left(1 + \frac{1}{\sqrt{2}} + \cdots + \frac{1}{\sqrt{n}}\right) > 2\sqrt{n + 1}$$
holds for every $n \in \mathbb{N}$.
10. Use mathematical induction to prove that
$$2\left(1 + \frac{1}{8} + \cdots + \frac{1}{n^3}\right) < 3 - \frac{1}{n^2}$$
holds for all $n \geq 2$.
11. The following Pascal segment is to compute the product of two integers m and n, where n is nonnegative. Use mathematical induction to prove that the segment is correct.

```
(*Initial state: The variables M, N, K, and P are of type
  INTEGER; the values of M and N are m and n, respec-
  tively, and n >= 0.*)
P := 0;
K := 0;
WHILE K < N DO
  BEGIN
    P := P + M;
    K := K + 1;
  END;
(*Final state: The value p of P is such that p = m * n.*)
```

| 3.6 | # THE STRONG FORM OF INDUCTION |

In some cases the method of induction used in the last section cannot be applied so nicely. To see this, consider the sequence of numbers $1, 1, 2, 3, 5, 8, 13, \ldots$. These numbers are known as the *Fibonacci numbers*. Letting f_n denote the nth Fibonacci number, where $n \geq 1$, we note that

$$\text{(a)} \quad f_1 = f_2 = 1$$

$$\text{(b)} \quad \text{For } n \geq 3, \ f_n = f_{n-2} + f_{n-1}$$

Suppose we wish to prove that $f_n < 2^n$ for every positive integer n. The statement of the problem suggests using induction; let's see what happens.

As usual, let S be the set of positive integers n for which the relation $f_n < 2^n$ holds. Clearly $1 \in S$. Let k be an arbitrary positive integer and assume $k \in S$; that is, assume $f_k < 2^k$. It must be shown that $k + 1 \in S$; that is, $f_{k+1} < 2^{k+1}$. What we do know is that $f_{k+1} = f_{k-1} + f_k$ if $k \geq 2$. It appears we are in a bit of a bind, as we have no direct information concerning f_{k-1}.

Problems such as this one can be handled using an alternate form of mathematical induction, called the *strong form of induction*. Like the PMI, the strong form of induction is equivalent to the principle of well-ordering. We demonstrate only that it is implied by the principle of well-ordering.

THEOREM 3.14   **Strong form of induction**

Let S be a set of positive integers such that

1. $1 \in S$
2. For any integer $n > 1$, if $k \in S$ for all integers k such that $1 \leq k < n$, then $n \in S$.

Then $S = \mathbb{N}$.

Proof   We proceed by contradiction and suppose S is a subset of $\mathbb{N}$ such that S satisfies conditions 1 and 2 of the theorem, but $S \neq \mathbb{N}$. Then $\mathbb{N} - S \neq \emptyset$ and $\mathbb{N} = (\mathbb{N} - S) \cup S$. Since $\mathbb{N} - S$ is a nonempty set of positive integers, it has a smallest element, say m.

Since $1 \in S$, we have $m > 1$, and since m is the smallest element of $\mathbb{N} - S$, the integers $1, 2, \ldots, m - 1$ must all belong to S. Thus condition 2 applies, with n replaced by m, to give that $m \in S$. But this contradicts the fact that $\mathbb{N} - S$ and S are disjoint sets; hence it follows that $S = \mathbb{N}$.   □

Example 3.17   We illustrate the strong form of induction by proving that the nth Fibonacci number f_n satisfies $f_n < 2^n$ for each $n \in \mathbb{N}$. As usual, let S be the set of positive integers such that $f_n < 2^n$.

1. Show $1 \in S$ and $2 \in S$. Since $f_1 = 1 < 2^1$ and $f_2 = 1 < 2^2$, we have $1 \in S$ and $2 \in S$. (The reason for anchoring the induction at both $n = 1$

and $n = 2$ is to allow us to make use of the relation $f_n = f_{n-1} + f_{n-2}$, which is valid for $n \geq 3$. Typically, when applying the strong form of induction, the induction must be anchored for several values.)

2. For an arbitrary integer $n \geq 3$, assume that $k \in S$ for all integers k such that $1 \leq k < n$; that is, $f_k < 2^k$ for $1 \leq k < n$.

3. We must show that $n \in S$, namely, that $f_n < 2^n$. We use the induction hypothesis and the relation $f_n = f_{n-2} + f_{n-1}$:

$$f_n = f_{n-2} + f_{n-1}$$
$$< 2^{n-2} + 2^{n-1} \qquad \text{(by IHOP)}$$
$$= 2^{n-2}(1 + 2)$$
$$= 2^{n-2}(3)$$
$$< 2^{n-2}(4)$$
$$= 2^n$$

This shows that $n \in S$ and therefore $S = \mathbb{N}$.   ∎

The strong form of induction may also be applied to prove that a statement $P(n)$ holds for all $n \geq n_0$, where n_0 is some fixed integer. The next corollary is similar to Corollary 3.13 in this respect, and its proof is left to Exercise 2.

COROLLARY 3.15   Let $n_0 \in \mathbb{Z}$ and let $M = \{n \in \mathbb{Z} \mid n \geq n_0\}$. Let S be a subset of M such that the following are true

1. $n_0 \in S$
2. For any $n > n_0$, if $k \in S$ for all integers k such that $n_0 \leq k < n$, then $n \in S$.

Then $S = M$.

Example 3.18   A professional football team may score a field goal for three points or a touchdown (with conversion) for seven points. (A safety, for two points, is also possible, but occurs rarely.) Use the strong form of induction to prove that, theoretically, it is possible for a football team to score some number of field goals and some number of touchdowns totaling n points for any integer $n \geq 12$.

Proof   Note that, if the team scores x field goals and y touchdowns, then the total number of points is $3x + 7y$. Hence what we are to prove is that for any integer $n \geq 12$, there exist nonnegative integers x and y (depending on n) such that

$$n = 3x + 7y$$

We proceed by induction on n; let $M = \{n \in \mathbb{N} \mid n \geq 12\}$ and let S be the set of all integers $n \in M$ for which there exist nonnegative integers x and y such that $n = 3x + 7y$.

1. It is possible to score 12 points with four field goals, 13 points with one touchdown and two field goals, and 14 points with two touchdowns. Thus $12 \in S$, $13 \in S$, and $14 \in S$. (We anchor the induction for $n = 12, 13, 14$, so that we may have $n > 14$ in the inductive step.)

2. Let n represent an arbitrary integer, $n > 14$, and assume that $k \in S$ for any integer k such that $12 \leq k < n$; that is, there exist nonnegative integers x and y such that $k = 3x + 7y$.

3. We must show that $n \in S$. Since $12 \leq n - 3 < n$, we have by the induction hypothesis that $n - 3 \in S$, and so there exist nonnegative integers x and y such that $n - 3 = 3x + 7y$. Thus

$$n = 3 + 3x + 7y = 3(x + 1) + 7y$$

and $x + 1$ and y are nonnegative integers. This shows that $n \in S$ and therefore $S = M$.   □

Example 3.19   Use the strong form of induction to prove the multiplication principle (Theorem 2.16):   If A and B are finite sets, then $|A \times B| = |A| \cdot |B|$.

Proof   We proceed by induction on $|A|$. To anchor the induction, it is easy to check that the result holds when $|A| = 0$ or $|A| = 1$. For an arbitrary integer $n > 1$, assume that the result holds whenever $0 \leq |A| < n$; that is, whenever A and B are finite sets and $|A| < n$, then $|A \times B| = |A| \cdot |B|$. Now let A and B be finite sets with $|A| = n$. We must show that $|A \times B| = |A| \cdot |B|$. Since $|A| = n > 1$, there exist nonempty disjoint subsets A_1 and A_2 of A such that $A = A_1 \cup A_2$. Then $1 \leq |A_1| < n$ and $1 \leq |A_2| < n$, so by the induction hypothesis, $|A_1 \times B| = |A_1| \cdot |B|$ and $|A_2 \times B| = |A_2| \cdot |B|$. Also recall from Chapter 2 the result $(A_1 \cup A_2) \times B = (A_1 \times B) \cup (A_2 \times B)$. This is a disjoint union, so by the addition principle,

$$\left|(A_1 \times B) \cup (A_2 \times B)\right| = |A_1 \times B| + |A_2 \times B|$$

Putting these results together, we have

$$
\begin{aligned}
|A \times B| &= \left|(A_1 \cup A_2) \times B\right| \\
&= \left|(A_1 \times B) \cup (A_2 \times B)\right| \\
&= |A_1 \times B| + |A_2 \times B| \\
&= |A_1| \cdot |B| + |A_2| \cdot |B| \\
&= (|A_1| + |A_2|) \cdot |B| \\
&= |A_1 \cup A_2| \cdot |B| \\
&= |A| \cdot |B|
\end{aligned}
$$

as was to be shown.   □

Example 3.20   The following Pascal segment is to compute the quotient $q = b \operatorname{DIV} a$ and the remainder $r = b \operatorname{MOD} a$ for a nonnegative dividend b and a positive divisor a. Use the strong form of induction to prove the segment is correct.

```
(*Initial state: The variables A, B, Q, and R are of type
 INTEGER; the values a and b of A and B, respectively, are
 such that a > 0 and b >= 0.*)
Q := 0;
R := B;
WHILE R >= A DO
  BEGIN
    Q := Q + 1;
    R := R - A
  END;
(*Final state: The values q and r of Q and R, respectively,
 are such that b = a * q + r and 0 <= r < a.*)
```

Proof   The loop invariant condition is that $b = a * q + r$ and $r \geq 0$. We claim that this condition holds whenever execution is at the top of the loop. This implies that the final state holds, since when the loop is exited, $r < a$ holds. It is fairly clear that $r \geq 0$ always holds (although this can be demonstrated rigorously), so we concentrate on our claim that $b = a * q + r$. To verify this, we could proceed by induction on q; however, to illustrate the strong form of induction, we proceed by induction on $b - r$. Note that $b - r$ is a nonnegative integer.

1. If $b - r = 0$, then $r = b$. Hence execution is at the top of the loop for the very first time. Thus $q = 0$, so that $b = a * q + r$ holds.

2. Let n be a positive integer, and assume that for every integer k, $0 \leq k < n$, if $b - r = k$ and execution is at the top of the loop, then $b = a * q + r$.

3. Suppose execution is at the top of the loop and $b - r = n$; we must show that $b = a * q + r$. Since $b - r = n$ and n is positive, the body of the loop has been executed at least once. Let's back up to the previous time execution was at the top of the loop; let $oldq$ and $oldr$ be the values of Q and R, respectively, at this time. Then $oldr = r + a > r$, and so $b - oldr < b - r$. Hence, by the induction hypothesis, $b = a * oldq + oldr$. Also note that $q = oldq + 1$. Therefore

$$b = a * oldq + oldr = a * (q - 1) + (r + a) = a * q + r$$

as we wished to show. This establishes our claim.   □

Exercises 3.6

1. Use the strong form of induction to prove the existence part of the fundamental theorem of arithmetic: Every positive integer $n \geq 2$ can be factored as a product of primes.

2. Prove Corollary 3.15.
3. Use the strong form of induction to prove that any integer $n \geq 24$ can be expressed as $n = 5x + 7y$, where x and y are nonnegative integers.
4. Define the sequence of numbers $t_1, t_2, t_3, \ldots$ by:
   (i) $t_1 = 1$, $t_2 = 2$, $t_3 = 3$
   (ii) $t_n = t_{n-3} + t_{n-2} + t_{n-1}$ for all $n \geq 4$
   Use the strong form of induction to prove that $t_n < 2^n$ for every $n \in \mathbb{N}$.
5. Use the strong form of induction to prove the correctness of Algorithm 3.2. (Hint: Let *rnew* denote the value of RNEW; use induction on $a - r$ to prove that whenever execution reaches the top of the loop, $\gcd(rnew, r) = \gcd(a, b)$.)
6. Let a, b, and c be positive integers with $2 \leq a < b$ and $\gcd(a, b) = 1$. Consider the Diophantine equation

$$ax + by = c \qquad\qquad (1)$$

(a) Discuss how to obtain a particular solution of (1) in integers x_1 and y_1.
(b) Let (x_1, y_1) be a particular solution. Show that (x, y) is a solution, where $x, y \in \mathbb{Z}$, if and only if $x = x_1 + bt$ and $y = y_1 - at$ for some integer t.
(c) Show that (1) has a solution in nonnegative integers x and y if and only if, for any particular solution (x_1, y_1), the interval $[-x_1/b, y_1/a]$ contains an integer.
(d) Show that the equation $ax + by = ab - a - b$ does not have a solution in nonnegative integers.
(e) Show that (1) has a solution in nonnegative integers if $c > ab - a - b$. (Hint: Write $c = ab - a - b + n$, where $n \in \mathbb{N}$, and use the strong form of induction on n. Anchor the induction by showing that (1) has a solution in nonnegative integers for $n \in \{1, 2, \ldots, a\}$. For $n \in \{1, 2, \ldots, a - 1\}$, let x_0 and y_0 be integers such that $n = ax_0 + by_0$. Note that y_0 may be chosen so that $1 \leq y_0 < a$. Then $c = a(x_0 + b - 1) + b(y_0 - 1)$. Show that $x_0 + b - 1 \geq 0$.)

CHAPTER PROBLEMS

Unless stated otherwise, all variables represent integers.

1. Find $b \operatorname{DIV} a$ and $b \operatorname{MOD} a$.
   (a) $a = 13$, $b = 100$ (b) $a = 13$, $b = -100$
   (c) $a = -13$, $b = 100$ (d) $a = -13$, $b = -100$
2. Give an example of a set A such that $\mathbb{N} \subset A \subset \mathbb{Z}$ and
   (a) A is well-ordered.
   (b) A is not well-ordered.
3. Prove for any integer m that $m^3 \operatorname{MOD} 7 \in \{0, 1, 6\}$.

4. Prove for any positive integer n that:
   (a) $(n^2 - n)$ is a multiple of 2.
   (b) $n^2 + 1$ is not a multiple of 4.
   (c) $n(n + 1)(2n + 1)$ is a multiple of 6.
5. Find the canonical factorization of
   (a) 24635975   (b) 1376375   (c) gcd(24635975, 1376375)
6. Assume $a \neq 0$ and prove that if $a \mid b$ and $a + b = c$, then $a \mid c$.
7. Assume $m > 1$. Show that if $m \mid (35n + 26)$ and $m \mid (7n + 3)$, then $m = 11$.
8. For $d \neq 0$, what is the relationship between the following?
   (a) $m \operatorname{DIV} d$ and $(m - d) \operatorname{DIV} d$
   (b) $m \operatorname{MOD} d$ and $(m - d) \operatorname{MOD} d$
9. Prove or disprove:
   (a) If $m \operatorname{MOD} 6 = 5$, then $m \operatorname{MOD} 3 = 2$.
   (b) If $m \operatorname{MOD} 3 = 2$, then $m \operatorname{MOD} 6 = 5$.
10. Let n be a positive integer and let m be an integer with $0 \leq m \leq 2^{n+1} - 1$.
    (a) Show that m is uniquely expressible in the form

    $$m = b_n \cdot 2^n + b_{n-1} \cdot 2^{n-1} + \cdots + b_1 \cdot 2^1 + b_0$$

    where each b_i is either 0 or 1. The representation $b_n b_{n-1} \cdots b_1 b_0$ is called the *binary representation* of m. (For example, the binary representation of 25 is 11001.)
    (b) Write a Pascal procedure DECTOBIN that inputs a nonnegative integer m and outputs the binary representation of m.
    (c) Write a Pascal procedure BINTODEC that inputs a bit string (a string of 0's and 1's) and outputs the nonnegative integer m having that string as its binary representation.
11. Find $d = \gcd(a, b)$ and integers s and t such that $d = as + bt$.
    (a) $a = 357, b = 629$ (b) $a = 1109, b = 4999$
12. How many integers between 1 and $6n$, inclusive, are relatively prime to 6? (Hint: Let $U = \{1, 2, \ldots, 6n\}$, $A = \{2, 4, \ldots, 6n\}$, and $B = \{3, 6, \ldots, 6n\}$; find $|A' \cap B'|$.)
13. Let n be both a perfect square and a perfect cube (for example, $64 = 8^2 = 4^3$). Show that $n \operatorname{MOD} 7 \in \{0, 1\}$.
14. Prove: If $n \in \mathbb{N}$, then $\gcd(a, a + n) \mid n$ for all $a \in \mathbb{Z}$.
15. Let c be a positive integer. Prove that $\gcd(ca, cb) = c \cdot \gcd(a, b)$.
16. Let $m = d_n \cdots d_1 d_0$ be the usual decimal representation of the positive integer m. Then

    $$m = d_n \cdot 10^n + \cdots + d_1 \cdot 10^1 + d_0$$

    where each $d_i \in \{0, 1, \ldots, 9\}$ and $d_n \neq 0$. Prove that m is a multiple of 9 if and only if $d_0 + d_1 + \cdots + d_n$ is a multiple of 9.
17. Prove: If $\gcd(a, b) = d$, then $\gcd(a/d, b/d) = 1$.

18. The *least common multiple* of integers a and b, denoted lcm(a, b), is defined as the smallest positive integer c such that $a \mid c$ and $b \mid c$. Suppose that

$$a = p_1^{\alpha_1} \cdot p_2^{\alpha_2} \cdots \cdot p_n^{\alpha_n}$$

$$b = p_1^{\beta_1} \cdot p_2^{\beta_2} \cdots \cdot p_n^{\beta_n}$$

where each p_i is prime and the α_i's and β_i's are nonnegative integers (see Exercise 3 in Exercises 3.4).
  (a) Give a formula for lcm(a, b) in terms of $p_1, p_2, \ldots, p_n$.
  (b) Use the result of part (a) to find lcm(4725, 9702).
  (c) Show that gcd(a, b) $\cdot$ lcm(a, b) $= ab$.
  (d) Prove: If $a \mid n$ and $b \mid n$, then lcm(a, b) $\mid n$.

19. If a, b, and c are positive integers such that $a^2 + b^2 = c^2$, then (a, b, c) is called a *Pythagorean triple*. Prove:
  (a) If (a, b, c) is a Pythagorean triple, then a or b is even.
  (b) If (a, b, c) is a Pythagorean triple and d is a positive integer, then (da, db, dc) is a Pythagorean triple.
  A Pythagorean triple is called *primitive* if gcd$(a, b) = 1$. (In this case it also happens that gcd$(a, c) = $ gcd$(b, c) = 1$.)
  (c) Write a Pascal program that inputs a positive integer n and outputs all primitive Pythagorean triples (a, b, c) such that
  $$1 < a < b < c \le n.$$

20. Given three integers a, b, and c, not all 0, we define their *greatest common divisor*, denoted gcd(a, b, c), to be the largest common divisor of all three.
  (a) Show that gcd(a, b, c) $=$ gcd(gcd(a, b), c).
  (b) Show that gcd(a, b, c) is the smallest positive integer d for which there exist integers x, y, and z such that $d = ax + by + cz$.
  (c) Prove or disprove: If gcd(a, b, c) $= 1$, then a, b, and c are pairwise relatively prime.
  (d) Suppose

$$a = p_1^{\alpha_1} \cdot p_2^{\alpha_2} \cdots \cdot p_n^{\alpha_n}$$

$$b = p_1^{\beta_1} \cdot p_2^{\beta_2} \cdots \cdot p_n^{\beta_n}$$

$$c = p_1^{\gamma_1} \cdot p_2^{\gamma_2} \cdots \cdot p_n^{\gamma_n}$$

where each p_i is prime and the α_i's, β_i's, and γ_i's are nonnegative integers. Give a formula for gcd(a, b, c) in terms of $p_1, p_2, \ldots, p_n$.

21. Let p be any prime. An integer a, where $1 < a < p$, is called a *primitive root* of p provided

$$\{ a, a^2 \text{ MOD } p, a^3 \text{ MOD } p, \ldots, a^{p-1} \text{ MOD } p \} = \{1, 2, 3, \ldots, p - 1\}$$

Find all primitive roots of:
  (a) 7   (b) 11

Let m be any integer and let a be a primitive root of p.

(c) Show that $\{m \text{ MOD } p, (m + a) \text{ MOD } p, (m + a^2) \text{ MOD } p, \dots,$
$(m + a^{p-1}) \text{ MOD } p\} = \{0, 1, 2, \dots, p - 1\}$.

22. Let m and n be positive integers and let $d = \gcd(m, n)$. Prove:

$$am \text{ MOD } n = bm \text{ MOD } n \rightarrow a \text{ MOD}\left(\frac{n}{d}\right) = b \text{ MOD}\left(\frac{n}{d}\right)$$

(Hence, if m and n are relatively prime, then $am \text{ MOD } n = bm \text{ MOD } n \rightarrow a \text{ MOD } n = b \text{ MOD } n$.)

23. Let m and n be positive integers such that $m \mid n$. Prove:

$$a \text{ MOD } n = b \text{ MOD } n \rightarrow a \text{ MOD } m = b \text{ MOD } m$$

24. Let a and b be positive integers and let p be prime.
   (a) Prove: $a^2 \text{ MOD } p = b^2 \text{ MOD } p \rightarrow (a \text{ MOD } p = b \text{ MOD } p$ or $a \text{ MOD } p = -b \text{ MOD } p)$.
   (b) Give an example to show the necessity of the requirement that p be prime in part (a).

25. Let

$$a = p_1^{\alpha_1} \cdot p_2^{\alpha_2} \cdot \dots \cdot p_n^{\alpha_n}$$
$$b = q_1^{\beta_1} \cdot q_2^{\beta_2} \cdot \dots \cdot q_n^{\beta_n}$$

be the canonical factorizations of positive integers a and b. What conditions must be satisfied by the α_i's and β_i's if
(a) a is a perfect square?
(b) b is a perfect cube?
(c) $a \mid b$?

26. Let a and b be integers and let m and n be positive integers such that m divides each of a, b, and n. Prove:

$$a \text{ MOD } n = b \text{ MOD } n \rightarrow \left(\frac{a}{m}\right) \text{MOD}\left(\frac{n}{m}\right) = \left(\frac{b}{m}\right) \text{MOD}\left(\frac{n}{m}\right)$$

27. Use mathematical induction to prove that the following statements hold for all $n \in \mathbb{N}$.
   (a) $1 \cdot 1 + 2 \cdot 2 \cdot 1 + 3 \cdot 3 \cdot 2 \cdot 1 + \dots + n \cdot n \cdot (n - 1) \cdots 1 = ((n + 1) \cdot n \cdot \dots \cdot 1) - 1$
   (b) $(2^n)^2 - 1$ is a multiple of 3.
   (c) $1^2 - 2^2 + \dots + (-1)^{n+1} n^2 = \dfrac{(-1)^{n+1} n(n + 1)}{2}$
   (d) $\dfrac{1}{1 \cdot 2} + \dfrac{1}{2 \cdot 3} + \dots + \dfrac{1}{n(n + 1)} = \dfrac{n}{n + 1}$
   (e) $1 \cdot 2 + 2 \cdot 3 + \dots + n(n + 1) = \dfrac{n(n + 1)(n + 2)}{3}$
   (f) $\dfrac{1}{\sqrt{1}} + \dfrac{1}{\sqrt{2}} + \dots + \dfrac{1}{\sqrt{n}} \le 2\sqrt{n} - 1$

28. Use induction on n to prove DeMorgan's laws for n sets $A_1, A_2, \ldots, A_n$:

(a) $(A_1 \cup A_2 \cup \cdots \cup A_n)' = (A_1' \cap A_2' \cap \cdots \cap A_n')$

(b) $(A_1 \cap A_2 \cap \cdots \cap A_n)' = (A_1' \cup A_2' \cup \cdots \cup A_n')$

29. Suppose you have an infinite supply of 8-cent and 13-cent stamps. What amounts of postage can you make? Use the strong form of induction to prove your answer.

30. Use induction on n to prove the extended version of Euclid's lemma: If $a_1, a_2, \ldots, a_n$ are integers and p is a prime such that $p \mid (a_1 a_2 \cdots a_n)$, then $p \mid a_i$ for some $i \in \{1, 2, \ldots, n\}$.

31. Let a and d be real numbers. Use induction to prove the following formula for the sum of a finite arithmetic series:

$$a + (a + d) + (a + 2d) + \cdots + (a + nd) = \frac{(n + 1)(2a + nd)}{2}$$

32. Use induction to prove that the number of primes is infinite. (Hint: For $n \in \mathbb{N}$, let $P(n)$ be the statement "There exist at least n primes.")

33. The following argument purports to prove that for any $n \in \mathbb{N}$ and for any set S of n cars, all cars in S get the same average number of miles per gallon of fuel. What is the flaw in the argument?

1. Clearly the result holds when $n = 1$.

2. Assume the result holds for an arbitrary integer $k \geq 1$; that is, all cars in any set of k cars get the same average number of miles per gallon.

3. We must show that the result holds when $n = k + 1$: that is, show that all cars in any set of $k + 1$ cars get the same average number of miles per gallon. Let S be a set of $k + 1$ cars. Choose subsets A and B of S such that $|A| = |B| = k$, $A \cap B \neq \emptyset$, and $A \cup B = S$. Then, by the induction hypothesis, all cars in A get the same average miles per gallon and all cars in B get the same average miles per gallon. Since there is at least one car in $A \cap B$, it follows that all cars in $S = A \cup B$ get the same average miles per gallon. Thus the result holds for $n = k + 1$ and therefore, by the PMI, the result holds for every $n \in \mathbb{N}$.

34. A method of finding all the primes up to some given positive integer n is known as the "sieve of Eratosthenes." Start with a list of the integers from 2 to n. The first number on the list, 2, is prime; output 2 and then delete all multiples of 2 from the list. The first number on the new list, 3, is prime; output 3 and then delete all multiples of 3 from the list. Continue this process until only prime numbers remain on the list, then output these. (At what point can we be certain that only primes remain on the list?) Implement the sieve of Eratosthenes as a Pascal program.

35. Let a and b be integers with $1 \leq a \leq b$, and let $r = b \bmod a$.

(a) Show that $2r < b$.

Let $P(b)$ be the statement that, if $1 < a < b$, then the number of divisions required by Algorithm 3.2 to compute $\gcd(a, b)$ is less than $2 \log_2 b$.

(b) Use the result of part (a) and the strong form of induction to prove that $P(b)$ holds for all integers $b \geq 3$.

(c) Use the result of part (b) to prove that, if $1 < a < b$, then the number of divisions required by Algorithm 3.2 to compute $\gcd(a, b)$ is less than $2 \log_2 a + 1$. Compare this bound with Lamé's bound, given in Section 3.3.

36. Let a and b be integers with $1 < a < b$.

(a) If a and b are both even, prove that $\gcd(a, b) = 2 \cdot \gcd(a/2, b/2)$.

(b) If a is even and b is odd, prove that $\gcd(a, b) = \gcd(a/2, b)$. (Or, if a is odd and b is even, then $\gcd(a, b) = \gcd(a, b/2)$.)

(c) If a and b are both odd, prove that $\gcd(a, b) = \gcd(a, b - a)$.

(d) Use the results of parts (a)–(c) to design an algorithm that inputs positive integers a and b and outputs $\gcd(a, b)$.

(e) Use the algorithm of part (d) to find $\gcd(1428, 2516)$.

37. Let n be a positive integer. The purpose of this problem is to develop efficient algorithms to do arithmetic modulo n; that is, given integers a and b with $0 \leq a < n$, $0 \leq b < n$, and a nonnegative integer m, we wish to compute $(a + b) \bmod n$, $ab \bmod n$, and $a^m \bmod n$.

(a) Write a Pascal function MOD_ADD that returns $(a + b) \bmod n$. (Note that if $a + b > n$, then $(a + b) \bmod n = a + b - n$.)

(b) Write a Pascal function MOD_MULTIPLY that returns $ab \bmod n$. Use the technique known as "repeated doubling." Compute the terms $a \bmod n, 2a \bmod n, 4a \bmod n, \ldots$ by doubling; then use the binary representation of b to determine which terms are needed for the final result. For example, $(a \cdot 101) \bmod n$ is computed as

$$\big(\big(\big(\big((a + (4a \bmod n))\bmod n\big) + (32a \bmod n)\big)\bmod n\big)$$
$$+ (64a \bmod n)\big)\bmod n$$

(c) Write a Pascal function MOD_EXPO that returns $a^m \bmod n$. Use the technique known as "repeated squaring." Compute the factors $a \bmod n, a^2 \bmod n, a^4 \bmod n, \ldots$ by squaring; then use the binary representation of m to determine which factors are needed for the final result. For example, $a^{101} \bmod n$ is computed as

$$\big(\big(\big(\big((a \cdot (a^4 \bmod n))\bmod n\big) \cdot (a^{32} \bmod n)\big)\bmod n\big)$$
$$\cdot (a^{64} \bmod n)\big)\bmod n$$

F O U R

Relations

INTRODUCTION

At this stage in our development, the student should have achieved some familiarity with the rudiments of set theory and the divisibility properties of integers. In this chapter we make use of the language of set theory to define the notion of a relation between two sets.

A rather nice example of a relation is provided by the idea of divisibility for positive integers. Given two positive integers a and b, we can ask whether a divides b; if so, we say that "a is related to b." It turns out that "divides" is a relation from $\mathbb{N}$ to $\mathbb{N}$. Notice that it is possible that a is related to b but b is not related to a; this happens, for example, when $a = 2$ and $b = 6$. Because of this, it makes good sense to single out the ordered pair (a, b) if $a \mid b$. Thus the relation determines a unique set of ordered pairs; namely,

$$R = \{(a, b) \mid a, b \in \mathbb{N} \wedge a \mid b\}$$

As another example, consider the sets P and L of all points and lines, respectively, in a given plane. Given a point p and a line l, we can ask whether p is incident with l (p is on l). Then "incidence" turns out to be a relation from P to L; in this case the relation determines the set

$$S = \{(p, l) \mid p \in P \wedge l \in L \wedge p \in l\}$$

What should be clear from these examples is that a relation from a set A to a set B determines a unique subset of the Cartesian product $A \times B$. This discussion indicates, in a certain sense, the general definition of the term *relation*.

DEFINITION 4.1

A *relation* from a set A to a set B is defined to be any subset of $A \times B$. If R is a relation from A to B and $(a, b) \in R$, then we say that *a is related to b* and write $a R b$.

Since the preceding definition involves two sets A and B, we sometimes refer to such a relation as a *binary relation*.

Example 4.1   Let

$$A = \{\text{Brinkerhoff, Chan, McKenna, Slonneger, Will, Yellen}\}$$

be the set of computer science instructors at a small college, and let

$$B = \{\text{CS105, CS260, CS261, CS360, CS450, CS460}\}$$

be the set of computer science courses offered next semester at that college. Then $A \times B$ gives all possible pairings of instructors and courses. Let the relation R from A to B be given by

$$R = \{(\text{Brinkerhoff, CS105}), (\text{Brinkerhoff, CS360}), (\text{Chan, CS260}),$$
$$(\text{Chan, CS360}), (\text{Chan, CS460}), (\text{McKenna, CS105}),$$
$$(\text{McKenna, CS260}), (\text{Slonneger, CS260}), (\text{Slonneger, CS261}),$$
$$(\text{Will, CS105}), (\text{Yellen, CS261}), (\text{Yellen, CS450})\}$$

Then R might tell us, for example, which instructors are assigned to teach which courses. ∎

Example 4.2   Let P be the set of primes and define a relation R from P to $\mathbb{N}$ by

$$p R n \leftrightarrow p \mid n$$

Find (a) all primes p such that $p R 126$, and (b) all n such that $3 R n$.

Solution   Since $126 = 2 \cdot 3 \cdot 3 \cdot 7$, for (a) we have $2 R 126$, $3 R 126$, and $7 R 126$. For (b), $3 R n$ if and only if $3 \mid n$; hence $3 R n$ if and only if n is a multiple of 3. Thus the set of positive integers to which 3 is related under R is $\{3, 6, 9, \ldots\}$.

If A and B are subsets of $\mathbb{R}$ and F is a relation from A to B, then F is a collection of ordered pairs of real numbers. Associated with each pair $(x, y) \in F$ there is a uniquely determined point in the xy-coordinate plane. In this special instance we refer to the set of all points so determined as the *graph* of F. In fact, it makes good sense to identify F with its graph.

Example 4.3   Define a relation F from $\mathbb{R}$ to the closed interval $[0, 1]$ by

$$x\,Fy \leftrightarrow y = \frac{1}{x^2 + 1}$$

Then we see that

$$F = \left\{ \left(x, \frac{1}{x^2 + 1} \right) \,\middle|\, x \in \mathbb{R} \right\}$$

The graph of F is shown in Figure 4.1. Notice that for all $x \in \mathbb{R}$,

$$0 < \frac{1}{x^2 + 1} \leq 1 \qquad \blacksquare$$

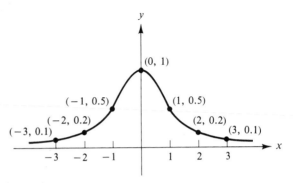

Figure 4.1 The graph of the relation $F = \left\{ (x, y) \mid y = \dfrac{1}{x^2 + 1} \right\}$

Example 4.4   Define a relation F from $\mathbb{R}$ to $\mathbb{R}$ by $x\,Fy \leftrightarrow x^2 + y^2 \leq 4$. Then the graph of F consists of all points (x, y) in the coordinate plane such that $x^2 + y^2 \leq 4$. Notice that F consists of all points that lie on or within the circle of radius 2 centered at the origin, as shown in Figure 4.2. $\blacksquare$

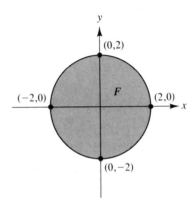

Figure 4.2 The graph of the relation $F = \{ (x, y) \mid x^2 + y^2 \leq 4 \}$

Suppose that A and B are finite sets, say, $A = \{a_1, a_2, \ldots, a_m\}$ and $B = \{b_1, b_2, \ldots, b_n\}$, and R is a relation from A to B. One way to represent R is to form a rectangular array having m rows and n columns in which the ith row corresponds to a_i, $1 \le i \le m$, and the jth column corresponds to b_j, $1 \le j \le n$. Specifically, if $a_i R b_j$, then we place a 1 in the position of the array corresponding to the ith row and jth column; if it is not the case that $a_i R b_j$, then we place a 0 in this position. We call this array the *matrix* of R. Denoting the element in the ith row and jth column of this array by r_{ij}, we have

$$r_{ij} = \begin{cases} 1 & \text{if} \quad a_i R b_j \\ 0 & \text{if} \quad \sim(a_i R b_j) \end{cases}$$

For example, the matrix of the relation R of Example 4.1 is

$$\begin{bmatrix} 1 & 0 & 0 & 1 & 0 & 0 \\ 0 & 1 & 0 & 1 & 0 & 1 \\ 1 & 1 & 0 & 0 & 0 & 0 \\ 0 & 1 & 1 & 0 & 0 & 0 \\ 1 & 0 & 0 & 0 & 0 & 0 \\ 0 & 0 & 1 & 0 & 1 & 0 \end{bmatrix}$$

where the rows correspond to Brinkerhoff, Chan, McKenna, Slonneger, Will, and Yellen, respectively, and the columns to CS105, CS260, CS261, CS360, CS450, and CS460, respectively.

In general, any rectangular array of numbers of the form

$$A = \begin{bmatrix} a_{11} & a_{12} & \cdots & a_{1n} \\ a_{21} & a_{22} & \cdots & a_{2n} \\ & & \vdots & \\ a_{m1} & a_{m2} & \cdots & a_{mn} \end{bmatrix}$$

is called an *m by n matrix* (written $m \times n$ matrix). For $1 \le i \le m$ and $1 \le j \le n$, we call a_{ij} the *(i, j)-entry* of the matrix. We use the customary shorthand notation $[a_{ij}]$ to denote A. Those students not already familiar with the elementary properties of matrices may refer to Appendix A for a self-contained treatment.

Example 4.5   In a certain tennis tournament involving five players, each contestant plays each of the others exactly once. If we denote the contestants by a_1, a_2, a_3, a_4, and a_5, then the results of the matches in the tournament can be represented as a relation R from the set $A = \{a_1, a_2, a_3, a_4, a_5\}$ to itself, where

$$a_i R a_j \leftrightarrow a_i \text{ beats } a_j$$

Suppose that the resulting relation is

$$R = \{(a_1, a_2), (a_1, a_3), (a_1, a_5), (a_2, a_3), (a_2, a_4),$$
$$(a_3, a_4), (a_3, a_5), (a_4, a_1), (a_5, a_2), (a_5, a_4)\}$$

Then the matrix of R is

$$\begin{bmatrix} 0 & 1 & 1 & 0 & 1 \\ 0 & 0 & 1 & 1 & 0 \\ 0 & 0 & 0 & 1 & 1 \\ 1 & 0 & 0 & 0 & 0 \\ 0 & 1 & 0 & 1 & 0 \end{bmatrix}$$

Note that, for $i \neq j$, exactly one of a_{ij} and a_{ji} is 1.   ∎

Example 4.6   Define a relation R from the set $A = \{2, 3, 7, 9, 14, 24, 27\}$ to itself by

$$a \, R \, b \leftrightarrow \gcd(a, b) > 1$$

Determine the relation R and its matrix.

Solution   The relation R is given by

$$R = \{(2,2), (2,14), (2,24), (3,3), (3,9), (3,24), (3,27), (7,7),$$
$$(7,14), (9,3), (9,9), (9,24), (9,27), (14,2), (14,7), (14,14),$$
$$(14,24), (24,2), (24,3), (24,9), (24,14), (24,24), (24,27),$$
$$(27,3), (27,9), (27,24), (27,27)\}$$

The matrix of R is then

$$\begin{bmatrix} 1 & 0 & 0 & 0 & 1 & 1 & 0 \\ 0 & 1 & 0 & 1 & 0 & 1 & 1 \\ 0 & 0 & 1 & 0 & 1 & 0 & 0 \\ 0 & 1 & 0 & 1 & 0 & 1 & 1 \\ 1 & 0 & 1 & 0 & 1 & 1 & 0 \\ 1 & 1 & 0 & 1 & 1 & 1 & 1 \\ 0 & 1 & 0 & 1 & 0 & 1 & 1 \end{bmatrix}$$

∎

Notice that the entries of the matrix in Example 4.6 satisfy the condition that $a_{ij} = a_{ji}$ for $1 \leq i, j \leq 7$. In general, an $m \times m$ matrix $A = [a_{ij}]$ is called a *symmetric matrix* provided $a_{ij} = a_{ji}$ for $1 \leq i \leq m$ and $1 \leq j \leq m$.

One final remark is in order regarding the matrix of a relation. If R is a relation from $A = \{a_1, a_2, \ldots, a_m\}$ to $B = \{b_1, b_2, \ldots, b_n\}$, and we are given the matrix M of R (with a_i corresponding to row i and b_j corresponding to column j) but are not given R explicitly, then we can determine the pairs of R from M. Thus, assuming some ordering of the elements in A and in B (that is, one knows which elements correspond to the rows and columns of M), one can think of M as a "matrix representation" of R.

The relations given in Examples 4.4, 4.5, and 4.6 are examples of relations from a set to itself. In general, if R is a relation from a set A to itself, then we call R a *relation on A*.

Suppose now that R is a relation on a finite set A. Another representation of R, a geometric one, is obtained using the following scheme. Each element of A corresponds to a point in the plane. If $a_1 R a_2$ for some $a_1, a_2 \in A$, where $a_1 \neq a_2$, then a directed simple curve is drawn from the point corresponding to a_1 to the point corresponding to a_2, as illustrated in Figure 4.3(a). (Here a simple curve is a curve that does not intersect itself, such as a line segment. A directed simple curve from a_1 to a_2 is a simple curve containing an arrow pointing to a_2.) If $a R a$ for some $a \in A$, then a directed simple closed curve, called a *loop*, is drawn from a to a, as shown in Figure 4.3(b). The resulting structure is referred to as the *directed graph* of R; the points are called *vertices* and the directed simple curves are called *directed edges* or *arcs*.

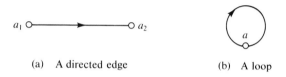

(a)   A directed edge   (b)   A loop

Figure 4.3

In general, the notion of a directed graph resides in that area of mathematics known as "graph theory." Graph theory is currently one of the more popular areas of mathematical research, not only because of the wealth of problems for study, but also because of the important applications of the subject to computer science and other fields. An introduction to graph theory, along with some of its applications, is provided in Chapter 7.

Example 4.7   Exhibit the directed graph of each of the following relations:

(a) the relation R of Example 4.5
(b) the relation T defined on the set $A = \{1, 2, 3, 4, 5, 6\}$ by $a T b \leftrightarrow \gcd(a, b) = 1$
(c) the relation S defined on the power set $\mathscr{P}(\{1, 2\})$ by $A S B \leftrightarrow A \subseteq B$

Solution   The directed graphs of R, T, and S are shown in Figure 4.4(a), (b), and (c), respectively.

One final observation: Suppose R is a relation on a set $A = \{a_1, a_2, \ldots, a_n\}$ and M and D are the matrix and directed graph of R,

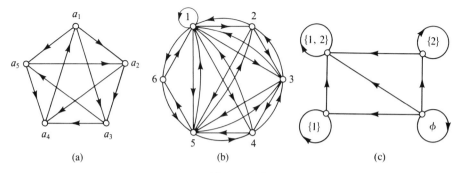

Figure 4.4

respectively. If $M = [m_{ij}]$, then we know that $m_{ij} = 1$ if and only if $a_i \, R \, a_j$. Also there is a directed edge from a_i to a_j in D if and only if $a_i \, R \, a_j$. Thus we see that $m_{ij} = 1$ if and only if there is a directed edge from a_i to a_j in D. We call M the *adjacency matrix* of D.

If R is a relation from a set A to a set B, then we know that R is some subset of $A \times B$. Thus R determines a subset of A, namely, the set of all first coordinates of the ordered pairs of R. Similarly, a subset of B is determined by R. These subsets, formally defined in Definition 4.2, are discussed in some detail in Chapter 5.

DEFINITION 4.2

Let R be a relation from the set A to the set B. The *domain* of R is the set dom R defined by

$$\text{dom } R = \{ a \in A \mid (a, b) \in R \text{ for some } b \in B \}$$

The *image* (or *range*) of R is the set im R defined by

$$\text{im } R = \{ b \in B \mid (a, b) \in R \text{ for some } a \in A \}$$

Example 4.8 In Example 4.1 dom $R = A$ and im $R = B$. In Example 4.2 dom $R = P$ and im $R = \mathbb{N} - \{1\}$. In Example 4.3 dom $F = \mathbb{R}$ and im $F = (0, 1]$. As another example, let A be the set of undergraduates at your school and let B be the set of undergraduate mathematics courses that your school offers. Define R from A to B by $a R b \leftrightarrow$ student a is taking course b this semester. Then dom R consists only of those students in A who are taking a mathematics course this semester. Since it is almost certainly the case that some students are not taking any mathematics courses this semester, it is probably true that dom R is a proper subset of A. Similarly, im R consists only of those mathematics courses being offered this semester. If it is the case that not every undergraduate mathematics course is offered in a given semester, then im R is a proper subset of B.

Exercises 4.1

1. Each of the following defines a relation from $\mathbb{R}$ to $\mathbb{R}$. Exhibit the graph of each relation.
    (a) $F = \{(x, y) \mid y = 2x + 1\}$
    (b) $F = \{(x, y) \mid y \leq 2x + 1\}$
    (c) $F = \{(x, y) \mid y = x^2\}$
    (d) $F = \{(x, y) \mid 4 > y > x^2\}$
    (e) $F = \{(x, y) \mid |x| + |y| = 4\}$
    (f) $F = \{(x, y) \mid x > 1 \wedge y < 2\}$
2. Let C be the set of classes you are taking this semester and let D be the set of days on which classes are held at your school. Define a relation R from C to D by $c\,R\,d \leftrightarrow$ class c meets on day d.
    (a) List the ordered pairs of R.
    (b) Exhibit the matrix of R.
    (c) Find the domain and image of R.
3. For each of the given relations on $\{1, 2, 3, 6\}$, determine the adjacency matrix and directed graph.
    (a) $\{(x, y) \mid x$ divides $y\}$
    (b) $\{(x, y) \mid x < y\}$
    (c) $\{(x, y) \mid x = y\}$
    (d) $\{(x, y) \mid x \neq y\}$
    (e) $\{(x, y) \mid x^2 \leq y\}$
    (f) $\{(x, y) \mid x + y$ is odd$\}$
4. Determine the adjacency matrix and directed graph of the relation R defined on $\{-3, -2, -1, 0, 1, 2, 3\}$ by $x\,R\,y \leftrightarrow 3 \mid (x - y)$.
5. Find the domain and image of each relation in Exercise 3.
6. Determine the adjacency matrix and directed graph of the relation R defined on $\{1, 2, 4, 5, 10, 20\}$ by $x\,R\,y \leftrightarrow (x < y \wedge x \mid y)$.
7. Let $A = \{a_1, a_2, \ldots, a_m\}$ and $B = \{b_1, b_2, \ldots, b_n\}$ be finite sets. A relation R from A to B can be represented in Pascal as a two-dimensional Boolean array; such an array can be declared as follows:

```
CONST M = 3; N = 4; (*or any positive integers*)
TYPE RELATION = ARRAY[1..M, 1..N] OF BOOLEAN;
VAR  R: RELATION;
```

The idea is that R[I,J] = TRUE if and only if $a_i\,R\,b_j$. (Note that R is actually the matrix representation of R, except that we are using the entries TRUE and FALSE rather than 0 and 1.)

Suppose that $A = \{1, 2, 3\}$ and $B = \{1, 2, 3, 4\}$. Give the representation of R if:
    (a) $R = \{(1, 1), (1, 3), (2, 4), (3, 1), (3, 2), (3, 4)\}$
    (b) $R = \varnothing$; that is, R is the *empty relation* from A to B
    (c) $R = A \times B$; that is, R is the *complete relation* from A to B
    (d) $R = \{(1, 1), (2, 2), (3, 3)\}$
    (e) $R = \{(x, y) \mid x \in A \wedge y \in B \wedge x \neq y\}$
    (f) $R = \{(x, y) \mid x \in A \wedge y \in B \wedge x < y\}$

4.2

PROPERTIES OF RELATIONS

A relation on a set A was defined in Section 4.1 as a relation from A to itself. In this section several key properties that a relation on a set may possess are introduced and some important examples are studied.

DEFINITION 4.3

If R is a relation on a set A, then

1. R is *reflexive* if $a R a$ for all $a \in A$.
2. R is *irreflexive* if $\sim(a R a)$ for all $a \in A$.
3. R is *symmetric* if $a R b \rightarrow b R a$ for all $a, b \in A$.
4. R is *antisymmetric* if $(a R b \wedge b R a) \rightarrow a = b$ for all $a, b \in A$.
5. R is *transitive* if $(a R b \wedge b R c) \rightarrow a R c$ for all $a, b, c \in A$.

Stated in terms of ordered pairs, a relation R on a set A is:

1. reflexive $\leftrightarrow (a, a) \in R$ for all $a \in A$
2. irreflexive $\leftrightarrow (a, a) \notin R$ for all $a \in A$
3. symmetric $\leftrightarrow (a, b) \in R \rightarrow (b, a) \in R$ for all $a, b \in A$
4. antisymmetric $\leftrightarrow ((a, b) \in R \wedge (b, a) \in R) \rightarrow a = b$
   for all $a, b \in A$
5. transitive $\leftrightarrow ((a, b) \in R \wedge (b, c) \in R) \rightarrow (a, c) \in R$
   for all $a, b, c \in A$

Moreover, making use of the partial contrapositive (Problem 13 in Chapter 1 Problems), we have that a relation R on a set A is antisymmetric if and only if the condition

$$((a, b) \in R \wedge a \neq b) \rightarrow (b, a) \notin R$$

holds for all $a, b \in A$. In other words, R is antisymmetric provided that, for any two distinct elements a and b of A, at most one of the ordered pairs (a, b) and (b, a) is an element of R.

Example 4.9   Let A be any nonempty set. Then the empty set $\emptyset$ is a relation on A, called the *empty relation*. It is clearly irreflexive. Is it symmetric? In other words, does it follow that, for all $a, b \in A$, $(a, b) \in \emptyset$ implies $(b, a) \in \emptyset$? Since $(a, b) \in \emptyset$ is false, this implication is true. So the relation is symmetric. Similarly, it is antisymmetric and transitive.

Next consider the relation

$$I_A = \{(a, a) \mid a \in A\}$$

which is called the *identity relation* on A. It is readily seen to be reflexive, symmetric, antisymmetric, and transitive, but it is not irreflexive. It should be noted, in fact, that a relation R on A is both symmetric and antisymmetric if and only if R is a subset I_A. To see the necessity of this, suppose $(a, b) \in R$. Then, since R is symmetric, $(b, a) \in R$. Also, since R is antisymmetric, $(a, b) \in R$ and $(b, a) \in R$ imply that $a = b$. So $(a, b) = (a, a) \in I_A$. Thus $R \subseteq I_A$.

Finally, consider the relation $S = A \times A$. We call this the *complete relation* on A. Notice that S is reflexive, symmetric, and transitive, but not irreflexive. It is antisymmetric if and only if $|A| = 1$. ∎

As a general comment, note that no relation on a nonempty set can be both reflexive and irreflexive.

Example 4.10 For each of these relations on $\{1, 2, 3, 4\}$, determine which of the properties from Definition 4.3 it satisfies.

(a) $R_1 = \{(1, 1), (1, 3), (1, 4), (2, 2), (3, 1), (3, 3), (3, 4), (4, 1), (4, 3), (4, 4)\}$
(b) $R_2 = \{(1, 1), (1, 2), (1, 3), (1, 4), (2, 2), (2, 3), (2, 4), (3, 3), (3, 4), (4, 4)\}$
(c) $R_3 = \{(1, 1), (1, 2), (2, 1), (2, 2), (2, 3), (3, 2), (3, 3), (3, 4)\}$
(d) $R_4 = \{(1, 3), (3, 1), (2, 4), (4, 2)\}$

Solution

(a) It is easy to check that R_1 is reflexive and symmetric; with some effort, it can also be checked that R_1 is transitive.
(b) Note that

$$R_2 = \{(x, y) \mid x \le y\}$$

The relation R_2 is reflexive, antisymmetric, and transitive.
(c) The relation R_3 satisfies none of the properties: it is not reflexive since $(4, 4) \notin R_3$; it is not irreflexive since, for instance, $(1, 1) \in R_3$; it is not symmetric, since $(3, 4) \in R_3$ but $(4, 3) \notin R_3$; it is not antisymmetric since, for instance, both $(1, 2) \in R_3$ and $(2, 1) \in R_3$; it is not transitive since, for instance, $(1, 2) \in R_3$ and $(2, 3) \in R_3$ but $(1, 3) \notin R_3$.
(d) The relation R_4 is irreflexive and symmetric. It is not transitive, since $(1, 3) \in R_4$ and $(3, 1) \in R_4$ but $(3, 3) \notin R_4$. ∎

Example 4.11 Define the relation R on $\mathbb{Z}$ by

$$a \, R \, b \leftrightarrow 3 \mid (a - b)$$

Show that R is reflexive, symmetric, and transitive.

Solution Since $3 \mid (a - a)$ for any integer a, the relation R is reflexive. Given any $a, b \in \mathbb{Z}$, if $3 \mid (a - b)$, then $3 \mid (b - a)$. It follows that R is symmetric. Also, by Theorem 3.2, part 3, if $3 \mid (a - b)$ and $3 \mid (b - c)$, then $3 \mid [(a - b) + (b - c)]$; that is, $3 \mid (a - c)$. This shows that R is transitive. The relation R is not antisymmetric, since, for instance, $6 \, R \, 9$ and $9 \, R \, 6$ but $6 \ne 9$. ∎

Example 4.12   Let U be a nonempty set and define the relation R on the power set $\mathcal{P}(U)$ of U by the following rule:

$$\text{For } A, B \in \mathcal{P}(U), \ A \, R \, B \leftrightarrow A \subseteq B.$$

Show that R is reflexive, antisymmetric, and transitive, but not symmetric.

Solution   Since $A \subseteq A$ for every $A \in \mathcal{P}(U)$, the relation R is reflexive. For any $A, B \in \mathcal{P}(U)$, if $A \subseteq B$ and $B \subseteq A$, then $A = B$. Hence R is antisymmetric. Also, by Theorem 2.1, if $A, B, C \in \mathcal{P}(U)$ with $A \subseteq B$ and $B \subseteq C$, then $A \subseteq C$. It follows that R is transitive. The relation R is not symmetric, for if $A, B \in \mathcal{P}(U)$ and A is a proper subset of B, then $A \subseteq B$ but $B \not\subseteq A$ (that is, $A \, R \, B$ and $\sim(B \, R \, A)$).   ■

Example 4.13   Consider the relation R from $A = \{$Brinkerhoff, Chan, McKenna, Slonneger, Will, Yellen$\}$ to $\{$CS105, CS260, CS261, CS360, CS450, CS460$\}$ defined in Example 4.1. Define a relation T on A by agreeing that two instructors who teach different sections of the same course are related. The relation T is symmetric but in this case is not transitive. This is because Brinkerhoff is related to Chan (both teach CS260) and Chan is related to Slonneger, but Brinkerhoff is not related to Slonneger.   ■

In the next example we examine how the reflexive, symmetric, and antisymmetric properties manifest themselves in the matrix and directed graph of a relation.

Example 4.14   Exhibit the matrix and directed graph of each of the relations given.

   (a) Let $U = \{1, 2\}$ and define the relation R on $\mathcal{P}(U)$ by $A \, R \, B \leftrightarrow A \subseteq B$.
   (b) Let $X = \{2, 3, 4, 5, 6\}$ and let R be the relation defined on X by

$$a \, R \, b \leftrightarrow \gcd(a, b) = 1$$

Solution   (a) With $\mathcal{P}(U) = \{\emptyset, \{1\}, \{2\}, \{1, 2\}\}$, we see that the matrix of R is

$$M = \begin{bmatrix} 1 & 1 & 1 & 1 \\ 0 & 1 & 0 & 1 \\ 0 & 0 & 1 & 1 \\ 0 & 0 & 0 & 1 \end{bmatrix}$$

The directed graph of R is shown in Figure 4.5(a). Observe that this relation is reflexive and antisymmetric. The reflexive property is revealed in the matrix M of R by the fact that each main diagonal entry of M is 1. It is reflected in the directed graph of R by the presence of a loop at every vertex. As a consequence of the antisymmetric property of R, we have, for $i \neq j$, that either the (i, j)-entry or the (j, i)-entry of M is 0. This is manifested in the directed graph of R by the fact that there is at most one arc joining distinct vertices.

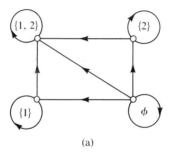 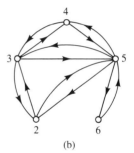

(a)   (b)

Figure 4.5

(b) In this case we readily see that the matrix M of R is

$$M = \begin{bmatrix} 0 & 1 & 0 & 1 & 0 \\ 1 & 0 & 1 & 1 & 0 \\ 0 & 1 & 0 & 1 & 0 \\ 1 & 1 & 1 & 0 & 1 \\ 0 & 0 & 0 & 1 & 0 \end{bmatrix}$$

The directed graph D of R is given in Figure 4.5(b). In this case, note that R is symmetric. In fact, the matrix M is a symmetric matrix and, for distinct vertices x and y of D, either both of the arcs joining a to b and b to a are present or there are no arcs joining a and b. The relation is also irreflexive. With M, this is given by the fact that each main diagonal entry of M is 0; with D, it is determined by the fact that D has no loops. ■

As indicated by the observations made in the preceding example, the following general remarks can be made regarding properties satisfied by a relation R on a finite set, its matrix M, and its digraph D:

1. R is reflexive $\leftrightarrow$ each main diagonal entry of M is 1 $\leftrightarrow$ there is a loop at each vertex of D
2. R is irreflexive $\leftrightarrow$ each main diagonal entry of M is 0 $\leftrightarrow$ there are no loops in D
3. R is symmetric $\leftrightarrow$ M is a symmetric matrix $\leftrightarrow$ either distinct vertices of D are joined by both possible arcs or there are no arcs joining them
4. R is antisymmetric $\leftrightarrow$ for $i \neq j$, not both the (i, j)-entry and the (j, i)-entry of M are 1 $\leftrightarrow$ there is at most one arc joining distinct vertices of D

If a relation R on a finite set A is symmetric, then for each pair of distinct, related elements a_1 and a_2 of A, the directed graph of R contains both an arc from a_1 to a_2 and an arc from a_2 to a_1. In addition, if R is irreflexive, then the associated directed graph has no loops. In this case, it is customary to replace each symmetric pair of arcs by a single undirected simple curve. We call such a curve an *edge* (or *undirected edge*). Figure 4.6 shows an edge joining vertices a_1 and a_2. The resulting structure is called the *graph* of the relation R.

$$a_1 \circ\!\!-\!\!\!-\!\!\!-\!\!\!-\!\!\!-\!\!\!-\!\!\!-\!\!\!-\!\!\!-\!\!\!-\!\!\circ\, a_2$$

Figure 4.6 An undirected edge

Example 4.15 Let $X = \{2, 3, 4, 5, 6\}$ and define the relation R on X by

$$a_1 \, R \, a_2 \leftrightarrow \gcd(a_1, a_2) = 1$$

As noted in Example 4.14, R is symmetric and irreflexive; the graph of R is given in Figure 4.7. ■

Graph theorists are not in total agreement as to the definition of a graph. Some interpret *graph* to mean a graph associated with a symmetric relation R on a finite set A, even if R is not irreflexive. In this case, if some element $a \in A$ is related to itself under R, then the graph of R contains an *undirected loop* at the vertex a.

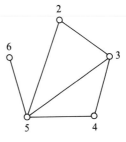

Figure 4.7

Exercises 4.2

1. For each of these relations on $\{1, 2, 3, 4, 5\}$, determine which of the properties (i) reflexive, (ii) irreflexive, (iii) symmetric, (iv) antisymmetric, and (v) transitive are satisfied.
   (a) $R_1 = \{(1, 1), (2, 2), (2, 3), (3, 2), (3, 3), (3, 4), (4, 3), (4, 4), (5, 5)\}$
   (b) $R_2 = \{(1, 2), (1, 4), (1, 5), (2, 4), (2, 5), (3, 4), (3, 5), (4, 5)\}$
   (c) $R_3 = \{(1, 3), (1, 5), (2, 4), (3, 1), (3, 5), (4, 2), (5, 1), (5, 3)\}$
   (d) $R_4 = R_3 \cup \{(1, 1), (2, 2), (3, 3), (4, 4), (5, 5)\}$
   (e) $R_5 = \{(1, 1), (1, 2), (2, 1), (2, 2), (4, 3), (5, 4)\}$
2. For each of the relations in Exercise 1, draw its graph or directed graph, as appropriate.
3. Follow the directions of Exercise 1 for these relations on $\{-2, -1, 0, 1, 2\}$.
   (a) $R_1 = \{(m, n) \mid m + n \le 6\}$ (b) $R_2 = \{(m, n) \mid n = m + 1\}$
   (c) $R_3 = \{(m, n) \mid m \le n^2\}$ (d) $R_4 = \{(m, n) \mid |m| = |n|\}$
4. For each of the relations in Exercise 3, draw its graph or directed graph, as appropriate.

5. For each of the following relations on $\mathbb{N} - \{1\}$, determine which of the properties (i) reflexive, (ii) irreflexive, (iii) symmetric, (iv) antisymmetric, and (v) transitive are satisfied.
   (a) $R_1 = \{(m, n) \mid \gcd(m, n) = 1\}$
   (b) $R_2 = \{(m, n) \mid \gcd(m, n) > 1\}$
   (c) $R_3 = \{(m, n) \mid m$ divides $n\}$
   (d) $R_4 = \{(m, n) \mid m + n$ is even$\}$
   (e) $R_5 = \{(m, n) \mid m + n$ is odd$\}$
   (f) $R_6 = \{(m, n) \mid mn$ is odd$\}$

6. Let R be a relation on a finite set $A = \{a_1, a_2, \ldots, a_n\}$. Characterize the transitive property in terms of the matrix and directed graph of R.

7. For each of these relations on $\mathbb{Z} - \{0\}$, determine which of the properties (i) reflexive, (ii) irreflexive, (iii) symmetric, (iv) antisymmetric, and (v) transitive are satisfied.
   (a) $R_1 = \{(m, n) \mid m \leq n\}$
   (b) $R_2 = \{(m, n) \mid m - n$ is even$\}$
   (c) $R_3 = \{(m, n) \mid m$ divides $n\}$
   (d) $R_4 = \{(m, n) \mid m - n$ is odd$\}$
   (e) $R_5 = \{(m, n) \mid 3$ divides $m - n\}$
   (f) $R_6 = \{(m, n) \mid mn > 0\}$

8. Let X be a nonempty set and define the relation R on the power set of X by $A \, R \, B \leftrightarrow A \cap B = \emptyset$. Clearly, R is symmetric.
   (a) Explain why R is neither reflexive nor irreflexive.
   (b) Show that R is not transitive.
   (c) Let $X = \{1, 2, 3\}$; draw the graph of R.

9. Each part of Figure 4.8 shows the directed graph of a relation on $\{1, 2, 3, 4\}$. For each of these relations, determine which of the properties (i) reflexive, (ii) irreflexive, (iii) symmetric, (iv) antisymmetric, and (v) transitive it satisfies.

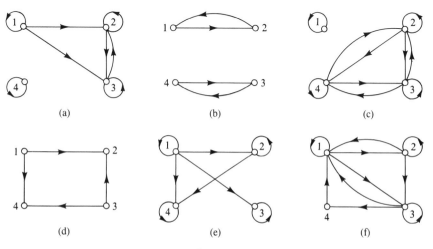

(a)   (b)   (c)

(d)   (e)   (f)

Figure 4.8

10. Let R_1 and R_2 be relations on the nonempty set A. Then each of $R_1 \cap R_2$, $R_1 \cup R_2$, and $R_1 - R_2$ is also a relation on A. (Why?) Indicate the truth-value of each of the following statements. (Optional: If the statement is true, prove it; if it is false, give a counterexample.)
   (a) If both R_1 and R_2 are reflexive, then $R_1 \cap R_2$ is reflexive.
   (b) If both R_1 and R_2 are reflexive, then $R_1 \cup R_2$ is reflexive.
   (c) If both R_1 and R_2 are reflexive, then $R_1 - R_2$ is irreflexive.
   (d) If both R_1 and R_2 are irreflexive, then $R_1 \cap R_2$ is irreflexive.
   (e) If both R_1 and R_2 are irreflexive, then $R_1 \cup R_2$ is irreflexive.
   (f) If both R_1 and R_2 are irreflexive, then $R_1 - R_2$ is irreflexive.
   (g) If both R_1 and R_2 are symmetric, then $R_1 \cap R_2$ is symmetric.
   (h) If both R_1 and R_2 are symmetric, then $R_1 \cup R_2$ is symmetric.
   (i) If both R_1 and R_2 are symmetric, then $R_1 - R_2$ is symmetric.
   (j) If both R_1 and R_2 are antisymmetric, then $R_1 \cap R_2$ is antisymmetric.
   (k) If both R_1 and R_2 are antisymmetric, then $R_1 \cup R_2$ is antisymmetric.
   (l) If both R_1 and R_2 are antisymmetric, then $R_1 - R_2$ is antisymmetric.
   (m) If both R_1 and R_2 are transitive, then $R_1 \cap R_2$ is transitive.
   (n) If both R_1 and R_2 are transitive, then $R_1 \cup R_2$ is transitive.
   (o) If both R_1 and R_2 are transitive, then $R_1 - R_2$ is transitive.

4.3 EQUIVALENCE RELATIONS

A relation R on the set A may have one or more of the properties of being reflexive, irreflexive, symmetric, antisymmetric, or transitive. Certain combinations of these properties lead to some important kinds of relations for which a general theory has been developed. For example, each of the following relations is reflexive, symmetric, and transitive:

1. "equals" on $\mathbb{Z}$
2. "having the same birthday" on a set of people
3. "congruence" on the set of triangles in a plane

DEFINITION 4.4

A relation R on a set A is called an *equivalence relation* provided it is reflexive, symmetric, and transitive.

Example 4.16 Let C be the set of currently enrolled undergraduate computer science majors at a certain university. Suppose that two students are related provided they have the same faculty advisor. Then it is easy to verify that this relation is reflexive, symmetric, and transitive and hence is an equivalence relation. ∎

Example 4.17 Let $S = \{1, 2, 3\}$. Determine which of the following relations on S are equivalence relations.

        (a) $R_1 = \{(1,2), (1,3), (2,1), (2,3)\ (3,1), (3,2)\}$
        (b) $R_2 = \{(1,1), (1,2), (2,1), (2,2), (2,3), (3,2), (3,3)\}$
        (c) $R_3 = \{(1,1), (1,2), (1,3), (2,2), (2,3), (3,3)\}$
        (d) $R_4 = \{(1,1), (1,2), (2,1), (2,2), (3,3)\}$

Solution   (a) It is clear that R_1 is symmetric and not reflexive. It is also not transitive. Let $a = 1$, $b = 2$, and $c = 1$ in the definition. Then we have $a R_1 b$ and $b R_1 c$, but a is not related to c under R_1.

    (b) It is evident that R_2 is both reflexive and symmetric. However, R_2 is not transitive because 1 is related to 2 and 2 is related to 3, but 1 is not related to 3.

    (c) The relation R_3 is reflexive and transitive but not symmetric.

    (d) The relation R_4 is reflexive, symmetric, and transitive.

    Hence, of these four relations, only R_4 is an equivalence relation on S. ∎

Consider the relation R of Example 4.11, where for integers a and b, we defined

$$a R b \leftrightarrow 3 \mid (a - b)$$

If $a R b$, then there is some integer n such that $3n = a - b$. Now suppose that $b \bmod 3 = r$ (recall that $r \in \{0, 1, 2\}$). Then there is some integer q such that $b = 3q + r$. Since $b = a - 3n$, we have

$$a - 3n = 3q + r$$

so

$$a = 3q + 3n + r$$
$$= 3(q + n) + r$$

This shows that $a \bmod 3 = r$. Hence, if $a R b$, then $a \bmod 3 = b \bmod 3$. In other words, a and b yield the same remainder upon division by 3. In this case we say that "a is congruent to b modulo 3" and, for this particular relation, represent $a R b$ with the notation $a \equiv b \pmod 3$. This relation is called *congruence modulo 3*. There is nothing special about the use of the integer 3 in this example; we can just as well use any positive integer m.

DEFINITION 4.5

Let m be any positive integer. Given integers a and b, we say that *a is congruent to b modulo m*, denoted $a \equiv b \pmod m$, provided $m \mid (a - b)$. This relation on $\mathbb{Z}$ is referred to as *congruence modulo m*.

THEOREM 4.1   For any $m \in \mathbb{N}$, the relation congruence modulo m is an equivalence relation on $\mathbb{Z}$.

We leave the proof of this theorem to Exercise 2. The student should note that, if a MOD $m = r$, then $a \equiv r(\text{mod } m)$, but not conversely. The relation of congruence modulo m gives rise to some interesting results in number theory and provides a number of very useful and fundamental examples in modern algebra.

Consider once again the relation congruence modulo 3 on $\mathbb{Z}$. The very use of the word *relation* suggests asking for the set of all "relatives" of a given fixed integer t; that is, we ask for the set of all those integers a for which $a \equiv t(\text{mod } 3)$. For any integer a, it has already been noted that a MOD 3 is exactly one of 0, 1, or 2, so it seems reasonable to consider the set of relatives of each of 0, 1, and 2. In particular, let

$$[0] = \{ a \in \mathbb{Z} \mid a \equiv 0(\text{mod } 3) \}$$

$$[1] = \{ a \in \mathbb{Z} \mid a \equiv 1(\text{mod } 3) \}$$

$$[2] = \{ a \in \mathbb{Z} \mid a \equiv 2(\text{mod } 3) \}$$

What are the elements of [0]? This is easily determined in the following string:

$$a \in [0] \leftrightarrow a \equiv 0(\text{mod } 3) \leftrightarrow 3 \mid a \leftrightarrow a = 3k \quad \text{for some } k \in \mathbb{Z}$$

Thus we see that $[0] = \{3k \mid k \in \mathbb{Z}\}$. In a similar fashion one obtains

$$[1] = \{3k + 1 \mid k \in \mathbb{Z}\} \quad \text{and} \quad [2] = \{3k + 2 \mid k \in \mathbb{Z}\}$$

Here are some relevant facts that should be observed about the sets [0], [1], and [2]:

1. [0], [1], and [2] are each nonempty.
2. [0], [1], and [2] are pairwise disjoint.
3. $[0] \cup [1] \cup [2] = \mathbb{Z}$.

Facts 2 and 3 follow from the division algorithm, since each $a \in \mathbb{Z}$ is uniquely expressible in the form $a = 3q + r$, where $r = 0$, 1, or 2. That is, each $a \in \mathbb{Z}$ belongs to exactly one of the sets [0], [1], or [2].

The preceding discussion can be applied to any equivalence relation.

DEFINITION 4.6

Let A be a nonempty set and let R be an equivalence relation on A. For each $a \in A$, the *equivalence class* of a is the set

$$[a] = \{ x \in A \mid x R a \}$$

Example 4.18 Consider again the equivalence relation of Example 4.16. Let F be the set of computer science faculty advisors, and for $f \in F$, let C_f denote the set of students in C who have f as an advisor. If $c \in C$ is a student, then $[c]$ is the set of students in C who have the same advisor as c; thus, if c has f for an advisor, then $[c] = C_f$. It follows that the sets $C_f, f \in F$, are the equivalence classes of this relation. Note that, since each student has a unique advisor, the sets C_f are pairwise disjoint and

$$\bigcup_{f \in F} C_f = C$$

Example 4.19 Find the equivalence classes of each of the following relations:

(a) the relation R_4 of Example 4.17
(b) the relation R_5 defined on $S = \{1, 2, 3, 4, 5\}$ by

$$R_5 = \{(1,1), (1,2), (1,3), (2,1), (2,2), (2,3), (3,1),$$
$$(3,2), (3,3), (4,4), (4,5), (5,4), (5,5)\}$$

(c) the relation R_6 (of Exercise 7(f) in Exercises 4.2) defined on $\mathbb{Z} - \{0\}$ by

$$R_6 = \{(m, n) \mid mn > 0\}$$

Solution   (a) For the equivalence relation R_4 of Example 4.17, observe that

$$[1] = [2] = \{1, 2\} \quad \text{and} \quad [3] = \{3\}$$

(b) One can verify in a routine manner that R_5 is an equivalence relation on S, with $[1] = [2] = [3] = \{1, 2, 3\}$ and $[4] = [5] = \{4, 5\}$.
(c) The relation R_6 is an equivalence relation with

$$[1] = [2] = \cdots = \mathbb{N} \quad \text{and} \quad [-1] = [-2] = \cdots = \{-1, -2, -3, \dots\} \quad \blacksquare$$

In each part of Example 4.19 we are given an equivalence relation R on a set A and we obtain (from the equivalence classes) a pairwise disjoint collection of nonempty subsets of A whose union is A. Such a collection is given a special name.

DEFINITION 4.7

Let A be a nonempty set. A set $\mathscr{P}$ of subsets of A is called a *partition* of A provided the following conditions hold:

1. $B \in \mathscr{P} \to B \neq \emptyset$
2. $B, C \in \mathscr{P} \to (B = C \vee B \cap C = \emptyset)$
3. $\bigcup_{B \in \mathscr{P}} B = A$

In this definition, conditions 1 and 2 say that $\mathscr{P}$ is a pairwise disjoint collection of nonempty subsets of A. Condition 3 says that the union of the subsets in $\mathscr{P}$ is A.

Example 4.20
(a) The relation congruence modulo 3 on $\mathbb{Z}$ yields the partition $\mathscr{P} = \{[0], [1], [2]\}$ of $\mathbb{Z}$, where $[r] = \{3k + r \mid k \in \mathbb{Z}\}$.
(b) The relation R_5 of Example 4.19(b) yields the partition $\mathscr{P} = \{\{1, 2, 3\}, \{4, 5\}\}$ of $\{1, 2, 3, 4, 5\}$.
(c) The relation R_6 of Example 4.19(c) yields the partition of $\mathbb{Z} - \{0\}$ into the set of positive integers and the set of negative integers. ∎

Example 4.21
The set of computer science courses offered at a certain college can be partitioned according to the level of the course as follows:

$A = $ the set of 100-level courses $= \{\text{CS105}, \text{CS125}\}$

$B = $ the set of 200-level courses $= \{\text{CS205}, \text{CS260}, \text{CS261}, \text{CS265}\}$

$C = $ the set of 300-level courses $= \{\text{CS340}, \text{CS350}, \text{CS360}, \text{CS361}, \text{CS380}\}$

$D = $ the set of 400-level courses $= \{\text{CS400}, \text{CS450}, \text{CS460}, \text{CS480}\}$

We can define a relation R on the set of computer science courses by saying that two courses at the same level are related. Then it can easily be verified that R is an equivalence relation and that A, B, C, and D are the equivalence classes of R. ∎

In reading the definitions of equivalence relation and partition, it is difficult to see any strong connection between the two concepts. However, as we have seen in the preceding examples, it is clear that they are very much related. The next result reveals this relationship.

THEOREM 4.2   **Fundamental theorem on equivalence relations**

Let A be a nonempty set. If R is an equivalence relation on A, then the set

$$\mathscr{P} = \{[a] \mid a \in A\}$$

is a partition of A. Conversely, if $\mathscr{P}$ is a partition of A, then the relation R defined on A by

$$a R b \leftrightarrow \exists C \in \mathscr{P}(a \in C \wedge b \in C)$$

is an equivalence relation on A.

Proof   We first prove that if R is an equivalence relation on A, then the collection $\mathscr{P}$ as defined in the statement of the theorem is a partition of A. First, it is clear that $[a]$ is nonempty for every $a \in A$; in particular, $a \in [a]$, since R is reflexive. Next, suppose that $[a], [b] \in \mathscr{P}$ and that $[a]$ and $[b]$ are not disjoint; then we must show that $[a] = [b]$. This can be accomplished by showing both $[a] \subseteq [b]$ and $[b] \subseteq [a]$. Since the proofs of these inclusions are similar, we consider only the proof that $[a] \subseteq [b]$. Let $x \in [a]$. Since $[a] \cap [b] \neq \emptyset$, there is some element $c \in A$ such that $c \in [a]$ and $c \in [b]$. Since R is an equivalence relation, it follows that $x R a$, $a R c$, and $c R b$. But $x R a$ and $a R c$ imply that $x R c$, and then $x R c$ and $c R b$ imply that $x R b$. Therefore $x \in [b]$, showing that $[a] \subseteq [b]$. So $\mathscr{P}$ is a pairwise disjoint collection. Last, if $a \in A$, then $a \in [a]$, so that $A \subseteq \bigcup_{c \in A} [c]$, and it follows that $A = \bigcup_{c \in A} [c]$. Thus $\mathscr{P}$ is a partition of A.

Next, suppose $\mathscr{P}$ is a partition of A and the relation R is defined as in the statement of the theorem. Notice that two elements of A are related if and only if they belong to the same set in the partition $\mathscr{P}$. It must be shown that R is an equivalence relation. To see that R is reflexive, let $a \in A$. Since $\mathscr{P}$ is a partition of A, there is some $C \in \mathscr{P}$ such that $a \in C$. Thus $a R a$. For the symmetric property, let $a, b \in A$ and assume $a R b$. Then there is some $C \in \mathscr{P}$ such that $a \in C$ and $b \in C$. But this is also the condition for $b R a$, so R is symmetric. Finally, suppose, for some $a, b, c \in A$, that $a R b$ and $b R c$. Then there are sets $D, E \in \mathscr{P}$ such that $a, b \in D$ and $b, c \in E$. It follows that $b \in D \cap E$, so D and E are not disjoint. Since $\mathscr{P}$ is a partition, it must be that $D = E$ and hence that $a R c$. This shows that R is transitive. Therefore R is an equivalence relation. ∎

To repeat, if R is an equivalence relation on a set A, then the set

$$\mathscr{P} = \{[a] \mid a \in A\}$$

of equivalence classes of R is a partition of A. Conversely, given a partition $\mathscr{C}$ of a set A, the partition $\mathscr{C}$ determines an equivalence relation R on A, where $a R b$ if and only if a and b belong to the same set in $\mathscr{C}$. We say that $\mathscr{C}$ *induces* an equivalence relation on A. In fact, it should be pointed out that if R is an equivalence relation on a set A and $\mathscr{P} = \{[a] \mid a \in A\}$, then the equivalence relation induced on A by $\mathscr{P}$ is precisely R. Thus, in a sense, the notions of equivalence relation and partition are the same.

Example 4.22 For each $c \in \mathbb{R}$, let L_c denote the line in the xy-plane with equation $y = x + c$. Then the set $\mathscr{L} = \{L_c \mid c \in \mathbb{R}\}$ is a partition of $\mathbb{R}^2 = \mathbb{R} \times \mathbb{R}$. Determine the equivalence relation on $\mathbb{R}^2$ induced by $\mathscr{L}$.

Solution Let $\sim$ denote the equivalence relation induced on $\mathbb{R}^2$ by $\mathscr{L}$ and let (a, b) and (r, s) be elements of $\mathbb{R}^2$. Then, according to the above discussion, $(a, b) \sim (r, s)$ if and only if (a, b) and (r, s) are points of the line L_c for some $c \in \mathbb{R}$. But this happens if and only if $b - a = c = s - r$. Hence the relation $\sim$ on $\mathbb{R}^2$ is given by

$$(a, b) \sim (r, s) \leftrightarrow b - a = s - r$$

If one starts with the equivalence relation $\sim$ on $\mathbb{R}^2$, given by $(a, b) \sim (r, s) \leftrightarrow b - a = s - r$, then the set of equivalence classes of $\sim$ is precisely $\mathscr{L}$.

Example 4.23   Each part gives a partition of the set $A = \{1, 2, 3, 4, 5, 6\}$. Find the equivalence relation on A induced by that partition.

(a) $\mathscr{P}_1 = \{\{1, 2\}, \{3, 4\}, \{5, 6\}\}$
(b) $\mathscr{P}_2 = \{\{1\}, \{2\}, \{3, 4, 5, 6\}\}$
(c) $\mathscr{P}_3 = \{\{1, 2, 3\}, \{4, 5, 6\}\}$

Solution   (a) $R_1 = \{(1, 1), (1, 2), (2, 1), (2, 2), (3, 3), (3, 4), (4, 3), (4, 4), (5, 5), (5, 6), (6, 5), (6, 6)\}$

(b) $R_2 = \{(1, 1), (2, 2), (3, 3), (3, 4), (3, 5), (3, 6), (4, 3), (4, 4), (4, 5), (4, 6), (5, 3), (5, 4), (5, 5), (5, 6), (6, 3), (6, 4), (6, 5), (6, 6)\}$

(c) $R_3 = \{(1, 1), (1, 2), (1, 3), (2, 1), (2, 2), (2, 3), (3, 1), (3, 2), (3, 3), (4, 4), (4, 5), (4, 6), (5, 4), (5, 5), (5, 6), (6, 4), (6, 5), (6, 6)\}$ ∎

In the special case of congruence modulo m on $\mathbb{Z}$, the implied partition is $\{[a] \mid a \in \mathbb{Z}\}$, where $[a] = \{c \in \mathbb{Z} \mid c \equiv a \pmod{m}\}$. We shall call $[a]$ the *residue class* of a; here *residue* is a synonym for "remainder." One obvious question that arises is, What are the residue classes of congruence modulo m? For any $a \in \mathbb{Z}$, recall that there is a unique integer r, $0 \le r \le m - 1$, such that a MOD $m = r$. Hence $a \equiv r \pmod{m}$, and we have $a \in [r]$. Thus every integer belongs to exactly one of the residue classes $[0], [1], \ldots$, $[m - 1]$. It follows that these classes are pairwise disjoint and that their union is $\mathbb{Z}$. Thus $\{[0], [1], \ldots, [m - 1]\}$ is a partition of $\mathbb{Z}$.

DEFINITION 4.8

The set

$$\mathbb{Z}_m = \{[0], [1], \ldots, [m - 1]\}$$

is called the *set of residue classes modulo m*.

Given $r \in \mathbb{Z}$, where $0 \le r \le m - 1$, what do the elements of the residue class $[r]$ look like? We proceed as we did earlier for the special case $m = 3$:

$$c \in [r] \leftrightarrow c \equiv r \pmod{m}$$

$$\leftrightarrow m \mid (c - r)$$

$$\leftrightarrow c - r = mq \quad \text{for some } q \in \mathbb{Z}$$

$$\leftrightarrow c = mq + r \quad \text{for some } q \in \mathbb{Z}$$

Hence $[r] = \{mq + r \mid q \in \mathbb{Z}\}$. For example, if $m = 4$, then the residue classes modulo 4 are:

$$[0] = \{4q \mid q \in \mathbb{Z}\} = \{\ldots, -8, -4, 0, 4, 8, \ldots\}$$
$$[1] = \{4q + 1 \mid q \in \mathbb{Z}\} = \{\ldots, -7, -3, 1, 5, 9, \ldots\}$$
$$[2] = \{4q + 2 \mid q \in \mathbb{Z}\} = \{\ldots, -6, -2, 2, 6, 10, \ldots\}$$
$$[3] = \{4q + 3 \mid q \in \mathbb{Z}\} = \{\ldots, -5, -1, 3, 7, 11, \ldots\}$$

With regard to congruence modulo m, it is very important to keep in mind that $[a] = [b] \leftrightarrow m \mid (a - b)$. Thus when $m = 4$, for instance, we see that

$$\cdots = [-6] = [-2] = [2] = [6] = [10] = \cdots$$

Each of the numbers $-6, -2, 2, 6, 10$, and so on is called a *representative* of the residue class [2]. This is because we may use any one of these integers to refer to the class. In some situations it might be desirable to use $[-6]$ instead of [2]. In general, with respect to congruence modulo m, each of the integers $mq + r$, $q \in \mathbb{Z}$, is a representative of the class $[r]$.

Exercises 4.3

1. Determine which of the relations defined in Exercise 1 of Exercises 4.2 are equivalence relations. Find the equivalence classes of those that are.
2. Prove Theorem 4.1.
3. Determine which of the relations defined in Exercise 3 of Exercises 4.2 are equivalence relations. Find the equivalence classes of those that are.
4. Let $m \in \mathbb{N}$ and let $a, b \in \mathbb{Z}$. Show that $a \equiv b \pmod{m}$ if and only if $a \text{ MOD } m = b \text{ MOD } m$.
5. Determine which of the relations defined in Exercise 5 of Exercises 4.2 are equivalence relations. Find the equivalence classes of those that are.
6. Each part gives a set S and a relation $\sim$ on S. Verify that $\sim$ is an equivalence relation and describe the equivalence classes into which $\sim$ partitions S.
   (a) S is the set of all polygons in a plane; for $A, B \in S$, $A \sim B \leftrightarrow$ A has the same number of sides as B.
   (b) $S = \mathbb{Z}$; for $a, b \in S$, $a \sim b \leftrightarrow |a - 2| = |b - 2|$.
   (c) S is the set of alumni of your university; for $a, b \in S$, $a \sim b \leftrightarrow$ a graduated the same year as did b.
7. Determine which of the relations defined in Exercise 7 of Exercises 4.2 are equivalence relations. Find the equivalence classes of those that are.
8. Which of the residue classes $[-22], [-12], [-6], [-3], [-2], [3], [6], [8], [39]$, and $[44]$ are equal in the following?
   (a) $\mathbb{Z}_2$　　(b) $\mathbb{Z}_3$　　(c) $\mathbb{Z}_5$　　(d) $\mathbb{Z}_9$
9. Determine which of the relations defined in Exercise 9 of Exercises 4.2 are equivalence relations. Find the equivalence classes of those that are.
10. Let R_1 and R_2 be relations on the nonempty set A. Then each of $R_1 \cap R_2$, $R_1 \cup R_2$, and $R_1 - R_2$ is also a relation on A. (See Exercise

10 of Exercises 4.2.) Indicate the truth-value of each of the following statements. (Optional: If the statement is true, prove it; if it is false, give a counterexample.)

(a) If R_1 and R_2 are equivalence relations, then so is $R_1 \cap R_2$.

(b) If R_1 and R_2 are equivalence relations, then so is $R_1 \cup R_2$.

(c) If R_1 and R_2 are equivalence relations, then so is $R_1 - R_2$.

11. Let A be a nonempty set and let B be a fixed subset of A. Define the relation $\sim$ on $\mathscr{P}(A)$ by

$$C \sim D \leftrightarrow C \cap B = D \cap B$$

(where C and D are subsets of A).

(a) Show that $\sim$ is an equivalence relation.

(b) In particular, if $A = \{1, 2, 3, 4, 5\}$, $B = \{1, 2, 5\}$, and $C = \{2, 4, 5\}$, find $[C]$.

12. How many distinct residue classes are there for the relation of congruence modulo m on $\mathbb{Z}$; that is, what is $|\mathbb{Z}_m|$?

13. Each part gives a partition of the set $A = \{1, 2, 3, 4, 5, 6\}$. Find the equivalence relation on A induced by that partition.

(a) $\mathscr{P}_1 = \{\{1\}, \{2\}, \{3\}, \{4\}, \{5\}, \{6\}\}$

(b) $\mathscr{P}_2 = \{\{1\}, \{2, 3\}, \{4, 5, 6\}\}$

(c) $\mathscr{P}_3 = \{\{1, 3, 5\}, \{2, 4, 6\}\}$

(d) $\mathscr{P}_4 = \{\{1, 2, 3, 4, 5, 6\}\}$

14. Let S denote the set of all points in the xy-plane, excluding the origin. For $r \in (0, \infty)$, let C_r denote the set of all points in S on the circle of radius r, centered at the origin. Let $\mathscr{C} = \{C_r \mid r \in (0, \infty)\}$.

(a) Give the equation of C_r.

(b) Show that $\mathscr{C}$ is a partition of S.

Let $\sim$ denote the equivalence relation on S induced by $\mathscr{C}$.

(c) Is $(2, 4) \sim (4, -2)$?   (d) Is $(3, 4) \sim (0, -5)$?

(e) Is $(2, 3) \sim (1, 4)$?

(f) State an analytic condition under which $(a, b) \sim (c, d)$.

15. Determine the number of equivalence relations on:

(a) $\{1\}$ (b) $\{1, 2\}$ (c) $\{1, 2, 3\}$ (d) $\{1, 2, 3, 4\}$

16. In Exercise 10 you observed that if R_1 and R_2 are both equivalence relations on a nonempty set A, then $R_1 \cap R_2$ is an equivalence relation on A. How are the equivalence classes for $R_1 \cap R_2$ determined from the equivalence classes for R_1 and those for R_2?

4.4   PARTIAL-ORDER RELATIONS

Consider the relation "is a subset of" or simply "$\subseteq$" on the power set $\mathscr{P}(U)$, where $U = \{1, 2, \ldots, 10\}$. This relation is reflexive and transitive but is clearly not symmetric. In fact, it is antisymmetric since, for any subsets A and B of U, $(A \subseteq B \wedge B \subseteq A) \rightarrow A = B$. Relations that are reflexive, antisymmetric, and transitive play an important role in mathematics and computer science and are addressed in this section.

DEFINITION 4.9

A relation R on a set A is called a *partial-order relation* provided R is reflexive, antisymmetric, and transitive. We also refer to R as a *partial ordering* of A and call (A, R) a *partially ordered set* (or *poset*).

Example 4.24 Consider the relation "divides" on $\mathbb{N}$. Show that divides is a partial ordering of $\mathbb{N}$. Is divides a partial ordering of $\mathbb{Z}$?

Solution We must verify that divides on $\mathbb{N}$ is reflexive, antisymmetric, and transitive. The reflexive property is obvious, and antisymmetry and transitivity follow from Theorem 3.2, parts 4 and 2, respectively. The relation divides is not a partial ordering of $\mathbb{Z}$ because it is not antisymmetric; for example, $-2 \mid 2$ and $2 \mid -2$, but $-2 \neq 2$. ∎

Example 4.25 Define the relation $\preccurlyeq$ on $A = [0, \infty) \times [0, \infty)$ by

$$(a, b) \preccurlyeq (c, d) \leftrightarrow (a^2 + b^2 \leq c^2 + d^2 \text{ and } b \leq d)$$

Show that $\preccurlyeq$ is a partial-order relation on A.

Solution Given any $(a, b) \in A$, the two inequalities $a^2 + b^2 \leq a^2 + b^2$ and $b \leq b$ are certainly valid, so $\preccurlyeq$ is reflexive. Suppose next that both

$$(a, b) \preccurlyeq (c, d) \quad \text{and} \quad (c, d) \preccurlyeq (a, b)$$

hold. Then we have the following inequalities:

$$a^2 + b^2 \leq c^2 + d^2 \qquad b \leq d \qquad c^2 + d^2 \leq a^2 + b^2 \qquad d \leq b$$

Thus we see immediately that $b = d$. This means that $a^2 \leq c^2$ and $c^2 \leq a^2$. Thus $a^2 = c^2$ and, since a and c are nonnegative real numbers, it follows that $a = c$. So $\preccurlyeq$ is antisymmetric. We leave it for the student to demonstrate that $\preccurlyeq$ is transitive as well. ∎

Perhaps the partial ordering most familiar to the student is the standard ordering "less than or equal to" or " $\leq$ " on the set of real numbers. It is the prototype of a partial-order relation and, for this reason, it is common to use the symbol $\preccurlyeq$ to denote a general partial ordering.

Some authors use the term *ordering* of a set A to mean a relation $\preccurlyeq$ on A that is transitive. If such a relation is also antisymmetric, then given $a \preccurlyeq b$ and $a \neq b$, we write "$a \prec b$" and say "a is less than b" (or "a precedes b"). The term *partial* is used to describe an ordering in which not every two distinct elements are necessarily related.

Example 4.26 Consider a set A on which a relation $\prec$ is defined that is irreflexive and transitive. If $a \preccurlyeq b$ is defined to mean that

$$a \prec b \quad \text{or} \quad a = b$$

show that $\preccurlyeq$ is a partial-order relation on A.

Solution   Since the statement $a \prec a$ or $a = a$ is clearly true for all $a \in A$, we see that $\preccurlyeq$ is reflexive. Next suppose that $a \preccurlyeq b$ and $b \preccurlyeq a$ both hold for some $a, b \in A$. Then the statements ($a \prec b$ or $a = b$) and ($b \prec a$ or $b = a$) both hold. If $a \prec b$ and $b \prec a$ were true, then the transitivity of $\prec$ would imply $a \prec a$, contradicting the irreflexive condition on $\prec$. Thus we must have $a = b$, showing that $\preccurlyeq$ is antisymmetric. Showing transitivity is left to Exercise 18. ∎

In Section 4.2 we discussed the use of a directed graph as a means of representing a relation on a finite set A. In case $(A, \preccurlyeq)$ is a poset, this method of representation can be simplified. First, since a partial ordering is understood to be reflexive, we omit the loop at each vertex. Second, we adopt the convention of omitting any directed edge that is implied by transitivity; in other words, there is a directed edge from a to b if and only if $a \prec b$ and there is no $c \in A$ with $a \prec c$ and $c \prec b$. Finally, since A is a poset, it is possible to construct its directed graph so that if $a \prec b$, then the vertex corresponding to b lies above the vertex corresponding to a. For this reason, we agree to use undirected edges, understanding that the orientation of all edges is from bottom to top. The resulting representation is then a graph and is called the *Hasse diagram* of the poset.

Example 4.27   (a) It follows from Example 4.12 that if U is any nonempty set, then $(\mathscr{P}(U), \subseteq)$ is a poset. The Hasse diagram of the poset $(\mathscr{P}(\{1, 2, 3\}), \subseteq)$ is shown in Figure 4.9(a). Note in this case that the Hasse diagram is similar to the subset diagram of $\mathscr{P}(\{1, 2, 3\})$.

(b) Let A denote the set of positive divisors of 24 and let $|$ denote the relation divides. Using the same arguments given in Example 4.24, we can readily check that $(A, |)$ is a poset. The Hasse diagram of $(A, |)$ is given in Figure 4.9(b).

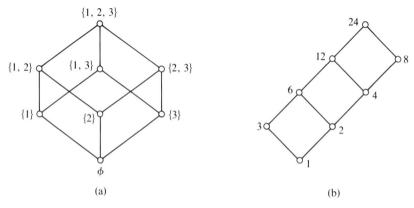

(a)   (b)

Figure 4.9 Two Hasse diagrams

DEFINITION 4.10

Let $(A, \preccurlyeq)$ be a poset and let $B \subseteq A$.

1. An element $x \in A$ is called a *lower bound* for B provided $x \preccurlyeq y$ for all $y \in B$. An element $z \in A$ is called the *greatest lower bound* for B provided the following hold:
   (a) z is a lower bound for B.
   (b) If x is any lower bound for B, then $x \preccurlyeq z$.
   We write $z = \text{glb}(B)$.
2. An element $x \in A$ is called an *upper bound* for B provided $y \preccurlyeq x$ for all $y \in B$. An element $z \in A$ is called the *least upper bound* for B provided the following hold:
   (a) z is an upper bound for B.
   (b) If x is any upper bound for B, then $z \preccurlyeq x$.
   We write $z = \text{lub}(B)$.

It should be noted, in the context of Definition 4.10, that $\text{glb}(B)$ and $\text{lub}(B)$, should they exist, are uniquely determined. This is suggested by the language used in the definition. For example, suppose that both z_1 and z_2 satisfy the conditions for $\text{glb}(B)$. Then both are lower bounds for B and, by part (b), we have both $z_1 \preccurlyeq z_2$ and $z_2 \preccurlyeq z_1$. Since $\preccurlyeq$ is antisymmetric, it follows that $z_1 = z_2$. So if $\text{glb}(B)$ exists, it is unique.

Example 4.28 Consider the poset whose Hasse diagram is given in Figure 4.10. Let $B = \{g, h, i\}$, $C = \{g, h\}$, and $D = \{e, f\}$. Find (if it exists):

(a) $\text{lub}(B)$ (b) $\text{lub}(C)$ (c) $\text{lub}(D)$ (d) $\text{glb}(D)$

Solution

(a) The set of upper bounds for B is $\{b, e\}$, and since $e \prec b$, we have that $e = \text{lub}(B)$.
(b) The set of upper bounds for C is $\{a, b, d, e\}$, but $\text{lub}(C)$ does not exist, since d and e are not comparable.
(c) The set D has no upper bounds, so $\text{lub}(D)$ does not exist.
(d) The elements j and k are the only lower bounds for D; since $k \prec j$, it follows that $j = \text{glb}(D)$.

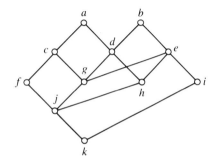

Figure 4.10 The poset in Example 4.10   ■

Example 4.29   Consider again the poset $(\mathscr{P}(U), \subseteq)$, where U is a nonempty set, and let $\mathscr{S}$ be a nonempty subset of $\mathscr{P}(U)$. Find:

(a) glb($\mathscr{S}$)   (b) lub($\mathscr{S}$)

Solution   (a) If C is a lower bound for $\mathscr{S}$, then $C \subseteq X$ for every $X \in \mathscr{S}$. But then $C \subseteq S$, where

$$S = \bigcap_{X \in \mathscr{S}} X$$

It is also the case that S is a lower bound for $\mathscr{S}$. Therefore glb($\mathscr{S}$) = S. In words, the greatest lower bound for the collection $\mathscr{S}$ is the intersection of all the sets in $\mathscr{S}$.

(b) We leave it to the student to show, in a similar fashion, that the least upper bound for the collection $\mathscr{S}$ is the union of all the sets in $\mathscr{S}$:

$$\text{lub}(\mathscr{S}) = \bigcup_{X \in \mathscr{S}} X \qquad\blacksquare$$

Notice that in the poset $(\mathscr{P}(U), \subseteq)$, every subset $\mathscr{S}$ of $\mathscr{P}(U)$ has both a least upper bound and a greatest lower bound. Posets in which every pair of distinct elements has both a greatest lower bound and a least upper bound are of special interest.

DEFINITION 4.11

If every two elements of a poset $(A, \preccurlyeq)$ possess both a greatest lower bound and a least upper bound, then the poset is called a *lattice*. Given x and y in A, we call glb($\{x, y\}$) the *meet* of x and y and denote it by $x \wedge y$. We call lub($\{x, y\}$) the *join* of x and y and denote it by $x \vee y$.

Thus it follows that the poset $(\mathscr{P}(U), \subseteq)$ is a lattice; for $X, Y \in \mathscr{P}(U)$,

$$X \wedge Y = X \cap Y \quad \text{and} \quad X \vee Y = X \cup Y$$

Example 4.30   Consider the set $\mathbb{N}$ of positive integers under the relation divides. If e is a lower bound of a and b, then $e \mid a$ and $e \mid b$, so e is a common divisor of a and b. Hence the meet of a and b is the greatest common divisor of a and b; that is,

$$a \wedge b = \gcd(a, b)$$

Similarly, if c is an upper bound of a and b, then $a \mid c$ and $b \mid c$, so c is a common multiple of a and b. It follows that the join of a and b is the least common multiple (see Problem 18 of Chapter 3 Problems) of a and b; that is,

$$a \vee b = \text{lcm}(a, b)$$

Since, for any two positive integers a and b, both $\gcd(a, b)$ and $\text{lcm}(a, b)$ exist, we may conclude that the poset $(\mathbb{N}, |)$ is a lattice.   ■

Example 4.31   It can be verified that the poset whose Hasse diagram is shown in Figure 4.11 is a lattice. For example, $b \vee c = a$, $b \wedge c = d$, $b \vee e = a$, and $b \wedge e = f$. It is left to Exercise 12 to find $x \vee y$ and $x \wedge y$ for the remaining values of x and y in $\{a, b, c, d, e, f\}$.   ■

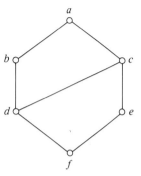

Figure 4.11

Example 4.32   Let $\mathcal{M}_n$ denote the set of $n \times n$ matrices with entries that are 0 or 1. Define a relation $\preccurlyeq$ on $\mathcal{M}_n$ as follows: for $A, B \in \mathcal{M}_n$, say $A = [a_{ij}]$ and $B = [b_{ij}]$, we define $A \preccurlyeq B$ provided $a_{ij} \leq b_{ij}$ for $1 \leq i \leq n$ and $1 \leq j \leq n$. Here we understand that 0 and 1 satisfy the natural order properties $0 \leq 0$, $0 < 1$, and $1 \leq 1$. It can be shown that $(\mathcal{M}_n, \preccurlyeq)$ is a lattice; do so in the case $n = 4$.

Solution   We leave it to the student to show that $(\mathcal{M}_4, \preccurlyeq)$ is a partially ordered set. To see that $(\mathcal{M}_4, \preccurlyeq)$ is a lattice, let A and B be any two elements of $\mathcal{M}_4$, say $A = [a_{ij}]$ and $B = [b_{ij}]$. If C is to serve as the join of A and B, then necessarily $A \preccurlyeq C$ and $B \preccurlyeq C$. Moreover, if $A \preccurlyeq X$ and $B \preccurlyeq X$ for some $X \in \mathcal{M}_4$, then necessarily $C \preccurlyeq X$. Thus if a given entry of A (or B) is 1, then the corresponding entry of C must also be 1. In addition, if the (i, j)-entries of A and B are both 0, then the (i, j)-entry of C must be 0. Hence C is the 4×4 matrix $[c_{ij}]$, where

$$c_{ij} = \max\{a_{ij}, b_{ij}\}$$

for $1 \leq i \leq 4$ and $1 \leq j \leq 4$. For example, the join of the matrices

$$A_1 = \begin{bmatrix} 1 & 0 & 0 & 1 \\ 0 & 1 & 0 & 0 \\ 1 & 1 & 0 & 0 \\ 0 & 0 & 1 & 0 \end{bmatrix} \quad \text{and} \quad B_1 = \begin{bmatrix} 0 & 1 & 1 & 0 \\ 0 & 0 & 1 & 1 \\ 1 & 1 & 1 & 0 \\ 0 & 0 & 0 & 1 \end{bmatrix}$$

is the matrix

$$C_1 = \begin{bmatrix} 1 & 1 & 1 & 1 \\ 0 & 1 & 1 & 1 \\ 1 & 1 & 1 & 0 \\ 0 & 0 & 1 & 1 \end{bmatrix}$$

If D is to be the meet of A and B, then we must have that both $D \preccurlyeq A$ and $D \preccurlyeq B$. Moreover, if $X \preccurlyeq A$ and $X \preccurlyeq B$ for some $X \in \mathcal{M}_4$, then we require $X \preccurlyeq D$. Thus if $D = [d_{ij}]$, then $d_{ij} \le a_{ij}$ and $d_{ij} \le b_{ij}$ for $1 \le i \le 4$ and $1 \le j \le 4$. Incorporating both required conditions, we see that

$$d_{ij} = \min\{a_{ij}, b_{ij}\}$$

For example, the meet of the matrices A_1 and B_1 is the matrix

$$D_1 = \begin{bmatrix} 0 & 0 & 0 & 0 \\ 0 & 0 & 0 & 0 \\ 1 & 1 & 0 & 0 \\ 0 & 0 & 0 & 0 \end{bmatrix}$$

■

As mentioned before, not every two elements in a poset are necessarily comparable. In $(\mathbb{N}, |)$, for example, neither of the integers 3 nor 8 divides the other; hence 3 and 8 are not comparable in this poset. However, in $(\mathbb{N}, \le)$ (here $\le$ has the usual interpretation "less than or equal to"), if $a, b \in \mathbb{N}$, then we know that either $a \le b$ or $b \le a$. This brings us to the next definition.

DEFINITION 4.12

A partial-order relation $\preccurlyeq$ on a set A is called a *total ordering* provided that, for any two elements $a, b \in A$, either $a \preccurlyeq b$ or $b \preccurlyeq a$.

We remark that if $\preccurlyeq$ is a total ordering on the set A, then the condition given in Definition 4.12 can be restated as follows:

For any two elements $a, b \in A$, precisely one of $a = b$, $a \prec b$, or $b \prec a$ holds.

This is sometimes called the *law of trichotomy*. A set A, together with a total ordering $\preccurlyeq$, is called a *totally ordered set*. In some settings a total ordering is also called a *linear*, *simple*, *complete*, or *full ordering*.

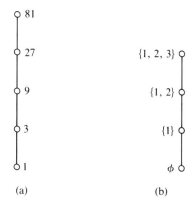

Figure 4.12 Totally ordered sets

Example 4.33

(a) Let A be the set of positive divisors of 24. The poset $(A, |)$, as shown in Figure 4.9(b), is not totally ordered; for example, 4 and 6 are not comparable.

(b) Let B be the set of positive divisors of 81. The poset $(B, |)$, as shown in Figure 4.12(a), is a totally ordered set.

(c) As shown in Figure 4.9(a), the poset $(\mathcal{P}(\{1, 2, 3\}), \subseteq)$ is not a totally ordered set.

(d) Let $\mathcal{B} = \{\emptyset, \{1\}, \{1, 2\}, \{1, 2, 3\}\}$. Then $(\mathcal{B}, \subseteq)$ is a totally ordered set; its Hasse diagram is shown in Figure 4.12(b). ∎

Exercises 4.4

1. Each part gives a relation on $A = \{a, b, c, d, e, f\}$. Determine whether the relation is a partial ordering. If it is, draw its Hasse diagram.

   (a) $R_1 = \{(f, f), (f, e), (f, d), (f, c), (f, b), (f, a), (e, e), (e, c), (e, a), (d, d), (d, b), (d, a), (c, c), (c, a), (b, b), (b, a), (a, a)\}$

   (b) $R_2 = \{(f, f), (f, d), (f, c), (f, b), (f, a), (e, e), (e, d), (e, c), (e, b), (e, a), (d, d), (d, c), (d, b), (d, a), (c, c), (c, b), (c, a), (b, b), (a, a)\}$

   (c) $R_3 = R_1 \cup \{(e, b), (d, c)\}$

   (d) $R_4 = \{(f, f), (f, e), (f, d), (f, c), (f, b), (f, a), (e, e), (e, d), (e, c), (e, b), (e, a), (d, d), (d, b), (d, a), (c, c), (c, b), (c, a), (b, b), (b, a), (a, a)\}$

   (e) $R_5 = \{(f, d), (f, b), (e, c), (e, a), (d, b), (c, a)\}$

   (f) $R_6 = \{(f, f), (e, e), (d, d), (c, c), (b, b), (a, a)\}$

   (g) $R_7 = R_5 \cup R_6$ **(h)** $R_8 = R_7 \cup \{(d, c)\}$

   (i) $R_9 = R_8 \cup \{(f, c)\}$ **(j)** $R_{10} = R_9 \cup \{(a, c)\}$

2. If $(A, \preceq)$ is a poset and $B \subseteq A$, show that $(B, \preceq)$ is a poset. We call $(B, \preceq)$ a *subposet* of $(A, \preceq)$.

3. Determine which of the relations defined in Exercise 9 of Exercises 4.2 are partial-order relations. For each such relation, draw the Hasse diagram.

4. Each part gives a subset S of $\mathbb{N}$. Draw the Hasse diagram of the (sub)poset $(S, |)$.
   - **(a)** $S = \{n \mid n \text{ divides } 30\}$
   - **(b)** $S = \{n \mid n \text{ divides } 45\}$
   - **(c)** $S = \{n \mid n \text{ divides } 36\}$
   - **(d)** $S = \{n \mid n \text{ divides } 36 \text{ or } n \text{ divides } 45\}$

5. Define the relation $\prec$ on $\mathbb{Z} - \{0\}$ by

$$m \prec n \leftrightarrow 2m \mid n$$

   - **(a)** Show that $\prec$ is irreflexive and transitive.
   - **(b)** Use the result of Example 4.26 to define a partial-order relation $\preccurlyeq$ on $\mathbb{Z} - \{0\}$.

6. Let n_0 be a fixed positive integer, and let $S_{n_0} = \{n \in \mathbb{N} \mid n \text{ divides } n_0\}$. Show that $(S_{n_0}, |)$ is a lattice. For $n_1, n_2 \in S$, what are $n_1 \wedge n_2$ and $n_1 \vee n_2$?

7. Define the relation $\preccurlyeq$ on $\mathbb{Z} \times \mathbb{Z}$ by

$$(a, b) \preccurlyeq (c, d) \leftrightarrow a \le c \text{ and } b \le d$$

   - **(a)** Show that $\preccurlyeq$ is a partial ordering of $\mathbb{Z} \times \mathbb{Z}$.
   - **(b)** Find glb($\{(-1, 4), (3, 2)\}$) and lub($\{(-1, 4), (3, 2)\}$).
   - **(c)** It turns out that $(\mathbb{Z} \times \mathbb{Z}, \preccurlyeq)$ is a lattice. For $(a, b), (c, d) \in \mathbb{Z}$, what are $(a, b) \wedge (c, d)$ and $(a, b) \vee (c, d)$?
   - **(d)** It also turns out that if $S \subseteq \mathbb{Z}$, then $(S \times S, \preccurlyeq)$ is a lattice. For $S = \{1, 2, 3\}$, draw the Hasse diagram of the lattice $(S \times S, \preccurlyeq)$.

8. Prove or disprove: If $(A, \preccurlyeq)$ is a lattice and B is a subset of A, then $(B, \preccurlyeq)$ is a lattice.

9. Define the relation $\preccurlyeq$ on $\mathbb{Z} \times \mathbb{Z}$ by

$$(a, b) \preccurlyeq (c, d) \leftrightarrow [a < c \text{ or } (a = c \text{ and } b \le d)]$$

   - **(a)** Show that $\preccurlyeq$ is a partial ordering of $\mathbb{Z} \times \mathbb{Z}$.
   - **(b)** Find glb($\{(-1, 4), (3, 2)\}$) and lub($\{(-1, 4), (3, 2)\}$).
   - **(c)** Find glb($\{(-1, 4), (-1, 7)\}$) and lub($\{(-1, 4), (-1, 7)\}$).
   - **(d)** Show that $\preccurlyeq$ is a total ordering of $\mathbb{Z} \times \mathbb{Z}$.
   - **(e)** For $S = \{1, 2, 3\}$, draw the Hasse diagram of the subposet $(S \times S, \preccurlyeq)$.

10. Prove or disprove: If $(A, \preccurlyeq)$ is a totally ordered set and B is a subset of A, then $(B, \preccurlyeq)$ is a totally ordered set.

11. Consider the poset $(\mathbb{Z} - \{0\}, \preccurlyeq)$, as defined in Exercise 5.
   - **(a)** Let $S_{24} = \{n \in \mathbb{N} \mid n \text{ divides } 24\}$. Draw the Hasse diagram of $(S_{24}, \preccurlyeq)$.
   - **(b)** Is $(S_{24}, \preccurlyeq)$ a lattice?
     Consider the subposet $(S, \preccurlyeq)$, where $S = \{1\} \cup \{2, 4, 6, 8, \dots\}$. It can be shown that $(S, \preccurlyeq)$ is a lattice.
   - **(c)** Find $16 \wedge 24$ and $16 \vee 24$.
   - **(d)** Find $20 \wedge 30$ and $20 \vee 30$.
   - **★(e)** For $m, n \in S$, what are $m \wedge n$ and $m \vee n$?

12. Complete Example 4.31 by finding $x \vee y$ and $x \wedge y$ for the remaining values of x and y in $\{a, b, c, d, e, f\}$.

13. Prove: If $(A, \preccurlyeq)$ is a totally ordered set, then $(A, \preccurlyeq)$ is a lattice.

14. Consider the lattice $(S_{n_0}, |)$ as defined in Exercise 6. For what values of n_0 is $(S_{n_0}, |)$ a totally ordered set?

15. Consider the lattice $(\mathcal{M}_2, \preccurlyeq)$ as defined in Example 4.32.
    (a) Draw the Hasse diagram of this lattice.
    (b) Find

$$\begin{bmatrix} 1 & 0 \\ 0 & 1 \end{bmatrix} \wedge \begin{bmatrix} 0 & 1 \\ 0 & 1 \end{bmatrix} \quad \text{and} \quad \begin{bmatrix} 1 & 0 \\ 0 & 1 \end{bmatrix} \vee \begin{bmatrix} 0 & 1 \\ 0 & 1 \end{bmatrix}$$

    (c) Find

$$\begin{bmatrix} 1 & 0 \\ 0 & 1 \end{bmatrix} \wedge \begin{bmatrix} 0 & 1 \\ 1 & 1 \end{bmatrix} \quad \text{and} \quad \begin{bmatrix} 1 & 0 \\ 0 & 1 \end{bmatrix} \vee \begin{bmatrix} 0 & 1 \\ 1 & 1 \end{bmatrix}$$

16. Let U be a nonempty set, and consider the poset $(\mathcal{P}(U), \subseteq)$. Let $\mathcal{S}$ be a nonempty subset of $\mathcal{P}(U)$. Show that

$$\text{lub}(\mathcal{S}) = \bigcup_{X \in \mathcal{S}} X$$

17. Consider the poset whose Hasse diagram is shown in Figure 4.13. Find, if it exists:
    (a) $\text{glb}(\{b, c, d\})$ (b) $\text{lub}(\{e, f, g\})$
    (c) $\text{glb}(\{b, d\})$ (d) $\text{lub}(\{h, i\})$
    (e) $\text{glb}(\{e, f, i\})$ (f) $\text{lub}(\{e, f, i\})$

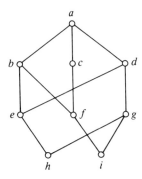

Figure 4.13

18. Verify the transitivity of the relation $\preccurlyeq$ defined in Example 4.26.

19. Let $V = \{a, b, c, d, e, f, g\}$ and define the partial-order relation A on V by

$$A = I_V \cup \{(g, f), (g, e), (g, d), (g, c), (g, b),$$
$$(g, a), (f, c), (f, a), (e, b), (e, a),$$
$$(d, c), (d, b), (d, a), (c, a), (b, a)\}$$

    (a) Draw the Hasse diagram of the poset (V, A).
    (b) Show that (V, A) is a lattice.

4.5

n-ARY RELATIONS

In this section we give a brief discussion of an extension of the idea of relation. Recall that a relation is also called a *binary* relation; this terminology emphasizes the fact that two sets are involved in the definition of a relation. There is a natural extension of this definition to any finite number of sets.

DEFINITION 4.13

Let $A_1, A_2, \ldots, A_n$ be sets, where $n \geq 2$. An *n-ary relation* among these sets is defined to be any subset of the Cartesian product $A_1 \times A_2 \times \cdots \times A_n$.

A relation among three sets is most often referred to as a *ternary relation*, whereas a relation among four sets is called a *quaternary relation*. If, in Definition 4.13, $A_1 = A_2 = \cdots = A_n = A$, then the relation is called an *n-ary relation on A*.

Example 4.34   The chairperson of the mathematics department at your college may be interested in the following sets:

A = the set of mathematics instructors

B = the set of mathematics courses to be offered next semester

C = the set of classrooms used for mathematics

D = the set of time slots during which courses meet (for example, MWF 2:00–3:00 or TTh 11:00–12:30)

In making up the teaching schedule, the chairperson might first come up with a binary relation R from A to B that assigns instructors to courses. Next it might be decided when and where each course is to meet; this would yield a ternary relation S among B, C, and D, where $(b, c, d) \in S$ if course b meets in location c during time slot d. The relations R and S could then be "joined" to yield the quaternary relation T among A, B, C, and D, where $(a, b, c, d) \in T$ means that instructor a is scheduled to teach course b in location c during time slot d. Notice that

$$(a, b, c, d) \in T \leftrightarrow [(a, b) \in R \wedge (b, c, d) \in S)]$$ ∎

Example 4.35   Define the ternary relation R on $\mathbb{N}$ by

$$(a, b, c) \in R \leftrightarrow c^2 = a^2 + b^2$$

In other words, a, b, and c are related provided they form a Pythagorean triple. (See Problem 19 in the Chapter 3 Problems.) Then, for instance, $(3, 4, 5) \in R$ and $(5, 12, 13) \in R$.   ∎

In the area of computer science known as database management, the main concern is the efficient manipulation of vast amounts of data. For example, suppose that a major university must maintain information concerning each student's registration. This could consist of several *n*-ary relations for various values of n. Here are several possibilities:

1. PDATA is a 5-ary relation whose 5-tuples have the form

   (NAME, HOME_ADDRESS, BIRTHDATE, SEX, IDNUMBER)

2. DEPT is a 4-ary relation with elements of the form

   (IDNUMBER, MAJOR, EARNED_HOURS, GPA)

3. ENROLLMENT is a ternary relation with elements of the form

   (NAME, IDNUMBER, COURSEID)

4. SCHEDULE is a 5-ary relation whose elements have the form

   (COURSEID, COURSENAME, TIME, LOCATION, INSTRUCTOR)

Thus, for the relation PDATA, let N be the set of all student names, A the set of all home addresses, B the set of all calendar dates, S the set $\{M, F\}$, and I the set of all student identification numbers. Then PDATA is a subset of

$$N \times A \times B \times S \times I$$

In the language of computer science, the set

$$D = \{\text{PDATA, DEPT, ENROLLMENT, SCHEDULE}\}$$

is called a *relational database*, and the relations in D are commonly referred to as *tables*. Indeed it is easy to imagine the relations in D being displayed in tabular form; for example, PDATA would be a table with headings NAME, HOME_ADDRESS, BIRTHDATE, SEX, and IDNUMBER. The rows of this table would contain the various 5-tuples of PDATA, with the value of each coordinate listed in a column under the appropriate heading. These columns are called *attributes* of the particular relation. Notice, for instance, that the relations of PDATA and ENROLLMENT have the attributes NAME and IDNUMBER in common.

In working with a relational database, four basic operations are commonly used. In the discussion that follows we describe these operations and demonstrate how they might be put to good use. To make life easy, we shall deal with two relations, R_1 and R_2, which are shown in Figures 4.14 and 4.15, respectively.

COURSEID	COURSENAME	TIME	LOCATION	INSTRUCTOR
CS105	Computing I	8	GEM 174	Straat
CS105	Computing I	9	GEM 174	Jones
CS105	Computing I	10	GEM 175	Moses
CS260	Data Structures	9	GEM 175	Straat
CS260	Data Structures	10	GEM 174	Lewis
CS340	Software Design	2	GEM 180	Lewis
MA122	Calculus I	11	GEM 180	Jones
MA122	Calculus I	1	GEM 180	Lewis
MA331	Algebra	10	GEM 179	Straat
MA350	Statistics	1	GEM 174	Moses

Figure 4.14 The relation R_1

COURSEID	TEXTAUTHOR	PREREQ	CREDITS
CS105	Allenton	CS102	3
CS260	Paston	CS105	3
CS340	Strong	CS261	3
MA122	Calvin	MA106	4
MA331	Seelo	MA231	3
MA350	Normal	MA231	3

Figure 4.15 The relation R_2

Now suppose that we would like to obtain a list of all courses being taught by Professor Straat, along with the text being used and the prerequisite for each course. Observe that R_1 contains the information about which courses Professor Straat is teaching (R_2 does not), while R_2 contains the text and prerequisite data (R_1 does not). Thus we must somehow make use of both R_1 and R_2 to obtain the desired information. We do this by applying the *relational operators* described next.

1. SELECT is an operation that derives a new relation S from a given relation R, where S consists of all elements of R that satisfy specified conditions. For example, the instruction

SELECT FROM R₁ WHERE INSTRUCTOR = `STRAAT`

produces the relation S consisting of all elements of R_1 whose INSTRUCTOR coordinate is Straat. The relation S is shown in Figure 4.16.

COURSEID	COURSENAME	TIME	LOCATION	INSTRUCTOR
CS105	Computing I	8	GEM 174	Straat
CS260	Data Structures	9	GEM 175	Straat
MA331	Algebra	10	GEM 179	Straat

Figure 4.16 The relation S

2. PROJECT forms a new relation P from a given relation R by extracting specified attributes (or columns) of R and, at the same time, removing any duplicate elements (rows) that may result from this extraction. For example, the instruction

PROJECT R₁ OVER LOCATION AND INSTRUCTOR

yields the relation P shown in Figure 4.17. (Note that the element (GEM 180, Lewis) is not listed twice.)

LOCATION	INSTRUCTOR
GEM 174	Straat
GEM 174	Jones
GEM 175	Moses
GEM 175	Straat
GEM 174	Lewis
GEM 180	Lewis
GEM 180	Jones
GEM 179	Straat
GEM 174	Moses

Figure 4.17 The relation P

3. JOIN forms a new relation as follows. Given an m-ary relation R and an n-ary relation S with common attribute A, the JOIN of R and S over A is an $(m + n - 1)$-ary relation J. The relation J consists of all $(m + n - 1)$-tuples obtained by taking elements $r \in R$ and $s \in S$ with common A-coordinate (the A-coordinates of r and s are equal), forming the $(n - 1)$-tuple s' by deleting the A-coordinate of s and then concatenating r with s'. This operation can also be applied over several attributes of R and S. For example, the instruction

```
JOIN R₁ AND R₂ OVER COURSEID
```

results in the relation J shown in Figure 4.18.

COURSEID	COURSENAME	TIME	LOCATION	INSTRUCTOR	TEXTAUTHOR	PREREQ	CREDITS
CS105	Computing I	8	GEM 174	Straat	Allenton	CS102	3
CS105	Computing I	9	GEM 174	Jones	Allenton	CS102	3
CS105	Computing I	10	GEM 175	Moses	Allenton	CS102	3
CS260	Data Structures	9	GEM 175	Straat	Paston	CS105	3
CS260	Data Structures	10	GEM 174	Lewis	Paston	CS105	3
CS340	Software Design	2	GEM 180	Lewis	Strong	CS261	3
MA122	Calculus I	11	GEM 180	Jones	Calvin	MA106	4
MA122	Calculus I	1	GEM 180	Lewis	Calvin	MA106	4
MA331	Algebra	10	GEM 179	Straat	Seelo	MA231	3
MA350	Statistics	1	GEM 174	Moses	Normal	MA231	3

Figure 4.18 The relation J

4. DIVISION is an operation that produces a new relation DIVIDE(R, S) from two relations R and S, where R and S satisfy

$$R \subseteq A_1 \times A_2 \times \cdots \times A_m \times B_1 \times B_2 \times \cdots \times B_n$$

and

$$S \subseteq B_1 \times B_2 \times \cdots \times B_n$$

for some sets $A_1, A_2, \ldots, A_m, B_1, B_2, \ldots, B_n$. Specifically, DIVIDE$(R, S)$ is the m-ary relation satisfying the condition

$$(a_1, a_2, \ldots, a_m) \in \text{DIVIDE}(R, S) \leftrightarrow$$

$$(a_1, a_2, \ldots, a_m, b_1, b_2, \ldots, b_n) \in R \;\forall (b_1, b_2, \ldots, b_n) \in S$$

This is an operation on relations R and S which requires that the n attributes of S be the same as the final sequence of n attributes of R. For example, let R be the relation of Figure 4.19 and suppose we wish to determine the names of all full professors who teach both CS105 and CS260. To do so we form the relation S of Figure 4.20 with attributes RANK and COURSEID.

INSTRUCTOR	RANK	COURSEID
Straat	Full Professor	CS105
Green	Full Professor	CS105
Straat	Full Professor	CS260
Jones	Assistant Prof.	MA122
Jones	Assistant Prof.	MA123
Green	Full Professor	CS260
Moses	Associate Prof.	CS260
Straat	Full Professor	MA337
Moses	Associate Prof.	CS105

Figure 4.19 The relation R

RANK	COURSEID
Full Professor	CS105
Full Professor	CS260

Figure 4.20 The relation S

Applying the division operation on R and S, we obtain the relation DIVIDE(R, S), consisting of the names of those full professors who teach both CS105 and CS260. The result is shown in Figure 4.21.

INSTRUCTOR
Straat
Green

Figure 4.21 The relation DIVIDE(R, S)

Let us go back to the problem posed earlier, that of determining all courses being taught by Professor Straat, along with the text and prerequisite for each course. We can get this information from the database by using the SELECT operator on the relation J of Figure 4.18. Simply use the instruction

```
SELECT FROM J WHERE INSTRUCTOR = `STRAAT´
```

to obtain the relation T of Figure 4.22. Then, to obtain the desired

COURSEID	COURSENAME	TIME	LOCATION	INSTRUCTOR	TEXTAUTHOR	PREREQ	CREDITS
CS105	Computing I	8	GEM 174	Straat	Allenton	CS102	3
CS260	Data Structures	9	GEM 175	Straat	Paston	CS105	3
MA331	Algebra	10	GEM 179	Straat	Seelo	MA231	3

Figure 4.22 The relation T

information, use the instruction

```
PROJECT T OVER COURSEID AND TEXTAUTHOR AND PREREQ
```

The result is the table of Figure 4.23.

COURSEID	TEXTAUTHOR	PREREQ
CS105	Allenton	CS102
CS260	Paston	CS105
MA331	Seelo	MA231

Figure 4.23

A relational database is one kind of "database management system." Such systems are used by many large corporations (such as airline companies, insurance companies, and banks) for organizing, storing, manipulating, and retrieving large amounts of data within a computing environment. In this context relational database management systems are often relatively efficient and easy to use.

Exercises 4.5

1. For the relations R_1 and R_2 of Figures 4.14 and 4.15, determine the relations produced by the following operations.
   (a) SELECT FROM R_1 WHERE TIME = 10
   (b) PROJECT R_2 OVER COURSEID AND TEXTAUTHOR
   (c) SELECT FROM R_2 WHERE CREDITS = 3
   (d) PROJECT R_1 OVER COURSEID AND COURSENAME AND TIME
   (e) SELECT FROM R_1 WHERE TIME $\neq$ 8 AND (INSTRUCTOR = `STRAAT` OR INSTRUCTOR = `MOSES`)

2. What might the following instruction produce, given the relations R_1 and R_2 of Figures 4.14 and 4.15?

```
SELECT FROM (PROJECT R₁ OVER COURSEID AND INSTRUCTOR)
WHERE INSTRUCTOR=`STRAAT`
```

3. Suppose the relation R_3, with attributes COURSEID, TIME, LOCATION, and ENROLLMENT, gives the enrollment (number of students) in each of the courses listed in Figure 4.14. Discuss how the operations SELECT, PROJECT, and JOIN can be used (with R_1 and R_3) to obtain the following information:
   (a) a relation Q_1 with attributes TIME and ENROLLMENT
   (b) a relation Q_2 giving the enrollment in each section of CS105
   (c) a ternary relation Q_3 with attributes COURSEID, INSTRUCTOR, and ENROLLMENT
   (d) a relation Q_4 giving the enrollment in each course taught by Professor Lewis
4. Discuss how the operation of division can be obtained from the operations SELECT and PROJECT.
5. Indicate how the relation of Figure 4.23 can be obtained from R_1 and R_2 by applying PROJECT, JOIN, and SELECT, in that order. Are there advantages or disadvantages to doing it this way, as opposed to the way it was done in this section?

CHAPTER PROBLEMS

1. Let R be a relation on a nonempty set A. Complete each of these statements.
   (a) R is not reflexive provided _____ .
   (b) R is not irreflexive provided _____ .
   (c) R is not symmetric provided _____ .
   (d) R is not antisymmetric provided _____ .
   (e) R is not transitive provided _____ .
2. Let R be a symmetric and transitive relation on a nonempty set A. Prove: If dom $R = A$, then R is reflexive.
3. Let R_1 and R_2 be relations on the nonempty set A. Indicate the truth-value of each of the following statements. (Optional: If the statement is true, prove it; if it is false, give a counterexample.)
   (a) If $R_1 \cap R_2$ is reflexive, then both R_1 and R_2 are reflexive.
   (b) If R_1 is reflexive, then $R_1 \cup R_2$ is reflexive.
   (c) If R_1 is reflexive and R_2 is irreflexive, then $R_1 - R_2$ is reflexive.
   (d) If R_1 is irreflexive, then $R_1 \cap R_2$ is irreflexive.
   (e) If $R_1 \cup R_2$ is irreflexive, then both R_1 and R_2 are irreflexive.
   (f) If R_1 is irreflexive, then $R_1 - R_2$ is irreflexive.
   (g) If $R_1 \cap R_2$ is symmetric, then both R_1 and R_2 are symmetric.
   (h) If $R_1 \cup R_2$ is symmetric, then both R_1 and R_2 are symmetric.
   (i) If R_1 is symmetric, then $R_1 - R_2$ is symmetric.
   (j) If R_1 is antisymmetric, then $R_1 \cap R_2$ is antisymmetric.
   (k) If $R_1 \cup R_2$ is antisymmetric, then both R_1 and R_2 are antisymmetric.

(l) If R_1 is antisymmetric, then $R_1 - R_2$ is antisymmetric.

(m) If $R_1 \cap R_2$ is transitive, then both R_1 and R_2 are transitive.

(n) If $R_1 \cup R_2$ is transitive, then both R_1 and R_2 are transitive.

4. Let $\sim$ be a reflexive relation on a nonempty set A. Prove that $\sim$ is an equivalence relation if and only if, given any elements a, b, and c of A (not necessarily distinct), the following condition holds:

$$(a \sim b \wedge a \sim c) \rightarrow b \sim c$$

5. For each of the following relations $\sim$ on $\mathbb{R} \times \mathbb{R}$, determine whether it is an equivalence relation. If it is, describe (geometrically) the equivalence class $[(a, b)]$.

   (a) $(a, b) \sim (c, d) \leftrightarrow a + d = b + c$

   (b) $(a, b) \sim (c, d) \leftrightarrow (a - 1)^2 + b^2 = (c - 1)^2 + d^2$

   (c) $(a, b) \sim (c, d) \leftrightarrow (a - c)(b - d) = 0$

   (d) $(a, b) \sim (c, d) \leftrightarrow |a| + |b| = |c| + |d|$

   (e) $(a, b) \sim (c, d) \leftrightarrow ab = cd$

6. For each of the following relations $\sim$ on $\mathbb{R}$, verify that $\sim$ is an equivalence relation and describe the equivalence classes of $\sim$.

   (a) $x \sim y \leftrightarrow \lfloor x \rfloor = \lfloor y \rfloor$, where $\lfloor x \rfloor$ denotes the largest integer m such that $m \le x$

   (b) $x \sim y \leftrightarrow \lfloor x + 0.5 \rfloor = \lfloor y + 0.5 \rfloor$ **(c)** $x \sim y \leftrightarrow |x| = |y|$

   (d) $x \sim y \leftrightarrow x - y \in \mathbb{Z}$ **(e)** $x \sim y \leftrightarrow x - y \in \mathbb{Q}$

7. Define the relation $\sim$ on $\mathbb{Z} \times \mathbb{N}$ by

$$(m_1, n_1) \sim (m_2, n_2) \leftrightarrow m_1 n_2 = n_1 m_2$$

   (a) Show that $\sim$ is an equivalence relation.

   (b) Describe the equivalence class $[(m, n)]$ in a nice way. (Hint: Think of (m, n) as the fraction m/n.)

8. Each part gives a set X, a relation $\preccurlyeq$ on X, and a subset Y of X. Verify that $\preccurlyeq$ is a partial ordering of X and draw the Hasse diagram of the subposet $(Y, \preccurlyeq)$.

   (a) $X = \mathbb{Z}$, $x \preccurlyeq y \leftrightarrow (x = y \vee |x| < |y|)$,
   $Y = \{-3, -2, -1, 0, 1, 2, 3\}$

   (b) $X = \mathbb{Z}$, $x \preccurlyeq y \leftrightarrow (x = y \vee x \bmod 4 < y \bmod 4)$,
   $Y = \{-4, -3, -2, -1, 0, 1, 2, 3, 4\}$

   (c) $X = \mathbb{Z} \times \mathbb{Z}$, $(a, b) \preccurlyeq (c, d) \leftrightarrow [(a, b) = (c, d) \vee a^2 + b^2 < c^2 + d^2]$, $Y = \{0, 1, 2\} \times \{0, 1, 2\}$

9. Let $A = \{(a, b) \mid a, b \in \mathbb{N} \wedge \gcd(a, b) = 1\}$. Define a relation $\preccurlyeq$ on A by

$$(a, b) \preccurlyeq (c, d) \leftrightarrow ad \le bc$$

   Show that $\preccurlyeq$ defines a total ordering of A.

10. Let $n_0 \in \mathbb{N}$ and let $S_{n_0} = \{n \in \mathbb{N} \mid n \text{ divides } n_0\}$. Draw the Hasse diagram for the lattice $(S_{n_0}, |)$, where:

    (a) $n_0 = 42$ **(b)** $n_0 = 54$ **(c)** $n_0 = 100$ **(d)** $n_0 = 126$

11. Let $\mathbb{Q}^+$ denote the set of positive rational numbers. Define the relation $\preccurlyeq$ on $\mathbb{Q}^+$ by

$$r \preccurlyeq s \leftrightarrow \frac{s}{r} \in \mathbb{N}$$

(a) Show that $\preccurlyeq$ is a partial ordering of $\mathbb{Q}^+$.
(b) Let $S = \{\frac{1}{6}, \frac{1}{3}, \frac{1}{2}, 1, 2, 3, 6\}$. Draw the Hasse diagram of the subposet $(S, \preccurlyeq)$.
(c) As shown in Problem 12, $(\mathbb{Q}^+, \preccurlyeq)$ is a lattice. Find each of these elements.

$$4 \wedge 6 \qquad 4 \vee 6 \qquad \frac{1}{6} \wedge \frac{1}{4} \qquad \frac{1}{6} \vee \frac{1}{4}$$

$$\frac{1}{4} \wedge 6 \qquad \frac{1}{4} \vee 6 \qquad \frac{5}{6} \wedge \frac{3}{4} \qquad \frac{5}{6} \vee \frac{3}{4}$$

12. Consider the poset $(\mathbb{Q}^+, \preccurlyeq)$ defined in Problem 11. The purpose of this problem is to show that this poset is a lattice.
(a) Let m and n be positive integers. Give a formula for each of these elements:

$$m \wedge n \qquad m \vee n \qquad \frac{1}{m} \wedge \frac{1}{n} \qquad \frac{1}{m} \vee \frac{1}{n}$$

★(b) Let a, b, c, and d be positive integers. Show that $(\mathbb{Q}^+, \preccurlyeq)$ is a lattice by finding general formulas for

$$\frac{a}{b} \wedge \frac{c}{d} \quad \text{and} \quad \frac{a}{b} \vee \frac{c}{d}$$

13. Define the relation $\prec$ on $\mathbb{N} - \{1\}$ by

$$m \prec n \leftrightarrow m^2 \mid n$$

(a) Show that $\prec$ is irreflexive and transitive.
(b) Apply the result of Example 4.26 to define a partial ordering $\preccurlyeq$ on $\mathbb{N} - \{1\}$. Extend this partial ordering to all of $\mathbb{N}$ by defining $1 \preccurlyeq n$ for all $n \in \mathbb{N}$.
(c) Let $S = \{1, 2, 3, 6, 4, 9, 16, 36, 81, 1296\}$. Draw the Hasse diagram of the subposets $(S, \preccurlyeq)$.
★(d) Prove that $(\mathbb{N}, \preccurlyeq)$ is a lattice.
(e) Find each of these elements:

$$45 \wedge 63 \qquad 45 \vee 63 \qquad 60 \wedge 84 \qquad 60 \vee 84$$

14. Let $(A, \preccurlyeq)$ be a lattice and let $a, b, c \in A$.
(a) Show that $\text{glb}(\{a, b, c\}) = (a \wedge b) \wedge c$.
(b) Show that $\text{lub}(\{a, b, c\}) = (a \vee b) \vee c$.

(These results can be generalized to provide a method for finding glb(B) and lub(B) for any finite subset B of A.)

(c) Verify the commutative laws: $a \wedge b = b \wedge a$ and $a \vee b = b \vee a$.

(d) Verify the associative laws:

$$(a \wedge b) \wedge c = a \wedge (b \wedge c) \quad \text{and} \quad (a \vee b) \vee c = a \vee (b \vee c)$$

(e) Verify the idempotent laws: $a \wedge a = a$ and $a \vee a = a$.

(f) Verify the absorption laws:

$$(a \wedge b) \vee a = a \quad \text{and} \quad (a \vee b) \wedge a = a$$

15. Define the relation $\sim$ on $\mathbb{Z}$ by

$$a \sim b \leftrightarrow 3 \,|\, (a + 2b)$$

(a) Show that $\sim$ is an equivalence relation.

(b) Determine the equivalence classes of $\sim$.

16. Each part gives a set X and a relation R on X. Determine which of the properties (i) reflexive, (ii) irreflexive, (iii) symmetric, (iv) antisymmetric, or (v) transitive R satisfies.

(a) $X = \mathscr{P}(\{1, 2, 3, 4\})$, $(A, B) \in R \leftrightarrow A \subseteq B \cup \{1\}$

(b) $X = \mathbb{Z}_7$, $([x], [y]) \in R \leftrightarrow [x - y] = 1$ or $[x - y] = 6$

(c) $X = \mathbb{Z}$, $(x, y) \in R \leftrightarrow |x - y| > 2$

(d) $X = \mathbb{Z}$, $(x, y) \in R \leftrightarrow xy \geq 0$

(e) $X = (0, 1)$, $(x, y) \in R \leftrightarrow xy \in \mathbb{Q}$

17. Consider the poset whose Hasse diagram is shown in Figure 4.24(a). Let $A = \{c, d, e\}$, $B = \{d, f\}$, and $C = \{e, f\}$. Find, if it exists:

(a) glb(A)   **(b)** lub(A)   **(c)** glb(B)

(d) lub(B)   **(e)** glb(C)   **(f)** lub(C)

18. Give an example of a relation on $\mathbb{Z}$ that is:

(a) irreflexive and symmetric   **(b)** irreflexive and transitive

(c) reflexive and symmetric but not transitive

(d) reflexive and transitive but not symmetric

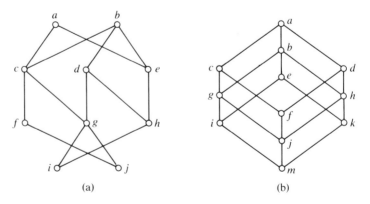

(a)   (b)

Figure 4.24

19. Consider the poset whose Hasse diagram is shown in Figure 4.24(b). It can be shown that this poset is a lattice. Find:

(a) $a \vee i$ (b) $c \wedge i$ (c) $g \vee k$ (d) $b \wedge d$

(e) $e \wedge f$ (f) $e \vee f$ (g) $d \wedge g$ (h) $a \wedge m$

20. Let R be a relation from A to B. The *converse* of R is the relation R^c from B to A defined for $a \in A$ and $b \in B$ by

$$(b, a) \in R^c \leftrightarrow (a, b) \in R$$

(a) For finite sets A and B, explain how the matrix representations of R and R^c are related.

(b) For a relation R on a finite set A, explain how the directed graph representations of R and R^c are related.

(c) Let R be a relation on A. Under what condition does $R = R^c$?

21. Let R be a relation on A. Define the relation R' on A by

$$(a, b) \in R' \leftrightarrow [(a, b) \in R \vee a = b]$$

that is, $R' = R \cup I_A$, where I_A denotes the identity relation on A.

(a) Show that R' is reflexive.

(b) Under what condition does $R' = R$?

(c) Prove: If R' is a reflexive relation on A and $R \subseteq R'$, then $R' \subseteq R'$. The relation R' is called the *reflexive closure* of R. Find the reflexive closure of each of these relations:

(d) the relation $R_1 = \{(1,1), (1,2), (2,1), (2,2), (3,4), (3,5), (4,3), (4,5), (5,3), (5,4)\}$ on $\{1, 2, 3, 4, 5\}$

(e) the relation $<$ on $\mathbb{Z}$

(f) the relation $\sim$ on $\mathbb{Z}$ defined by $m \sim n \leftrightarrow mn$ is odd

22. Let R be a relation on the nonempty set A and let R^c denote the converse of R. (See Problem 20.) Indicate the truth-value of each of the following statements. (Optional: If the statement is true, prove it; if it is false, give a counterexample.)

(a) If R is reflexive, then so is R^c.

(b) If R is irreflexive, then so is R^c.

(c) If R is antisymmetric, then so is R^c.

(d) If R is transitive, then so is R^c.

(e) If R is a partial-order relation, then so is R^c.

(f) If R is reflexive and transitive, then $R \cup R^c$ is an equivalence relation.

23. Let R be a relation on A. Define the relation R^s on A by

$$(a, b) \in R^s \leftrightarrow [(a, b) \in R \vee (b, a) \in R]$$

(a) Show that R^s is symmetric.

(b) Note that $R \subseteq R^s$. Under what condition does $R = R^s$?

(c) Show that $R^s = R \cup R^c$. (See Problem 20.)

(d) Prove: If R' is a symmetric relation on A and $R \subseteq R'$, then $R^s \subseteq R'$.

The relation R^s is called the *symmetric closure* of R. Find the symmetric closure of each of these relations:

(e) the relation $R_1 = \{(1,1),(1,2),(1,3),(1,4),(2,2),(3,3),(4,4)\}$ on $\{1,2,3,4\}$

(f) the relation $<$ on $\mathbb{Z}$

24. Let R_1 and R_2 be relations on a nonempty set A; suppose that R_1 is an equivalence relation and R_2 is a partial ordering.

(a) For example, suppose that $A = \mathbb{N}$, R_1 is the relation of congruence modulo 2, and R_2 is the relation of divides. Find $R_1 \cap R_2$ in this case.

(b) Show, in general, that $R_1 \cap R_2$ is a partial ordering of A.

25. Let R be a relation on A. The *transitive closure* of R is the relation R^t on A that is defined by the following properties: (i) R^t is transitive; (ii) $R \subseteq R^t$; (iii) if R' is a transitive relation on A and $R \subseteq R'$, then $R^t \subseteq R'$. Informally, R^t is the smallest transitive relation on A that contains R.

(a) Find the transitive closure of the relation $R_1 = \{(1,2),(2,1),(2,3),(3,4)\}$ on $\{1,2,3,4\}$.

(b) Find the transitive closure of the relation R_2 defined on $\mathbb{Z}$ by $(m, n) \in R_2 \leftrightarrow n = m + 1$.

(c) Find the transitive closure of the relation "is the father of" on your family tree.

(d) Define the relation R^2 on A by

$$(a, b) \in R^2 \leftrightarrow [(a, b) \in R \vee \exists c \in A[(a, c) \in R \wedge (c, b) \in R]]$$

In view of Problems 21 and 23, it is tempting to define R^t by $R^t = R^2$. However, show that R^2 is not transitive in general.

26. Write a Pascal program that inputs a positive integer n and a relation R on $\{1, 2, \ldots, n\}$ and then performs the following jobs:

(a) It determines whether R is reflexive; if not, it finds R^r. (See Problem 21.)

(b) It determines whether R is symmetric; if not, it finds R^s. (See Problem 23.)

(c) It determines whether R is transitive; if not, it finds R^t. (See Problem 25.)

(Note: A well-known algorithm for finding the transitive closure of a relation is "Warshall's algorithm." We consider Warshall's algorithm in Chapter 7 in the context of finding the "reachability matrix" for a given directed graph.)

27. Let R be a relation on A. Prove or disprove each of these statements.

(a) If R is symmetric and transitive, then so is R^r. (See Problem 21.)

(b) If R is reflexive and transitive, then so is R^s. (See Problem 23.)

(c) If R is reflexive and symmetric, then so is R^t. (See Problem 25.)

(d) The relation $((R^r)^s)^t$ is an equivalence relation on A.

(e) If R is reflexive and antisymmetric, then R^t is a partial ordering of A.

28. Write a Pascal program that inputs a positive integer n and a relation R on $\{1, 2, \ldots, n\}$ and finds the relation $((R^r)^s)^t$, which is the smallest equivalence relation containing R. (Refer to Problems 26 and 27.)

29. Let R be a relation on A and let S be a relation on B. Define the relations T_1 and T_2 on $A \times B$ as follows:

$$(a_1, b_1)\, T_1\, (a_2, b_2) \leftrightarrow (a_1\, R\, a_2 \wedge b_1\, S\, b_2)$$

$$(a_1, b_1)\, T_2\, (a_2, b_2) \leftrightarrow [(a_1 \neq a_2 \wedge a_1\, R\, a_2) \vee (a_1 = a_2 \wedge b_1\, S\, b_2)]$$

Indicate the truth-value of each of the following statements. (Optional: If the statement is true, prove it; if it is false, give a counterexample.)

(a) If both R and S are reflexive, then T_1 is reflexive.

(b) If S is reflexive, then T_2 is reflexive.

(c) If both R and S are irreflexive, then T_1 is irreflexive.

(d) If S is irreflexive, then T_2 is irreflexive.

(e) If both R and S are symmetric, then T_1 is symmetric.

(f) If both R and S are symmetric, then T_2 is symmetric.

(g) If both R and S are antisymmetric, then T_1 is antisymmetric.

(h) If both R and S are antisymmetric, then T_2 is antisymmetric.

(i) If both R and S are transitive, then T_1 is transitive.

(j) If both R and S are transitive, then T_2 is transitive.

30. A lattice $(A, \preccurlyeq)$ is called a *distributive lattice* provided the distributive laws hold: $\forall\, a, b, c \in A$,

$$a \vee (b \wedge c) = (a \vee b) \wedge (a \vee c)$$

and

$$a \wedge (b \vee c) = (a \wedge b) \vee (a \wedge c)$$

Determine whether each of these lattices is a distributive lattice.

(a) $(\mathscr{P}(U), \subseteq)$ **(b)** the lattice of Figure 4.11   **(c)** $(\mathbb{N}, |)$

31. An element m in a poset $(A, \preccurlyeq)$ is called a *minimal element* provided

$$\forall\, a \in A (a \preccurlyeq m \rightarrow a = m)$$

In other words, m is a minimal element provided there does not exist an element $a \in A$ such that $a \prec m$. Similarly, m is called a *maximal element* provided

$$\forall\, a \in A (m \preccurlyeq a \rightarrow a = m)$$

For example, in the poset of Figure 4.10, a and b are maximal elements and k is a minimal element.

(a) Find any maximal or minimal elements in the poset of Figure 4.24(a).

(b) Find the maximal and minimal elements of the poset $(\mathscr{P}(U), \subseteq)$.

(c) Find the minimal elements of the poset $(\mathbb{N} - \{1\}, |)$.

(d) Show that $(\mathbb{N}, |)$ has no maximal element.

(e) Show that every nonempty finite poset $(A, \preccurlyeq)$ has both a minimal and a maximal element. Given $A = \{a_1, a_2, \ldots, a_n\}$, describe an algorithm for finding a minimal (or maximal) element.

32. Given a nonempty finite poset $(A, \preccurlyeq)$, it is possible to *embed* the partial ordering $\preccurlyeq$ in a total ordering $<=$. That is, we can define a total ordering $<=$ on A such that, if $a, b \in A$ with $a \preccurlyeq b$, then $a <= b$. We can then write $A = \{a_1, a_2, \ldots, a_n\}$ with a_1 preceding a_2, a_2 preceding a_3, and so on. The process of finding such a total ordering is called *sorting*. One method of sorting is called the *topological sort*. This method makes use of the fact (see Problem 31) that every nonempty finite poset contains a minimal element. Choose a minimal element m of A and set $a_1 = m$. Next, for a_2, choose a minimal element of the poset $A - \{a_1\}$. Continue this process until the sort is completed.

(a) Perform a topological sort of the poset whose Hasse diagram is shown in Figure 4.10.

(b) Implement the topological sort as a Pascal program that inputs a partial-order relation R on some subset of the letters $\{a, b, \ldots, z\}$ and performs a topological sort.

33. Let $(A, \preccurlyeq)$ be a poset. A *chain* in $(A, \preccurlyeq)$ is a nonempty subset C of A such that $(C, \preccurlyeq)$ is totally ordered. If C is finite, then we may write $C = \{c_1, c_2, \ldots, c_n\}$ with $c_1 \prec c_2 \prec \cdots \prec c_n$; the number $n - 1$ is called the *length* of the chain C. As an example, in the poset of Figure 4.10, $C = \{j, g, e, b\}$ is a chain of length three.

(a) Find all chains of length four in the poset of Figure 4.10.

(b) Find a chain of maximum length in the poset of Figure 4.24(a).

(c) Find a chain of maximum length in the poset $(\mathscr{P}(U), \subseteq)$, where $U = \{1, 2, \ldots, n\}$.

(d) Prove: If $a_1 \prec a_2 \prec \cdots \prec a_n$ is a chain of maximum length in $(A, \preccurlyeq)$, then a_1 is a minimal element (and a_n is a maximal element). (See Problem 31.)

(e) Let the maximum length of a chain in $(A, \preccurlyeq)$ be $n \geq 1$. Prove: If M is the set of maximal elements in A, then the maximum length of a chain in $(A - M, \preccurlyeq)$ is $n - 1$.

34. An *antichain* in a poset $(A, \preccurlyeq)$ is a subset B of A such that no two distinct elements of B are comparable. Symbolically, B is an antichain provided

$$\forall\, a, b \in B (a \preccurlyeq b \rightarrow a = b)$$

For example, $B = \{d, e, f\}$ is an antichain in the poset of Figure 4.10. For this poset:

(a) Find an antichain having four elements.

(b) Find the largest antichain that contains the element j.

(c) Show that no antichain has five elements.

Let $(A, \preceq)$ be a nonempty poset.

(d) Show that the set of minimal (or maximal) elements of A is an antichain.

(e) Let $a_1 \prec a_2 \prec \cdots \prec a_n$ be a chain of maximum length in A, and for $1 \leq i \leq n - 1$, define $B_i = \{b \in A \mid a_i \preceq b \prec a_{i+1}\}$. Show that each B_i is an antichain.

35. Let A denote the ASCII character set. Then A contains the letters A, B, ..., Z, the digits $0, 1, 2, \ldots, 9$, and other special characters such as), $+$, :, $=$. A *character string* is a sequence of characters from A; we denote a character string by writing the sequence of characters within quotes. For example, "FREDONIA" and "93 MAIN ST." are character strings. The *length* of a character string is the number of characters it contains (not counting the enclosing quotes). Thus the length of "FREDONIA" is 8 and the length of "93 MAIN ST." is 11 (blanks are characters, too). The unique string of length 0 is called the *empty string* and is denoted "".

Comparison of two character strings is made possible by the *collating sequence* of A, by which A is totally ordered. Let us denote this ordering by $\leq$. In the ASCII character set, A < B < $\cdots$ < Z < 0 < 1 < $\cdots$ < 9. In this problem we consider two possible ways of extending the ordering of the ASCII character set A to an ordering of all character strings over A.

The first method is called *lexicographic ordering* and is denoted $\leq_l$. In this ordering the empty string precedes any nonempty string, while for two nonempty strings $s_1 = "a_1 a_2 \cdots a_m"$ and $s_2 = "b_1 b_2 \cdots b_n"$, we have $s_1 \leq_l s_2$ provided either $m \leq n$ and $a_i = b_i$, $1 \leq i \leq m$, or provided, proceeding from left to right, i is the first value where $a_i \neq b_i$ and $a_i < b_i$.

(a) Verify that $\leq_l$ is a total ordering.

(b) Find character strings $s_1, s_2, s_3, \ldots$ such that $s_2 <_l s_1$, $s_3 <_l s_2$, and so on. (As usual, we write $s_1 <_l s_2$ to mean $s_1 \leq_l s_2$ and $s_1 \neq s_2$.)

The second method of ordering character strings is called *standard ordering* and is denoted $\leq_s$. For two strings s_1 and s_2, we have $s_1 \leq_s s_2$ provided the length of s_1 is less than the length of s_2 or s_1 and s_2 have the same length and $s_1 \leq_l s_2$.

(c) Verify that $\leq_s$ is a total ordering.

(d) Given any nonempty set of character strings S, show that S has a smallest element. Compare with part (b).

36. Let $(A, \preceq)$ be a finite poset. Prove: If the maximum length of a chain in $(A, \preceq)$ is n, then A can be partitioned into $n + 1$ antichains. (Hint:

Proceed by induction on n; let M be the set of maximal elements in A and consider $A - M$.)

37. Let $(A, \preccurlyeq)$ be a finite totally ordered set. Let $S = A \times A \times \cdots \times A$ be the n-fold Cartesian product of A with itself. Using Problem 35 as a guide,
   (a) define a "lexicographic" total ordering $\preccurlyeq_l$ on S;
   (b) define a "standard" total ordering $\preccurlyeq_s$ on S.

38. Let $V = \{a, b, c, d, e, f, g, h, i, j\}$ and define the partial-order relation A on V by $A = I_V \cup \{(j, i), (j, h), (j, g), (j, f), (j, e), (j, d), (j, c), (j, b), (j, a), (i, f), (i, d), (i, c), (i, a), (h, e), (h, d), (h, b), (h, a), (g, c), (g, b), (g, a), (f, c), (f, a), (e, b), (e, a), (d, a), (c, a), (b, a)\}$.
   (a) Draw the Hasse diagram of the poset (V, A).
   (b) Show that (V, A) is a lattice.

39. Define the relation $\sim$ on the set $\mathbb{R}^+$ of positive real numbers by

$$x \sim y \leftrightarrow \frac{x}{y} \in \mathbb{Q}^+$$

(where $\mathbb{Q}^+$ denotes the set of positive rational numbers). Show that $\sim$ is an equivalence relation and discuss the equivalence classes of $\sim$.

F I V E

Functions

5.1 INTRODUCTION

Recall that a relation R from a set A to a set B is any subset of $A \times B$. The domain of R is the set

$$\text{dom } R = \left\{ a \in A \mid (a, b) \in R \text{ for some } b \in B \right\}$$

and the image of R is the set

$$\text{im } R = \left\{ b \in B \mid (a, b) \in R \text{ for some } a \in A \right\}$$

We have already encountered two very special types of relations, namely equivalence relations and partial-order relations. These are well motivated by some rather classical examples: "equals" is an equivalence relation on $\mathbb{N}$ and "is less than or equal to" is a partial ordering of $\mathbb{N}$.

In this chapter we study what is probably the most fundamental type of relation used in mathematics.

DEFINITION 5.1

A relation f from a set A to a set B is called a *function from A to B* provided the following conditions hold:

1. dom $f = A$
2. No two distinct ordered pairs in f have the same first coordinate.

We denote the fact that f is a function from A to B by writing $f: A \to B$. The set A is called the *domain* of the function f, and B is called the *codomain* of f. If $A = B$, then we call f a *function on A*.

We remark that condition 2 of the definition of a function can be stated symbolically as follows:

$$[(x, y) \in f \wedge (x, z) \in f] \to y = z$$

Example 5.1 Determine which of the following relations are functions and find the image of each function.
(a) The relation f_1 on $\{1, 2, 3\}$ given by $f_1 = \{(1, 2), (2, 1), (3, 2)\}$.
(b) The relation f_2 on $\{1, 2, 3\}$ given by $f_2 = \{(1, 1), (1, 3), (2, 3), (3, 1)\}$.
(c) The relation f_3 on $\mathbb{R}$ defined by $f_3 = \{(x, y) \mid x^2 + y^2 = 4\}$.
(d) The relation f_4 on $\mathbb{Z}$ defined by $f_4 = \{(m, n) \mid n = 2m + 1\}$.
(e) The relation f_5 defined on your family tree by

$$(x, y) \in f_5 \leftrightarrow y \text{ is the (biological) father of } x$$

Solution (a) The relation f_1 is a function; im $f_1 = \{1, 2\}$.
(b) The relation f_2 is not a function because $(1, 1)$ and $(1, 3)$ both belong to f_2.
(c) The relation f_3 is not a function because, for instance, $(0, 2) \in f_3$ and $(0, -2) \in f_3$.
(d) This relation is a function; im f_4 is precisely the set of odd integers.
(e) This relation is a function, since each person has a unique father. The image of f_5 consists of those people in your family who are (or were) fathers. ∎

Suppose now that f is a function from A to B. Then conditions 1 and 2 in Definition 5.1 can be replaced by the following single condition:

Given $x \in A$, there is a unique $y \in B$ such that $(x, y) \in f$.

We call y the *image of x under f* and write $y = f(x)$; this equation is read "y equals f of x." Thus $(x, y) \in f$ if and only if $y = f(x)$, and it should then be observed that

$$f = \{(x, y) \mid x \in A \text{ and } y = f(x)\} = \{(x, f(x)) \mid x \in A\}$$

It should be mentioned that $f(x)$ is also referred to as the *value of f at x*.
It is common to refer to the equation $y = f(x)$ as the "defining equation" of f. Indeed it is common to define a function by writing: "Define $f: A \to B$ by $y = f(x)$." In some sense, this equation can be viewed as a rule specifying how to compute the image of a given $x \in A$. In fact, it is not uncommon to see the following statement as a definition of a function $f: A \to B$:

3. A function from A to B is a rule f that associates with each $x \in A$ a unique element $y \in B$.

This statement is, at best, imprecise. The reason for this lies in the lack of a precise meaning for the term "rule." Condition 3 simply relies on the reader's intuition as to what the term "rule" means.

In view of the foregoing discussion it is clear that, given $f: A \rightarrow B$, the uniqueness of the image of each element of A under f is a key feature in the definition of a function. Consider the following attempt at defining a function from $\mathbb{Z}_3$ to $\mathbb{Z}_6$: Define $g: \mathbb{Z}_3 \rightarrow \mathbb{Z}_6$ by $g([a]_3) = [a]_6$. Here $[a]_3$ denotes an arbitrary element of $\mathbb{Z}_3$ and $[a]_6$ denotes an arbitrary element of $\mathbb{Z}_6$. The question is, Do we really have a function? In particular, is each $[a]_3$ in $\mathbb{Z}_3$ associated with exactly one element of $\mathbb{Z}_6$? The problem is that there are infinitely many ways to represent a given element of $\mathbb{Z}_3$; for example, $[2]_3 = [5]_3$. In this particular case, we see that $g([2]_3) = [2]_6$ and $g([5]_3) = [5]_6$, but $[2]_6 \neq [5]_6$. Thus we are forced to conclude that g associates with $[2]_3$ at least two different elements of $\mathbb{Z}_6$. Therefore g is not a function.

In general terms, what went wrong with the previous example is that an attempt was made to define a function g from a set A to a set B, where the elements of A can be represented by more than one "name." For instance, $[2]_3$ and $[5]_3$ are different names for the same element of $\mathbb{Z}_3$. If g is to be a function, then the image of each element of A must be independent of the name chosen to represent it. If this happens, then we emphasize the property by stating that g is *well-defined*.

Example 5.2   Define $g: \mathbb{Z}_7 \rightarrow \mathbb{Z}_7$ by $g([a]) = [4a]$. Show that g is well-defined.

Solution   Suppose that $[a] = [b]$ in $\mathbb{Z}_7$. We must show that $g([a]) = g([b])$. Since $[a] = [b]$, it follows that $7 \mid (a - b)$. Hence $7 \mid 4(a - b)$; that is, $7 \mid (4a - 4b)$. Thus $[4a] = [4b]$, or $g([a]) = g([b])$, and it follows that g is well-defined.   ∎

If f is a function whose domain and range are both subsets of $\mathbb{R}$, then associated with each ordered pair $(x, y) \in f$ there is a uniquely determined point (x, y) in the xy-coordinate plane. In Chapter 4 the set of all points so determined was called the *graph* of f. Here we also speak of the graph of such a function f. We assume the reader has had considerable experience with graphing functions.

Example 5.3   Sketch the graphs of the following functions.

(a) $f: \mathbb{R} \rightarrow \mathbb{R}$; $f(x) = 4 - 2x$
(b) $f: (0, \infty) \rightarrow (0, \infty)$; $f(x) = 1/x$
(c) $f: \mathbb{R} \rightarrow [-4, \infty)$; $f(x) = x^2 - 2x - 3$

Solution   The graphs are shown in Figure 5.1(a), (b) and (c).   ∎

If X and Y are subsets of $\mathbb{R}$ and $f: X \rightarrow Y$ is given, then the property
$$[(x, y_1) \in f \wedge (x, y_2) \in f] \rightarrow y_1 = y_2$$

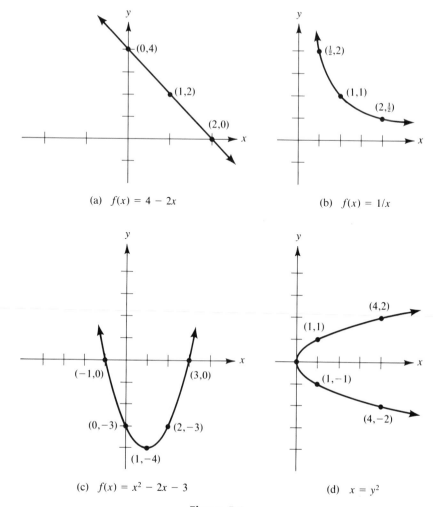

Figure 5.1

can be interpreted geometrically. It implies that each vertical line intersects the graph of f in at most one point. Thus, if we are given a set of points in a coordinate plane, then we can determine whether the associated set of ordered pairs is a function by applying this "vertical line test."

Example 5.4   It is easy to apply the vertical line test to each of the graphs in Figure 5.1. Those in parts (a) through (c) are seen to determine functions, which of course we know from the previous example. Figure 5.1(d) is the graph of the relation $f = \{(x, y) \mid y \in \mathbb{R} \wedge x = y^2\}$. Note that any vertical line $x = a^2$, where $a \neq 0$, intersects the graph of f in two points—namely, (a^2, a) and $(a^2, -a)$. This shows that this relation is not a function.   ∎

Exercises 5.1

1. Let $S = \{1, 2, 3, 4\}$. Determine which of these relations on S are functions.

   (a) $f_1 = \{(1, 2), (3, 4), (4, 1)\}$ (b) $f_2 = \{(1, 3), (2, 3), (3, 3), (4, 3)\}$
   (c) $f_3 = \{(1, 1), (2, 2), (3, 3), (3, 4), (4, 4)\}$
   (d) $f_4 = \{(1, 3), (2, 4), (3, 1), (4, 2)\}$

2. Determine which of the following alleged "functions" are really functions on $\mathbb{Z}_6$; that is, determine which of them are well-defined.

   (a) $g_1([a]) = [a + 1]$ (b) $g_2([a]) = [2a]$
   (c) $g_3([a]) = [a \text{ DIV } 2]$ (d) $g_4([a]) = [-a]$

3. Determine which of the following relations are functions and find the images of those that are.

   (a) the relation f on $\mathbb{R}$ defined by $(x, y) \in f \leftrightarrow y = x^2$
   (b) the relation g on $\mathbb{R}$ defined by $(x, y) \in g \leftrightarrow |x - y| = 2$
   (c) the relation h from $\mathbb{N} - \{1\}$ to the set P of primes defined by

   $$(n, p) \in h \leftrightarrow p \text{ is the smallest prime factor of } n$$

   (d) the relation f defined on your family tree by $(x, y) \in f \leftrightarrow y$ is an uncle of x
   (e) the relation g defined on your family tree by

   $$(x, y) \in g \leftrightarrow y \text{ is the eldest son of } x\text{'s paternal grandfather}$$

4. Determine which of these alleged "functions" are well-defined.

   (a) $f_1: \mathbb{Z}_{12} \to \mathbb{Z}_8; \ f_1([n]) = [n]$ (b) $f_2: \mathbb{Z}_8 \to \mathbb{Z}_{12}; \ f_2([n]) = [n + 2]$
   (c) $f_3: \mathbb{Z}_{12} \to \mathbb{Z}_8; \ f_3([n]) = [2n]$ (d) $f_4: \mathbb{Z}_8 \to \mathbb{Z}_{12}; \ f_4([n]) = [3n]$
   (e) $f_5: \mathbb{Z}_{12} \to \mathbb{Z}_6; \ f_5([n]) = [n]$ (f) $f_6: \mathbb{Z}_6 \to \mathbb{Z}_{12}; \ f_6([n]) = [n + 2]$
   (g) $f_7: \mathbb{Z}_{12} \to \mathbb{Z}_6; \ f_7([n]) = [2n]$ (h) $f_8: \mathbb{Z}_6 \to \mathbb{Z}_{12}; \ f_8[(n]) = [3n]$

5. Recalling that $M_3(\mathbb{Z})$ is the set of all 3 by 3 matrices over $\mathbb{Z}$, define a relation g on $M_3(\mathbb{Z})$ by $(A, B) \in g$ if and only if $B = 2A$. Show that g is a function and find im g.

5.2

ONE-TO-ONE AND ONTO FUNCTIONS

Given a function $f: A \to B$, it should be emphasized that, for each $a \in A$, there is exactly one $b \in B$ for which $(a, b) \in f$. However, it need not be the case that for each $d \in B$ there is exactly one $c \in A$ such that $(c, d) \in f$. In fact, for some $d \in B$ there may be no element $c \in A$ such that $(c, d) \in f$, or there may be several elements $c \in A$ with $(c, d) \in f$. For example, consider the function $f: \mathbb{R} \to \mathbb{R}$ defined by $f(x) = x^2 + 1$. It is clear that $f(x) \geq 1$ for all $x \in \mathbb{R}$, so, for instance, there is no $x \in \mathbb{R}$ for which $f(x) = 0$. On the other hand, it is readily observed that $f(1) = f(-1) = 2$; actually, $f(a) = f(-a) = a^2 + 1$ for all $a \in \mathbb{R}$.

Some functions $f: A \to B$ satisfy the property that for each $b \in B$ there is at most one $a \in A$ such that $f(a) = b$. This condition may be rephrased as follows:

For $a_1, a_2 \in A$, if $f(a_1) = f(a_2)$, then $a_1 = a_2$.

For example, consider the function $f: \mathbb{R} \to \mathbb{R}$ defined by $f(x) = 3x + 5$. To show that the aforementioned condition holds, suppose that $f(a_1) = f(a_2)$ for some $a_1, a_2 \in \mathbb{R}$. Then $3a_1 + 5 = 3a_2 + 5$, from which it easily follows that $a_1 = a_2$.

DEFINITION 5.2

A function $f: A \to B$ is called *one-to-one* provided that, for all $a_1, a_2 \in A$, the following condition holds:

$$f(a_1) = f(a_2) \to a_1 = a_2$$

As just indicated, the condition stated in the definition that a function be one-to-one is a workable one. It is sometimes convenient to use the condition in the form:

For all $a_1, a_2 \in A$, $a_1 \neq a_2 \to f(a_1) \neq f(a_2)$.

This is obtained by taking the contrapositive of the implication in the definition.

When is the function $f: A \to B$ not one-to-one? Taking the negation of the condition in Definition 5.2, we obtain

f is not one-to-one $\leftrightarrow$ for some $a_1, a_2 \in A$, $a_1 \neq a_2 \wedge f(a_1) = f(a_2)$

Example 5.5   Determine which of the following functions are one-to-one.

(a) $f_1: \{1, 2, 3\} \to \{1, 2, 3\}$; $f_1(1) = 2$, $f_1(2) = 2$, $f_1(3) = 1$
(b) $f_2: \{1, 2, 3\} \to \{1, 2, 3\}$; $f_2(1) = 3$, $f_2(2) = 2$, $f_2(3) = 1$
(c) $f: \mathbb{Z} \to \mathbb{Z}$; $f(m) = m - 1$
(d) $g_1: \mathbb{Z} \to \mathbb{Z}$; $g_1(m) = 3m + 1$
(e) $h: \mathbb{Z} \to \mathbb{N}$; $h(m) = |m| + 1$
(f) $p: \mathbb{Q} - \{1\} \to \mathbb{Q}$; $p(r) = r/(1 - r)$

Solution   (a) Here f_1 is not one-to-one since $f_1(1) = f_1(2) = 2$.

(b) This function f_2 is one-to-one since no two elements of $\{1, 2, 3\}$ have the same image under f_2.

(c) This function f maps each integer to its predecessor, and it is easily seen to be one-to-one.

(d) If $g_1(m_1) = g_1(m_2)$, where $m_1, m_2 \in \mathbb{Z}$, then $3m_1 + 1 = 3m_2 + 1$, which implies that $m_1 = m_2$. So g_1 is one-to-one.

(e) If $h(m_1) = h(m_2)$, then $|m_1| + 1 = |m_2| + 1$, which implies that $|m_1| = |m_2|$. But this does not imply that $m_1 = m_2$, which leads one to suspect that h is not one-to-one. Indeed, if we let $m_1 = -1$ and $m_2 = 1$, then we see that $h(-1) = h(1) = 2$, which shows that h is not one-to-one.

(f) Let $r_1 \neq 1$ and $r_2 \neq 1$ be rational numbers. Then

$$p(r_1) = p(r_2) \rightarrow \frac{r_1}{1 - r_1} = \frac{r_2}{1 - r_2}$$

$$\rightarrow r_1(1 - r_2) = r_2(1 - r_1)$$

$$\rightarrow r_1 - r_1 r_2 = r_2 - r_2 r_1$$

$$\rightarrow r_1 = r_2$$

This shows that the function p is one-to-one. ∎

Suppose we have a function $f: X \rightarrow Y$, where both X and Y are subsets of $\mathbb{R}$. How can we determine from the graph of f whether f is one-to-one? Well, if f is not one-to-one, then there exist $x_1 \neq x_2$ such that $f(x_1) = f(x_2)$. Letting $y_1 = f(x_1)$, we have the two distinct points (x_1, y_1) and (x_2, y_1) that are on the graph of f and that are also on the horizontal line $y = y_1$. Conversely, if some horizontal line intersects the graph of f in more than one point, then f is not one-to-one. This yields the "horizontal line test:"

The function f is one-to-one if and only if every horizontal line intersects the graph of f in at most one point.

Example 5.6 Apply the horizontal line test to determine which of the functions defined in Example 5.3 are one-to-one. (See Figure 5.1.)

Solution
(a) The function $f: \mathbb{R} \rightarrow \mathbb{R}$ defined by $f(x) = 4 - 2x$ is one-to-one by the horizontal line test.

(b) The function $f: (0, \infty) \rightarrow (0, \infty)$ defined by $f(x) = 1/x$ is also seen to be one-to-one.

(c) The function $f: \mathbb{R} \rightarrow [-4, \infty)$ defined by $f(x) = x^2 - 2x - 3$ fails the horizontal line test. For example, the line $y = 0$ (the x-axis) intersects the graph in the points $(-1, 0)$ and $(3, 0)$. Thus, f is not one-to-one. ∎

We caution the reader that the horizontal line test, as well as the vertical line test mentioned in Section 5.1, apply to a function f only when both the domain and codomain of f are subsets of $\mathbb{R}$.

Given a function $f: A \rightarrow B$, what can be said about im f? Knowing nothing else, we can say only that im $f \subseteq B$. One extreme possibility is provided by choosing a fixed element $b \in B$ and defining $g: A \rightarrow B$ by $g(a) = b$ for all $a \in A$. In this case, im $f = \{b\}$, and g is called a *constant function* (the value of g is constant at b). The other extreme is the case of a function $f: A \rightarrow B$ for which im $f = B$.

> **DEFINITION 5.3**
>
> A function $f: A \rightarrow B$ is called *onto* provided im $f = B$.

Observe that a function $f: A \rightarrow B$ is onto provided, for each $b \in B$, there exists an $a \in A$ such that $f(a) = b$. This condition provides a very common method for proving that a given function $f: A \rightarrow B$ is onto. Choose an arbitrary element $b \in B$, set $f(a) = b$, and then attempt to solve this equation for a in terms of b. If such a solution exists and is in A, then f is onto. On the other hand, if for some $b \in B$ there is no solution in A to the equation $f(a) = b$, then f is not onto.

Example 5.7   Determine which of the following functions are onto.

(a) $f_1: \{1, 2, 3\} \rightarrow \{1, 2, 3\}$; $f_1(1) = 2$, $f_1(2) = 2$, $f_1(3) = 1$
(b) $f_2: \{1\,2, 3\} \rightarrow \{1, 2, 3\}$; $f_2(1) = 3$, $f_2(2) = 2$, $f_2(3) = 1$
(c) $f: \mathbb{Z} \rightarrow \mathbb{Z}$; $f(m) = m - 1$
(d) $g_1: \mathbb{Z} \rightarrow \mathbb{Z}$; $g_1(m) = 3m + 1$
(e) $h: \mathbb{Z} \rightarrow \mathbb{N}$; $h(m) = |m| + 1$
(f) $p: \mathbb{Q} - \{1\} \rightarrow \mathbb{Q}$; $p(r) = r/(1 - r)$
(g) $g_2: \mathbb{Q} \rightarrow \mathbb{Q}$; $g_2(r) = 3r + 1$

Solution

(a) Here f_1 is not onto since im $f_1 = \{1, 2\} \neq \{1, 2, 3\}$.
(b) The function f_2 is onto since im $f_2 = \{1, 2, 3\}$, the codomain of f_2.
(c) The function f is onto, since for any $n \in \mathbb{Z}$,

$$f(n + 1) = (n + 1) - 1 = n$$

(d) In this case, we find im g_1 directly and decide if g_1 is onto. Thus we have

$$\text{im } g_1 = \{g_1(m) \mid m \in \mathbb{Z}\} = \{3m + 1 \mid m \in \mathbb{Z}\}$$
$$= \{\ldots, -5, -2, 1, 4, 7, \ldots\} \neq \mathbb{Z}$$

So g_1 is not onto.
(e) For $n \in \mathbb{N}$,

$$h(m) = n \leftrightarrow |m| + 1 = n$$
$$\leftrightarrow |m| = n - 1$$
$$\leftrightarrow m = n - 1 \quad \text{or} \quad m = 1 - n$$

Thus h is onto; in fact, each $n \in \mathbb{N} - \{1\}$ is the image of two integers, namely, $n - 1$ and $1 - n$.
(f) For $s \in \mathbb{Q}$,

$$p(r) = s \leftrightarrow \frac{r}{1 - r} = s$$
$$\leftrightarrow r = s(1 - r)$$
$$\leftrightarrow r + rs = s$$
$$\leftrightarrow r(1 + s) = s$$
$$\leftrightarrow r = \frac{s}{1 + s}$$

Thus, if $s \neq -1$, then $r = s/(1 + s) \in \mathbb{Q} - \{1\}$ and $p(r) = s$. However, there does not exist an $r \in \mathbb{Q} - \{1\}$ such that $p(r) = -1$. Therefore im $p = \mathbb{Q} - \{-1\}$, and the function p just misses being onto.

(g) Note that this function g_2 has the same rule as the function g_1 of part (d), but the domain and codomain have been changed from $\mathbb{Z}$ to $\mathbb{Q}$. Let's see what happens. Let $s \in \mathbb{Q}$; we wish to find $r \in \mathbb{Q}$ such that $g_2(r) = s$. Now

$$g_2(r) = s \leftrightarrow 3r + 1 = s \leftrightarrow r = \frac{s - 1}{3}$$

This shows that g_2 is onto; for each rational number s, the image of the rational number $r = (s - 1)/3$ is s. This example illustrates the important point that "ontoness" for a given function depends not only on the defining equation of the function, but on the domain and codomain as well. ∎

We have seen examples of functions that are one-to-one and not onto, and the reverse possibility, functions that are onto but not one-to-one. Under what conditions does the existence of one condition imply the other? One very important case is supplied by the following theorem.

THEOREM 5.1 Let A and B be finite sets with $|A| = |B|$ and let f be a function from A to B. Then f is one-to-one if and only if f is onto.

Proof Let $|A| = |B| = m$ and suppose $A = \{a_1, a_2, \ldots, a_m\}$.

We first assume that f is one-to-one and show that f is onto. The image of f is the set

$$\text{im } f = \{f(a) \mid a \in A\}$$
$$= \{f(a_1), f(a_2), \ldots, f(a_m)\}$$

and we know that im $f \subseteq B$. If we can show $|\text{im } f| = m$, then we will have both im $f \subseteq B$ and $|\text{im } f| = |B|$. We may then conclude that im $f = B$. To show that $|\text{im } f| = m$, it suffices to show that $f(a_1), f(a_2), \ldots, f(a_m)$ are distinct. Suppose $f(a_i) = f(a_j)$ for some i and j. Since f is one-to-one, $f(a_i) = f(a_j)$ implies that $a_i = a_j$, and hence that $i = j$. This shows that $f(a_1), f(a_2), \ldots, f(a_m)$ are distinct, thus proving that im $f = B$.

Next we assume that f is onto and show that f is one-to-one. Since f is onto, im $f = B$. Then $\{f(a_1), f(a_2), \ldots, f(a_m)\} = B$ and $|B| = m$, so it must be that $f(a_1), f(a_2), \ldots, f(a_m)$ are distinct. Hence $a_i \neq a_j$ implies that $f(a_i) \neq f(a_j)$, which shows that f is one-to-one. □

Theorem 5.1 can be nicely applied in various situations. For example, consider the function $f: \mathbb{Z}_{30} \to \mathbb{Z}_{30}$ defined by $f([a]) = [7a]$. (We leave it to the student to show that f is well-defined.) The following steps show that

f is one-to-one:

$$f([a]) = f([b]) \rightarrow [7a] = [7b]$$

$$\rightarrow 7a \equiv 7b \pmod{30}$$

$$\rightarrow 30 \mid (7a - 7b)$$

$$\rightarrow 30 \mid 7(a - b)$$

$$\rightarrow 30 \mid (a - b) \quad (\text{since } \gcd(7, 30) = 1)$$

$$\rightarrow [a] = [b]$$

Thus f is one-to-one. We now get that f is onto with no additional work; just apply Theorem 5.1!

We inject a word of caution with regard to the proper application of Theorem 5.1. In order to apply this theorem to a function $f: A \rightarrow B$, it must be made certain that A and B are finite sets and that $|A| = |B|$.

Let $A = \{a_1, a_2, \ldots, a_n\}$ and suppose that $f: A \rightarrow A$ is a one-to-one function. From this it follows that the ordered n-tuple $(f(a_1), f(a_2), \ldots, f(a_n))$ is simply an ordered arrangement of the elements of A. Indeed, it seems clear that any one-to-one function on A can be thought of as selecting the elements of A in some order, or as simply "permuting" the elements of A.

DEFINITION 5.4

Let A be a nonempty set. A function $f: A \rightarrow A$ is called a *permutation* of A provided f is both one-to-one and onto.

If A is a finite set, then it is altogether natural to ask for the number of permutations of A. Certainly it is more revealing, and perhaps more pertinent, to ask for a method of generating the actual permutations of A. For the time being, we consider only the former question and postpone the latter until Chapter 6. The formula for the number of permutations of a finite set is a direct consequence of our next theorem. Recall that if n is a nonnegative integer, then n-factorial is given by

$$n! = \begin{cases} 1 & \text{if} \quad n = 0 \\ 1 \cdot 2 \cdots n & \text{if} \quad n \geq 1 \end{cases}$$

Recursively, n-factorial is given by

$$n! = \begin{cases} 1 & \text{if} \quad n = 0 \\ n \cdot (n - 1)! & \text{if} \quad n \geq 1 \end{cases}$$

THEOREM 5.2   Let A and B be nonempty finite sets with $|A| = r$, $|B| = m$, and $r \leq m$. Then the number of one-to-one functions from A to B is given by

$$\frac{m!}{(m-r)!}$$

Proof   We proceed by induction on m and let S be the set of all positive integers m for which the result holds. We denote by $P(m, r)$ the number of one-to-one functions from an r-element set to an m-element set. If $m = 1$, then necessarily $r = 1$, and it is clear that $P(1, 1) = 1 = 1!/(1-1)!$. So $1 \in S$. Assume $k \in S$ for some arbitrary integer $k \geq 1$. Thus the induction hypothesis is the statement $P(k, r) = k!/(k-r)!$ for all r, where $1 \leq r \leq k$.

We must now show that $k + 1 \in S$. To see this, let $A = \{a_1, a_2, \ldots, a_r\}$, let $|B| = k + 1$, with $r \leq k + 1$, and let $f: A \rightarrow B$ be any one-to-one function from A to B. Consider $f(a_1)$; there are $k + 1$ choices for this image. Once $f(a_1)$ is selected, distinct images $f(a_2), f(a_3), \ldots, f(a_r)$ must be selected in $B - \{f(a_1)\}$. But this is equivalent to choosing a one-to-one function from $A - \{a_1\}$ to $B - \{f(a_1)\}$, which can be done in $P(k, r-1)$ ways (by the induction hypothesis). Hence, by the multiplication principle, f can be selected in

$$(k+1)P(k, r-1) = \frac{(k+1) \cdot k!}{(k-(r-1))!}$$

$$= \frac{(k+1)!}{((k+1)-r)!}$$

$$= P(k+1, r)$$

ways. So $k + 1 \in S$, and it follows that $P(m, r) = m!/(m-r)!$ for all $m \in \mathbb{N}$.   □

Applying Theorem 5.2 in the case $B = A$ yields the following corollary.

COROLLARY 5.3   For any $n \in \mathbb{N}$, the number of permutations of a set A of cardinality n is $n!$.

Example 5.8   As shown above, the function $f: \mathbb{Z}_{30} \rightarrow \mathbb{Z}_{30}$ defined by $f([a]) = [7a]$ is a permutation of $\mathbb{Z}_{30}$. From Examples 5.5(c) and 5.7(c), it follows that the function $f: \mathbb{Z} \rightarrow \mathbb{Z}$ defined by $f(m) = m - 1$ is a permutation of $\mathbb{Z}$. In Example 5.7(g) we showed that the function $g_2: \mathbb{Q} \rightarrow \mathbb{Q}$ defined by $g_2(r) = 3r + 1$ is onto. It is not hard to show that this function is also one-to-one. Therefore g_2 is a permutation of $\mathbb{Q}$.   ■

Suppose now that we are given $f\colon A \to B$; we know that im $f = \{f(a) \mid a \in A\}$. Thus it seems natural to write im $f = f(A)$. More generally, if $C \subseteq A$, then we define the *image of C under f* to be the set

$$f(C) = \{f(c) \mid c \in C\}$$

Turn the situation around: given $D \subseteq B$, how should we refer to those elements $a \in A$ for which $f(a) \in D$? In a sense, seeking such elements in A has the effect of reversing the direction of f. In other words, the action proceeds from D to A. Formally, we define the *inverse image of D under f* to be the set

$$f^{-1}(D) = \{a \in A \mid f(a) \in D\}$$

Example 5.9   Let g be the permutation of $\mathbb{Q}$ defined by $g(r) = 3r + 1$. Find the following sets:

    (a) $g(\mathbb{Z})$ (b) $g(E)$, where E is the set of even integers
    (c) $g^{-1}(\mathbb{N})$ (d) $g^{-1}(D)$, where D is the set of odd integers

Solution   (a) If $m \in \mathbb{Z}$, then $g(m) = 3m + 1$. Thus $g(\mathbb{Z}) = \{3m + 1 \mid m \in \mathbb{Z}\} = \{\ldots, -5, -2, 1, 4, 7, \ldots\}$.
    (b) We have

$$k \in g(E) \leftrightarrow k = g(2m) \quad \text{for some } m \in \mathbb{Z}$$
$$\leftrightarrow k = 3(2m) + 1 \quad \text{for some } m \in \mathbb{Z}$$
$$\leftrightarrow k = 6m + 1 \quad \text{for some } m \in \mathbb{Z}$$

Thus $g(E) = \{6m + 1 \mid m \in \mathbb{Z}\} = \{\ldots, -11, -5, 1, 7, 13, \ldots\}$.
    (c) Here we have

$$r \in g^{-1}(\mathbb{N}) \leftrightarrow 3r + 1 = n \quad \text{for some } n \in \mathbb{N}$$
$$\leftrightarrow r = \frac{n-1}{3} \quad \text{for some } n \in \mathbb{N}$$

Hence $g^{-1}(\mathbb{N}) = \{(n-1)/3 \mid n \in \mathbb{N}\} = \{0, \frac{1}{3}, \frac{2}{3}, 1, \frac{4}{3}, \frac{5}{3}, \ldots\}$.
    (d) Similarly,

$$r \in g^{-1}(D) \leftrightarrow 3r + 1 = 2m + 1 \quad \text{for some } m \in \mathbb{Z}$$
$$\leftrightarrow r = \frac{2m}{3} \quad \text{for some } m \in \mathbb{Z}$$

Hence $g^{-1}(D) = \{2m/3 \mid m \in \mathbb{Z}\} = \{\ldots, \frac{-4}{3}, \frac{-2}{3}, 0, \frac{2}{3}, \frac{4}{3}, \ldots\}$.   ■

THEOREM 5.4   Given $f\colon A \to B$, let C and D be subsets of A and let E and F be subsets of B. Then the following relationships hold.

    1. (a) $f(C \cup D) = f(C) \cup f(D)$
        (b) $f^{-1}(E \cup F) = f^{-1}(E) \cup f^{-1}(F)$
    2. (a) $f(C \cap D) \subseteq f(C) \cap f(D)$
        (b) $f^{-1}(E \cap F) = f^{-1}(E) \cap f^{-1}(F)$

3. (a) $f(C) - f(D) \subseteq f(C - D)$
   (b) $f^{-1}(E) - f^{-1}(F) = f^{-1}(E - F)$
4. (a) If $C \subseteq D$, then $f(C) \subseteq f(D)$.
   (b) If $E \subseteq F$, then $f^{-1}(E) \subseteq f^{-1}(F)$.

Proof   We prove 1(a) and 3(b). The proofs of the remaining parts are left to Exercise 2.

The proof that $f(C \cup D) = f(C) \cup f(D)$ proceeds as follows:

$$y \in f(C \cup D) \leftrightarrow \exists x \in C \cup D (y = f(x))$$

$$\leftrightarrow [\exists x \in C(y = f(x))] \vee [\exists x \in D(y = f(x))]$$

$$\leftrightarrow y \in f(C) \vee y \in f(D)$$

$$\leftrightarrow y \in f(C) \cup f(D)$$

Thus $f(C \cup D) = f(C) \cup f(D)$.

The proof that $f^{-1}(E) - f^{-1}(F) = f^{-1}(E - F)$ proceeds along similar lines:

$$x \in f^{-1}(E) - f^{-1}(F) \leftrightarrow \left(x \in f^{-1}(E) \wedge x \notin f^{-1}(F) \right)$$

$$\leftrightarrow f(x) \in E \wedge f(x) \notin F$$

$$\leftrightarrow f(x) \in E - F$$

$$\leftrightarrow x \in f^{-1}(E - F)$$

Hence $f^{-1}(E) - f^{-1}(F) = f^{-1}(E - F)$.   □

There are several additional properties involving the notions of image and inverse image. Some of these are addressed in the exercises.

Exercises 5.2

1. Determine which of the following functions are one-to-one.
   (a) $f: \mathbb{Z} \to \mathbb{N}$; $f(m) = m^2 + 1$ (b) $g: \mathbb{R} \to \mathbb{R}$; $g(x) = x^3$
   (c) $h: \mathbb{R} \to \mathbb{R}$; $h(x) = x^3 - x$ (d) $p: \mathbb{Q} \to \mathbb{R}$; $p(r) = 2^r$
   (e) the "cardinality function" n from $\mathscr{P}(A)$, where A is a nonempty finite set, to $\{0, 1, \ldots, |A|\}$: for $B \in \mathscr{P}(A)$, $n(B) = |B|$
   (f) the "complement function" c on $\mathscr{P}(A)$: for $B \in \mathscr{P}(A)$, $c(B) = A - B$

2. Prove the remaining parts of Theorem 5.4:
   - (a) part 1(b)   (b) part 2(a)   (c) part 2(b)
   - (d) part 3(a)   (e) part 4(a)   (f) part 4(b)
3. Determine which of the functions in Exercise 1 are onto.
4. List all the one-to-one functions from $\{1,2\}$ to $\{1,2,3,4\}$.
5. Each part gives sets A and B and a function from A to B. Determine whether the function is one-to-one.
   - (a) $A = \{1,2,3,4\}$, $B = \{1,2,3\}$; $f_1(1) = 2$, $f_1(2) = 3$, $f_1(3) = f_1(4) = 1$
   - (b) $A = \{1,2,3\}$, $B = \{1,2,3,4\}$; $f_2(1) = 3$, $f_2(2) = 2$, $f_2(3) = 1$
   - (c) $A = B = \{1,2,3,4\}$; $f_3(1) = f_3(3) = 2$, $f_3(2) = f_3(4) = 1$
   - (d) $A = B = \{1,2,3,4\}$; $f_4(1) = 3$, $f_4(2) = 4$, $f_4(3) = 1$, $f_4(4) = 2$
   - (e) $A = B = \mathbb{Z}$; $f_5(m) = -m$
   - (f) $A = B = \mathbb{Z}$; if $m \geq 0$, $f_6(m) = 2m$, if $m < 0$, $f_6(m) = 3m$
   - (g) $A = B = \mathbb{N}$; if n is odd, $f_7(n) = (n + 1)/2$, if n is even, $f_7(n) = n/2$
   - (h) $A = B = \mathbb{N}$; if n is odd, $f_8(n) = n + 1$, if n is even, $f_8(n) = n - 1$
6. List all the permutations of:
   - (a) $\{1\}$ (b) $\{1,2\}$ (c) $\{1,2,3\}$
7. Determine which of the functions in Exercise 5 are onto.
8. Let A and B be nonempty finite sets with $|A| = r$ and $|B| = m$. What can be said about the possibility that a function $f: A \rightarrow B$ is one-to-one or onto if
   - (a) $r < m$?   (b) $m < r$?
9. Determine which of the following functions are one-to-one.
   - (a) $f_1: \mathbb{Z}_{10} \rightarrow \mathbb{Z}_{10}$; $f_1([n]) = [3n]$ (b) $f_2: \mathbb{Z}_{10} \rightarrow \mathbb{Z}_{10}$; $f_2([n]) = [5n]$
   - (c) $f_3: \mathbb{Z}_{36} \rightarrow \mathbb{Z}_{36}$; $f_3([n]) = [3n]$ (d) $f_4: \mathbb{Z}_{36} \rightarrow \mathbb{Z}_{36}$; $f_4([n]) = [5n]$
   - (e) $f_5: \mathbb{Z}_{10} \rightarrow \mathbb{Z}_{10}$; $f_5([n]) = [n + 3]$
   - (f) $f_6: \mathbb{Z}_{10} \rightarrow \mathbb{Z}_{10}$; $f_6([n]) = [n + 5]$
   - (g) $f_7: \mathbb{Z}_{12} \rightarrow \mathbb{Z}_8$; $f_7([n]) = [2n]$ (h) $f_8: \mathbb{Z}_8 \rightarrow \mathbb{Z}_{12}$; $f_8([n]) = [3n]$
   - (i) $f_9: \mathbb{Z}_6 \rightarrow \mathbb{Z}_{12}$; $f_9([n]) = [2n]$ (j) $f_0: \mathbb{Z}_{12} \rightarrow \mathbb{Z}_{36}$; $f_0([n]) = [6n]$
10. Determine which of the following functions on $\mathbb{Z}_{12}$ are permutations. In each case find the image of $C = \{[1], [5], [7], [11]\}$ and the inverse image of $D = \{[4], [8]\}$.
    - (a) $f_1([n]) = [2n]$ (b) $f_2([n]) = [4n]$ (c) $f_3([n]) = [5n]$
11. Determine which of the functions in Exercise 9 are onto.
12. Give an example of a function on $\mathbb{N}$ that is:
    - (a) neither one-to-one nor onto   (b) one-to-one but not onto
    - (c) onto but not one-to-one   (d) both one-to-one and onto
13. Each part refers to the corresponding function defined in Exercise 5. For $C \subseteq A$ and $D \subseteq B$, find the image of C and the inverse image of D.
    - (a) $C = \{1,2\} = D$ (b) $C = \{1,3\}$, $D = \{2,4\}$
    - (c) $C = \{1,3\}$, $D = \{1\}$ (d) $C = \{2\} = D$
    - (e) $C = \{2m \mid m \in \mathbb{Z}\}$, $D = \mathbb{N}$ (f) $C = \mathbb{N}$, $D = \{2m \mid m \in \mathbb{Z}\}$
    - (g) $C = D = \{2n \mid n \in \mathbb{N}\}$ (h) $C = D = \{2n \mid n \in \mathbb{N}\}$

| 5.3 | **INVERSE FUNCTIONS AND COMPOSITION** |

Suppose that $f: A \to B$ is both one-to-one and onto. Since f is onto, given any $b \in B$, there is an element $a \in A$ such that $f(a) = b$. Moreover, since f is one-to-one, this element a is uniquely determined. Thus for each $b \in B$, there is exactly one $a \in A$ such that $b = f(a)$. We can then define a new function $g: B \to A$ as follows: For each $b \in B$,

$$g(b) = a \leftrightarrow f(a) = b$$

In other words, $g(b)$ is that unique element $a \in A$ for which $f(a) = b$.

DEFINITION 5.5

Let $f: A \to B$ be one-to-one and onto. The function $g: B \to A$ defined by

$$g(b) = a \leftrightarrow f(a) = b$$

for each $b \in B$ is called the *inverse function* of f and is denoted by f^{-1}.

The situation of a function $f: A \to B$ and its inverse $f^{-1}: B \to A$ is depicted in Figure 5.2, where $f(a) = b$. Let $f: A \to B$ be one-to-one and onto and let $D \subseteq B$. At this point we have two interpretations for the notation $f^{-1}(D)$: one is the inverse image of D under f, and the other is the image of D under the function f^{-1}. In Exercise 12 the student is asked to show that these two sets are equal. It should be pointed out, however, that if f is not both one-to-one and onto, then only the former interpretation for $f^{-1}(D)$ applies.

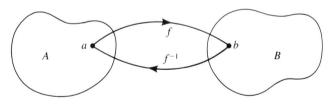

Figure 5.2

THEOREM 5.5 If $f: A \to B$ is one-to-one and onto, then the inverse function $f^{-1}: B \to A$ is also one-to-one and onto.

Proof We first show that f^{-1} is one-to-one. Suppose $f^{-1}(b_1) = f^{-1}(b_2) = a$ for some $b_1, b_2 \in B$ and $a \in A$. Then, by definition, $b_1 = f(a)$ and $b_2 = f(a)$. Since f is a function, it follows that $b_1 = b_2$. Thus f^{-1} is one-to-one. Next we show that f^{-1} is onto. Let $a \in A$ and suppose $f(a) = b$. Then, again by definition, $a = f^{-1}(b)$, so f^{-1} is onto. ☐

As a special case, Theorem 5.5 tells us that, for any set X, if $f: X \to X$ is a permutation of X, then f^{-1} is also a permutation of X.

If $f: A \to B$ is one-to-one and onto, how do we find the inverse function $f^{-1}: B \to A$? For example, given $f: \mathbb{R} \to \mathbb{R}$ defined by $f(x) = 3x - 7$, how can we find f^{-1}? Given $r \in \mathbb{R}$, we want to find $x \in \mathbb{R}$ such that $f^{-1}(r) = x$. This means that $r = f(x)$ or, in particular, that $r = 3x - 7$. With a bit of elementary algebra, one finds that $x = (r + 7)/3$; hence

$$f^{-1}(r) = \frac{r + 7}{3}$$

Example 5.10   In Examples 5.5 and 5.7, it was shown that the function $p: \mathbb{Q} - \{1\} \to \mathbb{Q} - \{-1\}$ defined by $p(r) = r/(1 - r)$ is one-to-one and onto. Find p^{-1}.

Solution   If $x \in \mathbb{Q} - \{-1\}$ and $p^{-1}(x) = r$, where $r \in \mathbb{Q} - \{1\}$, then $p(r) = x$. Thus, $r/(1 - r) = x$, or $r = (1 - r)x$. Solving for r, we obtain $r = x/(x + 1)$. Therefore $p^{-1}: \mathbb{Q} - \{-1\} \to \mathbb{Q} - \{1\}$ is given by

$$p^{-1}(x) = \frac{x}{x + 1}$$ ∎

Example 5.11   In the last section it was shown that the function $f: \mathbb{Z}_{30} \to \mathbb{Z}_{30}$, defined by $f([a]) = [7a]$, is a permutation of $\mathbb{Z}_{30}$. Find f^{-1}.

Solution   We have that

$$f^{-1}([b]) = [a] \leftrightarrow f([a]) = [b]$$
$$\leftrightarrow [7a] = [b]$$
$$\leftrightarrow 30 \,|\, (7a - b)$$
$$\leftrightarrow 30q = 7a - b$$

for some $q \in \mathbb{Z}$. In order to find $f^{-1}([b])$, we must rewrite the equation $30q = 7a - b$ in the form $30q' = a - sb$ for some integers q' and s. It will then follow that $f^{-1}([b]) = [a] = [sb]$. Recall that, since $\gcd(7, 30) = 1$, there exist integers s and t such that $7s + 30t = 1$; in fact, we know how to find s and t. Using the Euclidean algorithm, we find that $7(13) + 30(-3) = 1$. Thus, continuing from above, we have that

$$30q = 7a - b \leftrightarrow 30q \cdot 13 = (7 \cdot 13)a - 13b$$
$$\leftrightarrow 30q \cdot 13 = (1 + 3 \cdot 30)a - 13b$$
$$\leftrightarrow 30(13q - 3a) = a - 13b$$
$$\leftrightarrow [a] = [13b]$$

Therefore $f^{-1}: \mathbb{Z}_{30} \to \mathbb{Z}_{30}$ is defined by $f^{-1}([b]) = [13b]$. ∎

There are various ways in which two functions may be combined to produce a third function. One of the most common and important of these is called *composition*. Suppose $f: A \to B$ and $g: B \to C$ are given. Then, for any $a \in A$, there is associated a unique $b \in B$ such that $f(a) = b$. For this element b, there is a unique $c \in C$ such that $c = g(b) = g(f(a))$. Hence, with each $a \in A$, there is associated a unique element $c \in C$, where $c = g(f(a))$. This association allows us to define a new function $h: A \to C$ by $h(a) = g(f(a))$. This situation is depicted in Figure 5.3.

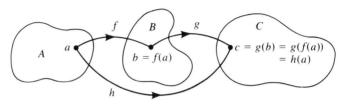

Figure 5.3

DEFINITION 5.6

Given $f: A \to B$ and $g: B \to C$, we define the *composite function* $g \circ f$: $A \to C$ as follows: For each $a \in A$,

$$(g \circ f)(a) = g(f(a))$$

Example 5.12   Let E be the set of even integers, let $f: \mathbb{Z} \to E$ be defined by $f(m) = 2m$, and let $g: E \to \mathbb{N}$ be defined by $g(n) = |n|/2 + 1$. Then $g \circ f: \mathbb{Z} \to \mathbb{N}$ is defined by

$$(g \circ f)(m) = g(f(m)) = g(2m) = \frac{|2m|}{2} + 1 = |m| + 1 \qquad \blacksquare$$

Example 5.13   Let $f: \mathbb{R} - \{0\} \to \mathbb{R} - \{1\}$ and $g: \mathbb{R} - \{1\} \to \mathbb{R} - \{2\}$ be given by $f(x) = (x + 1)/x$ and $g(x) = 3x - 1$. Then $g \circ f: \mathbb{R} - \{0\} \to \mathbb{R} - \{2\}$ is defined by

$$(g \circ f)(x) = g(f(x)) = g\left(\frac{x + 1}{x}\right)$$

$$= \frac{3(x + 1)}{x} - 1$$

$$= \frac{3(x + 1)}{x} - \frac{x}{x}$$

$$= \frac{2x + 3}{x} \qquad \blacksquare$$

Example 5.14   Let $f: \mathbb{Z} \to \mathbb{Z}$ and $g: \mathbb{Z} \to \mathbb{Z}$ be defined by $f(x) = x + 3$ and $g(x) = -x$. Find (a) $f \circ g: \mathbb{Z} \to \mathbb{Z}$ and (b) $g \circ f: \mathbb{Z} \to \mathbb{Z}$.

Solution   (a) $(f \circ g)(x) = f(g(x)) = f(-x) = -x + 3$
(b) $(g \circ f)(x) = g(f(x)) = g(x + 3) = -(x + 3) = -x - 3$

Notice that $f \circ g \neq g \circ f$, and that f, g, $f \circ g$, and $g \circ f$ are all permutations of $\mathbb{Z}$.   ■

In what follows, we shall be interested in knowing when certain functions are "equal." What does this mean? After all, if f and g are functions, then each is a set of ordered pairs, and perhaps equality should simply mean that these sets are equal. This condition would certainly imply that the two functions have the same domain and image, although not necessarily the same codomain. We shall not insist that the codomains be the same, but instead keep in mind that a function is a relation (a set of ordered pairs), and we already know when two sets are equal.

DEFINITION 5.7

The functions $f: A \to B$ and $g: C \to D$ are *equal* if and only if $A = C$ and $f(a) = g(a)$ for all $a \in A$.

For instance, the functions $f_1: \mathbb{R} \to \mathbb{R}$ and $f_2: \mathbb{R} \to [0, \infty)$ defined by $f_1(x) = f_2(x) = x^2$ are equal by our definition.

There are several interesting results that involve the composition of functions and the properties onto and one-to-one. For example, suppose $f: A \to B$ is one-to-one and onto; thus $f^{-1}: B \to A$ exists. Given any $a \in A$, if $f(a) = b$, then we know that $f^{-1}(b) = a$ and hence that $f^{-1}(f(a)) = a$. So the composite function

$$f^{-1} \circ f: A \to A$$

satisfies the property $(f^{-1} \circ f)(a) = a$ for all $a \in A$. In a similar fashion, we can determine that $f \circ f^{-1}: B \to B$ satisfies the condition $(f \circ f^{-1})(b) = b$ for every $b \in B$. Note that each of $f \circ f^{-1}$ and $f^{-1} \circ f$ is a function of the type $h: X \to X$, where $h(x) = x$ for all $x \in X$.

DEFINITION 5.8

For any nonempty set A, the function $i_A: A \to A$ defined by $i_A(a) = a$ for all $a \in A$ is called the *identity function on* A.

In view of the preceding discussion, if $f: A \rightarrow B$ is one-to-one and onto, then $f^{-1} \circ f = i_A$ and $f \circ f^{-1} = i_B$. A very basic and easily verified property of identity functions is contained in the following theorem.

THEOREM 5.6   Given any function $f: A \rightarrow B$, we have $i_B \circ f = f$ and $f \circ i_A = f$.

The proof of Theorem 5.6 is left to Exercise 2.

THEOREM 5.7   Given $f: A \rightarrow B$ and $g: B \rightarrow C$, the following hold:

1. If f and g are both one-to-one, then $g \circ f$ is one-to-one.
2. If f and g are both onto, then $g \circ f$ is onto.

Proof   First consider the proof of part 1. Suppose, for $a_1, a_2 \in A$, that $(g \circ f)(a_1) = (g \circ f)(a_2)$. Then, by the definition of composition, $g(f(a_1)) = g(f(a_2))$. Since g is one-to-one, we have $f(a_1) = f(a_2)$. Then, since f is also one-to-one, $a_1 = a_2$. Hence $g \circ f$ is one-to-one.

Next consider the proof of part 2. To prove that $g \circ f$ is onto, begin with an arbitrary element $c \in C$. Since g is onto, there is an element $b \in B$ such that $g(b) = c$. Since $b \in B$ and since f is onto, there is an element $a \in A$ for which $f(a) = b$. Thus $(g \circ f)(a) = g(f(a)) = g(b) = c$, and it follows that $g \circ f$ is onto.   □

COROLLARY 5.8   If $f: A \rightarrow B$ and $g: B \rightarrow C$ are one-to-one and onto, then the function $g \circ f: A \rightarrow C$ is one-to-one and onto. In particular, if f and g are permutations of the set A, then $g \circ f$ is a permutation of A.

For each of the statements 1 and 2 in Theorem 5.7, the converse is false (see Exercises 4 and 6). However, a partial converse does hold.

THEOREM 5.9   Given $f: A \rightarrow B$ and $g: B \rightarrow C$, the following hold:

1. If $g \circ f$ is one-to-one, then f is one-to-one.
2. If $g \circ f$ is onto, then g is onto.

Proof   We prove part 1, leaving part 2 to Exercise 10. Suppose then that $f(a_1) = f(a_2)$ for some $a_1, a_2 \in A$. Then, since $f(a_1) \in B$, we have that $g(f(a_1)) = g(f(a_2))$ or $(g \circ f)(a_1) = (g \circ f)(a_2)$. Since $g \circ f$ is given to be one-to-one, we may conclude that $a_1 = a_2$. This shows that the function f is one-to-one.   □

Given functions $f: A \rightarrow B$, $g: B \rightarrow C$, and $h: C \rightarrow D$, notice that $h \circ g$ is a function from B to D and $g \circ f$ is a function from A to C. Thus $(h \circ g) \circ f$ and $h \circ (g \circ f)$ are both functions from A to D. In fact, they are equal functions!

THEOREM 5.10   Given $f: A \rightarrow B$, $g: B \rightarrow C$, and $h: C \rightarrow D$, we have

$$(h \circ g) \circ f = h \circ (g \circ f)$$

Proof   Both $(h \circ g) \circ f$ and $h \circ (g \circ f)$ have domain A. To show equality, we must show that the two functions have the same value at each $a \in A$. Now

$$
\begin{aligned}
[(h \circ g) \circ f](a) &= (h \circ g)[f(a)] \\
&= h(g(f(a))) \\
&= h[(g \circ f)(a)] \\
&= [h \circ (g \circ f)](a)
\end{aligned}
$$

Therefore, by the definition of equality of functions, $(h \circ g) \circ f = h \circ (g \circ f)$.

□

Exercises 5.3

1. Find the inverse of each of the following functions.
   (a) $f_1 : \mathbb{Q} \to \mathbb{Q}$; $f_1(r) = 4r + 2$
   (b) $f_2 : \mathbb{R} - \{1\} \to \mathbb{R} - \{1\}$; $f_2(x) = x/(x - 1)$
   (c) $f_3 : \mathbb{Z}_{12} \to \mathbb{Z}_{12}$; $f_3([x]) = [5x]$
   (d) $f_4 : \mathbb{Z}_{39} \to \mathbb{Z}_{39}$; $f_4([x]) = [5x + 2]$
   (e) $f_5 : \mathbb{Z} \to \mathbb{Z}$; $f_5(m) = m + 1$
   (f) $f_6 : \mathbb{Z} \to \mathbb{N}$; $f_6(m) = \begin{cases} 2m + 1, & m \geq 0 \\ -2m, & m < 0 \end{cases}$
   (g) $f_7 : \{1,2,3,4\} \to \{1,2,3,4\}$; $f_7(1) = 4$, $f_7(2) = 1$, $f_7(3) = 2$, $f_7(4) = 3$
   (h) $f_8 : \{1,2,3,4\} \to \{1,2,3,4\}$; $f_8(1) = 3$, $f_8(2) = 4$, $f_8(3) = 1$, $f_8(4) = 2$

2. Prove Theorem 5.6.

3. Find $g \circ f$.
   (a) $f : \mathbb{Z} \to \mathbb{N}$; $f(m) = |m| + 1$, $g : \mathbb{N} \to (0, \infty)$; $g(n) = 1/n$
   (b) $f : \mathbb{R} \to (0, 1)$; $f(x) = 1/(x^2 + 1)$, $g : (0, 1) \to (0, 1)$; $g(x) = 1 - x$
   (c) $f : \mathbb{R} - \{2\} \to \mathbb{R} - \{0\}$; $f(x) = 1/(x - 2)$, $g : \mathbb{R} - \{0\} \to \mathbb{R} - \{0\}$; $g(x) = 1/x$
   (d) $f : \mathbb{R} \to [1, \infty)$; $f(x) = x^2 + 1$, $g : [1, \infty) \to [0, \infty)$; $g(x) = \sqrt{x - 1}$
   (e) $f : \mathbb{Q} - \{10/3\} \to \mathbb{Q} - \{3\}$; $f(r) = 3r - 7$, $g : \mathbb{Q} - \{3\} \to \mathbb{Q} - \{2\}$; $g(r) = 2r/(r - 3)$
   (f) $f : \mathbb{Z} \to \mathbb{Z}_5$; $f(m) = [m]$, $g : \mathbb{Z}_5 \to \mathbb{Z}_5$; $g([m]) = [m + 1]$
   (g) $f : \mathbb{Z}_8 \to \mathbb{Z}_{12}$; $f([m]) = [3m]$, $g : \mathbb{Z}_{12} \to \mathbb{Z}_6$; $g([m]) = [2m]$
   (h) $f : \{1,2,3,4\} \to \{1,2,3,4\}$; $f(1) = 4$, $f(2) = 1$, $f(3) = 2$, $f(4) = 3$, $g : \{1,2,3,4\} \to \{1,2,3,4\}$; $g(1) = 3$, $g(2) = 4$, $g(3) = 1$, $g(4) = 2$

4. Give an example of sets A, B, and C and of functions $f : A \to B$ and $g : B \to C$ such that $g \circ f$ and f are both one-to-one, but g is not.

5. Given the permutations f and g, find f^{-1}, g^{-1}, $f \circ g$, $(f \circ g)^{-1}$, and $g^{-1} \circ f^{-1}$.
   (a) $f : \mathbb{Z} \to \mathbb{Z}$; $f(m) = m + 1$, $g : \mathbb{Z} \to \mathbb{Z}$; $g(m) = 2 - m$
   (b) $f : \mathbb{Z}_7 \to \mathbb{Z}_7$; $f([m]) = [m + 3]$, $g : \mathbb{Z}_7 \to \mathbb{Z}_7$; $g(m) = [2m]$
   (c) $f : \{1,2,3,4\} \to \{1,2,3,4\}$; $f(1) = 4$, $f(2) = 1$, $f(3) = 2$, $f(4) = 3$, $g : \{1,2,3,4\} \to \{1,2,3,4\}$; $g(1) = 3$, $g(2) = 4$, $g(3) = 1$, $g(4) = 2$

(d) f: $\{1, 2, 3, 4\} \rightarrow \{1, 2, 3, 4\}$; $f(1) = 2$, $f(2) = 4$, $f(3) = 3$, $f(4) = 1$,
g: $\{1, 2, 3, 4\} \rightarrow \{1, 2, 3, 4\}$; $g(1) = 1$, $g(2) = 3$, $g(3) = 4$, $g(4) = 2$

(e) f: $\mathbb{Q} \rightarrow \mathbb{Q}$; $f(r) = 4r$, g: $\mathbb{Q} \rightarrow \mathbb{Q}$; $g(r) = (r - 3)/2$

(f) f: $\mathbb{Q} - \{1\} \rightarrow \mathbb{Q} - \{1\}$; $f(r) = 2r - 1$, g: $\mathbb{Q} - \{1\} \rightarrow \mathbb{Q} - \{1\}$;
$g(r) = r/(r - 1)$

6. Give an example of sets A, B, and C and of functions f: $A \rightarrow B$ and g: $B \rightarrow C$ such that $g \circ f$ and g are both onto, but f is not.

7. For the permutations f and g given in Exercise 5, find $g \circ f$ and $f^{-1} \circ g^{-1}$.

8. Define the functions f and g on your family tree by $f(x) =$ the father of x and $g(x) =$ the eldest child of the father of x. Describe these functions.

(a) $f \circ f$ (b) $f \circ g$ (c) $g \circ f$ (d) $g \circ g$

9. Let f: $A \rightarrow B$ and g: $B \rightarrow C$ be both one-to-one and onto functions. Prove that $(g \circ f)^{-1} = f^{-1} \circ g^{-1}$.

10. Prove Theorem 5.9, part 2.

11. Let f: $A \rightarrow B$ be one-to-one and onto. Prove that $(f^{-1})^{-1} = f$.

12. Let f: $A \rightarrow B$ be one-to-one and onto and let $D \subseteq B$. Let C_1 be the inverse image of D under f and let C_2 be the image of D under f^{-1}. Show that $C_1 = C_2$.

13. Let f: $A \rightarrow B$ and g: $B \rightarrow A$ be one-to-one and onto functions. Prove: If $g \circ f = i_A$ (or $f \circ g = i_B$), then $g = f^{-1}$.

5.4

BIG-O NOTATION

In implementing an algorithm on a computer, one of the major concerns is the running time of the algorithm. For instance, suppose we are using an algorithm to sort a list of n integers in nonincreasing order. How fast the algorithm runs is dependent on many factors, some being the input size n, the computer being used, the operating system, and the particular implementation of the algorithm. We are only interested in running time as a function $f(n)$ of n (one can assume other factors are constant).

Of special significance is how $f(n)$ increases as n increases, or the order of magnitude of $f(n)$ for large n. For example, suppose that $f(n) = 3n^3 + 6n^2 + 7n$. Then, for $n \geq 7$, we have

$$3n^3 + 6n^2 + 7n \leq 3n^3 + 6n^2 + n^2$$

$$\leq 3n^3 + 7n^2$$

$$\leq 3n^3 + n^3$$

$$\leq 4n^3$$

So the running time of the algorithm is no more than $4n^3$. Thus if a list of 100 integers can be sorted in two seconds, then a list of 300 integers can be sorted in no more than $2 \cdot 3^3 = 54$ seconds. We would say that $f(n)$ has order of magnitude n^3 or grows on the order of n^3. A special notation is used to express such facts.

DEFINITION 5.9

Given functions $f: \mathbb{N} \to \mathbb{R}$ and $g: \mathbb{N} \to \mathbb{R}$, we write $f(n) = O(g(n))$ if there is some constant C and some positive integer n_0 such that

$$|f(n)| \le C|g(n)|$$

for all $n \ge n_0$. We read $f(n) = O(g(n))$ as $f(n)$ *is big-O of* $g(n)$.

In other words, $f(n) = O(g(n))$ provided, for large enough values of n, $|f(n)|$ is bounded above by some constant multiple of $|g(n)|$. Hence we now see that $3n^3 + 6n^2 + 7n = O(n^3)$.

Example 5.15   Show each of the following:

(a) $6n + 5 = O(n)$ (b) $(n^2 + 8)(n + 1) = O(n^3)$
(c) $1 + 2 + \cdots + n = O(n^2)$

Solution   (a) Observing that $6n + 5 \le 7n$ for $n \ge 5$, we have $6n + 5 = O(n)$.
(b) In this case we see that, for $n \ge 3$,

$$(n^2 + 8)(n + 1) \le (n^2 + n)(2n)$$
$$= 2n^3 + 2n^2$$
$$\le 2n^3 + n^3$$
$$= 3n^3$$

Hence $(n^2 + 8)(n + 1) = O(n^3)$.

(c) Simply notice that $1 + 2 + \cdots + n \le n + n + \cdots + n$ (n times) for all $n \ge 1$. So we quickly have $1 + 2 + \cdots + n = O(n^2)$. ∎

Notice, in Example 5.15, that each of the given functions has order of magnitude n^k for some $k \in \mathbb{N}$. In each case the power of k used is the smallest possible (for large enough values of n). For instance, observe that $(n^2 + 8)(n + 1) = O(n^r)$ for all $r \ge 3$, and $(n^2 + 8)(n + 1) \neq O(n^r)$ if $r < 3$.

In general, if $f(n) = O(g(n))$, then we would like $g(n)$ to have a simple form and be as small as possible. In estimates of the running time of an algorithm, some functions that are commonly used are the following: 1, n, n^2, n^3, $\log_2 n$, $n \log_2 n$, and 2^n. A comparison of the relative sizes of these functions is provided by their graphs, shown in Figure 5.4. It should be emphasized that other functions, such as $n^2/\log_2 n$, $\sqrt{n}$, and $\sqrt{n} \log_2 n$, arise naturally in estimating running times.

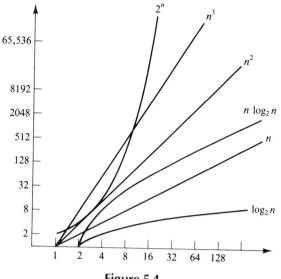

Figure 5.4

Example 5.16   Prove: $\log_2 n < n < 2^n$ for all $n \in \mathbb{N}$.

Solution   We proceed by induction to show that $n < 2^n$ for all $n \in \mathbb{N}$. Let S be the set of positive integers for which the result holds. Clearly $1 \in S$. Assume $k \in S$ for an arbitrary $k \geq 1$. Thus our induction hypothesis is: $k < 2^k$. Using this, we observe that

$$k + 1 < 2^k + 1 < 2^k + 2^k = 2 \cdot 2^k = 2^{k+1}$$

Hence $k + 1 \in S$. Therefore $S = \mathbb{N}$.

To obtain the remaining inequality $\log_2 n < n$, we use the fact that the function $f(x) = \log_2 x$ is increasing on $(0, \infty)$; thus if $0 < a < b$, then $\log_2 a < \log_2 b$. Hence, from the fact that $n < 2^n$ for all $n \in \mathbb{N}$, we get $\log_2 n < \log_2 2^n$, or $\log_2 n < n$, for all $n \in \mathbb{N}$.   ∎

Example 5.17   From the list: $1, \log_2 n, n, n \log_2 n, n^2, 2^n$, find the smallest big-O estimate for each of the given expressions.

(a) $\dfrac{\log_2 n}{n}$ (b) $n + 3\log_2 n$ (c) $\dfrac{n^2 + 3}{n}$

Solution   (a) We know that $\log_2 n < n$ for all $n \geq 1$. Thus

$$\frac{\log_2 n}{n} < 1$$

for all $n \geq 1$. So we have

$$\frac{\log_2 n}{n} = O(1)$$

(b) Since $\log_2 n < n$ for all $n \geq 1$, it is clear that $n + 3\log_2 n = O(n)$. Using the result of Exercise 9, we see that there do not exist r, $n_0 \in \mathbb{N}$ such that $n < r\log_2 n$ for all $n > n_0$. Hence $n + 3\log_2 n \neq O(\log_2 n)$.

(c) Note that

$$\frac{n^2 + 3}{n} = n + \frac{3}{n} < 2n$$

for all $n \geq 1$. Thus

$$\frac{n^2 + 3}{n} = O(n) \qquad \blacksquare$$

Exercises 5.4

1. From the list of commonly used functions, find the smallest big-O estimate for each of the expressions given.
   (a) $\log_2(\log_2 n)$ (b) $(1 + 3\log_2 n)/n^2$
   (c) $(n^4 + 2n^3 - 6n + 3)/(3n^2 + 4)$ (d) $n^3 + 2^n$

2. For each part, determine the smallest n_0 such that $f(n) < g(n)$ for all $n > n_0$. Then prove $f(n) < g(n)$ for all $n > n_0$.
   (a) $f(n) = n^2$, $g(n) = 2^n$
   (b) $f(n) = 6n^2 + 7n + 5$, $g(n) = 5n^3 + 3$
   (c) $f(n) = 3n + n\log_2 n$, $g(n) = 7n^2$

3. Prove: If r is any positive integer and $c_0, c_1, c_2, \ldots, c_r$ are any real constants, with $c_r \neq 0$, then

$$c_0 + c_1 n + c_2 n^2 + \cdots + c_r n^r = O(n^r)$$

4. Show that

$$1^2 + 2^2 + \cdots + n^2 = O(n^3)$$

5. Prove: If $f(n) = O(g(n))$ and $g(n) = O(h(n))$, then $f(n) = O(h(n))$.

6. Give a big-O estimate for $f(n) = (2n + 3)(\log_2(n^2 + 2) + 2n^3)$.

7. Give a big-O estimate for $g(n) = \log_2(n!)$.

8. Prove: If $f_1(n) = O(g(n))$ and $f_2(n) = O(h(n))$, then $f_1(n)f_2(n) = O(g(n)h(n))$.

9. Prove: If $r \in \mathbb{N}$, there exists an $n_0 \in \mathbb{N}$ such that $n^r < 2^n$ for all $n > n_0$.

5.5   SET EQUIVALENCE

Recall that a set A is finite provided A consists of m elements for some nonnegative integer m. We called m the cardinal number of A. In this section we extend the concept of cardinal number to infinite sets in a rather natural, albeit naive, manner. In what follows we agree to set $I_m = \{1, 2, \ldots, m\}$ for each $m \in \mathbb{N}$, $I_0 = \emptyset$, and, as usual, assume the sets under consideration are subsets of some universal set U. Also remember that a function $f: A \to B$ is a subset of $A \times B$ satisfying the condition:

For each $a \in A$, there is a unique $b \in B$ such that $(a, b) \in f$.

Thus, in particular, if $A = \emptyset$, then $f = \emptyset$ is a function from A to B.

In this section we frequently deal with functions $f: A \to B$ that are one-to-one and onto.

DEFINITION 5.10

A function $f: A \to B$ that is one-to-one and onto is called a *bijection*.

In what follows, we use the term *bijection* in place of the more cumbersome wording "one-to-one and onto." It is hoped that the student has gained sufficient familiarity with the one-to-one and onto concepts so that the use of term *bijection* poses no confusion. In other words, when we refer to a bijection $f: A \to B$, the student should automatically think of f as being a one-to-one and onto function, and then have a good understanding of the properties satisfied by f.

THEOREM 5.11 A set A is finite if and only if there is a bijection $f: I_m \to A$ for some nonnegative integer m.

Proof Assume A is finite. If $A = \emptyset$, then there is a (unique) bijection $f: I_0 \to A$, namely $f = \emptyset$. If $A \neq \emptyset$, say $A = \{a_1, a_2, \ldots, a_m\}$, define $f: I_m \to A$ by $f(i) = a_i$ for $i = 1, 2, \ldots, m$. Then f is easily seen to be a bijection.

Conversely, assume there is a bijection $f: I_m \to A$ for some nonnegative integer m. If $m = 0$, then necessarily $f = \emptyset$. Since f is onto, it must be that $A = \emptyset$. Suppose next that $m > 0$. Since f is onto, we have $A = \operatorname{im} f = \{f(a_1), f(a_2), \ldots, f(a_m)\}$, and since f is one-to-one, it follows that $f(a_1), f(a_2), \ldots, f(a_m)$ are distinct. Hence $|A| = m$. $\qquad\square$

COROLLARY 5.12 Let A and B be finite sets. Then $|A| = |B|$ if and only if there is a bijection from A to B.

Proof If $|A| = |B| = m$, then there are bijections $f: I_m \to A$ and $g: I_m \to B$. Since f is a bijection, $f^{-1}: A \to I_m$ exists and, by Theorem 5.5, is a bijection. By Corollary 5.8, $g \circ f^{-1}: A \to B$ is a bijection.

Conversely, suppose there is a bijection $h: A \to B$. Let $|A| = m$; then there is a bijection $f: I_m \to A$. But then $h \circ f: I_m \to B$ is a bijection, and so $|B| = m$. $\qquad\square$

In order to arrive at some reasonable idea of cardinal number for infinite sets, it is essential that we be able to decide when two such sets have the same cardinal number. Corollary 5.12 tells us that two nonempty finite sets have the same cardinal number provided there is a bijection from one to the other. It is precisely this criterion that we use in deciding when two infinite sets have the same cardinal number.

DEFINITION 5.11

Two sets A and B are called *set equivalent* provided there is a bijection f from A to B. We write $A \sim B$.

Thus two finite sets A and B have the same cardinal number if and only if $A \sim B$. Moreover, a set A is finite if and only if $A \sim I_m$ for some nonnegative integer m.

In order to gain some insight into the infinite case, consider the sets $\mathbb{N}$ and $A = \{1, 4, 9, 16, \dots \}$. Should these sets have the same cardinal number (whatever that number is)? On the one hand, we see clearly that $A \subset \mathbb{N}$, so that the set $\mathbb{N}$ appears to be "larger" than A or seems to have more elements than does A. On the other hand, we have a criterion that can be applied, in general, to any two sets. In other words, we wish to determine if $\mathbb{N} \sim A$. The obvious candidate for a bijection from $\mathbb{N}$ to A is the mapping $f \colon \mathbb{N} \to A$ given by $f(n) = n^2$. The function f is plainly onto, so we need only show that it is one-to-one. Let n_1, $n_2 \in \mathbb{N}$ and assume $f(n_1) = f(n_2)$; that is, $n_1^2 = n_2^2$. Since n_1, $n_2 \in \mathbb{N}$, it follows that $n_1 = n_2$. Hence $\mathbb{N} \sim A$ and, placing full faith in our criterion, we consider A and $\mathbb{N}$ to have the same cardinal number. This example serves to point out an interesting property of infinite sets—namely, that an infinite set is equivalent to a proper subset of itself. In fact, this property is characteristic of infinite sets. We return to this later. For now, we formally adopt the following definition.

DEFINITION 5.12

Two sets A and B have the same cardinal number (or cardinality) provided $A \sim B$.

Example 5.18   Show these set equivalences:

(a) $(-1, 1) \sim \mathbb{R}$ (b) $\mathbb{Z} \sim \mathbb{N}$

Solution   (a) Define $f \colon (-1, 1) \to \mathbb{R}$ by

$$
f(x) = \begin{cases}
1 - \dfrac{1}{x} & \text{if } \ 0 < x < 1 \\[2mm]
0 & \text{if } \ x = 0 \\[2mm]
-1 - \dfrac{1}{x} & \text{if } \ -1 < x < 0
\end{cases}
$$

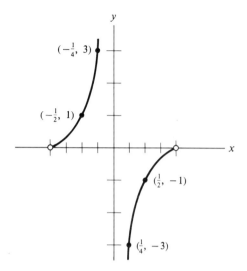

Figure 5.5

The graph of f is shown in Figure 5.5. The horizontal line test shows that f is a one-to-one function. To see that f is onto, let $r \in \mathbb{R}$ and consider the three cases $r > 0$, $r = 0$, and $r < 0$. If $r > 0$, then set $-1 - 1/x = r$ and solve for x. We obtain $x = -1/(r + 1)$. Note that $-1/(r + 1) < 0$ and $f(-1/(r + 1)) = r$. Clearly $f(0) = 0$. For $r < 0$, set $1 - 1/x = r$ and solve for x. We obtain $x = 1/(1 - r)$. Here note that $1/(1 - r) > 0$ and $f(1/(1 - r)) = r$. Thus f is onto, and we have established that $(-1, 1) \sim \mathbb{R}$.

(b) Define $g: \mathbb{Z} \to \mathbb{N}$ as follows:

$$g(m) = \begin{cases} 2m + 1 & \text{if} \quad m \geq 0 \\ -2m & \text{if} \quad m < 0 \end{cases}$$

We observe that g maps the nonnegative integers to the odd positive integers and the negative integers to the even positive integers, so that g is onto. Suppose $g(m_1) = g(m_2)$. Then clearly m_1 and m_2 have the same sign. If $m_1 \geq 0$ and $m_2 \geq 0$, then we have $2m_1 + 1 = 2m_2 + 1$, from which it follows that $m_1 = m_2$. The case for $m_1 < 0$ and $m_2 < 0$ is handled similarly. Therefore g is one-to-one. ∎

Definition 5.12 seems quite reasonable in one way and seems a bit mystifying in another. It tells us when two sets have the same cardinal number, even though we have no concept of cardinal number for infinite sets. Perhaps what is truly important is not any special symbol or mark used to represent the cardinal number of an infinite set, but being able to compare two cardinal numbers. Understanding this, we denote the cardinal number of a set A by $|A|$. Thus, given two sets A and B, we should be able to decide if $|A| = |B|$, $|A| < |B|$, or $|B| < |A|$. Of course, the key to deciding whether $|A| = |B|$ is Definition 5.12. In trying to decide whether or not $|A| < |B|$, we must keep in mind that we are comparing the sizes of two sets and that it is some relationship between the two sets that must tell us if $|A| < |B|$. The key to this lies in the next definition.

DEFINITION 5.13

If A and B are sets such that A is set equivalent to some subset of B, then we say B *dominates* A and we write $A \preccurlyeq B$. We also write $|A| \leq |B|$.

We remark that $A \preccurlyeq B$ provided there is a function $f: A \to B$ such that f is one-to-one. In such a case, $A \sim \text{im } f \subseteq B$.

Following Definition 5.13, it makes good sense to define $|A| < |B|$ provided $|A| \leq |B|$ and $|A| \neq |B|$; equivalently, $|A| < |B|$ provided $A \preccurlyeq B$ and it is not the case that $A \sim B$. In addition, the relation $\leq$ (on cardinal numbers) should be a partial-order relation if one is ever to develop a reasonable arithmetic for cardinal numbers. In order to determine this, consider the relation $\preccurlyeq$ on the set of all subsets of some universal set U. Is $\preccurlyeq$ a partial-order relation? In other words, if $\preccurlyeq$ is defined on the set of all subsets of U, then is $\preccurlyeq$ reflexive, antisymmetric, and transitive? One can reasonably verify the reflexive and transitive properties—that is,

1. $A \preccurlyeq A$ for all $A \in \mathscr{P}(U)$.
2. For A, B, C in $\mathscr{P}(U)$, if $A \preccurlyeq B$ and $B \preccurlyeq C$, then $A \preccurlyeq C$.

However, antisymmetry fails. To see this, consider the special case $U = \mathbb{R}$. Compare the subsets $\mathbb{N}$ and $E = \{2, 4, 6, \ldots\}$. If we define $f: \mathbb{N} \to E$ by $f(n) = 2n$, then f is one-to-one, and so $\mathbb{N} \preccurlyeq E$. On the other hand, the mapping $g: E \to \mathbb{N}$, given by $g(m) = m$, is one-to-one, so that $E \preccurlyeq \mathbb{N}$. Hence we have both $\mathbb{N} \preccurlyeq E$ and $E \preccurlyeq \mathbb{N}$, but $E \neq \mathbb{N}$. Thus $\preccurlyeq$ is not a partial-order relation. Still, it can be concluded that $|E| = |\mathbb{N}|$. This is a consequence of the following well-known and important result. Its proof is lengthy and intricate and is omitted.

THEOREM 5.13   **(Schroeder–Bernstein)**

Given sets A and B, if $A \preccurlyeq B$ and $B \preccurlyeq A$, then $A \sim B$.

Thus, even though $\preccurlyeq$ is not a partial-order relation on the power set of a given universal set U, it does follow that $\leq$ is a partial-order relation on cardinal numbers. Hence, as alluded to above, it now makes sense to consider an arithmetic for cardinal numbers. We address the sum and product operations in the chapter problems. Finally, it should be mentioned that the relation $\leq$ is a total ordering on cardinal numbers; thus, for any subsets A and B of U, either $|A| \leq |B|$ or $|B| \leq |A|$.

Example 5.19   Let $\mathbb{Q}^+$ denote the set of positive rational numbers. Show that $\mathbb{Q}^+ \sim \mathbb{N}$ by showing that $\mathbb{Q}^+ \preccurlyeq \mathbb{N}$ and $\mathbb{N} \preccurlyeq \mathbb{Q}^+$.

Solution   Define $f: \mathbb{N} \to \mathbb{Q}^+$ by $f(n) = n$ for all $n \in \mathbb{N}$. It is readily seen that f is one-to-one. Hence $\mathbb{N} \preccurlyeq \mathbb{Q}^+$.

To show that $\mathbb{Q}^+ \preccurlyeq \mathbb{N}$, recall that each $r \in \mathbb{Q}^+$ is uniquely expressible in the form $r = p/q$, where p and q are relatively prime positive integers. In other words, r can be written in lowest terms as p/q. We call p/q the *simple form* of r. Then define $g: \mathbb{Q}^+ \to \mathbb{N}$ by $g(r) = 2^p 3^q$, where p/q is the simple form of r. Since each element of $\mathbb{Q}^+$ has a unique simple form, we see that g is indeed a function (g is well-defined). To see that g is one-to-one, assume $g(r) = g(s)$ for some $r, s \in \mathbb{Q}^+$, say $g(r) = 2^p 3^q$ and $g(s) = 2^m 3^n$ (so p/q and m/n are the simple forms of r and s, respectively). So $2^p 3^q = 2^m 3^n$. Now these are canonical factorizations of the same positive integer so, by the uniqueness of such factorizations, we have that $p = m$ and $q = n$. Hence it follows that $p/q = m/n$; that is, $r = s$. Thus g is one-to-one, and we conclude that $\mathbb{Q}^+ \preccurlyeq \mathbb{N}$.

By the Schroeder–Bernstein theorem it follows that $\mathbb{Q}^+ \sim \mathbb{N}$.   ∎

We now turn our attention to some fundamental properties of infinite sets under set equivalence.

DEFINITION 5.14

A set A is called *countable* provided either A is finite or $A \sim \mathbb{N}$. We call A *countably infinite* or *denumerable* provided $A \sim \mathbb{N}$.

Example 5.20   In Example 5.18 it was shown that $\mathbb{Z} \sim \mathbb{N}$. Thus $\mathbb{Z}$ is countably infinite. In Example 5.19 we showed that $\mathbb{Q}^+ \sim \mathbb{N}$, and so $\mathbb{Q}^+$ is a countably infinite set. In fact, it follows from our next result that $\mathbb{Q}$ is countably infinite.   □

THEOREM 5.14   The union of two disjoint denumerable sets is denumerable.

Proof   Let A and B be disjoint denumerable sets and let $f: \mathbb{N} \to A$ and $g: \mathbb{N} \to B$ be bijections. Define $h: \mathbb{N} \to A \cup B$ as follows:

$$h(n) = \begin{cases} f\left(\dfrac{n+1}{2}\right) & \text{if } n \text{ is odd} \\[2ex] g\left(\dfrac{n}{2}\right) & \text{if } n \text{ is even} \end{cases}$$

Then h is a bijection (see Exercise 2) and we obtain $\mathbb{N} \sim A \cup B$. So $A \cup B$ is denumerable.   □

COROLLARY 5.15 If $A_1, A_2, \ldots, A_m$ are pairwise disjoint denumerable sets, then $A_1 \cup A_2 \cup \cdots \cup A_m$ is denumerable.

The proof of Corollary 5.15 is left to Exercise 4.

As mentioned previously, Example 5.19 and Theorem 5.14 can be used to demonstrate that $\mathbb{Q}$ is denumerable.

THEOREM 5.16 The set $\mathbb{Q}$ of rational numbers is denumerable.

Proof Let $\mathbb{Q}^+$ denote the set of positive rationals and $\mathbb{Q}^-$ the set of negative rationals. By Example 5.19, $\mathbb{Q}^+$ is a denumerable set. Also it is readily seen that $\mathbb{Q}^+ \sim \mathbb{Q}^-$, so $\mathbb{Q}^-$ is denumerable. Hence, by Theorem 5.14, $\mathbb{Q}^+ \cup \mathbb{Q}^-$ is a denumerable set. It then follows that $\mathbb{Q} = \mathbb{Q}^+ \cup \mathbb{Q}^- \cup \{0\}$ is denumerable. □

One question that is begging for an answer at this point is: Is there a smallest cardinality that an infinite set can have? The answer is provided by the following result.

THEOREM 5.17 If A is an infinite set, then $\mathbb{N} \preccurlyeq A$.

Proof We must show that there is a one-to-one function $f: \mathbb{N} \to A$. We proceed by induction (the strong form) to show that $f(n)$ can be defined, for each $n \in \mathbb{N}$, so that f is one-to-one. Let S be the set of all positive integers on which f has been defined. Since $A \neq \emptyset$, we can choose an element from A; let it be $f(1)$. Thus $1 \in S$. For an arbitrary $n > 1$, assume $k \in S$ for $1 \leq k < n$. Thus $f(1), f(2), \ldots, f(n-1)$ are defined so as to be distinct elements of A. Since A is infinite, the set $T = A - \{f(1), f(2), \ldots, f(n-1)\}$ is nonempty. Hence we can choose $f(n)$ in T. Then $f(1), f(2), \ldots, f(n)$ are distinct and we obtain $n \in S$. Therefore, by the PMI, $S = \mathbb{N}$. So a one-to-one function $f: \mathbb{N} \to A$ can be defined and we then have $\mathbb{N} \preccurlyeq A$. □

We would be less than honest if we did not confess to some rather heavy hand waving in the preceding proof. We made use of a very famous axiom, the "axiom of choice," in choosing infinitely many elements from the set A. To be precise about it, we have given no recipe or formula for the function f in this proof. We tacitly assume that we can choose an infinite set of elements, even though we prescribe no method for doing so. Of the many equivalent forms of the axiom of choice, one form in which our application is perhaps most evident is the following:

> **Axiom of Choice** Given any nonempty collection $\mathscr{C}$ of nonempty sets, there is a function g defined from $\mathscr{C}$ to $\cup \mathscr{C}$ (the union of the elements of $\mathscr{C}$), such that $g(B) \in B$ for all $B \in \mathscr{C}$.

The function g is called a "choice function," and such a function was defined on the collection

$$\{A, A - \{f(1)\}, A - \{f(1), f(2)\}, \ldots\}$$

in Theorem 5.17.

COROLLARY 5.18   Any infinite subset of a denumerable set is denumerable.

Proof   Let A be a denumerable set and B any infinite subset of A. Since $B \subseteq A$, we have $B \preccurlyeq A$. And since A is denumerable, there is a bijection $f: A \to \mathbb{N}$. As B is infinite, $\mathbb{N} \preccurlyeq B$, and there is a one-to-one function $g: \mathbb{N} \to B$ by Theorem 5.17. Then the function $g \circ f: A \to B$ is one-to-one, so we have $A \preccurlyeq B$. Hence, by the Schroeder–Bernstein theorem, $B \sim A$. Thus B is denumerable.   □

Since Corollary 5.18 uses Theorem 5.13, it depends on the axiom of choice. Actually, one can prove the corollary without appealing to the axiom of choice. In other words, one can construct a one-to-one function $f: \mathbb{N} \to B$. The next example serves to illustrate this in a special case.

Example 5.21   Show that any subset of $\mathbb{N}$ is countable.

Solution   Let $A \subseteq \mathbb{N}$. If A is finite, then certainly A is countable. Assume A is infinite. We construct a bijection $f: \mathbb{N} \to A$. By the principle of well-ordering (PWO), A has a smallest element, say a_1. Define $f(1) = a_1$. Since A is infinite, $A - \{a_1\} \neq \emptyset$, so this set has a smallest element, say a_2. Define $f(2) = a_2$. Suppose, for an arbitrary $k \geq 2$, that $a_1, a_2, \ldots, a_k$ have been so selected and we have defined $f(i) = a_i$ for $i = 1, 2, \ldots, k$. Again, since A is infinite, $A - \{a_1, a_2, \ldots, a_k\} \neq \emptyset$, so this set has a smallest element, say a_{k+1}. Define $f(k + 1) = a_{k+1}$. Hence (using induction) we have constructed a one-to-one function $f: \mathbb{N} \to A$, where $f(n) = a_n$ for each $n \in \mathbb{N}$. Moreover, $a_1 < a_2 < a_3 < \cdots$. We now claim that f is onto. Let $x \in A$. If $x \notin A - \{a_1\}$, then $x = a_1$. Otherwise $a_1 < x$ and, by the PWO, there is a smallest positive integer m such that $a_m < x \leq a_{m+1}$, so that $x \in A - \{a_1, a_2, \ldots, a_m\}$. But since a_{m+1} is the smallest element of the set $A - \{a_1, a_2, \ldots, a_m\}$, we must have $x = a_{m+1}$. So f is onto.   ∎

Another important corollary to Theorem 5.17 is a result that was alluded to earlier. Some authors use it as the definition of a finite set. We state it as a characterization of infinite sets, leaving the proof to Exercise 8.

COROLLARY 5.19   A set is infinite if and only if it is set equivalent to a proper subset of itself.

Needless to say, one can consider many more interesting properties of denumerable sets. Some of these are addressed in the exercises. For now we consider another teasing question: Are all infinite sets denumerable? As shown by the next result, the answer is no; an infinite set that is not denumerable is called *nondenumerable*.

THEOREM 5.20   The open interval $(0, 1)$ is a nondenumerable set.

Proof   The proof depends on the fact that each number x in the interval $(0, 1)$ is uniquely expressible in the decimal form

$$x = 0.x_1x_2x_3\ldots$$

where the decimal expansion is nonterminating. Thus, for example, the terminating decimal expansion 0.4 is not allowed; instead we use the repeating decimal expansion $0.3999\ldots$. With this convention, two numbers in the interval $(0, 1)$ are equal if and only if their decimal expansions are identical.

We proceed by contradiction and assume that $(0, 1)$ is denumerable; say $f: \mathbb{N} \to (0, 1)$ is a bijection. Suppose that

$$f(1) = 0.a_{11}a_{12}a_{13}\ldots$$

$$f(2) = 0.a_{21}a_{22}a_{23}\ldots$$

$$\vdots$$

$$f(n) = 0.a_{n1}a_{n2}a_{n3}\ldots$$

$$\vdots$$

Consider the number $z = 0.z_1z_2z_3\ldots$, defined by

$$z_m = \begin{cases} 1 & \text{if} \quad a_{mm} \neq 1 \\ 2 & \text{if} \quad a_{mm} = 1 \end{cases}$$

Then $0.z_1z_2z_3\ldots$ is the nonterminating decimal expansion of z and, for each $n \in \mathbb{N}$, $z \neq f(n)$, since $z_n \neq a_{nn}$. Thus we have arrived at a contradiction, and it follows that the interval $(0, 1)$ in nondenumerable.   □

According to convention the cardinal number of $\mathbb{N}$ is denoted by $\aleph_0$, pronounced "aleph null," and the cardinal number of $\mathbb{R}$ is denoted by c or by $\aleph$ (aleph). The symbol $\aleph$ is a member of the Hebrew alphabet. Theorem 5.17 tells us that

$$\aleph_0 < c$$

Before ending this section, we consider a pair of interesting questions concerning these cardinal numbers:

1. Does there exist a cardinal number d such that $\aleph_0 < d < c$?
2. Is there a largest infinite cardinal number?

The first question turns out to be profound and vexing. For what it is asking about is the existence of a subset C of $\mathbb{R}$ for which $\mathbb{N} \prec C$ and $C \prec \mathbb{R}$. This problem was studied at length by the German mathematician Georg Cantor in the late 19th century. It was Cantor who first considered a theory of infinite cardinal numbers in a formal way; for example, he was the first to show that $\mathbb{N}$ is set equivalent to the set of positive rational numbers. He conjectured that there is no cardinal number strictly between $\aleph_0$ and c, as all attempts to find a subset of $\mathbb{R}$ with cardinal number strictly between $\aleph_0$ and c led to sets with cardinal number $\aleph_0$ or c. Cantor's conjecture came to be known as the *continuum hypothesis*: If we think of the real line as the continuum, then the cardinal number of the continuum is the immediate successor of $\aleph_0$ in the list of infinite cardinal numbers. In 1938 it was shown by Kurt Gödel that the continuum hypothesis is consistent with the standard axioms of set theory; in other words, no contradiction results from assuming the continuum hypothesis is true. In 1963 another landmark result came from the pen of Paul J. Cohen, who showed that, in fact, one could not prove the continuum hypothesis from the standard axioms. What this means is that the continuum hypothesis can be taken as false (or true) and no inconsistency can arise.

The second question is more manageable, and the answer is provided by the following result, due to Cantor.

THEOREM 5.21   For any set A, we have $A \prec \mathscr{P}(A)$.

Proof   To see that $A \preceq \mathscr{P}(A)$, define a function $f: A \to \mathscr{P}(A)$ by $f(x) = \{x\}$. This function is clearly one-to-one, so $A \preceq \mathscr{P}(A)$. To show $A \prec \mathscr{P}(A)$, we proceed by contradiction. Thus, suppose that $A \sim \mathscr{P}(A)$, with a bijection $g: A \to \mathscr{P}(A)$. Let

$$B = \{x \in A \mid x \notin g(x)\}$$

So $B \subseteq A$, and hence $B \in \mathscr{P}(A)$. Since g is onto, there must be an element $x_0 \in A$ for which $g(x_0) = B$. Now either $x_0 \in B$ or $x_0 \notin B$; consider each case. If $x_0 \in B$, then we have $x_0 \in g(x_0)$. But, by the definition of B, $x_0 \in B$ implies $x_0 \notin g(x_0)$, and since $g(x_0) = B$, this is impossible. On the other hand, if $x_0 \notin B$, then $x_0 \notin g(x_0)$, and the definition of B tells us that $x_0 \in B$, another impossibility. Hence we have arrived at a contradiction, and the proof is complete. $\qquad\square$

As a consequence of Cantor's theorem, there can be no largest cardinal number. For if A is any infinite set, then $|A| < |\mathscr{P}(A)|$. In particular, we see that

$$\mathbb{N} \prec \mathscr{P}(\mathbb{N}) \prec \mathscr{P}(\mathscr{P}(\mathbb{N})) \prec \cdots$$

Exercises 5.5

1. Show that $(0, 1) \sim (-1, 1)$.
2. Show that the function h in the proof of Theorem 5.14 is a bijection.
3. Let a and b be real numbers with $a < b$. Show that $(0, 1) \sim (a, b)$.
4. Prove Corollary 5.15.
5. Show that $[0, 1) \sim [0, \infty)$.
6. Show that set equivalence is an equivalence relation on the power set of a given universal set U.
7. Given real numbers a, b, c, and d such that $a < b$ and $c < d$, show that $(a, b) \sim (c, d)$.
8. Prove Corollary 5.19.
9. Given real numbers a, b, and c such that $b < c$, show that $[b, c) \sim [a, \infty)$.
10. Prove that the set of all circles in the xy-plane with rational radii and whose centers have rational coordinates is denumerable.
11. Given any nonempty set A, denote by 2^A the set of all functions from A to the set $\{0, 1\}$. Prove that $\mathscr{P}(A) \sim 2^A$. Notice the similarity to the well-known result: If $|A| = n$, then $|\mathscr{P}(A)| = 2^n$.
12. Let X and Y be sets. Prove that, if $X - Y \sim Y - X$, then $X \sim Y$.
13. Prove that the set of all irrational numbers is nondenumerable.
14. Let A and B be sets. Prove: If there exists a function f from A onto B, then $B \preccurlyeq A$.
15. Let $A = \{ f: [0, 1] \to [0, 1] \mid x \text{ irrational} \to f(x) = 0 \}$. Determine the cardinal number of A.
16. A complex number is called an *algebraic integer* provided it is a solution of a polynomial equation with integer coefficients. Let $\mathbb{A}$ denote the set of algebraic integers.
    (a) Show that $\mathbb{Q}$ is a proper subset of $\mathbb{A}$.
    (b) Use the fact that a polynomial equation of degree n has at most n distinct roots to prove that the set $\mathbb{A}$ is denumerable.
17. Let A, B, A_1, and B_1 be sets with A and B disjoint and A_1 and B_1 disjoint. If $A \sim A_1$ and $B \sim B_1$, then prove that $A \cup B \sim A_1 \cup B_1$.
18. Let A, B, A_1, and B_1 be sets with $A \sim A_1$ and $B \sim B_1$. Prove that $A \times B \sim A_1 \times B_1$.
19. Make use of the function $f: (-1, 1) \to \mathbb{R}$ defined by $f(x) = x/(1 - x^2)$ to show that $(-1, 1) \sim \mathbb{R}$.

CHAPTER PROBLEMS

1. Give an example of a function $f: [0, 1] \to [0, 1]$ that is:
   (a) both one-to-one and onto   (b) one-to-one but not onto
   (c) onto but not one-to-one   (d) neither one-to-one nor onto
2. Repeat Problem 1 for $f: E \to \mathbb{N}$, where E is the set of even integers.

3. For each function, determine whether it is one-to-one and whether it is onto.
   (a) $f: \mathbb{Q} \to \mathbb{Q}$; $f(x) = x^3$
   (b) $g: \mathbb{Q} - \{1/2\} \to \mathbb{Q} - \{3/2\}$; $g(x) = 3x/(2x - 1)$
   (c) $h: \mathbb{Z} \to \mathbb{N}$; $h(m) = |m - 4| + 1$ (d) $p: \mathbb{Z} \to \mathbb{N}$; $p(m) = m^2$
   (e) $t: \mathbb{Z}_{149} \to \mathbb{Z}_{149}$; $t([n]) = [6n]$ (f) $q: \mathbb{N} \to \mathbb{N}$; $q(n) = n^2$

4. Let $f: X \to Y$, where X and Y are subsets of $\mathbb{R}$. The function f is said to be *increasing* provided the condition

$$x_1 < x_2 \to f(x_1) < f(x_2)$$

holds for all $x_1, x_2 \in X$. Similarly, the function f is said to be *decreasing* provided the condition

$$x_1 < x_2 \to f(x_1) > f(x_2)$$

holds for all $x_1, x_2 \in X$. If f is either increasing or decreasing, then we say that f is *monotonic*.
   (a) Prove that a monotonic function is one-to-one.
   (b) Apply the result of part (a) to show that $f: \mathbb{R} \to \mathbb{R}$ defined by $f(x) = x^3 + x - 2$ is one-to-one.

5. Determine $f \circ g$ and $g \circ f$.
   (a) $f: \mathbb{R} \to \mathbb{R}$; $f(x) = x^2 + x$, $g: \mathbb{R} \to \mathbb{R}$; $g(x) = 3x + 4$
   (b) $f: (0, \infty) \to (0, \infty)$; $f(x) = x/(x^2 + 1)$, $g: (0, \infty) \to (0, \infty)$; $g(x) = 1/x$

6. Let X and Y be nonempty sets. Prove: A function $f: X \to Y$ is one-to-one if and only if the condition

$$f(A \cap B) = f(A) \cap f(B)$$

holds for all subsets A and B of X.

7. Show that there are infinitely many pairs of distinct functions f and g on $\mathbb{Q}$ such that none of f, g, and $f \circ g$ is the identity function on $\mathbb{Q}$ and $f \circ g = g \circ f$.

8. Let A and B be nonempty sets and let $f: A \to B$. Prove these results.
   (a) f is onto $\leftrightarrow \forall D \subseteq B[f(f^{-1}(D)) = D]$
   (b) f is one-to-one $\leftrightarrow \forall C \subseteq A[f^{-1}(f(C)) = C]$

9. Let f, g and h be functions on $\mathbb{Z}$ defined as follows:

$$f(m) = m + 1 \qquad g(m) = 2m \qquad h(m) = \begin{cases} 0 & \text{if } m \text{ is even} \\ 1 & \text{if } m \text{ is odd} \end{cases}$$

Determine the following composite functions:
   (a) $f \circ g$ (b) $g \circ f$ (c) $f \circ h$ (d) $h \circ f$
   (e) $g \circ h$ (f) $h \circ g$ (g) $g \circ g$ (h) $h \circ f \circ g$

10. Given a nonempty set X, let $\mathscr{R}(X)$ denote the set of real-valued functions with domain X:

$$\mathscr{R}(X) = \{f \mid f: X \to \mathbb{R}\}$$

For $f, g \in \mathscr{R}(X)$, we define functions $f + g$ and $f \cdot g$ in $\mathscr{R}(X)$, called the *sum* and *product* of f and g, respectively, as follows: For $x \in X$,

$$(f + g)(x) = f(x) + g(x)$$
$$(f \cdot g)(x) = f(x) \cdot g(x)$$

Determine the functions $f + g$ and $f \cdot g$ in each of the following.
(a) $X = \mathbb{N}$, $f(x) = x^2$ and $g(x) = 2x + 1$
(b) $X = \mathbb{Q}$, $f(x) = x/3$ and $g(x) = 3x + 2$
(c) $X = \mathbb{R} - \{0\}$, $f(x) = (x^2 + 1)/x$ and $g(x) = x/(x^2 + 1)$
(d) $X = \mathbb{R}$, $f(x) = x^2 - 2x + 3$ and $g(x) = -x^2 + 2x - 3$

11. Let A denote the ASCII character set, $\mathscr{S}$ the set of character strings over A, and $\mathscr{S}^+$ the set of nonempty strings. Also, for $x, y \in \mathscr{S}$, let $x \& y$ denote the catenation of x with y. We define the following functions.

    (i) For a fixed $\alpha \in A$, the functions $p_\alpha: \mathscr{S} \to \mathscr{S}$ and $s_\alpha: \mathscr{S} \to \mathscr{S}$ are defined by

$$p_\alpha(x) = \text{``}\alpha\text{''} \& x \quad \text{and} \quad s_\alpha(x) = x \& \text{``}\alpha\text{''}$$

    (ii) The function $r: \mathscr{S} \to \mathscr{S}$ is defined as follows: $r(x)$ is the string obtained by reversing x.
    (iii) The function $t: \mathscr{S}^+ \to \mathscr{S}$ is defined as follows: $t(x)$ is the string obtained by deleting the first character of x.

For example, if $\alpha = $ 'S' and $x = $ "TAB", then $p_\alpha(x) = $ "STAB", $s_\alpha(x)$ $= $ "TABS", $r(x) = $ "BAT", and $t(x) = $ "AB".
(a) Determine which of these functions are one-to-one.
(b) Find the image of each function and determine which are onto.
(c) Describe those strings x for which $r(x) = x$.
(d) Show that $r \circ p_\alpha = s_\alpha \circ r$.   (e) Describe $p_\alpha \circ t$ and $t \circ p_\alpha$.

12. Let U be a nonempty universal set. For $A \subseteq U$, define the function $\chi_A: U \to \{0, 1\}$ by

$$\chi_A(x) = \begin{cases} 0 & \text{if} \quad x \notin A \\ 1 & \text{if} \quad x \in A \end{cases}$$

The function χ_A is called the *characteristic function* of A. For $A, B \in \mathscr{P}(U)$, let $C = A \cap B$, $D = A \cup B$, and $E = A - B$. Prove that the following hold for all $x \in U$.
(a) $\chi_C(x) = \chi_A(x) \cdot \chi_B(x)$
(b) $\chi_D(x) = \chi_A(x) + \chi_B(x) - [\chi_A(x) \cdot \chi_B(x)]$
(c) $\chi_U(x) = 1$ (d) $\chi_\emptyset(x) = 0$
(e) $\chi_{B'}(x) = 1 - \chi_B(x)$ (f) $\chi_E(x) = \chi_A(x)[1 - \chi_B(x)]$

13. Let X be a nonempty set and let f be a function on X. Define $f^1 = f$, $f^2 = f \circ f$, $f^3 = f \circ f \circ f$, and so on. (In general, for $n > 1$, $f^n = f \circ f^{n-1}$.) For this problem, let i denote the identity function on $\mathbb{Z}$. Give an example of a function $f: \mathbb{Z} \to \mathbb{Z}$ such that:
    (a) $f \neq i$ but $f^2 = i$ (b) $f^2 \neq i$ but $f^3 = i$
    (c) Generalize parts (a) and (b); for each $n > 1$, give an example of a function $f: \mathbb{Z} \to \mathbb{Z}$ such that $f^{n-1} \neq i$ but $f^n = i$.

14. (Addition of cardinal numbers) If a and b are cardinal numbers, then the *sum* $a + b$ is the cardinal number of $A \cup B$, where A and B are any two disjoint sets whose cardinal numbers are, respectively, a and b. Prove the following:
    (a) The definition of the sum $a + b$ of the cardinal numbers a and b does not depend on the choice of the disjoint sets A and B whose cardinal numbers are a and b, respectively. (See Exercise 17 in Exercises 5.5.)
    (b) $\aleph_0 + \aleph_0 = \aleph_0$
    (c) $c + c = c$

15. For each function in Problem 3 that is both one-to-one and onto, find its inverse.

16. (Product of cardinal numbers) If a and b are cardinal numbers, then the *product* $a \cdot b$ is the cardinal number of $A \times B$, where A and B are any two sets whose cardinal numbers are, respectively, a and b. Prove the following:
    (a) The definition of the product $a \cdot b$ of the cardinal numbers a and b does not depend on the choice of the sets A and B whose cardinal numbers are a and b, respectively. (See Exercise 18 of Exercises 5.5.)
    (b) $\aleph_0 \cdot \aleph_0 = \aleph_0$
    (c) $c \cdot c = c$

17. Let A and B be nonempty sets, and consider a function $f: A \to B$ that is onto. Define a relation $\sim$ on A by

$$x \sim y \leftrightarrow f(x) = f(y)$$

    (a) Show that $\sim$ is an equivalence relation on A.
    Recall that $[a]$ denotes the equivalence class containing the element a. Let $\mathscr{P}$ be the partition of A induced by $\sim$. Define $q: A \to \mathscr{P}$ by $q(a) = [a]$, and define $\bar{f}: \mathscr{P} \to B$ by $\bar{f}([a]) = f(a)$.
    (b) Show that the function $\bar{f}$ is well-defined; that is, if $[a_1] = [a_2]$, then $\bar{f}([a_1]) = \bar{f}([a_2])$.
    (c) Show that $\bar{f}$ is one-to-one and onto.
    (d) Show that $f = \bar{f} \circ q$.
    Thus, every onto mapping (f) may be "factored" as the composition of a "quotient mapping" (q) and a one-to-one, onto mapping ($\bar{f}$). This result is the set-theoretic analogue of the "fundamental morphism theorem" of abstract algebra.

18. (A problem for those who have had calculus) Let $\mathscr{F}$ denote the set of functions defined on $\mathbb{R}$, let $\mathscr{C} = \{ f \in \mathscr{F} \mid f \text{ is continuous} \}$, and let $\mathscr{D} = \{ f \in \mathscr{F} \mid f \text{ is differentiable} \}$. Consider the function $\Delta \colon \mathscr{D} \to \mathscr{F}$ that maps each function $f \in \mathscr{D}$ to its derivative f' in $\mathscr{F}$; that is, $\Delta(f) = f'$.
    (a) Is the function Δ one-to-one?
    (b) Let $f \in \mathscr{D}$. What is $\Delta^{-1}(\{ f' \})$?
    (c) Show that $\mathscr{C} \subseteq \operatorname{im} \Delta$. In fact, it can be shown that $\mathscr{C} \subset \operatorname{im} \Delta$. For example, consider f defined by

$$f(x) = \begin{cases} x^2 \sin \dfrac{1}{x} & \text{if } x \neq 0 \\ 0 & \text{if } x = 0 \end{cases}$$

It can be shown that f is differentiable but that f' is not continuous at 0.

19. Define $f \colon \mathbb{Z}_{119} \to \mathbb{Z}_{119}$ by $f([x]) = [15x]$. Show that f is a permutation of $\mathbb{Z}_{119}$ and find f^{-1}.

20. Define $g \colon \{1, 2, \ldots, 100\} \to \{1, 2, \ldots, 100\}$ by

$$g(n) = \begin{cases} \dfrac{n}{2} & \text{if } n \text{ is even} \\ n + 1 & \text{if } n \text{ is odd} \end{cases}$$

Determine whether g is a permutation of $\{1, 2, \ldots, 100\}$.

21. Define $f \colon \mathbb{Q} - \{1/4\} \to \mathbb{Q} - \{0\}$ by $f(x) = 1 - 4x$ and the function $g \colon \mathbb{Q} - \{0\} \to \mathbb{Q} - \{3/2\}$ by $g(x) = (3x - 1)/(2x)$. Determine each of the functions $g \circ f$, $(g \circ f)^{-1}$, f^{-1}, g^{-1}, and $f^{-1} \circ g^{-1}$.

22. Define $f \colon \mathbb{Q} - \{1\} \to \mathbb{Q} - \{2\}$ by $f(x) = (5x^3 + 7)/6$ and $g \colon \mathbb{Q} - \{2\} \to \mathbb{Q} - \{1\}$ by $g(x) = x/(x - 2)$. Determine any of these functions that exist: $g \circ f$, $(g \circ f)^{-1}$, $f \circ g$, and $(f \circ g)^{-1}$.

23. Let m and n be positive integers, and let $A = \{0, 1, \ldots, m - 1\}$, $B = \{0, 1, \ldots, n - 1\}$, and $C = \{0, 1, \ldots, mn - 1\}$. Construct a bijection $f \colon A \times B \to C$.

24. Given that $g_1(n) \geq 0$ and $g_2(n) \geq 0$ for all $n \in \mathbb{N}$ and both $f_1(n) = O(g_1(n))$ and $f_2(n) = O(g_2(n))$, prove that $f_1(n) + f_2(n) = O(g_1(n) + g_2(n))$.

25. Prove: If $f_1(n) = O(g_1(n))$ and $f_2(n) = O(g_2(n))$, then $f_1(n)f_2(n) = O(g_1(n)g_2(n))$.

26. Given functions: $f \colon \mathbb{N} \to [0, \infty)$ and $g \colon \mathbb{N} \to [0, \infty)$, we write $f(n) = \Theta(g(n))$ provided there exist positive constants C_1 and C_2 and some positive integer n_0 such that

$$C_1 g(n) \leq f(n) \leq C_2 g(n)$$

for all $n \geq n_0$. We read $f(n) = \Theta(g(n))$ as $f(n)$ is *theta* of $g(n)$.
    (a) Show that $f(n) = \Theta(g(n))$ if and only if both $f(n) = O(g(n))$ and $g(n) = O(f(n))$.
    (b) Show that the relation "is theta of" is an equivalence relation on the set of functions from $\mathbb{N}$ to $[0, \infty)$.

27. Each part gives two functions $f(n)$ and $g(n)$ from $\mathbb{N}$ to $[0, \infty)$. Choose the statement $f(n) = O(g(n))$ or $f(n) = \Theta(g(n))$ that most accurately describes the relationship between f and g.

   (a) $f(n) = 3n^4 + 5n^3 + 7n + 2$, $g(n) = n^4$

   (b) $f(n) = \sqrt{n^2 + 3}$, $g(n) = n$

   (c) $f(n) = 4n^{1.9} + 15n + 9$, $g(n) = n^2$

   (d) $f(n) = 3^n + n^3$, $g(n) = 3^n$

   (e) $f(n) = \begin{cases} n & \text{if } n \text{ is prime} \\ 1 & \text{otherwise} \end{cases}$, $g(n) = n$

   ★**(f)** $f(n) = n^{99}$, $g(n) = 2^{\sqrt{n}}$

   (g) $f(n) = 5n \log_2 n + 4n + 3 \log_2 n + 2$, $g(n) = n \log_2 n$

   ★**(h)** $f(n) = 2^{\sqrt{n}}$, $g(n) = (1.01)^n$

   ★**(i)** $f(n) = \log_2(2^n + n^2)$, $g(n) = n$

28. Prove: If $f_1(n) = \Theta(g(n))$ and $f_2(n) = O(g(n))$, then $f_1(n) + f_2(n) = \Theta(g(n))$.

29. Let $f(n) = 2^n$ and $g(n) = n!$. Show that $f(n) = O(g(n))$.

S I X

Combinatorial Mathematics

INTRODUCTION

Although it was originally a source of recreation in solving puzzles and playing games, combinatorial mathematics is currently one of the more active areas of mathematical research. In addition, it has provided important applications to fields such as computer science, operations research, statistics, and the social and physical sciences.

In Chapter 2 we touched on the addition principle and the principle of inclusion–exclusion. These are combinatorial laws that have to do with counting the number of elements in a union of sets. In a large measure combinatorial mathematics deals with arrangements of elements that satisfy specified conditions. Typically, the elements belong to sets that are either finite or countably infinite, and so combinatorial mathematics falls within the area of discrete mathematics.

When we deal with arrangements of a specified type, several questions generally arise:

1. Existence. Do such arrangements exist?
2. Enumeration or classification. How many valid arrangements of the specified type are there? Can they be classified in some special way?
3. Algorithms. Is there a definite method for constructing such arrangements?

In addition to the above questions, one is interested in the possibility of generalizations. In other words, does the problem under consideration suggest other related problems?

The following examples serve to illustrate some of the above ideas.

Example 6.1   Seven generals store top-secret documents in a safe. They want to be able to open the safe only when a majority of the group is present, so they propose locking the safe with a number of different locks. Each general will be given keys to certain of the locks, with the possibility that two of them may have keys to the same lock. How many locks (at least) must there be and how many keys (at least) must each general possess?

Discussion   Let $G_1, G_2, \ldots, G_7$ denote the generals. Since only a majority of the generals can open the safe, it will be impossible for any three of them to open it. Hence, given any three generals, there must be a lock to which none of them has a key. Can two different subsets of three lack a key to the same lock? For example, suppose that both G_1, G_2, G_3 and G_2, G_3, G_4 have no key to the same lock. Then $G_1, G_2 G_3, G_4$ will have no key to that lock, contrary to the given condition that any majority can open the safe. Thus there must be at least as many different locks as there are ways to choose a three-element subset of the seven generals. In considering the number of keys each general must possess, we focus our attention on one of the generals, say G_7. If we add G_7 to a subset S consisting of three of the other generals, then the resulting set of four must be able to open the safe. However, as noted above, there is a lock to which the members of S have no key. So G_7 must have a key to this lock. Moreover, since no two different subsets of three lack a key to the same lock, G_7 must have at least as many keys as there are ways to choose a three-element subset from a set of six elements.

The general problem of determining the number of k-element subsets of a set consisting of n elements is addressed later in this chapter.   ∎

Example 6.2   In Chapter 5 the notion of a permutation of a set was introduced as a one-to-one function of the set onto itself. In particular, if A is a finite set, say $|A| = n$, then we saw that a permutation A could be viewed as an ordered selection of the elements of A or as an ordered n-tuple whose coordinates are the elements of A (in some order). It was shown that there are $n!$ permutations of a set with n elements. For example, let $A = \{1, 2, 3\}$. Then $(1, 2, 3)$, $(1, 3, 2)$, $(2, 3, 1)$, $(2, 1, 3)$, $(3, 1, 2)$, and $(3, 2, 1)$ are the $3! = 6$ permutations of A. In general, for an arbitrary finite set A, does there exist a systematic way to obtain all the permutations of A? This question is addressed in a later section. In Chapter 8 permutations are encountered again, from an algebraic point of view.   ∎

Example 6.3   In how many ways can one distribute 37 plain donuts among five children? If we label the children $c_1, c_2, \ldots, c_5$, then let x_i denote the number of donuts c_i receives, for $i = 1, 2, 3, 4, 5$. We see that each x_i is a nonnegative integer and

$$x_1 + x_2 + \cdots + x_5 = 37$$

As an obvious generalization, consider the problem of determining the number of solutions to the equation $x_1 + x_2 + \cdots + x_n = m$, where each x_i is a nonnegative integer, for $i = 1, 2, \ldots, n$ and $m > 0$.   ∎

Exercises 6.1

1. List the permutations of each set.
    (a) $\{1\}$ (b) $\{1, 2\}$ (c) $\{1, 2, 3\}$ (d) $\{1, 2, 3, 4\}$
2. Let n be a positive integer and let $A = \{1, 2, \ldots, n\}$. We define an ordering $\preccurlyeq$ on the permutations of A as follows: For two permutations $\pi_1 = (a_1, a_2, \ldots, a_n)$ and $\pi_2 = (b_1, b_2, \ldots, b_n)$ of A, we define $\pi_1 \preccurlyeq \pi_2$ provided either $\pi_1 = \pi_2$, or $a_1 < b_1$, or $a_1 = b_1$, $a_2 = b_2, \ldots, a_{i-1} = b_{i-1}$, and $a_i < b_i$ for some i, $1 < i \leq n$. This ordering is known as *lexicographic ordering*.
    (a) Show that $\preccurlyeq$ is a total ordering of the set of permutations of A.
    (b) List the permutations of $\{1, 2, 3, 4\}$ in lexicographic order.
    (c) For $n = 7$, what permutation is the immediate successor of $(1, 2, 3, 4, 6, 7, 5)$ (in lexicographic order)?
    (d) For $n = 7$, what permutation is the immediate successor of $(1, 6, 2, 7, 5, 4, 3)$?
3. List the three-element subsets of each set.
    (a) $\{1, 2, 3, 4, 5\}$ (b) $\{1, 2, 3, 4, 5, 6\}$
4. Let n and A be as in Exercise 2, and let k be an integer, where $1 \leq k \leq n$. We define an ordering $\preccurlyeq$, also called *lexicographic ordering*, on the k-element subsets of A as follows: Let $S_1 = \{a_1, a_2, \ldots, a_k\}$ and $S_2 = \{b_1, b_2, \ldots, b_k\}$ be two k-element subsets of A, and assume the elements are listed so that $a_1 < a_2 < \cdots < a_k$ and $b_1 < b_2 < \cdots < b_k$; then $S_1 \preccurlyeq S_2$ provided either $S_1 = S_2$, or $a_1 < b_1$, or $a_1 = b_1$, $a_2 = b_2, \ldots, a_{i-1} = b_{i-1}$, and $a_i < b_i$ for some i, $1 < i \leq k$.
    (a) Show that $\preccurlyeq$ is a total ordering of the k-element subsets of A.
    (b) List the three-element subsets of $\{1, 2, 3, 4, 5, 6\}$ in lexicographic order.
    (c) For $n = 7$, what four-element subset is the immediate successor of $\{1, 3, 4, 5\}$ (in lexicographic order)?
    (d) For $n = 7$, what four-element subset is the immediate successor of $\{2, 4, 6, 7\}$?
5. Let n and A be as in Exercises 2 and 4. Let us extend the definition of $\preccurlyeq$ given in Exercise 4 to a "lexicographic ordering" of all the subsets of A, namely, to $\mathcal{P}(A)$. First we agree that the empty subset precedes any nonempty subset. Then, for two nonempty sets A and B, with $A \neq B$, let x be the smallest element of the symmetric difference $(A - B) \cup (B - A)$. If $x \in A$, then $A \prec B$; otherwise, $B \prec A$. (For two subsets A and B the notation $A \preccurlyeq B$ means that either $A \prec B$ or $A = B$.)
    (a) Show that $\preccurlyeq$ is a total ordering of $\mathcal{P}(A)$.
    (b) List the subsets of $\{1, 2, 3, 4\}$ in lexicographic order.
    (c) For $n = 7$, what subset is the immediate successor of $\{1, 3, 4, 5\}$ (in lexicographic order)?

(d) For $n = 7$, what subset is the immediate successor of $\{1, 2, 3, 4, 5, 6, 7\}$?

(e) For $n = 7$, what subset is the immediate successor of $\{1, 7\}$?

6.2

BASIC COUNTING TECHNIQUES

In Section 2.4 we stated both the addition principle and the multiplication principle. Each was proved for the case of two sets and stated in general form. Hence, for finite sets A and B, we proved the following special cases:

Addition Principle If A and B are disjoint, then $|A \cup B| = |A| + |B|$

Multiplication Principle $|A \times B| = |A| \cdot |B|$

Here we prove each in general form and address further applications.

THEOREM 6.1   **(Addition Principle)**

If $A_1, A_2, \ldots, A_n$ are pairwise disjoint finite sets, then

$$|A_1 \cup A_2 \cup \cdots \cup A_n| = |A_1| + |A_2| + \cdots + |A_n|$$

Proof   We proceed by induction on n and let S be the set of positive integers for which the statement holds. Clearly $1 \in S$ and, as was established in Section 2.4, we have that $2 \in S$. Assume $k \in S$ for an arbitrary integer $k \geq 2$; thus the addition principle is assumed true for any pairwise disjoint collection of k finite sets. To show $k + 1 \in S$, let $A_1, A_2, \ldots, A_{k+1}$ be any $k + 1$ pairwise disjoint finite sets. Then $A_1 \cup A_2 \cup \cdots \cup A_k$ and A_{k+1} are disjoint finite sets and hence

$$
\begin{aligned}
|A_1 \cup A_2 \cup \cdots \cup A_k \cup A_{k+1}| &= \left|(A_1 \cup A_2 \cup \cdots \cup A_k) \cup A_{k+1}\right| \\
&= |A_1 \cup A_2 \cup \cdots \cup A_k| + |A_{k+1}| \\
&= |A_1| + |A_2| + \cdots + |A_k| + |A_{k+1}|
\end{aligned}
$$

The second equality follows since $2 \in S$, while the third equality follows from the induction hypothesis. Consequently $k + 1 \in S$, and we obtain $S = \mathbb{N}$. ☐

Example 6.4   An experiment consists of rolling a pair of dice. In how many ways can a seven or a ten result?

Solution   We associate with the result of rolling the dice an ordered pair (r, s), where r represents the value obtained on the first die and s the value obtained on the second die. Let A denote the set of ordered pairs that yield a result of seven and let B denote the set of ordered pairs yielding a result of ten. Thus

$$A = \{(1,6), (2,5), (3,4), (4,3), (5,2), (6,1)\}$$

and

$$B = \{(4,6), (5,5), (6,4)\}$$

Then $A \cup B$ is the set of ordered pairs associated with a result of seven or ten, and the solution to the problem is $|A \cup B|$. Since A and B are disjoint sets, we have

$$|A \cup B| = |A| + |B| = 6 + 3 = 9 \qquad \blacksquare$$

Example 6.5   An experiment consists of flipping an ordinary coin four times in succession and noting the resulting sequence of four outcomes. In how many ways can one obtain one, two, or three heads?

Solution   As stated, the result of the experiment consists of a sequence of four terms, each term being an H (head) or a T (tail) (for example, HTHH is a possibility). Let E_k be the set of sequences associated with exactly k heads, where $0 \le k \le 4$. Thus

$$E_1 = \{HTTT, THTT, TTHT, TTTH\}$$
$$E_2 = \{HHTT, HTHT, HTTH, THHT, THTH, TTHH\}$$
$$E_3 = \{HHHT, HHTH, HTHH, THHH\}$$

Then $E_1 \cup E_2 \cup E_3$ is the set of sequences yielding a result of one, two, or three heads, and the solution to the problem is $|E_1 \cup E_2 \cup E_3|$. Since E_1, E_2, and E_3 are pairwise disjoint, we obtain

$$|E_1 \cup E_2 \cup E_3| = |E_1| + |E_2| + |E_3|$$
$$= 4 + 6 + 4$$
$$= 14 \qquad \blacksquare$$

THEOREM 6.2   **(Multiplication Principle)**

If $A_1, A_2, \ldots, A_n$ are finite sets, then

$$|A_1 \times A_2 \times \cdots \times A_n| = |A_1| \cdot |A_2| \cdot \cdots \cdot |A_n|$$

Proof   Proceed by induction on n and let S be the set of positive integers for which the result holds. We already have that $1, 2 \in S$. Assume $k \in S$ for an arbitrary $k \ge 2$. Then the statement of the theorem holds for any k finite sets. To show that $k + 1 \in S$, let $A_1, A_2, \ldots, A_k, A_{k+1}$ be finite sets and consider the mapping

$$f: (A_1 \times A_2 \times \cdots \times A_k) \times A_{k+1} \to A_1 \times A_2 \times \cdots \times A_k \times A_{k+1}$$

given by

$$f(((a_1, a_2, \ldots, a_k), a_{k+1})) = (a_1, a_2, \ldots, a_k, a_{k+1})$$

This mapping is both one-to-one and onto (see Exercise 14). Hence, by Corollary 5.12, we have

$$\left|(A_1 \times A_2 \times \cdots \times A_k) \times A_{k+1}\right| = \left|A_1 \times A_2 \times \cdots \times A_k \times A_{k+1}\right|$$

Since $2 \in S$ and $k \in S$, we have

$$\left|(A_1 \times A_2 \times \cdots \times A_k) \times A_{k+1}\right| = \left|A_1 \times A_2 \times \cdots \times A_k\right| \cdot \left|A_{k+1}\right|$$
$$= |A_1| \cdot |A_2| \cdots \cdot |A_k| \cdot |A_{k+1}|$$

Hence it follows that $k + 1 \in S$. Therefore $S = \mathbb{N}$ and the proof is complete.   □

The multiplication principle can be stated in a more pedestrian and perhaps more meaningful manner as follows.

Multiplication Principle Suppose that an experiment consists of making an ordered selection of n objects such that:

(1) The first object can be selected in m_1 ways.
(2) For each choice of the first object, the second object can be selected in m_2 ways.
(3) For each choice of the first two objects, the third object can be chosen in m_3 ways.
$$\vdots$$
(n) For each choice of the first $n - 1$ objects, the nth object can be chosen in m_n ways.

Then the n objects can be chosen together in a total of $m_1 \cdot m_2 \cdots \cdot m_n$ ways.

Example 6.6   In how many ways can a sequence of four cards be chosen at random from a standard deck of 52 cards (a) with replacement? (b) without replacement?

Solution   (a) There are 52 choices for the first card in the sequence. Since the first card chosen is put back into the deck before the choice of the second card is made, there are also 52 choices for the second card. Similarly, there are 52 choices for the third card and 52 choices for the fourth card. Therefore, by the multiplication principle, a sequence of four cards may be chosen, with replacement, in $52 \cdot 52 \cdot 52 \cdot 52 = 52^4$ ways.

(b) There are 52 choices for the first card in the sequence. Since the first card chosen is not replaced in the deck, there are then only 51 choices for the second card. Similarly, there are 50 choices for the third card and 49 choices for the fourth card. Therefore, by the multiplication principle, a sequence of four cards may be chosen, without replacement, in $52 \cdot 51 \cdot 50 \cdot 49$ ways.

Example 6.7   How many integers n are there such that both $5000 < n < 10000$ and the digits of n are distinct?

Solution   If n is such an integer, then n contains four digits and the thousands digit must be one of 5, 6, 7, 8, 9. Thus there are five choices for the thousands digit. Once this digit has been selected, we can choose the hundreds digit in any one of nine ways. (Note that 0 can be used for this digit.) Once these two digits have been selected, there are eight choices for the tens digit. Finally, once the first three digits have been selected, there remain seven choices for the units digit. Thus, by the multiplication principle, there are $5 \cdot 9 \cdot 8 \cdot 7$ total ways to choose n.   ∎

Another very basic and well-used counting principle in combinatorics is the "pigeonhole principle." It is quite simple to state and easy to understand; indeed, it seems so straightforward that the student may feel confident of proving it. (If so, try it now, using a proof by contradiction.)

THEOREM 6.3   **(Pigeonhole Principle)**

If $n + 1$ objects are placed in n boxes, then some box contains at least two of the objects.

This result follows from a more general version that is presented later in this section. The importance of the pigeonhole principle rests in its deep generalizations and in its application to a wide variety of difficult and sometimes tricky problems. Incidentally, note that it is possible to place n objects into n boxes with exactly one object in each box.

Example 6.8   Given a group of n women and their husbands, how many people must be chosen from this group of $2n$ people in order to guarantee that the set of those selected contains a married couple?

Solution   Think of associating the n married couples with boxes labeled 1 to n, so that whenever a member is selected from the set of $2n$ people, then that person is "placed" into his or her associated box. Thus the question reduces to asking for the smallest number of members that can be placed in the n boxes in order that some box contains two members. Clearly, n does not suffice; however, by the pigeonhole principle, $n + 1$ works.   ∎

Example 6.9   In any set of 21 integers chosen from $\{1, 2, \ldots, 40\}$, show that there must be two members such that one is divisible by the other.

Solution   Let $x_1, x_2, \ldots, x_{21}$ denote the 21 integers chosen. We can express each x_i, $1 \leq i \leq 21$, in the form

$$x_i = 2^{m_i} \cdot b_i$$

where b_i is odd. Consider the odd integers $b_1, b_2, \ldots, b_{21}$. There are exactly 20 odd integers between 1 and 40; hence, by the pigeonhole principle, at least two of $b_1, b_2, \ldots, b_{21}$ are equal. So $b_r = b_s$ for some r and s, where

$r \neq s$. Then we have

$$x_r = 2^{m_r} \cdot b_r \quad \text{and} \quad x_s = 2^{m_s} \cdot b_s$$

Since $x_r \neq x_s$, we have $m_r \neq m_s$; without loss of generality, we can assume $m_r < m_s$. But then $2^{m_r} \mid 2^{m_s}$ and consequently, $x_r \mid x_s$. Is it possible to choose 20 integers from the list so that no number is divisible by any other? ∎

An interesting and perhaps discomforting feature of the preceding problem is that it is not at all clear from the statement that an application of the pigeonhole principle is called for. In many such problems, it is only after performing some clever maneuver that one sees an opening for its use. Let's try another.

Example 6.10   Show: If five points are chosen on, or within, an equilateral triangle of side length 2, then at least two of the points are at a distance at most 1 from each other.

Solution   Partition the given triangular region into four subregions by joining the midpoints of the sides by line segments. The resulting "triangulated" region is shown in Figure 6.1. If the five points are placed in the given region, then the pigeonhole principle implies that one of four subregions must contain at least two of the points. Observe that the distance between two such points is at most one. ∎

Figure 6.1

As alluded to earlier, the pigeonhole principle admits to various generalizations, some with rather profound applications in combinatorial theory and problems. The following theorem represents a natural (next-in-line) extension.

THEOREM 6.4   **(Pigeonhole Principle — Strong Form)**

Let n, r, and d be positive integers. If $n(r - 1) + d$ objects are placed into n boxes, then some box contains at least r of the objects.

Proof   Proceed by contradiction and assume that each box contains at most $r - 1$ of the objects. Then the total number of objects contained in the n boxes would be at most $n(r - 1)$, thus contradicting the hypothesis that

$n(r - 1) + d$ objects (with $d > 0$) are placed into the boxes. Hence some box contains at least r objects. □

Theorem 6.3 is an immediate consequence of Theorem 6.4—simply set $r = 2$ and $d = 1$. The following result is also a consequence of Theorem 6.4; its proof is left to Exercise 16.

COROLLARY 6.5 Given n positive integers $m_1, m_2, \ldots, m_n$ whose average

$$(m_1 + m_2 + \cdots + m_n)/n$$

exceeds the integer $r - 1$, then $m_i \geq r$ for some i, $1 \leq i \leq n$.

Example 6.11 A violinist practiced for a total of 110 hours over a period of 12 days. Show that he practiced at least 19 hours on some pair of consecutive days. Assume that he practiced a whole number of hours on each day.

Solution Number the days $1, 2, \ldots, 12$ and consider the subsets $\{1, 2\}, \{3, 4\}, \ldots,$ $\{11, 12\}$ associated with six pairs of consecutive days. Since these subsets partition the 12-day period, the 110 hours of practice can be distributed among them. And since $110 = 6 \cdot 18 + 2$, the pigeonhole principle (strong form) implies that some consecutive 2-day period contains at least 19 hours. ■

Example 6.12 Suppose that the numbers $1, 2, \ldots, 10$ are randomly placed in ten locations on a circle; Figure 6.2 shows one possible assignment. Show that the sum of the numbers occupying some set of three consecutive locations is more than 16.

Solution Number the locations $1, 2, \ldots, 10$ and let a_i be the number assigned to location i, $1 \leq i \leq 10$. There are ten sums to consider, namely

$$a_1 + a_2 + a_3, a_2 + a_3 + a_4, \ldots, a_9 + a_{10} + a_1, a_{10} + a_1 + a_2$$

and each a_i appears in exactly three of the sums. Hence the total of these

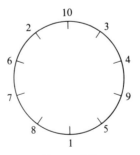

Figure 6.2

sums is given by

$$\sum_{i=1}^{10} 3a_i = 3 \cdot \sum_{i=1}^{10} a_i$$

$$= 3 \cdot \sum_{i=1}^{10} i$$

$$= 3 \cdot \frac{(10)(11)}{2}$$

$$= 165$$

Thus the average value of the sums is $165/10 = 16.5 > 16$, and so, by Corollary 6.5, one of the sums has value at least 17. ∎

Exercises 6.2

1. Find the number of bit strings of length 1, 2, 3, or 4.
2. The American Standard Code for Information Exchange, or ASCII, defines a sequence of 128 characters. Thirty-three of these characters are control characters and 95 are graphic characters, including the space, the 10 digits, the 52 uppercase and lowercase letters, and 32 special characters such as $*$, $/$, and $=$. Note that $128 = 2^7$, so that each ASCII character may be encoded as a bit string of length 7; for example, A is encoded as 1000001. Suppose we wish to represent each ASCII character by a ternary string, where each digit is 0, 1, or 2. What is the minimum length required for such a string?
3. Looking at the standard telephone dial, note that each of the digits 2 through 9 corresponds to three letters.
   (a) In the "good old days," a local telephone number consisted of two letters followed by five digits, such as EM6-3415. How many such telephone numbers are there?
   (b) Today a local number is simply a sequence of seven digits, with the restriction that the first digit cannot be 0 or 1. A full telephone number prefixes the local number with a three-digit area code; the area code may not begin with 0 or 1 and must have a 0 or 1 as the second digit. How many full telephone numbers are there?
4. How many "different" logical expressions are there using (any of) n logical variables $p_1, p_2, \ldots, p_n$? (Two expressions that are logically equivalent are not considered to be "different.")
5. How many students must a university have in order to guarantee that some two students have the same three initials?
6. Let p_1, p_2, p_3, p_4 be distinct primes and let a_1, a_2, a_3, a_4 be positive integers. We are interested in counting the number of distinct positive

divisors of

$$n = p_1^{a_1} \cdot p_2^{a_2} \cdot p_3^{a_3} \cdot p_4^{a_4}$$

(a) How many distinct positive divisors does $3^5 \cdot 7^4 \cdot 11^3 \cdot 17^8$ have?

(b) Answer the general question.

7. Argue that, in a class of 20 students, there are two students who are acquainted with the same number of students in the class.

8. Argue that, in a round-robin tournament with ten teams, either there is a unique team with i wins, $0 \leq i \leq 9$, or there are two teams with the same number of wins.

9. A drawer contains 14 socks in 7 matching, but unsorted, pairs. A mathematician randomly selects socks from the drawer, one at a time (without replacement). How many socks must be selected to guarantee that a matching pair is chosen?

10. Each of 16 students has a positive whole number of dollars and together the students have 30 dollars. For each n, $1 \leq n \leq 29$, show that some students together have n dollars.

11. Sarah campaigns for the State Assembly, visiting 45 towns in 30 days. If Sarah visits a positive whole number of towns each day, show that there must exist some period of consecutive days during which Sarah visits exactly 14 towns.

12. Given a sequence $(x_1, x_2, \ldots, x_n)$ of n integers, show that there exist i and j with $1 \leq i \leq j \leq n$ such that $(x_i + x_{i+1} + \cdots + x_j) \operatorname{MOD} n = 0$.

13. If $(x_0, x_1, \ldots, x_9)$ and $(y_0, y_1, \ldots, y_9)$ are binary sequences such that

$$x_0 + x_1 + \cdots + x_9 = 5 = y_0 + y_1 + \cdots + y_9$$

show that some "cycle permutation" of the y sequence matches the x sequence in at least five positions. (A *cycle permutation* of the sequence $(y_0, y_1, \ldots, y_9)$ is a sequence $(y_i, y_{i+1}, \ldots, y_{i+9})$ for some i, $0 \leq i \leq 9$, where the subscripts are taken modulo 10.)

14. Show that the function f defined in the proof of Theorem 6.2 is one-to-one and onto.

15. Let A and B be nonempty finite sets with $|A| = m$ and $|B| = n$. Let f be a function from A to B. Prove:

(a) If $m > n$, then f is not one-to-one.

(b) For some $k \in \mathbb{N}$, if $m > kn$, then there is some $b \in B$ having at least $k + 1$ preimages under f.

16. Prove Corollary 6.5. (In fact, one can prove a more general result: Given n real numbers $x_1, x_2, \ldots, x_n$ whose average exceeds the real number t, then $x_i > t$ for some i, $1 \leq i \leq n$.)

17. A certain make of combination lock has the integers 0 through 49 arranged in a circle. A combination for opening the lock consists of a sequence of three of these numbers, $(7, 32, 18)$, for example. How many combinations are there? How many combinations (x_1, x_2, x_3) have the property that $x_1 \neq x_2$ and $x_2 \neq x_3$?

| 6.3 | ## PERMUTATIONS AND COMBINATIONS |

In Chapter 5 a permutation of a nonempty set A is defined to be any function $f: A \to A$ that is both one-to-one and onto. In case A is a finite set, say $A = \{x_1, x_2, \ldots, x_n\}$, then it is pointed out that a permutation of A can be thought of as an ordered n-tuple $(f(x_1), f(x_2), \ldots, f(x_n))$, or an ordered selection, of the elements of A. The number of permutations of an n-element set was determined in Chapter 5. We repeat the result here and give an alternate proof.

THEOREM 6.6   The number of permutations of an n-element set is n-factorial.

Proof   Let A be an n-element set. We apply the multiplication principle to count the number of ordered n-tuples $(a_1, a_2, \ldots, a_n)$, where $a_1, a_2, \ldots, a_n$ are the elements of A. We see directly that:

(1) There are n choices for the first coordinate a_1.
(2) Once a_1 is chosen, there are $n - 1$ choices for the second coordinate a_2.
(3) Once a_1 and a_2 are chosen, there are then $n - 2$ choices for the third coordinate a_3.
$\vdots$
(n) Once $a_1, a_2, \ldots, a_{n-1}$ are chosen, there is then one choice for the nth coordinate a_n.

By the multiplication principle, there are $1 \cdot 2 \cdot 3 \cdots (n - 1) \cdot n = n!$ permutations of A in all.   □

In the succeeding pages we make use of the following generalization of the notion of permutation.

DEFINITION 6.1

Let A be a finite set with $|A| = n \geq 1$. For $1 \leq r \leq n$, an *r-permutation* of A is any r-tuple $(a_1, a_2, \ldots, a_r)$, where $a_1, a_2, \ldots, a_r$ are distinct elements of A. We denote the number of r-permutations of an n-element set by $P(n, r)$.

Note that this definition includes the case of a permutation of A (let $r = n$): a permutation of an n-element set A is also called an n-permutation of A. In this case we agree to write $P(n)$ in place of $P(n, n)$.

As indicated for permutations, we can think of an r-permutation of a finite set A as an ordered selection of r distinct elements from A. For instance, suppose $A = \{1, 2, 3, 4\}$ and $r = 3$, so that we seek the 3-permutations of A. If we select, in order, $a_1 = 2$, $a_2 = 1$, and $a_3 = 4$, then we obtain the 3-permutation $(2, 1, 4)$. How many 3-permutations of A are there? Well, there are four choices for the first element a_1, then three choices for a_2, and then two choices for a_3. By the multiplication principle, there are $4 \cdot 3 \cdot 2 = 24$ choices in all; that is, the number of 3-permutations of A is 24. We list them here:

$$(1,2,3), (1,3,2), (2,1,3), (2,3,1), (3,1,2), (3,2,1)$$
$$(1,2,4), (1,4,2), (2,1,4), (2,4,1), (4,1,2), (4,2,1)$$
$$(1,3,4), (1,4,3), (3,1,4), (3,4,1), (4,1,3), (4,3,1)$$
$$(2,3,4), (2,4,3), (3,2,4), (3,4,2), (4,2,3), (4,3,2)$$

In general, how many r-permutations of an n-element set A are there? One way to answer this question is to consider the fact that an r-permutation of A can be thought of as a one-to-one function $f \colon X_r \to A$, where $X_r = \{1, 2, \ldots, r\}$. After all, if f is one-to-one, then $f(1), f(2), \ldots, f(r)$ are distinct elements of A. Thus, if $(a_1, a_2, \ldots, a_r)$ is an r-permutation of A, then the associated function f is given by $f(i) = a_i$ for $1 \leq i \leq r$. How many such functions are there? This question is answered by Theorem 5.2, which we repeat here as Theorem 6.7.

THEOREM 6.7 If X and Y are nonempty sets with $|X| = r$, $|Y| = n$, and $r \leq n$, then the number of one-to-one functions from X to Y is $n!/(n-r)!$.

Noting that

$$\frac{n!}{(n-r)!} = \frac{n(n-1) \cdots (n-r+1)(n-r) \cdots (2)(1)}{(n-r) \cdots (2)(1)}$$
$$= n(n-1) \cdots (n-r+1)$$

we can restate Theorem 6.7 as follows:

THEOREM 6.8 For each $n \in \mathbb{N}$ and $1 \leq r \leq n$,

$$P(n, r) = n(n-1) \cdots (n-r+1)$$
$$= \frac{n!}{(n-r)!}$$

Theorem 6.8 can be proved using induction and the multiplication principle for two sets. It can also be proved using the general version of the multiplication principle (Theorem 6.2). Perhaps it is instructive to run through the argument.

In making an ordered selection of r objects from a set of n distinct objects, the first object can be chosen in n ways. Once chosen, there are then $n - 1$ ways to choose the second object; this done, there are $n - 2$ ways to choose the third object, and so on, until the rth object is chosen, which can be done in $n - (r - 1) = n - r + 1$ ways. Thus, by the multiplication principle, the r objects can be chosen in $n(n - 1) \cdots (n - r + 1)$ ways, and the result of Theorem 6.8 follows.

Example 6.13 How many five-letter words can be formed from the letters a, b, c, d, e, f, g, and h if no letter is repeated? How many of these include the letter a?

Solution In the first question, a five-letter word is a 5-permutation of the given letters. Hence there are

$$P(8,5) = \frac{8!}{(8 - 5)!}$$
$$= 8 \cdot 7 \cdot 6 \cdot 5 \cdot 4$$
$$= 6720$$

such words.

In the second question, we insist that the letter a be present. There are five possible positions for it and, for each such position, the associated word can be completed by choosing a four-letter word from the letters b, c, d, e, f, g, and h. Hence, by the multiplication principle, there are

$$5 \cdot P(7,4) = 5 \cdot \frac{7!}{(7 - 4)!}$$
$$= 5 \cdot 7 \cdot 6 \cdot 5 \cdot 4$$
$$= 4200$$

such five-letter words. ∎

Example 6.14 In how many ways can 10 couples be chosen from 10 boys and 16 girls if a couple is to consist of a boy and a girl?

Solution The solution of the problem can be determined by lining up the ten boys and then choosing a girl for each boy. Thus, if we represent the ten boys by $b_1, b_2, \ldots, b_{10}$, then we wish to fill in the second coordinates of the pairs

$$(b_1, g_1), (b_2, g_2), \ldots, (b_{10}, g_{10})$$

with 10 of the 16 girls, $g_1, g_2, \ldots, g_{10}$. This can be done in $P(16, 10)$ ways. ∎

In choosing an r-permutation of a set A, the order in which the elements are chosen is important. For example, the telephone numbers 673-1245 and 673-1542 are different. However, there are times when, in making a series of choices, the order in which the choices are made is unimportant. For instance, consider the process of being dealt a five-card poker hand; what matters is the *set* of five cards in the final hand, not the order in which they were dealt.

Suppose we wish to determine the number of possible poker hands, or the number of poker hands that contain exactly three aces. Since a poker hand can be thought of as a five-element subset of the set of 52 cards, we need to determine the number of five-element subsets of a 52-element set. Similarly, counting the poker hands containing exactly three aces involves finding the number of three-element subsets of a four-element set (the set of aces), as well as the number of two-element subsets of a 48-element set (the set of cards that are not aces).

In general, if S is an n-element set, then any r-element subset of S is called an *r-combination* of S. We denote the number of r-combinations of S by $C(n, r)$ or $\binom{n}{r}$. Sometimes $C(n, r)$ is referred to as the *number of combinations of n things taken r at a time*.

To help us determine a formula for $C(n, r)$, we consider an example. Let's list the 3-combinations of $A = \{1, 2, 3, 4\}$, and next to each subset list the 3-permutations of A using the elements from that subset:

$$\{1, 2, 3\}: \quad (1, 2, 3), (1, 3, 2), (2, 1, 3), (2, 3, 1), (3, 1, 2), (3, 2, 1)$$
$$\{1, 2, 4\}: \quad (1, 2, 4), (1, 4, 2), (2, 1, 4), (2, 4, 1), (4, 1, 2), (4, 2, 1)$$
$$\{1, 3, 4\}: \quad (1, 3, 4), (1, 4, 3), (3, 1, 4), (3, 4, 1), (4, 1, 3), (4, 3, 1)$$
$$\{2, 3, 4\}: \quad (2, 3, 4), (2, 4, 3), (3, 2, 4), (3, 4, 2), (4, 2, 3), (4, 3, 2)$$

Note that each of the four 3-combinations of A yields six 3-permutations of A. Thus we have that

$$P(4, 3) = 24 = 6 \cdot 4 = P(3) \cdot C(4, 3) = 3! \cdot C(4, 3)$$

This example suggests a general relationship between $P(n, r)$ and $C(n, r)$.

THEOREM 6.9   For $n \in \mathbb{N}$ and $1 \leq r \leq n$, we have

$$P(n, r) = r! \cdot C(n, r)$$

Thus, for $0 \leq r \leq n$,

$$C(n, r) = \frac{n!}{r!(n-r)!}$$

Proof   Consider any r-element subset of an n-element set A. Such a subset gives rise to $r!$ r-permutations of A, since the r elements in the subset can be ordered in $r!$ ways. Moreover, every r-permutation of A is so determined. Thus we obtain the relation

$$P(n, r) = r! \cdot C(n, r)$$

which holds for $1 \leq r \leq n$. Applying Theorem 6.8 then yields

$$C(n, r) = \frac{P(n, r)}{r!} = \frac{n!}{r!(n-r)!}$$

for $1 \leq r \leq n$. For $r = 0$, it follows easily that $C(n, 0) = \dfrac{n!}{0!n!} = 1$. $\square$

Example 6.15   Determine the number of distinct poker hands. How many of these contain exactly three aces?

Solution   The number of distinct poker hands is the same as the number of 5-combinations of a 52-element set; that is,

$$C(52,5) = \frac{52!}{(5!)(47!)} = 2,598,960$$

To determine the number of poker hands containing exactly three aces, we employ the multiplication principle. There are $C(4,3)$ ways to choose three of the four aces and, once chosen, there are then $C(48,2)$ ways to choose the remaining two cards from the 48 cards that are not aces. Thus, by the multiplication principle, there are $C(4,3) \cdot C(48,2)$ ways to obtain a poker hand with exactly three aces.   ■

Example 6.16   A lot of 100 items from a manufacturing process is known to contain 10 defective items (and 90 items that are nondefective). A sample of seven items is to be selected at random and checked. How many samples contain:

(a) exactly three defective items?
(b) at least one defective item?

Solution   (a) If the sample contains exactly three defective items, then it contains exactly four nondefective items. There are $C(10,3)$ ways to choose exactly three defective items and $C(90,4)$ ways to choose exactly four nondefective items. Hence, by the multiplication principle, there are $C(10,3) \cdot C(90,4)$ ways to choose a sample with exactly three defective items.

(b) The best way to answer this question is to consider the complement of the set of samples containing at least one defective item. This complement is the set of samples containing no defective items; since there are 90 nondefective items, there are exactly $C(90,7)$ samples containing no defective items. And since there are $C(100,7)$ samples in all, it follows that there are $C(100,7) - C(90,7)$ samples that contain at least one defective item.   ■

We have defined the power set $\mathscr{P}(A)$ of a set A to be the set of all subsets of A. Suppose $|A| = n$. Since $C(n,r)$ is the number of r-element subsets of A, we have that

$$|\mathscr{P}(A)| = C(n,0) + C(n,1) + \cdots + C(n,n) = \sum_{r=0}^{n} C(n,r)$$

In Chapter 3 we proved that $|\mathscr{P}(A)| = 2^n$; hence we have the following result.

THEOREM 6.10 For each positive integer n,

$$C(n,0) + C(n,1) + \cdots + C(n,n) = 2^n$$

Example 6.17 The purpose of this example is to point out an error that is frequently made in counting problems. Consider a box containing seven distinct colored balls: three blue, two red, and two green. A subset of three of the balls is to be selected at random. How many such subsets contain at least two blue balls?

Incorrect solution We want a subset that contains at least two blue balls. Start by choosing two blue balls; since there are three blue balls to choose from this may be done in $C(3,2)$ ways. Now any one of the remaining five balls may be chosen as the third ball in the subset. Thus, by the multiplication principle, the number of subsets containing at least two blue balls is

$$C(3,2) \cdot C(5,1) = 3 \cdot 5 = 15$$

Correct solution We first solve the problem by brute force. The number of balls was purposely kept small in this problem to allow a listing of all the 3-combinations with at least two blue balls. We denote the set of balls by

$$\{ b_1, b_2, b_3, r_1, r_2, g_1, g_2 \}$$

where b_1, b_2, and b_3 are the blue balls. Also let x denote any one of the four nonblue balls. Then the 3-combinations with at least two blue balls are $\{ b_1, b_2, b_3 \}$, the four subsets of the form $\{ b_1, b_2, x \}$, the four subsets of the form $\{ b_1, b_3, x \}$, and the four subsets of the form $\{ b_2, b_3, x \}$. Hence there are 13 such subsets, not 15. We can also obtain the correct answer by properly applying our counting techniques. If a 3-combination is to contain at least two blue balls, then either it contains exactly two blue balls or it contains exactly three blue balls. There are $C(3,2) \cdot 4 = 12$ 3-combinations with exactly two blue balls. And there is just one 3-combination with exactly three blue balls. Thus, by the addition principle, there are $12 + 1 = 13$ three-element subsets that contain at least two blue balls. (The student is encouraged to find the flaw in the incorrect solution.) ■

Example 6.18 **(Circular Permutations)**

A *circular r-permutation* of n distinct objects is an ordered placement of r of the objects in r equally spaced positions on a circle. Two such permutations are considered *equal* if one can be obtained from the other by a suitable rotation of the circle about its center. For example, consider the circular 4-permutations in Figure 6.3. The permutations in (a) and (b) are equal, while that in (c) is not equal to the others. Determine the number of circular r-permutations of n distinct objects.

Solution First choose the r objects to be placed in the r positions. This can be done in $C(n,r)$ ways. Suppose the objects selected are $a_1, a_2, \ldots, a_r$. These r objects must now be placed in the r positions of the circle. In how many

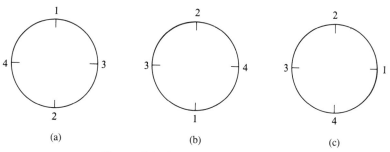

Figure 6.3 Circular permutations

different ways can the first object a_1 be positioned? Well, because of the definition of equality for circular r-permutations, the positions (all r of them) for this first choice are all equivalent. Hence there is only one way to select a position for a_1. Once this has been done, we must then choose positions for the remaining $r - 1$ objects. But this is equivalent to making an ordered selection of $r - 1$ distinct objects, and this can be done in $(r - 1)!$ ways. Hence, the total number of circular r-permutations of n distinct objects is

$$C(n, r) \cdot (r - 1)!$$ ■

Incidentally, a circular n-permutation of n distinct objects is simply called a *circular permutation* of the n objects. As an application, consider the following question: In how many ways can five men and five women be seated at a circular table if no two neighbors are to be of the same sex? To answer the question, consider seating the men first; they must be seated in alternate seats. This amounts to choosing a circular permutation of five objects, and this can be done in 4! ways. Once this has been done, we then place the five women in the remaining five seats, and this can be done in 5! ways. Thus the total number of ways to seat the group is $(4!)(5!) = 2880$.

We now return to the problem that was posed in Example 6.3 of Section 6.1. The general form of the problem that arose reads as follows:

Given positive integers m and n, determine the number of nonnegative integer solutions to the equation

$$x_1 + x_2 + \cdots + x_n = m$$

Consider the special case

$$x_1 + x_2 + x_3 + x_4 = 8$$

One possible solution to this equation is given by $x_1 = 1$, $x_2 = 2$, $x_3 = 3$, and $x_4 = 2$. This solution corresponds to a partitioning of the string 11111111 of eight 1's into the four substrings 1, 11, 111, 11. This partitioned list can be more conveniently displayed as a single string 1/11/111/11 of 1's and /'s (slashes). The solution $x_1 = 0$, $x_2 = 4$, $x_3 = 0$, and $x_4 = 4$, corresponds to the string /1111//1111. In fact, any solution to the given equation corresponds to a uniquely determined string of eight 1's and three /'s (try some yourself). Conversely, any such string of eight 1's and three /'s corresponds to a uniquely determined solution of the equation. Hence the number of solutions of the equation in nonnegative integers is then equal to the number of different sequences of eight 1's and three /'s. How many of these are there? Well, each such sequence consists of 11 terms (eight 1's and three /'s) and is determined once positions have been chosen for the eight 1's. In how many different ways can one choose eight positions out of a possible 11? This is just $C(11, 8)$. Hence the answer to the question is $C(11, 8)$.

Using an argument similar to that given above, one can prove that the answer to the question posed in Example 6.3 is given by the following result. Its proof is left to Exercise 27.

THEOREM 6.11   Let m and n be positive integers. The number of solutions to the equation

$$x_1 + x_2 + \cdots + x_n = m$$

where $x_1, x_2, \ldots, x_n$ are nonnegative integers, is $C(n + m - 1, m)$.

The above theorem can be phrased in a more mundane fashion as follows:

> Suppose we are given n distinct objects, each available in large quantities (for example, seven varieties of donuts, each available in large quantities). Then the number of ways to select m of the objects, with repetitions allowed, is $C(n + m - 1, m)$.

In many problems one is asked to determine the number of positive integer solutions to an equation of the form shown in Theorem 6.11. In this case we are given n distinct objects, each available in large quantities, and wish to determine the number of ways to choose m of the objects with the proviso that at least one of each object is chosen. Notice that the condition $m \geq n$ must hold for the problem to make sense. We can handle this case with Theorem 6.11. First choose one each of the n objects (this can be done in only one way) and then choose $m - n$ of the n objects in any fashion. By Theorem 6.11, this can be done in precisely $C(n + (m - n) - 1, m - n) = C(m - 1, m - n)$ ways. Hence we have the following result.

THEOREM 6.12   Let m and n be positive integers with $m \geq n$. The number of solutions to the equation

$$y_1 + y_2 + \cdots + y_n = m$$

where $y_1, y_2, \ldots, y_n$ are positive integers, is $C(m - 1, m - n)$.

Example 6.19 A bakery sells seven kinds of donuts.

(a) How many ways are there to choose 12 donuts?
(b) How many ways are there to choose 12 donuts if there must be at least one donut of each kind?

Solution What matters here is the number of donuts of each kind chosen. Thus, let x_i be the number of donuts of kind i chosen, where $1 \leq i \leq 7$. Then $x_1 + x_2 + \cdots + x_7 = 12$.

(a) Here each x_i is a nonnegative integer, so we apply Theorem 6.11. Hence we get that there are $C(7 + 12 - 1, 12) = C(18, 12)$ ways to choose 12 donuts.

(b) In this part, each x_i is a positive integer, so Theorem 6.12 applies to yield that there are $C(12 - 1, 12 - 7) = C(11, 5)$ ways to choose 12 donuts. ∎

Example 6.20 How many ways are there to place nine identical rings on the four fingers of your right hand?

Solution One can liken this problem to that of distributing nine identical objects into four different boxes (the boxes corresponding to the four fingers of your right hand). Assuming the four fingers are labeled f_1, f_2, f_3, and f_4, let x_i denote the number of rings that are placed on f_i, where $1 \leq i \leq 4$. Then the answer to the problem is precisely the number of nonnegative integer solutions to the equation $x_1 + x_2 + x_3 + x_4 = 9$. By Theorem 6.11, this number is $C(4 + 9 - 1, 9) = C(12, 9)$. ∎

Exercises 6.3

1. Seven runners are entered in the mile race at a track meet. Different trophies are awarded to the first, second, and third place finishers. In how many ways might the trophies be awarded?
2. If Tom has 12 horses and Bill has 15 horses, in how many ways can they trade 8 horses with each other?
3. In the game of Mastermind (see Exercise 14 in Exercises 2.4), how many codes use exactly
   (a) 1 color?   (b) 2 colors?   (c) 3 colors?   (d) 4 colors?
4. Let n be an integer, where $n \geq 2$, and suppose the canonical factorization of n is

$$n = p_1^{a_1} \cdot p_2^{a_2} \cdot \cdots \cdot p_k^{a_k}$$

(a) Give a formula (in terms of the a_i's) for the number of positive factors of n.
(b) Show that n has an odd number of positive factors if and only if n is a perfect square.

5. Ron has n friends who enjoy playing bridge, and he is able to invite a different subset of three of them over to his home every Wednesday night for 2 years. How large (at least) is n?

6. How many four-digit numbers can be formed from the digits 1, 2, 3, 4, 5, 6? How many of these have distinct digits?

7. Solve the problem of the seven generals from Example 6.1.

8. For $0 \le k \le n$, consider the identity $C(n, k) = C(n, n - k)$.
   (a) Verify this identity using Theorem 6.9.
   (b) Give a combinatorial justification for this identity.

9. The Department of Mathematics and Computer Science has four full professors, eight associate professors, and three assistant professors. A four-person search committee is to be selected. How many ways are there to choose the committee under these conditions?
   (a) At least one full professor is chosen.
   (b) The committee contains at least one professor of each rank.
   (c) The committee contains exactly one assistant professor or it contains exactly two associate professors.

10. In how many ways can 12 distinct homework problems be distributed among 20 students if each student gets at most one problem?

11. A university is to interview six candidates for the position of provost; Dr. Deming is a member of the search committee.
   (a) After interviewing the candidates, Dr. Deming is to rank the candidates from first choice to sixth choice. In how many ways can this be done?
   (b) Suppose that, rather than ranking all the candidates, Dr. Deming is to give a first choice, a second choice, and a third choice. In how many ways can this be done?
   (c) Suppose that, rather than ranking the candidates, Dr. Deming is to indicate a set of three "acceptable" candidates. In how many ways can this be done?
   (d) Suppose that, rather than ranking the candidates, Dr. Deming is to indicate a set of "acceptable" candidates. (This is called "approval voting.") In how many ways can this be done?

12. How many permutations of $\{1, 2, 3, 4, 5\}$ meet the following conditions?
   (a) 1 is in the first position.
   (b) 1 is in the first position and 2 is in the second position.
   (c) 1 is in the first position or 2 is in the second position.
   (d) 1 is not in the first position and 2 is not in the second position.
   (e) 1 is not in the first position or 2 is not in the second position.

13. How many binary strings of length 16 contain exactly six 1's?

14. In how many ways can a nonempty subset of people be chosen from six men and four women if the subset has equal numbers of men and women?

15. How many five-card poker hands have at least one card from each suit?

16. How many five-digit numbers formed from the digits 1, 2, 3, 4, and 5,
   (a) have the 1 before the 5?
   (b) contain "12" or "21"?

(c) contain "12" or "34"?

(d) contain neither "12" nor "23"?

17. In how many ways can seven distinct keys be put on a key ring? (Assume the orientation of the individual keys is not important.)

18. Five married couples are to be seated at a round table. How many ways are there to seat these ten people under the following conditions?
   (a) There are no restrictions.
   (b) Each man is seated next to two women.
   (c) Each husband must be seated next to his wife.
   (d) The five men are seated consecutively.

19. How many five-card poker hands contain the following?
   (a) exactly one pair (not two pairs or three of a kind)
   (b) exactly three of a rank (not four of a rank or a full house)
   (c) a full house (three of one rank, a pair of another)
   (d) a straight flush (for example, 7, 8, 9, 10, jack, all the same suit)
   (e) a straight (for example, 7, 8, 9, 10, jack, not all the same suit)
   (f) a flush (all the same suit, but not a straight)

20. If the U.S. Senate has 52 Democrats and 48 Republicans, in how many ways can a committee of six Democrats and four Republicans be formed? How many ways are there to form this committee if one of the Democrats is to be appointed chair and one of the Republicans is to be appointed vice chair of the committee?

21. A pizza may be ordered with any of the following extras: extra cheese, pepperoni, mushrooms, sausage, green peppers, onions, or anchovies.
   (a) How many kinds of pizza may be ordered?
   (b) How many kinds of pizza with exactly three extras may be ordered?
   (c) If a "vegetarian pizza" may not contain pepperoni, sausage, or anchovies, how many kinds of vegetarian pizza are there?

22. A member of the Association for Computing Machinery (ACM) may belong to any of 33 Special Interest Groups (SIGs). The ACM wishes to assign each member a code number indicating the SIGs to which that member belongs. How many code numbers are needed?

23. In how many ways can one choose ten coins from piles of pennies, nickels, dimes, and quarters?

24. In how many ways can one choose an unordered group of eight numbers between 1 and 25 inclusive, if repetitions are allowed?

25. (a) How many ways are there to distribute 15 identical candy canes to seven children if each child is to receive at least one cane?
   (b) How many ways are there to distribute 15 identical candy canes and ten identical gumballs to five children?

26. Eight identical balls are to be placed into six different boxes $B_1, B_2, \ldots, B_6$.
   (a) In how many ways can this be done?
   (b) In how many ways can it be done if the boxes B_1 and B_2 are to remain empty?
   (c) In how many ways can it be done if at least two of the boxes are to remain empty?

(d) In how many ways can it be done if the boxes B_1 and B_2 are to contain, collectively, at most two of the objects?

27. Complete the proof of Theorem 6.11.

28. Write a Pascal function COMBO(N, K) that takes integers N and K with $0 \leq K \leq N$ and returns the value $C(N, K)$.

29. The purpose of this exercise is to describe a procedure to generate the permutations of $\{1, 2, \ldots, n\}$ in lexicographic order (see Exercise 2 of Exercises 6.1). Let's call the procedure PERMS(N). PERMS uses an array PI[1], PI[2], ..., PI[N] to hold the current permutation and, for convenience, sets PI[0] = 0. Also let PERMS contain two local procedures to help it perform its task—INITIALIZE simply initializes PI to the identity permutation (which is first in lexicographic order) and NEXTPERM updates PI to the next permutation in lexicographic order. The body of PERMS is then easy to write:

```
BEGIN (*PERMS*)
  DONE := FALSE;
  INITIALIZE;
  WHILE NOT DONE DO
    BEGIN
      output PI;
      NEXTPERM;
    END;
END; (*PERMS*)
```

(a) Write the procedure INITIALIZE.

(b) Write the procedure NEXTPERM. Here is a description of the algorithm:

Step 1. Set M equal to N − 1. While PI[M] > PI[M + 1], decrement M by 1.

Step 2. If M = 0, set DONE equal to TRUE and stop; otherwise, set K equal to N; while PI[M] > PI[K], decrement K by 1.

Step 3. Interchange the values of PI[M] and PI[K].

Step 4. For J = M + 1 to (M + N) DIV 2, interchange the values of PI[J] and PI[M + N + 1 − J].

(c) Trace NEXTPERM with PI = $(1, 2, 3, 4, 6, 7, 5)$.

(d) Trace NEXTPERM with PI = $(1, 6, 2, 7, 5, 4, 3)$.

(e) If M = 0 in NEXTPERM, what is PI? What happens if, instead of stopping when M = 0, we change Steps 2 and 3 of the algorithm as follows?

Step 2′. Set K equal to N. While PI[M] > PI[K], decrement K by 1.

Step 3′. If M > 0, interchange the values of PI[M] and PI[K].

30. Given a permutation $\pi = (a_1, a_2, \ldots, a_n)$ of $\{1, 2, \ldots, n\}$, the *lexicographic rank* of π is an integer between 0 and $n! - 1$ giving the number of predecessors of π in the list of all permutations of $\{1, 2, \ldots, n\}$ in lexicographic order.

(a) Given an integer m, $0 \leq m \leq n! - 1$, show that m is uniquely expressed as

$$\sum_{k=1}^{n-1} d_k k!$$

where d_k is an integer, $0 \leq d_k \leq k$. (The representation $d_{n-1} \cdots d_2 d_1$ is called the *factorial representation* of m. For instance,

$$80 = 3 \cdot 4! + 1 \cdot 3! + 0 \cdot 2! + 2 \cdot 1!$$

so the factorial representation of 80 is 3102.)

(b) Write a Pascal function PERM_TO_RANK that returns the lexicograhic rank of a given permutation. (Hint: Let r be the lexicographic rank of $\pi = (a_1, a_2, \ldots, a_n)$, and let $d_{n-1} \cdots d_2 d_1$ be the factorial representation of r; then a_1 determines d_{n-1}, and then a_2 determines $d_{n-2}, \ldots$, and then a_{n-1} determines d_1.)

(c) Write a Pascal function RANK_TO_PERM that, given integers R and N with $0 \leq R \leq N! - 1$, returns the permutation of $\{1, 2, \ldots, N\}$ with rank R.

31. The purpose of this exercise is to describe a procedure to generate the k-combinations of $\{1, 2, \ldots, n\}$ in lexicographic order (see Exercise 4 of Exercises 6.1). Let's call the procedure SUBSETS(N, K). SUBSETS uses an array S[1], S[2], \ldots, S[K] to hold the current subset, where S[1] < S[2] < $\cdots$ < S[K]. As with PERMS in Exercise 29, let's have SUBSETS contain two local procedures to help it perform its task—INITIALIZE initializes S to the subset $\{1, 2, \ldots, K\}$ (which is first in lexicographic order) and NEXTSUBSET updates S to the next subset in lexicographic order. SUBSETS also has two local variables H and M—S[H] is the smallest value in S to be modified by NEXTSUBSET, and M is the new value of S[H]. The variables H and M are initialized by INITIALIZE to the value of K. (To avoid trivial cases, let us assume that $0 < K < N$.)

(a) The body of SUBSETS is analogous to the body of PERMS, as shown in Exercise 29; write it.

(b) Write the procedure INITIALIZE.

(c) Write the procedure NEXTSUBSET. Here is a description of the algorithm:

Step 1. If M < N − K + H, then set H equal to K, set M equal to S[K] + 1, and skip to Step 3.

Step 2. If H = 1, then set DONE equal to TRUE and stop; otherwise, decrement H by 1 and set M equal to S[H] + 1.

Step 3. For each J, H $\leq$ J $\leq$ K, set S[J] equal to M + J − H.

(d) Trace NEXTSUBSET for N = 7 and S = $\{1, 3, 4, 5\}$. (Note: The predecessor of S is $\{1, 2, 6, 7\}$, so H = 2 and M = 3.)

(e) Trace NEXTSUBSET for N = 7, and S = $\{2, 4, 6, 7\}$. (Note: The predecessor of S is $\{2, 4, 5, 7\}$, so H = 3 and M = 6.)

(f) If H = 1 and M = N − K + 1 in NEXTSUBSET, what is S? What happens if, instead of stopping in this case, we change Step 2 of the algorithm as follows?

> **Step 2′.** If H = 1, then set M equal to 1; otherwise, decrement H by 1 and set M equal to S[H] + 1.

32. Given a *k*-combination $S = \{a_1, a_2, \ldots, a_k\}$ of $\{1, 2, \ldots, n\}$, the *lexicographic rank* of *S* is an integer between 0 and $C(n, k) - 1$ giving the number of predecessors of *S* in the list of all the *k*-combinations of $\{1, 2, \ldots, n\}$ in lexicographic order.

(a) Write a Pascal function COMBO_TO_RANK that returns the lexicographic rank of a given *k*-combination. (Hint: Let $S = \{a_1, a_2, \ldots, a_k\}$, where $1 \le a_1 < a_2 < \cdots < a_k \le n$; based on the value a_1, give a lower bound on the number of *k*-combinations of $\{1, 2, \ldots, n\}$ that follow *S* in lexicographic order.)

(b) Write a Pascal function RANK_TO_COMBO that, given integers R, K, and N with $0 < K < N$ and $0 \le R \le C(N, K) - 1$, returns the K-combination of $\{1, 2, \ldots, N\}$ with lexicographic rank R.

33. The purpose of this exercise is to describe a procedure to generate the subsets of $\{1, 2, \ldots, n\}$ in lexicographic order (see Exercise 5 of Exercises 6.1). Let's call the procedure SUBSETS(N). SUBSETS uses an array S[1], S[2], …, S[N] to hold the current subset and a variable K to hold the cardinality of the current subset, where always $1 \le S[1] < S[2] < \cdots < S[K]$. As with PERMS in Exercise 29, let SUBSETS contain two local procedures to help it perform its task—INITIALIZE initializes S to the empty subset (which is first in lexicographic order) and NEXTSUBSET updates S to the next subset in lexicographic order.

(a) The body of SUBSETS is analogous to the body of PERMS, as shown in Exercise 29; write it.

(b) Write the procedure INITIALIZE.

(c) Write the procedure NEXTSUBSET. Here is a description of the algorithm:

> **Step 1.** If K = 0, then set K equal to 1, set M equal to 0, and skip to Step 4.
>
> **Step 2.** If S[K] < N, then set M equal to S[K], increment K by 1, and skip to Step 4.
>
> **Step 3.** Decrement K by 1; if K > 0, then set M equal to S[K]; otherwise, set DONE equal to TRUE and stop.
>
> **Step 4.** Set S[K] equal to M + 1.

(d) Trace NEXTSUBSET for N = 7 and S = {1, 3, 4, 5}.

(e) Trace NEXTSUBSET for N = 7 and S = {1, 2, 3, 4, 5, 6, 7}.

(f) Trace NEXTSUBSET for N = 7 and S = {1, 7}.

(g) The given algorithm for NEXTSUBSET sets DONE equal to TRUE when S = {N}, since this is the last subset in lexicographic order. What is S changed to in this case?

34. Given a subset $S = \{a_1, a_2, \ldots, a_k\}$ of $\{1, 2, \ldots, n\}$, the *lexicographic rank* of S is an integer between 0 and $2^n - 1$ giving the number of predecessors of S in the list of all the subsets of $\{1, 2, \ldots, n\}$ in lexicographic order.

(a) Write a Pascal function SUBSET_TO_RANK that returns the lexicographic rank of a given subset. (Hint: Let $S = \{a_1, a_2, \ldots, a_k\}$, where $1 \le a_1 < a_2 < \cdots < a_k \le n$; based on the value a_1, give a lower bound on the number of subsets of $\{1, 2, \ldots, n\}$ that follow S in lexicographic order.)

(b) Write a Pascal function RANK_TO_SUBSET that, given integers R and N with

$$0 \le R \le 2^N - 1$$

returns the subset of $\{1, 2, \ldots, N\}$ with lexicographic rank R.

6.4 THE BINOMIAL THEOREM

Consider the expression $(x + y)^4$. Its expansion can be determined by considering the product

$$(x + y)(x + y)(x + y)(x + y)$$

An arbitrary term in the expansion is obtained by choosing an x or a y in each of the four factors, thus leading to $2 \cdot 2 \cdot 2 \cdot 2 = 2^4$ terms in all. Each term in the expansion has the form $x^{4-k}y^k$, where $0 \le k \le 4$. For example, the term x^3y is obtained by choosing a y in one of the four factors and then choosing an x in each of the remaining three factors. Or, to put it another way, x^3y is obtained by choosing a y in exactly one of the four factors. Choosing a y in one of the four factors corresponds to choosing one factor out of the four, and this can be done in precisely $C(4, 1) = 4$ ways. Hence the term x^3y appears four times in the expansion of $(x + y)^4$; that is, the coefficient of x^3y in the expansion is 4. What is the coefficient of x^2y^2 in the expansion? In keeping with our discussion of x^3y, we see that the coefficient of x^2y^2 is exactly the number of ways of choosing two factors out of the four. Hence the coefficient of x^2y^2 is $C(4, 2) = 6$. Using this combinatorial line of reasoning, we obtain

$$(x + y)^4 = C(4, 0)x^4 + C(4, 1)x^3y + C(4, 2)x^2y^2 + C(4, 3)xy^3 + C(4, 4)y^4$$

$$= x^4 + 4x^3y + 6x^2y^2 + 4xy^3 + y^4$$

The student is invited to check this by expanding $(x + y)^4$ in the familiar manner. At any rate, it seems that we are not only prepared to conjecture a formula for $(x + y)^n$, but we also have a good idea of how to prove it.

THEOREM 6.13   **(Binomial Theorem)**

For each of $n \in \mathbb{N}$ and any $x, y \in \mathbb{R}$,

$$(x + y)^n = \sum_{k=0}^{n} C(n, k) x^{n-k} y^k$$

Proof   A combinatorial argument is given. Consider $(x + y)^n$ in the form

$$(x + y)^n = (x + y)(x + y) \cdots (x + y)$$

with n factors present on the right-hand side. An arbitrary term in the expansion of $(x + y)^n$ is obtained by choosing an x or a y in each of the above n factors. Thus each such term has the form $x^{n-k} y^k$ for some k, where $0 \le k \le n$. In particular, $x^{n-k} y^k$ can be obtained by first choosing a y in k of the n factors and then choosing an x in the remaining $n - k$ factors. The number of ways to choose a y in exactly k of the n factors is the same as the number of ways to choose k factors out of n. Hence the number of times $x^{n-k} y^k$ appears in the expansion of $(x + y)^n$ is precisely $C(n, k)$. So we have the relation

$$(x + y)^n = C(n, 0) x^n + C(n, 1) x^{n-1} y + \cdots +$$
$$C(n, k) x^{n-k} y^k + \cdots + C(n, n) y^n$$
$$= \sum_{k=0}^{n} C(n, k) x^{n-k} y^k \qquad \square$$

The above theorem can also be proved by induction on n (see Exercise 2).

COROLLARY 6.14   For each $n \in \mathbb{N}$ and any real number x,

$$(1 + x)^n = \sum_{k=0}^{n} C(n, k) x^k$$

In view of the above results, one frequently refers to the numbers $C(n, k)$ as the *binomial coefficients*. Notice that setting $x = 1$ in the expansion formula for $(1 + x)^n$ yields the familiar result (Theorem 6.10)

$$2^n = C(n, 0) + C(n, 1) + \cdots + C(n, n)$$

If we set $x = -1$ in $(1 + x)^n$, we obtain

$$0 = C(n, 0) - C(n, 1) + \cdots + (-1)^k C(n, k) + \cdots + (-1)^n C(n, n)$$

One can also use other methods to yield some interesting identities.

Example 6.21   Prove that

$$n \cdot 2^{n-1} = \sum_{k=1}^{n} kC(n, k)$$

for all $n \in \mathbb{N}$.

Solution   One possibility is to try induction on n; however, we shall employ a different technique. Consider the identity

$$(1 + x)^n = C(n, 0) + C(n, 1)x + C(n, 2)x^2 + \cdots +$$
$$C(n, k)x^k + \cdots + C(n, n)x^n$$

Using a little calculus, we differentiate both sides of this identity with respect to x and obtain the identity

$$n(1 + x)^{n-1} = C(n, 1) + 2C(n, 2)x + \cdots + kC(n, k)x^{k-1} +$$
$$\cdots + nC(n, n)x^{n-1}$$

Setting $x = 1$, we see that

$$n2^{n-1} = C(n, 1) + 2C(n, 2) + \cdots + kC(n, k) + \cdots + nC(n, n) =$$
$$\sum_{k=1}^{n} kC(n, k) \qquad \blacksquare$$

In what follows we shall use the convention that $C(n, r) = 0$ if either $r > n$ or $r < 0$. The binomial coefficient $C(n, r)$, $0 \le r \le n$, can be conveniently displayed, for successive values of n, in the following triangular array:

$$C(0, 0)$$
$$C(1, 0) \qquad C(1, 1)$$
$$C(2, 0) \qquad C(2, 1) \qquad C(2, 2)$$
$$C(3, 0) \qquad C(3, 1) \qquad C(3, 2) \qquad C(3, 3)$$
$$C(4, 0) \qquad C(4, 1) \qquad C(4, 2) \qquad C(4, 3) \qquad C(4, 4)$$
$$\vdots$$

or

$$1$$
$$1 \qquad 1$$
$$1 \qquad 2 \qquad 1$$
$$1 \qquad 3 \qquad 3 \qquad 1$$
$$1 \qquad 4 \qquad 6 \qquad 4 \qquad 1$$
$$\vdots$$

The preceding array is called *Pascal's triangle*, after the 17th-century French mathematician Blaise Pascal. Although he did not discover the triangle, he did derive several of its interesting properties. If we take a careful look at it, we can see some interesting patterns. For example, notice that the array is symmetric with respect to the column containing $C(0,0)$; that is, for $0 \le r \le n$, it appears that

$$C(n,r) = C(n, n-r)$$

The proof of this identity is addressed in Exercise 8 of Exercises 6.3. Also, for $n \ge 2$ and $0 < r < n$, it appears that $C(n, r)$ is the sum of the two terms that appear to its immediate right and left in the preceding row. What we have found is the following very useful identity.

THEOREM 6.15   **(Pascal's Identity)**

For each integer $n \ge 2$ and for $0 < r < n$,

$$C(n, r) = C(n-1, r-1) + C(n-1, r)$$

This identity can be proved using Theorem 6.9 or it can be proved combinatorially; its proof is left to Exercise 4.

Another interesting identity can be produced by looking at the sum of the elements on a diagonal that slopes upward from left to right (starting in a fixed row); for example, starting with row 5, observe that

$$C(0,0) + C(1,0) + C(2,0) + C(3,0) + C(4,0) = 5 = C(5,1)$$
$$C(1,1) + C(2,1) + C(3,1) + C(4,1) = 10 = C(5,2)$$
$$C(2,2) + C(3,2) + C(4,2) = 10 = C(5,3)$$

Can the student conjecture the general form of this identity? Here are both the conjecture and its proof.

THEOREM 6.16   For each $n \in \mathbb{N}$ and for $0 \le r \le n$,

$$C(r,r) + C(r+1, r) + C(r+2, r) + \cdots + C(n,r) = C(n+1, r+1)$$

Proof   We proceed by induction on n and let S be the set of positive integers n for which the result holds. To see that $1 \in S$, we must test the result for $r = 0$ and $r = 1$. For $r = 0$, we see that $C(0,0) + C(1,0) = C(2,1)$ holds, while for $r = 1$, note that $C(1,1) = C(2,2)$. So $1 \in S$.

Assume $k \in S$ for some arbitrary $k \ge 1$. Hence our induction hypothesis is that

$$C(r,r) + C(r+1, r) + \cdots + C(k, r) = C(k+1, r+1)$$

for $0 \le r \le k$. To show that $k + 1 \in S$, we must show that

$$C(r,r) + C(r+1, r) + \cdots + C(k, r) + C(k+1, r) = C(k+2, r+1)$$

for $0 \le r \le k + 1$. This statement is clearly true if $r = k + 1$. Using the induction hypothesis and Pascal's identity, we obtain, for $0 \le r \le k$,

$$C(r, r) + C(r + 1, r) + \cdots + C(k, r) + C(k + 1, r) =$$
$$C(k + 1, r + 1) + C(k + 1, r) =$$
$$C(k + 2, r + 1)$$

Hence $k + 1 \in S$. So $S = \mathbb{N}$ and our argument is complete. □

Example 6.22 Evaluate the sum

$$\sum_{k=3}^{n} k(k - 1)(k - 2)$$

Solution We apply Theorem 6.16 to arrive at the value of this sum. Consider the kth term, where $k \ge 4$:

$$k(k - 1)(k - 2) = \frac{(1)(2) \cdots (k - 3)(k - 2)(k - 1)k}{(1)(2) \cdots (k - 3)}$$
$$= \frac{k!}{(k - 3)!}$$
$$= 3! \cdot \frac{k!}{3!(k - 3)!}$$
$$= 3!C(k, 3)$$

Note that $k(k - 1)(k - 2) = 3!C(k, 3)$ also holds when $k = 3$. Thus

$$\sum_{k=3}^{n} k(k - 1)(k - 2) = \sum_{k=3}^{n} 3!C(k, 3)$$
$$= 3! \sum_{k=3}^{n} C(k, 3)$$
$$= 3!(C(3, 3) + C(4, 3) + \cdots + C(n, 3))$$
$$= 3!C(n + 1, 4)$$

by Theorem 6.16. ∎

Further identities and applications are explored in the exercises.

Exercises 6.4

1. Use Pascal's identity (Theorem 6.15) to evaluate:
(a) $C(4, 2)$ (b) $C(6, 3)$
2. Prove the binomial theorem (Theorem 6.13) by induction on n.

3. Use the binomial theorem to expand:

    (a) $(x + y)^5$ (b) $(x + y)^6$ (c) $(x + 3y)^7$

4. Prove Pascal's identity.

    (a) Give an algebraic proof, using Theorem 6.9.

    (b) Use a combinatorial argument. (Hint: Partition the k-element subsets of $\{1, 2, \ldots, n\}$ into two classes—those that contain n and those that do not—and apply the addition principle.)

5. Give a combinatorial proof of Theorem 6.16. (Hint: Consider choosing an $(r + 1)$-element subset of $\{1, 2, \ldots, n + 1\}$; partition these subsets into $n - r + 1$ classes according to the largest element chosen, and apply the addition principle.)

6. Prove the following property (called the *unimodal property*) of the binomial coefficients:

    (a) $C(n, k) \le C(n, k + 1)$ for $0 \le 2k < n$

    (b) $C(n, k) \ge C(n, k + 1)$ for $n \le 2k \le 2n - 2$

    (c) $C(n, k) = C(n, k + 1)$ if and only if $2k = n - 1$

7. Find the coefficient of:

    (a) $x^{11}y^4$ in the expansion of $(x + y)^{15}$

    (b) x^6y^4 in the expansion of $(2x + y)^{10}$

    (c) x^3y^5 in the expansion of $(3x - 2y)^8$

    (d) x^5 in the expansion of $(1 + x + x^2)(1 + x)^6$

8. For $n \in \mathbb{N}$, prove that

$$C(n, 0)^2 + C(n, 1)^2 + \cdots + C(n, n)^2 = C(2n, n)$$

    (a) Use the binomial theorem.

    (b) Give a combinatorial proof based on the fact that each n-element subset of $\{1, 2, \ldots, 2n\}$ is the union of a k-element subset of $\{1, 2, \ldots, n\}$ and an $(n - k)$-element subset of $\{n + 1, n + 2, \ldots, 2n\}$ for some k, $0 \le k \le n$; partition according to the value of k and apply the addition principle.

9. Use an appropriate identity to simplify each of these expressions.

    (a) $C(12, 4) + C(12, 5)$

    (b) $C(5, 5) + C(6, 5) + C(7, 5) + \cdots + C(11, 5)$

    (c) $C(10, 0) + C(10, 1) + C(10, 2) + \cdots + C(10, 10)$

    (d) $1 \cdot C(9, 1) + 2 \cdot C(9, 2) + 3 \cdot C(9, 3) + \cdots + 9 \cdot C(9, 9)$

10. Consider Pascal's triangle with n rows; let $f(n)$ be the sum of the numbers that lie on the median from the lower left-hand corner (of row n) to the opposite side. For instance, $f(6) = 1 + 4 + 3 = 8$ and $f(7) = 1 + 5 + 6 + 1 = 13$.

    (a) Express $f(n)$ as a sum of binomial coefficients.

    (b) Do you recognize the sequence $(f(1), f(2), f(3), \ldots)$?

11. Let n and k be integers with $0 \le k < n$.

    (a) Give an algebraic proof (using Theorem 6.9) of the identity

$$(n - k)C(n, k) = nC(n - 1, k)$$

(b) Give a combinatorial proof of the same identity. (Hint: Let $A = \{1, 2, \ldots, n\}$; how many ways are there to select a pair (x, B), where $B \subseteq A$, $|B| = k < n$, and $x \in A - B$?)

12. For n an integer, $n > 1$, prove the identity

$$\sum_{k=1}^{n} (-1)^{k-1} k C(n, k) = 0$$

13. Use the identity of Exercise 11 (and the fact that $C(k, k) = 1$) to compute:

(a) $C(4, 2)$ **(b)** $C(6, 3)$

14. Let $n, k \in \mathbb{N}$.

(a) Find integers a, b, and c such that

$$k^3 = aC(k, 3) + bC(k, 2) + cC(k, 1)$$

(b) Using part (a) and appropriate identities, find a formula for

$$1^3 + 2^3 + \cdots + n^3$$

15. For $n, k \in \mathbb{N}$, consider the identity

$$C(n, 0) + C(n + 1, 1) + C(n + 2, 2) + \cdots + C(n + k, k) = $$
$$C(n + k + 1, k)$$

(a) Give a combinatorial proof of this identity. (Hint: Consider choosing a k-element subset of $\{1, 2, \ldots, n + k + 1\}$; partition these subsets according to the largest element not chosen, and apply the addition principle.)

(b) Prove the identity by induction on k.

(c) Prove the identity by induction on n.

16. Let A be an n-element set, where $n > 0$. Show that the number of subsets of A with even cardinality is the same as the number of subsets of A with odd cardinality. What is this number?

17. Let n be a positive integer and let r be a real number.

(a) Use Corollary 6.14 to show that

$$3^n = C(n, 0)2^0 + C(n, 1)2^1 + \cdots + C(n, n)2^n$$

(b) Generalize part (a) to give a formula for the sum

$$C(n, 0)r^0 + C(n, 1)r^1 + \cdots + C(n, n)r^n$$

18. (For those who have had calculus) Here is an outline of an alternate proof of Corollary 6.14. We know that $f(x) = (1 + x)^n$ is a polynomial of degree n, say $g(x) = a_0 + a_1 x + a_2 x^2 + \cdots + a_n x^n$, and suppose $f(x) = g(x)$ for all $x \in \mathbb{R}$. We wish to determine the coefficients $a_0, a_1, a_2, \ldots, a_n$. To determine a_0, use the fact that $f(0) = g(0)$; to determine a_1, use the fact that $f'(0) = g'(0)$; to determine a_2, use the fact that $f''(0) = g''(0)$, and so on.

19. Find a formula for

$$C(n, 0) + 3C(n, 1) + 5C(n, 2) + \cdots + (2n + 1)C(n, n)$$

6.5   ## PRINCIPLE OF INCLUSION – EXCLUSION

In Section 2.4 the principle of inclusion–exclusion was addressed for two sets in the form of Corollary 2.15, which we restate here as Theorem 6.17.

THEOREM 6.17   If A and B are subsets of a finite set S, then

$$|A' \cap B'| = |S| - (|A| + |B|) + |A \cap B|$$

Now we wish to consider the general situation, where $A_1, A_2, \ldots, A_n$ are subsets of the finite set S, and derive a formula for $|A_1' \cap A_2' \cap \cdots \cap A_n'|$. Our approach is to think of a subset A of S as being specified or defined by a property P:

$$A = \{x \in S \mid x \text{ has property } P\}$$

In fact, any subset of S can be so described. With this interpretation, suppose that P and Q are properties and that $A = \{x \in S \mid x \text{ has } P\}$ and $B = \{x \in S \mid x \text{ has } Q\}$. Then notice that $A' \cap B'$ is the set of all elements in S having neither of the properties P or Q. The next example serves to illustrate the case for three subsets.

Example 6.23   Suppose that three men go to a restaurant and, upon entering, check their hats. In how many ways can their hats be returned to them so that no man gets his own hat?

Solution   Let m_1, m_2, m_3 denote the three men and let P_i, $1 \le i \le 3$, be the property that m_i receives his own hat. Let S be the collection of all ways that the three hats can be returned to them and, for $1 \le i \le 3$, let $A_i = \{x \in S \mid x \text{ has } P_i\}$. We then seek $|A_1' \cap A_2' \cap A_3'|$. But this is just $|S| - |A_1 \cup A_2 \cup A_3|$ and, using Exercise 6 of Exercises 2.4 (a formula for $|A_1 \cup A_2 \cup A_3|$), we see that

$$|A_1' \cap A_2' \cap A_3'| = |S| - (|A_1| + |A_2| + |A_3|) +$$
$$(|A_1 \cap A_2| + |A_1 \cap A_3| + |A_2 \cap A_3|) - |A_1 \cap A_2 \cap A_3|$$

and that

$$|S| = 3! = 6 \qquad |A_1| = |A_2| = |A_3| = 2! = 2$$
$$|A_1 \cap A_2| = |A_1 \cap A_3| = |A_2 \cap A_3| = 1$$
$$|A_1 \cap A_2 \cap A_3| = 1$$

Hence

$$|A_2' \cap A_2' \cap A_3'| = 6 - 3 \cdot 2 + 3 \cdot 1 - 1 = 2 \qquad \blacksquare$$

In general, suppose $P_1, P_2, \ldots, P_n$ are properties that each of the objects in S may or may not possess. For each $i = 1, 2, \ldots, n$, let

$$A_i = \{x \in S \mid x \text{ has } P_i\}$$

With this description of the A_i's, we see that

$$\bigcup_{i=1}^{n} A_i = \{ x \in S \mid x \text{ has at least one of the properties } P_i,\ i = 1, 2, \ldots, n \}$$

and

$$\bigcap_{i=1}^{n} A_i' = \{ x \in S \mid x \text{ has none of the properties } P_i,\ i = 1, 2, \ldots, n \}$$

We wish to determine $\left| \bigcap_{i=1}^{n} A_i' \right|$.

THEOREM 6.18   **(Principle of Inclusion–Exclusion)**

In keeping with the above notation, we have

$$\left| \bigcap_{i=1}^{n} A_i' \right| = |S| - \sum_{i=1}^{n} |A_i| + \sum |A_i \cap A_j| - \sum |A_i \cap A_j \cap A_k| + \cdots \tag{1}$$

$$+ (-1)^r \sum |A_{i_1} \cap A_{i_2} \cap \cdots \cap A_{i_r}| + \cdots + (-1)^n |A_1 \cap A_2 \cap \cdots \cap A_n|$$

where, for $2 \leq r \leq n$, the sums

$$\sum |A_{i_1} \cap A_{i_2} \cap \cdots \cap A_{i_r}|$$

are taken over all r-combinations $\{ i_1, i_2, \ldots, i_r \}$ of $\{ 1, 2, \ldots, n \}$.

Proof   To prove the theorem, we show that each $x \in S$ contributes the same amount to both sides of (1). We distinguish the following cases:

Case 1. x has none of the properties $P_1, P_2, \ldots, P_n$. In this case, $x \in A_1' \cap A_2' \cap \cdots \cap A_n'$, so x contributes 1 to the left-hand side of (1). Also the facts $x \in S$ and $x \notin A_i$ for $i = 1, 2, \ldots, n$ imply that the contribution of x to the right-hand side of (1) is 1.

Case 2. x has exactly k of the properties $P_1, P_2, \ldots, P_n$, where $1 \leq k \leq n$. Then $x \notin A_1' \cap A_2' \cap \cdots \cap A_n'$, so the contribution of x to the left-hand side of (1) is 0. To see the contribution of x to the right-hand side, we examine each sum on the right-hand side. Since $x \in S$, there is a contribution of 1 to $|S|$. Also, since x belongs to exactly k of the sets $A_1, A_2, \ldots, A_n$, the contribution of x to $\sum |A_i|$ must be k. Consider next $\sum |A_i \cap A_j|$, where the sum is over all 2-combinations of $\{ 1, 2, \ldots, n \}$. The number of times x is counted in this sum is precisely the number of ways to choose two of the k subsets to which x belongs; this is just $C(k, 2)$. More generally, consider $\sum |A_{i_1} \cap A_{i_2} \cap \cdots \cap A_{i_r}|$, where the sum is over all r-combinations $\{ i_1, i_2, \ldots, i_r \}$ of $\{ 1, 2, \ldots, n \}$, $3 \leq r \leq k$. How many times is x counted in this sum? We see that the answer is precisely the number of ways of choosing r of the k subsets that contain x, and this is $C(k, r)$. For $r > k$

notice that, no matter how one chooses r subsets $A_{i_1}, A_{i_2}, \ldots, A_{i_r}$, we have $x \notin A_{i_1} \cap A_{i_2} \cap \cdots \cap A_{i_r}$. Hence the total contribution of x to the right-hand side of (1) is

$$C(k,0) - C(k,1) + C(k,2) - C(k,3) + \cdots + (-1)^k C(k,k)$$

In Section 6.4, as an immediate application of Corollary 6.14, we proved this sum is 0. So the contribution of x to the right-hand side of (1) is 0, the same as its contribution to the left-hand side.

This completes the argument for Case 2 and, as a consequence, establishes the theorem. □

COROLLARY 6.19 The number of objects in S that possess at least one of the properties $P_1, P_2, \ldots, P_n$ is

$$\left| \bigcup_{i=1}^{n} A_i \right| = \sum_{i=1}^{n} |A_i| - \sum |A_i \cap A_j| + \cdots +$$

$$(-1)^{r-1} \sum |A_{i_1} \cap A_{i_2} \cap \cdots \cap A_{i_r}| + \cdots +$$

$$(-1)^{n-1} |A_1 \cap A_2 \cap \cdots \cap A_n|$$

where, for $2 \le r \le n$, the sums $\sum |A_{i_1} \cap A_{i_2} \cap \cdots \cap A_{i_r}|$ are taken over all r-combinations $\{i_1, i_2, \ldots, i_r\}$ of $\{1, 2, \ldots, n\}$.

The proof of Corollary 6.19 is left to Exercise 14.

Example 6.24 How many ways are there to select a five-card poker hand from a standard deck such that the hand contains at least one card of each suit?

Solution Let S be the set of all five-card poker hands. For a given $x \in S$ we define properties P_1, P_2, P_3, and P_4 as follows:

$$P_1: \quad x \text{ has no hearts}$$
$$P_2: \quad x \text{ has no spades}$$
$$P_3: \quad x \text{ has no diamonds}$$
$$P_4: \quad x \text{ has no clubs}$$

For $i = 1, 2, 3, 4$, let $A_i = \{x \in S \mid x \text{ has property } P_i\}$. Then $B = A_1' \cap A_2' \cap A_3' \cap A_4' = \{x \in S \mid x \text{ does not have } P_i \text{ for } 1 \le i \le 4\}$. In other words, B is the set of five-card poker hands that contain at least one card of each suit. Thus we seek $|B|$. By the principle of inclusion–exclusion, we have

$$|B| = |S| - \sum_{i=1}^{4} |A_i| + \sum |A_i \cap A_j|$$

$$- \sum |A_i \cap A_j \cap A_k| + |A_1 \cap A_2 \cap A_3 \cap A_4|$$

where the second sum on the right-hand side is taken over the six 2-combinations $\{i, j\}$ of $\{1, 2, 3, 4\}$ and the third sum is taken over the four 3-combinations $\{i, j, k\}$ of $\{1, 2, 3, 4\}$. Notice that

$$|A_i| = C(39, 5) \qquad |A_i \cap A_j| = C(26, 5) \qquad |A_i \cap A_j \cap A_k| = C(13, 5)$$

for any $1 \le i < j < k \le 4$; moreover, $|A_1 \cap A_2 \cap A_3 \cap A_4| = 0$. Hence

$$|B| = C(52, 5) - 4 \cdot C(39, 5) + 6 \cdot C(26, 5) - 4 \cdot C(13, 5) + 0 \quad \blacksquare$$

Example 6.25   At a restaurant, n men check their hats at the coat-check counter. When the evening is ended, each man returns to the coat-checker to retrieve his hat. If the hats are mixed up and returned to the men in random fashion, in how many ways can it happen that no man gets his own hat?

Solution   Let S be the set of all possible ways for the n men to receive the n hats, one hat to a man. We see that $|S| = n!$. In order to set up an application of the principle of inclusion–exclusion, we assume the n men to be named say, $m_1, m_2, \ldots, m_n$. For each i, $1 \le i \le n$, let P_i be the property that m_i receives his own hat, and let

$$A_i = \{x \in S \mid x \text{ has } P_i\}$$

Then

$$A_1' \cap A_2' \cap \cdots \cap A_n' = \{x \in S \mid x \text{ does not have } P_i \text{ for } 1 \le i \le n\}$$

In other words, $A_1' \cap A_2' \cap \cdots \cap A_n'$ is the set of ways of returning the n hats so that no man gets his own hat. In order to apply the principle of inclusion–exclusion, let r be an integer, $1 \le r \le n$, and let $\{i_1, i_2, \ldots, i_r\}$ be an r-combination of $\{1, 2, \ldots, n\}$. We need to determine

$$|A_{i_1} \cap A_{i_2} \cap \cdots \cap A_{i_r}|$$

This is simply the number of ways to return the hats so that the men $m_{i_1}, m_{i_2}, \ldots, m_{i_r}$ receive their own hats, namely $(n - r)!$. Hence

$$\left| \bigcap_{i=1}^{n} A_i' \right| = |S| - \sum_{i=1}^{n} |A_i| + \cdots + (-1)^r \sum |A_{i_1} \cap A_{i_2} \cap \cdots \cap A_{i_r}| + \cdots +$$

$$(-1)^n |A_1 \cap A_2 \cap \cdots \cap A_n|$$

$$= n! - n \cdot (n - 1)! + C(n, 2)(n - 2)! - C(n, 3)(n - 3)! + \cdots +$$

$$(-1)^r C(n, r)(n - r)! + \cdots + (-1)^n n!$$

$$= n! \left(1 - 1 + \frac{1}{2!} - \frac{1}{3!} + \cdots + \frac{(-1)^r}{r!} + \cdots + \frac{(-1)^n}{n!} \right)$$

$$= n! \left(\frac{1}{0!} - \frac{1}{1!} + \frac{1}{2!} - \frac{1}{3!} + \cdots + \frac{(-1)^r}{r!} + \cdots + \frac{(-1)^n}{n!} \right)$$

$$= n! \sum_{k=0}^{n} \frac{(-1)^k}{k!} \qquad \blacksquare$$

The preceding example is interesting in the sense that we can view the elements of S as permutations of n objects. With this interpretation, the problem asks for the number of permutations of n objects that leave no element "fixed"; that is, if X is a nonempty set with n elements, we seek the number of one-to-one functions $f: X \rightarrow X$ such that $f(x) \neq x$ for all $x \in X$. Such a permutation is called a *derangement* and we say that the n elements of X are *deranged*. We denote the number of derangements of n objects by D_n; thus we see from the preceding example that

$$D_n = n! \sum_{k=0}^{n} \frac{(-1)^k}{k!}$$

Example 6.26 Determine the number of permutations of the set $X = \{1, 2, \ldots, n\}$ in which at least one element is fixed.

Solution If S represents the set of all permutations of X and T represents the set of derangements of X, then $S - T$ is the set of all permutations of X that leave at least one element of X·fixed. And

$$|S - T| = |S| - |T| = n! - D_n \qquad \blacksquare$$

Example 6.27 Use the principle of inclusion–exclusion to determine the number of functions from $\{1, 2, \ldots, m\}$ onto $\{1, 2, \ldots, n\}$, where $m \geq n$.

Solution Let S be the set of all functions from $\{1, 2, \ldots, m\}$ to $\{1, 2, \ldots, n\}$; so $|S| = n^m$. For $1 \leq i \leq n$, let

$$A_i = \{f \in S \mid f^{-1}(\{i\}) = \emptyset\}$$

Notice that our problem is then to find $|A_1' \cap A_2' \cap \cdots \cap A_n'|$. In order to apply the principle of inclusion–exclusion, let r be an integer, $1 \leq r \leq n$, and let $\{i_1, i_2, \ldots, i_r\}$ be an r-combination of $\{1, 2, \ldots, n\}$. We need to determine

$$|A_{i_1} \cap A_{i_2} \cap \cdots \cap A_{i_r}|$$

This is simply the number of functions f from $\{1, 2, \ldots, m\}$ to $\{1, 2, \ldots, n\}$ such that $f^{-1}(\{i_k\}) = \emptyset$ for $1 \leq k \leq r$. If this is the case, then there are $n - r$ choices for each image $f(x)$. Thus

$$|A_{i_1} \cap A_{i_2} \cap \cdots \cap A_{i_r}| = (n - r)^m$$

Hence

$$\left| \bigcap_{i=1}^{n} A_i' \right| = |S| - \sum_{i=1}^{n} |A_i| + \cdots + (-1)^r \sum |A_{i_1} \cap A_{i_2} \cap \cdots \cap A_{i_r}| + \cdots +$$

$$(-1)^n |A_1 \cap A_2 \cap \cdots \cap A_n| =$$

$$n^m - C(n, 1)(n - 1)^m + C(n, 2)(n - 2)^m - C(n, 3)(n - 3)^m + \cdots +$$

$$(-1)^r C(n, r)(n - r)^m + \cdots + (-1)^n C(n, n)(n - n)^m$$

$$= \sum_{k=0}^{n-1} (-1)^k C(n, k)(n - k)^m$$

For example, when $m = 5$ and $n = 3$, we obtain that the number of functions from $\{1, 2, 3, 4, 5\}$ onto $\{1, 2, 3\}$ is

$$3^5 - 3 \cdot 2^5 + 3 \cdot 1^5 = 243 - 96 + 3 = 150 \qquad \blacksquare$$

Exercises 6.5

1. A bakery sells seven kinds of donuts. How many ways are there to choose one dozen donuts if no more than three donuts of any kind are used? (Hint: Let x_i be the number of donuts chosen of kind i, where $1 \le i \le 7$. Then you are to find the number of 7-tuples $(x_1, x_2, \ldots, x_7)$ of integers such that $x_1 + x_2 + \cdots + x_7 = 12$ and $0 \le x_i \le 3$. Let $S = \{(x_1, x_2, \ldots, x_7) \mid x_1 + x_2 + \cdots + x_7 = 12$ and $x_i \ge 0\}$ and, for $1 \le i \le 7$, let $A_i = \{(x_1, x_2, \ldots, x_7) \in S \mid x_i > 3\}$. Apply the principle of inclusion–exclusion to find $|A_1' \cap A_2' \cap \cdots \cap A_7'|$.)

2. Find the number of integer solutions to $x_1 + x_2 + x_3 + x_4 + x_5 + x_6 = 20$, where $1 \le x_i \le 4$, for $i = 1, 2, 3, 4, 5, 6$.

3. Find the number of nine-digit sequences that contain each of the odd digits 1, 3, 5, 7, and 9.

4. How many five-card poker hands contain a jack, a queen, and a king?

5. Given five distinct pairs of gloves, in how many ways can each of five students select two gloves at random so that no student gets a matching pair? (Hint: Label the gloves g_{ij}, $1 \le i \le 5$, $j = 1, 2$, where g_{i1} and g_{i2} match, and label the students s_1, s_2, s_3, s_4, s_5. Then each of the five students is to select two of the gloves (a 2-combination) and you can assume that s_1 selects first, s_2 second, and so on. Let S be the set of all such ordered selections; show that $|S| = 10!/(2! \cdot 2! \cdot 2! \cdot 2! \cdot 2!)$. For each k, $1 \le k \le 5$, let A_k be the set of those elements in S such that s_k gets a matching pair of gloves. Find $|A_1' \cap A_2' \cap A_3' \cap A_4' \cap A_5'|$.)

6. How many nine-letter sequences formed from the letters A, A, A, B, B, B, C, C, C have the three A's, or the three B's, or the three C's, consecutive?

7. How many ways are there to distribute 30 identical cans of beer among five fraternity brothers if no brother gets more than seven beers?

8. Given n identical objects of type 1, n identical objects of type 2, ..., n identical objects of type t, how many sequences of these nt objects have all the objects of some type appearing consecutively?

9. How many permutations of the ten digits $0, 1, 2, \ldots, 9$ contain at least one of the subsequences $0, 1, 2$, or $3, 4$, or $5, 6$, or $7, 8, 9$?

10. Let n be an odd positive integer.
    (a) How many functions f on $\{1, 2, \ldots, n\}$ have the property that $f(2k - 1) = 2k - 1$ for each k, $2 \le 2k \le n + 1$?
    (b) How many permutations f of $\{1, 2, \ldots, n\}$ have the property that $f(2k - 1) \ne 2k - 1$ for each k, $2 \le 2k \le n + 1$?

11. In how many ways can n married couples be seated at a circular dinner table so that no husband is seated next to his wife? Answer for the

following:

(a) $n = 2$ (b) $n = 3$ (c) $n = 4$

(d) an arbitrary positive integer n (See Exercise 18 in Exercises 6.3).

12. Find the number of integers between 1 and 10^{30} that are neither perfect squares nor perfect cubes nor perfect fifth powers.

13. Ron has eight friends who enjoy playing bridge and wishes to invite a different subset of three of them over to his home for a game every Wednesday evening for five consecutive weeks. In how many ways can this be done so that each friend receives at least one invitation?

14. Prove Corollary 6.19.

15. Find the number of permutations of $\{1, 2, \ldots, 7\}$ in which exactly three numbers are fixed.

16. Find the number of permutations of $\{1, 2, \ldots, n\}$ in which exactly k numbers are fixed ($0 \leq k \leq n$).

17. Prove the identity

$$n! = C(n,0)D_n + C(n,1)D_{n-1} + \cdots + C(n,n)D_0 = \sum_{k=0}^{n} C(n,k)D_k$$

where D_0 is defined to be 1.

6.6   RECURRENCE RELATIONS

In Section 3.6 we introduced the Fibonacci sequence 1, 1, 2, 3, 5, 8, ... and denoted the nth term of the sequence by f_n for $n \geq 1$. In this section we wish to look at the sequence again; however, we prefer to view it more directly as a function $f: \mathbb{N} \to \mathbb{N}$, where $f(n) = f_n$ for $n \in \mathbb{N}$. Given this form, we showed in Section 3.6 that

1. $f(1) = f(2) = 1$
2. $f(n) = f(n-1) + f(n-2)$ for $n \geq 3$

Thus, for $n \geq 3$, $f(n)$ is expressed in terms of $f(n-1)$ and $f(n-2)$. It is an example of a "recursively defined function."

DEFINITION 6.2

A function $f: \mathbb{N} \to \mathbb{R}$ is said to be *recursively* (or *inductively*) *defined* provided, for some positive integer n_0, the following hold:

1. The values of $f(1), f(2), \ldots, f(n_0)$ are known.
2. For $n > n_0$, $f(n)$ is defined in terms of $f(1), f(2), \ldots, f(n-1)$.

We call $f(1), f(2), \ldots, f(n_0)$ the *initial values* of f and refer to the equation describing $f(n)$ in terms of $f(1), f(2), \ldots, f(n-1)$ as a *recurrence relation* for f.

In other words, given a sequence of numbers $g(1), g(2), \ldots, g(n), \ldots,$ any equation that describes $g(n)$ in terms of its predecessors in the sequence, and that holds for all integers n greater than some fixed positive integer n_0, is called a *recurrence relation* for g. In some cases, we are interested in finding a recurrence relation for a function g defined on the set $\mathbb{N} \cup \{0\}$ of nonnegative integers (g is the sequence of numbers $(g(0), g(1), \ldots)$. In such cases, we permit Definition 6.3 to be stretched a little so as to include the possibility of $\mathbb{N} \cup \{0\}$ as a domain. For example, recall that n-factorial is defined for all $n \geq 0$ as follows:

1. $0! = 1$
2. $n! = 1 \cdot 2 \cdots (n-1) \cdot n$

In Section 6.3 we used the notation $P(n)$ to denote the number of permutations of an n-element set, for each $n \geq 1$. So $P(n) = n!$ for $n \in \mathbb{N}$ and it seems entirely appropriate to define $P(0) = 0!$. Given this, can we define the function P recursively? Realizing that $n! = n \cdot (n-1)!$ for $n \geq 1$, we can define $P: \mathbb{N} \cup \{0\} \to \mathbb{N}$ recursively by

1. $P(0) = 1$
2. $P(n) = n \cdot P(n-1)$ for $n \geq 1$

The identity in condition 2 is the associated recurrence relation.

Example 6.28 Let $d(n)$ denote the number of derangements of $S = \{1, 2, \ldots, n\}$. Then $d(1) = 0$ and $d(2) = 1$; show that $d(n)$ satisfies the recurrence relation

$$d(n) = (n-1)[d(n-1) + d(n-2)]$$

for $n \geq 3$.

Solution We could proceed using the formula

$$d(n) = n!\left(\frac{1}{0!} - \frac{1}{1!} + \cdots + \frac{(-1)^n}{n!}\right)$$

that was proved in Section 6.5; however, we leave this route to Exercise 2 and provide a combinatorial argument instead. Let f denote an arbitrary derangement of S; we wish to determine the number of possibilities for f. In viewing f as a permutation of S with no fixed elements, notice that there are $n - 1$ choices for the image of 1, say $f(1) = a$, where $2 \leq a \leq n$. For each such choice of a, we must determine the number of ways to complete f. We distinguish two cases that partition the set of possibilities, and then apply the addition and multiplication principles.

Case 1. $f(a) = 1$. Here we have $f(1) = a$ and $f(a) = 1$, so f maps $S - \{1, a\}$ to itself. Hence, to complete f, one must choose a derangement of $S - \{1, a\}$. This can be done in $d(n-2)$ ways.

Case 2. $f(a) \neq 1$. In this case f must be completed so that

$$f(b) \neq b \quad \text{for } b \in \{2, 3, \ldots, n\} - \{a\}$$

and $f(a) \neq 1$. But the number of ways of doing this is the same as the number of ways of choosing a derangement of $\{2, 3, \ldots, n\}$ (simply think of 1 in the inequality $f(a) \neq 1$ as playing the role of a). Hence there are $d(n-1)$ ways to complete f in this case.

It follows from the addition and multiplication principles that

$$d(n) = (n-1)[d(n-1) + d(n-2)] \qquad \blacksquare$$

Example 6.29   Let $g(n)$ be the number of k-element subsets of an n-element set, where k is a fixed positive integer and $n \geq k$. (So $g(n) = C(n, k)$.) We note that $g(k) = 1$. Find a recurrence relation for $g(n)$, where $n > k$.

Solution   Let $A = \{1, 2, \ldots, n\}$, where $n > k$. We begin by considering the k-element subsets of $A - \{x\}$, where x is an arbitrary (but fixed) element of A. Since $A - \{x\}$ consists of $n - 1$ elements, it has $g(n-1)$ subsets of cardinality k. Summing over all n elements that can be x gives a total of $n \cdot g(n-1)$ subsets of cardinality k. (The case for $n = 5$ and $k = 3$ is illustrated at the end of this example.) However, these $n \cdot g(n-1)$ subsets are not distinct. So we must determine the number of times each k-element subset appears. Well, a given k-element subset B appears as a subset of $A - \{x\}$ whenever $x \in A - B$. So there are $n - k$ choices for x. (As illustrated for $n = 5$ and $k = 3$, the subset $B = \{1, 2, 3\}$ appears twice in the list, for $x = 4$ and $x = 5$.) Therefore the number of distinct k-element subsets of A is $n \cdot g(n-1)/(n-k)$, which yields the recurrence relation

$$g(n) = \frac{ng(n-1)}{n-k}$$

(Note that this is the identity derived in Exercise 11 of Exercises 6.4.) The 3-element subsets of $\{1, 2, 3, 4, 5\}$ are listed in tabular form below. Note that each subset appears twice.

$x = 1$	$x = 2$	$x = 3$	$x = 4$	$x = 5$
$\{2, 3, 4\}$	$\{1, 3, 4\}$	$\{1, 2, 4\}$	$\{1, 2, 3\}$	$(1, 2, 3)$
$(2, 3, 5)$	$(1, 3, 5)$	$\{1, 2, 5\}$	$\{1, 2, 5\}$	$\{1, 2, 4\}$
$\{2, 4, 5\}$	$\{1, 4, 5\}$	$\{1, 4, 5\}$	$\{1, 3, 5\}$	$\{1, 3, 4\}$
$\{3, 4, 5\}$	$\{3, 4, 5\}$	$\{2, 4, 5\}$	$\{2, 3, 5\}$	$\{2, 3, 4\}$

$\blacksquare$

Example 6.30   Consider character strings of length $n \geq 1$ consisting entirely of the letters A and B. Call those strings that do not contain two consecutive A's "good." Let $s(n)$ be the number of good strings of length n. Then $s(1) = 2$ (A and B) and $s(2) = 3$ (AB, BA, and BB). Find a recurrence relation for $s(n)$, where $n \geq 3$.

Solution Let x denote a good string of length n. If the first character of x is B, then the second through nth characters of x form one of the $s(n-1)$ good strings of length $n-1$. On the other hand, if the first character of x is A, then necessarily the second character of x is B and the third through nth characters form one of the $s(n-2)$ good strings of length $n-2$. It follows by the addition principle that

$$s(n) = s(n-1) + s(n-2)$$

Note that the recurrence relation for $s(n)$ is the same as that for the Fibonacci numbers; only the initial values differ. ∎

Example 6.31 (Towers of Hanoi) We are given three pegs, labeled peg 1, peg 2, and peg 3, planted on a board. In addition, we are given n circular disks of different diameters, each with a hole drilled through its center. The disks are labeled $d_1, d_2, \ldots, d_n$ in order of decreasing size (d_1 is the largest and d_n the smallest) and then placed on peg 1 in the order $d_1, d_2, \ldots, d_n$ (see Figure 6.4).

The n disks are to be transferred from peg 1 to one of the other pegs, say peg 3, subject to the following rules:

1. Each move consists of transferring the top disk from one peg to another peg.
2. At no time can a larger disk be placed on top of a smaller one.

Find the initial value and a recurrence relation for the number of moves $h(n)$ required to accomplish this.

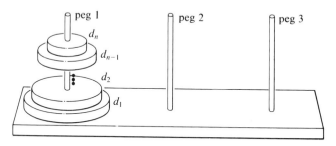

Figure 6.4

Solution Clearly $h(1) = 1$. Suppose $n > 1$; then in order to transfer d_1 to a disk-free peg 3, we must first transfer $d_2, d_3, \ldots, d_n$ from peg 1 to peg 2. According to our notational convention, this requires $h(n-1)$ moves. Once this is accomplished, d_1 can be transferred from peg 1 to peg 3 in one move. At this point, the situation is as depicted in Figure 6.5, with $h(n-1)+1$ moves having taken place.

Finally, $h(n-1)$ moves are required to transfer $d_2, d_3, \ldots, d_n$ from peg 2 to peg 3. So the desired transfer of all n disks can be accomplished in $[h(n-1)+1] + h(n-1) = 2h(n-1)+1$ moves. Thus the initial value

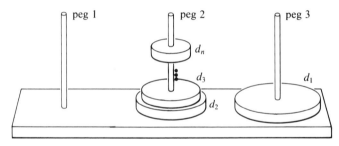

Figure 6.5

and recurrence relation are given by:

1. $h(1) = 1$
2. $h(n) = 2h(n-1) + 1$ for $n \geq 2$ ∎

Suppose we wish to find an explicit formula for $h(n)$ in the problem of the Towers of Hanoi. Knowing the initial value and the recurrence relation for $h(n)$ from the preceding example, we may calculate successive values of the function h as follows:

$$h(1) = 1$$

$$h(2) = 2h(1) + 1 = 3$$

$$h(3) = 2h(2) + 1 = 7$$

$$h(4) = 2h(3) + 1 = 15$$

$$h(5) = 2h(4) + 1 = 31$$

and so on.

This method of obtaining successive values of a recursively defined function is called *iteration*. Iteration is easily implemented in most programming languages using a counter-controlled loop; with a computer, values of a function like $h(n)$ can be found for many values of n. To illustrate, the following Pascal segment uses iteration (a FOR loop) to output values of $h(n)$ for $n = 1, 2, \ldots, 25$.

```
BEGIN
 WRITELN('VALUE OF N    VALUE OF H(N)');
 N := 1; (*initialize N*)
 H := 1; (*initialize H*)
 WRITELN(N:10, H:17);
 FOR N := 2 TO 25 DO
  BEGIN
   H := 2 * H + 1 (*use the recurrence relation to compute
                    next value of H(N)*)
   WRITELN(N:10, H:17)
  END
END.
```

Upon inspection of several values of $h(n)$, perhaps the student can guess an explicit formula for $h(n)$ in terms of n. It certainly appears that $h(n) = 2^n - 1$ is a good guess. Can we in fact prove such a formula holds for each $n \in \mathbb{N}$? What method of proof should be employed? Induction, of course!

Example 6.32   Use mathematical induction to prove that $h(n) = 2^n - 1$ for all $n \in \mathbb{N}$.

Solution   As usual, let S be the set of all positive integers n for which $h(n) = 2^n - 1$.

1. Show that $1 \in S$. We have already observed that $h(1) = 1 = 2^1 - 1$, so $1 \in S$.
2. Assume $k \in S$ for some arbitrary $k \geq 1$. The induction hypothesis (IHOP) is that $h(k) = 2^k - 1$.
3. Show that $k + 1 \in S$; that is, show that $h(k + 1) = 2^{k+1} - 1$. Observe that

$$
\begin{aligned}
h(k + 1) &= 2h(k) + 1 && \text{(by the recurrence relation)} \\
&= 2(2^k - 1) + 1 && \text{(by IHOP)} \\
&= 2^{k+1} - 1
\end{aligned}
$$

Hence $k + 1 \in S$, and it follows that $S = \mathbb{N}$. Thus, in the Towers of Hanoi problem, $2^n - 1$ moves are required to transfer the n disks from peg 1 to peg 3.   ■

Example 6.33   Find an explicit formula for the function $g(n)$ defined recursively by $g(1) = 1$ and $g(n) = g(n - 1) + 2n - 1$, $n \geq 2$.

Solution   The first several values of $g(n)$, $n \geq 2$, are computed as follows:

$$
\begin{aligned}
g(2) &= g(1) + 2 \cdot 2 - 1 = 4 \\
g(3) &= g(2) + 2 \cdot 3 - 1 = 9 \\
g(4) &= g(3) + 2 \cdot 4 - 1 = 16
\end{aligned}
$$

It seems reasonable at this stage to conjecture that $g(n) = n^2$.

In fact, it is easily seen that $g(n) = n^2$ satisfies the recurrence relation $g(n) = g(n - 1) + 2n - 1$ and the initial condition $g(1) = 1$. What we would like to show is that $g(n) = n^2$ is *the only function* from $\mathbb{N}$ to $\mathbb{N}$ that satisfies this recurrence relation and initial condition. In other words, we must prove that if $g: \mathbb{N} \to \mathbb{N}$ has the properties:

1. $g(1) = 1$
2. If $g(n) = g(n - 1) + 2n - 1$ for all $n \geq 2$,
   then $g(n) = n^2$. The proof is by induction and the details are left to Exercise 4.   ■

Example 6.34   Let $r(n)$ denote the number of "comparisons" required to sort a list of 2^n numbers using the "merge sort" technique. Assume $r(0) = 0$. For $n > 0$, a merge sort is performed recursively by partitioning the list of 2^n numbers into two sublists of 2^{n-1} numbers each, sorting each of these lists, and then "merging" them into a single sorted list. Assume that it takes k comparisons to merge two sorted sublists into a single sorted list of k numbers. Find a recurrence relation for $r(n)$, where $n \geq 1$.

Solution   It takes $r(n-1)$ comparisons to sort each sublist of 2^{n-1} numbers, then 2^n comparisons to merge these sublists into a single sorted list. Thus a recurrence relation for $r(n)$ is given by

$$r(n) = 2r(n-1) + 2^n$$

for $n \geq 1$.

It can be shown by induction that $r(n) = n2^n$. On the other hand, if $m = 2^n$ and $r'(m) = r(n)$, then notice that

$$r'(m) = r(n) = n2^n = nm = mn = m \log_2 m$$

The interpretation of this is that the merge sort takes (approximately) $m \log_2 m$ comparisons to sort a list of m numbers.   ∎

General methods exist for solving certain types of recurrence relations. Some of these are discussed in Section 6.8.

Exercises 6.6

1. For a fixed positive integer n and for $0 \leq m \leq n$, let $k(m)$ be the number of m-element subsets of an n-element set. Then $k(0) = 1$. Find a recurrence relation for $k(m)$ in terms of $k(m-1)$, $1 \leq m \leq n$. (Hint: See Chapter Problem 39.)

2. Prove the recurrence relation $d(n) = (n-1)[d(n-1) + d(n-2)]$ of Example 6.28 by using the formula

$$d(n) = n!\left(\frac{1}{0!} - \frac{1}{1!} + \cdots + \frac{(-1)^n}{n!} \right)$$

3. Consider character strings of length $n \geq 0$ over the alphabet $\{A, B, C\}$. Find initial values and a recurrence relation for each of these functions.
   (a) $h_1(n)$ is the number of strings (of length n) not containing the substring AA.
   (b) $h_2(n)$ is the number of strings not containing either of the substrings AA or AB.
   (c) $h_3(n)$ is the number of strings not containing the substring BBB.
   ★(d) $h_4(n)$ is the number of strings not containing the substring AB.
4. Give the induction proof needed to complete Example 6.33.
5. Let $t(n)$ be the number of ways in which $2n$ tennis players can be paired to play n matches. (For example, if $n = 3$ and the players are numbered $1, 2, 3, 4, 5, 6$, then one possible pairing is to have 1 play 2, 3 play 4, and 5 play 6; another possibility is to have 1 play 6, 2 play 5,

and 3 play 4.) Find $t(1)$ and a recurrence relation for $t(n)$, $n \geq 2$. (Hint: Number the players $1, 2, \ldots, 2n$; once the opponent for player 1 has been decided, in how many ways can the remaining $2n - 2$ players be paired?)

6. At the beginning of year 1, Moe invests $1000 in a savings account that pays 10% annual interest, payable at the end of each year. At the beginning of subsequent years, Moe deposits an additional $100 to his account. Let $v(n)$ be the value of the account at the beginning of year n, where $n \geq 1$. Then $v(1) = 1000$, $v(2) = 1000 + 100 + 100 = 1200$, $v(3) = 1200 + 120 + 100 = 1420$, $v(4) = 1420 + 142 + 100 = 1662$, and so on. Find a recurrence relation for $v(n)$, $n \geq 2$.

7. A wizard must climb an infinite staircase whose steps are numbered with the positive integers. At each step, the wizard may go up one stair or two. Let $w(n)$ be the number of ways for the wizard to reach step number n; then $w(1) = 1$, $w(2) = 2$, $w(3) = 3$, and so on. Find a recurrence relation for $w(n)$, $n \geq 3$. (Hint: The set of ways for the wizard to reach step n may be partitioned into two sets according to whether, at his last step, the wizard went up one stair or two.)

8. Let $r(n)$ be the number of regions into which a plane is divided by n lines in general position (each pair of lines intersect at a point and no point is on more than two lines). Then $r(1) = 2$, $r(2) = 4$, $r(3) = 7$, and so on. Find a recurrence relation for $r(n)$, $n \geq 2$.

9. For $n \geq 3$, label the vertices of an n-sided convex polygon with the integers $1, 2, \ldots, n$. Let $c(n)$ be the number of ways to partition the interior of an n-gon into $n - 2$ triangular regions by adding $n - 3$ diagonals that do not intersect in the interior. Then $c(3) = 1$, $c(4) = 2$, $c(5) = 5$ (see Figure 6.6), and so on. Find a recurrence relation for $c(n)$, $n \geq 4$. (Hint: There is a triangle with vertices 1, 2, and k for some k, $3 \leq k \leq n$; partition according to the value of k.)

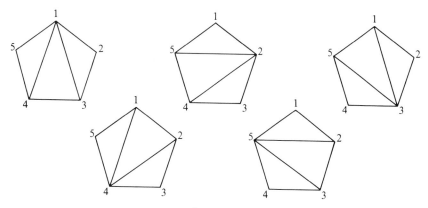

Figure 6.6

10. Let $s(m)$ denote the number of comparisons required to sort a list of m numbers using a "selection sort" technique. Assume $s(1) = 0$. For $m \geq 2$, a selection sort of m numbers can be described recursively as follows. First find the smallest of the m numbers and place it at the beginning of the list. Assume that this step requires $m - 1$ comparisons. Then sort the other $m - 1$ numbers.

(a) Find a recurrence relation for $s(m)$.

(b) Show that $s(m) = m(m - 1)/2$. Compare this with the result of Example 6.34.

11. Let $b(n)$ be the number of bees in the nth previous generation of a male bee. Assume that a male bee is produced asexually from a single female parent, while a female bee has both a male parent and a female parent. So $b(1) = 1$, $b(2) = 2$, $b(3) = 3$, and so on. Find a recurrence relation for $b(n)$, $n \geq 2$.

12. Moe obtains a home mortgage of $40,000 for which the annual interest rate is 10%. Moe wishes to pay off the mortgage in yearly installments of $6000 (except for the last payment, which may be less than $6000). Let $p(n)$ be the amount owed by Moe at the end of year n, and, for convenience, define $p(0) = 40,000$. Then $p(1) = 40,000 + 4000 - 6000 = 38,000$, $p(2) = 38,000 + 3800 - 6000 = 35,800$, and so on. Find a recurrence relation for $p(n)$, $n \geq 1$.

13. Consider a single-elimination tournament involving 2^n players. Suppose that each player receives prize money in the amount of $100m$ dollars, where m is the number of players in the subtournament won by that player. (Thus a player who loses in the first round gets $100, a player who loses in the second round gets $200, a player who loses in the third round gets $400, and so on.) Let $p(n)$ be the total amount of prize money awarded; then $p(1) = 300$. Find a recurrence relation for $p(n)$, $n \geq 2$.

14. Consider the following game played by a single player on a bit string $b_1 b_2 \ldots b_n$ of length n. At each move, the player switches the value of some bit b_i ($b_i := 1 - b_i$) subject to the following rules: (i) b_1 may be switched; (ii) b_2 may be switched only if $b_1 = 1$; (iii) for $2 < i \leq n$, b_i may be switched only if $b_{i-1} = 1$ and $b_k = 0$ for $1 \leq k < i - 1$. Let $f(n)$ be the number of moves required to change $b_1 b_2 \ldots b_n$ from all 0's to the bit string with exactly one 1, in position n. Then $b(1) = 1$; find a recurrence relation for $b(n)$, $n \geq 2$.

15. Let $g(n)$ be the number of ways to seat n married couples at a round table for dinner subject to the condition that no husband sits next to his wife. Then $g(1) = 0$, $g(2) = 2$, $g(3) = 24$, and so on. Find a recurrence relation for $g(n)$, $n \geq 3$. (See Exercise 18 of Exercises 6.3.)

6.7

RECURSIVE ALGORITHMS

It has already been observed how an iterative program can be written to find successive values of a recursively defined function. What is especially intriguing is that many high-level languages, like Pascal, allow one to implement recursion. In other words, just as a recursive function is defined in terms of itself, one can write subprograms that invoke themselves. As a result, such subprograms are called *recursive*. In Pascal, both procedures and functions can be recursive. Since we have seen several examples of recursively defined functions, it seems appropriate to first see how such functions can be implemented in Pascal.

For example, consider $P(n) = n!$, which was defined recursively in Section 6.7 by

    1. $P(0) = 1$
    2. $P(n) = n \cdot P(n - 1)$ for $n \geq 1$

A recursive Pascal implementation for this function is given next.

```
FUNCTION FACTORIAL(N: INTEGER): INTEGER;
BEGIN
 IF N = 0 THEN
  FACTORIAL := 1
 ELSE
  FACTORIAL := N * FACTORIAL(N-1)
END;
```

Note how closely the body of the Pascal function resembles the recursive definition of $P(n)$. Suppose now that, in a main program that has defined FACTORIAL, a reference is made to FACTORIAL(3). Then calls to FACTORIAL are made and assignments are executed as indicated by the following steps:

1. FACTORIAL(3) is referenced; $3 \neq 0$, so

```
FACTORIAL(3) := 3 * FACTORIAL(2)
```

2. FACTORIAL(2) is referenced; $2 \neq 0$, so

```
FACTORIAL(2) := 2 * FACTORIAL(1)
```

3. FACTORIAL(1) is referenced; $1 \neq 0$, so

```
FACTORIAL(1) := 1 * FACTORIAL(0)
```

4. FACTORIAL(0) is referenced; here 1 is calculated as the value of FACTORIAL(0).
5. The value 1 is returned as the value of FACTORIAL(0) for the expression in Step 3, and so $1*1 = 1$ is calculated as the value of FACTORIAL(1).

6. The value 1 is returned as the value of FACTORIAL(1) for the expression in Step 2, and so $2 * 1 = 2$ is calculated as the value of FACTORIAL(2).
7. The value 2 is returned as the value of FACTORIAL(2) for the expression in Step 1, and so $3 * 2 = 6$ is calculated as the value of FACTORIAL(3).
8. The value 6 is returned to the main program as the value of FACTORIAL(3).

The preceding sequence of steps can be represented schematically as follows:

```
FACTORIAL(3)
   ↓
FACTORIAL(3) := 3 * FACTORIAL(2) = 3 * 2 = 6
                     ↓
            FACTORIAL(2) := 2 * FACTORIAL(1) = 2 * 1 = 2
                                 ↓
                        FACTORIAL(1) := 1 * FACTORIAL(0) = 1 * 1 = 1
```

In Exercise 8 the student is asked to write the function FACTORIAL iteratively, using a FOR loop. It is of some interest to compare the two functions from the standpoint of "computational complexity," whereby an algorithm is analyzed as to its memory requirements and execution time. Although we do not attempt to do this, we urge the student to experiment with the two methods. For example, each function could be called for various values of N and a graph of (average) execution time versus N could be constructed.

Example 6.35   Write a recursive Pascal function FIB that calculates Fibonacci numbers. Produce an outline that traces the calculation of FIB(5).

Solution   Recall that the Fibonacci sequence is given recursively by

1. $f(1) = f(2) = 1$
2. $f(n) = f(n - 1) + f(n - 2)$ for $n \geq 3$

Using the model provided previously by the function FACTORIAL, we write the function FIB as follows:

```
FUNCTION FIB(N: INTEGER): INTEGER;
 BEGIN
  IF (N = 1) OR (N = 2) THEN
   FIB := 1
  ELSE
   FIB := FIB(N-1) + FIB(N-2)
 END;
```

The calculation of FIB(5) is traced schematically as follows:

```
FIB(5)
  ↓
FIB(5) := FIB(4) + FIB(3) = 2 + 3 = 5
            ↓         ↓
            ↓         FIB(3) := FIB(2) + FIB(1) = 1 + 1 = 2
          FIB(4) := FIB(3) + FIB(2) = 2 + 1 = 3
                      ↓
                    FIB(3) := FIB(2) + FIB(1) = 1 + 1 = 2     ■
```

In Exercise 6 the student is asked to write an iterative version of FIB and to compare the execution times of the two versions.

Example 6.36 Write a Pascal function REVERSE that inputs an integer $N \geq 0$ and finds the reversal of N; that is, the integer obtained by reversing the digits of N. (For example, if N = 1870, then 781 is to be output.) Use recursive Pascal functions

$$\text{LENGTH}(N): \text{to compute the number of digits in } N$$

$$\text{POWER}(X, M): \text{to compute } X \text{ to the power } M$$

The function REVERSE should also be recursive.

Solution We first seek a recursive function R defined on the set of nonnegative integers and whose value $R(n)$ is the reversal of n. If $0 \leq n \leq 9$, then clearly $R(n) = n$. To see what happens for $n \geq 10$, consider $n = 1873$. We want $R(1873) = 3781$; thus notice that

$$R(1873) = R(187) + 3 \cdot 10^3$$

Letting $L(n)$ denote the number of digits in n, we see that

$$R(1873) = R(187) + 3 \cdot 10^{L(1873)-1}$$

In general, we have

1. $R(n) = n$ if $0 \leq n \leq 9$
2. $R(n) = R(n \text{ DIV } 10) + (n \text{ MOD } 10) \cdot 10^{L(n)-1}$ if $n \geq 10$

We now see the need for the Pascal functions LENGTH and POWER (for computing $10^{L(n)-1}$). The function LENGTH will be within our grasp if we can recursively define the function $L(n)$, for the number of digits in a nonnegative integer n. This is given by

$$L(n) = 1 \quad \text{if } 0 \leq n \leq 9$$

$$L(n) = L(n \text{ DIV } 10) + 1 \quad \text{for } n \geq 10$$

The function POWER is derived from the usual rule for exponentiation; in particular, for any positive number x,

$$x^0 = 1$$

$$x^n = x^{n-1} \cdot x \quad \text{for } n \geq 1$$

and depends on both n and x. We can now see how equation 2 will be given in the Pascal function REVERSE:

$$\text{REVERSE} := \text{REVERSE}(N \text{ DIV } 10) + (N \text{ MOD } 10) * \text{POWER}(10, \text{LENGTH}(N) - 1)$$

We are now prepared to write the REVERSE function.

```
FUNCTION REVERSE(N: INTEGER): INTEGER;
 FUNCTION LENGTH(M: INTEGER): INTEGER;
 BEGIN
   IF M DIV 10 = 0 THEN
    LENGTH := 1
   ELSE
    LENGTH := LENGTH(M DIV 10) + 1
 END;
 FUNCTION POWER(X,M: INTEGER): INTEGER;
  BEGIN
   IF M = 0 THEN
    POWER := 1
   ELSE
    POWER := POWER(X, M - 1) * X
  END;
BEGIN (*REVERSE*)
  IF N DIV 10 = 0 THEN
   REVERSE := N
  ELSE
   REVERSE := REVERSE(N DIV 10)+(N MOD 10)*POWER(10, LENGTH(N)-1)
END;                                                          ■
```

As was noted at the outset of this section, recursive functions are often more readily implemented in Pascal than are recursive procedures. It seems evident from the preceding examples that, when presented with a recursively defined function, one can code a corresponding recursive Pascal function without much difficulty. How procedures are implemented recursively is a slightly different story. In general, a recursive subprogram deals with a problem that can be reduced to a simpler instance of itself. In fact, the problem must be reducible to a case that is easily solved. We call this the *anchor case*. In contrast to a function, which returns a value to a calling program, a procedure may call for output before transferring control to a calling program.

Consider the example of the Towers of Hanoi (see Example 6.31). We already know a recursive function $h(n)$ for the number of moves required to transfer n disks from peg 1 to peg 3:

$$h(1) = 1$$

$$h(n) = 2h(n - 1) + 1$$

It is fairly easy to write a recursive Pascal function for h (see Exercise 9). What if we want a listing of the actual moves required to transfer n disks from peg 1 to peg 3? For instance, when $n = 1$, our output might look like

```
MOVE DISK FROM PEG 1 TO PEG 3
```

This is the anchor case. For $n = 3$, our output should be the following:

```
MOVE DISK FROM PEG 1 TO PEG 3
MOVE DISK FROM PEG 1 TO PEG 2
MOVE DISK FROM PEG 3 TO PEG 2
MOVE DISK FROM PEG 1 TO PEG 3
MOVE DISK FROM PEG 2 TO PEG 1
MOVE DISK FROM PEG 2 TO PEG 3
MOVE DISK FROM PEG 1 TO PEG 3
```

For an arbitrary value of n, this problem can be solved as follows (recall that the disks are labeled $d_1, d_2, \ldots, d_n$ in order of decreasing size):

1. List the moves required to transfer $d_2, d_3, \ldots, d_n$ from peg 1 to peg 2 with the aid of peg 3.
2. List the move transferring d_1 from peg 1 to peg 3.
3. List the moves required to transfer $d_2, d_3, \ldots, d_n$ from peg 2 to peg 3 with the aid of peg 1.

Notice that steps 1 and 3 call for the solution of the same problem with one less disk; thus a recursive procedure seems appropriate. If we call this procedure MOVES, then what should its parameter list be? Judging from steps 1 and 3, we need to access the number of disks being moved and the three pegs in some order. For example, if we wish to list the moves required to transfer n disks from peg 1 to peg 3, using peg 2 as auxiliary, then we would make the call

```
MOVES(N,1,3,2)
```

Since the procedure is recursive, it calls itself with possibly different orderings of the pegs, so it is important to denote the parameters representing the pegs in some general (and suggestive) manner. We use the following names:

START: starting peg

END: ending peg

AUX: auxiliary peg

with the obvious connotations.

With the aid of steps 1–3, we may now write the procedure using the established notation.

```
PROCEDURE MOVES(N,START,END,AUX:INTEGER);
(*This procedure outputs the sequence of moves required to
solve the puzzle of the ``Towers of Hanoi'' for N disks,
where the disks start on peg START, are to end up on peg
END, and peg AUX is to be used as the auxiliary peg.*)
BEGIN
 IF N=1 THEN
  WRITELN(`MOVE DISK FROM PEG', START:1, `TO PEG' END:1)
 ELSE
  BEGIN
   MOVES(N - 1, START, AUX, END);
   WRITELN(`MOVE DISK FROM PEG', START:1, `TO PEG' END:1);
   MOVES(N - 1, AUX, END, START)
  END
END;
```

Now suppose the call MOVES(3, 1, 2, 3) is made. The following steps indicate the recursive calls made to MOVES, along with the output of each call:

(a) MOVES(3, 1, 3, 2) is called and $3 \neq 1$, so the following statements are executed:

```
MOVES(2,1,2,3);
WRITELN(`MOVE DISK FROM PEG', 1, `TO PEG', 3);
MOVES(2,2,3,1);
```

(b) The first of the above instructions, MOVES(2, 1, 2, 3), is executed. Since $2 \neq 1$, the following statements are executed:

```
MOVES(1,1,3,2);
WRITELN(`MOVE DISK FROM PEG', 1, `TO PEG', 2);
MOVES(1,3,2,1);
```

(c) The three instructions in (b) result in the following output;

```
MOVE DISK FROM PEG 1 TO PEG 3
MOVE DISK FROM PEG 1 TO PEG 2
MOVE DISK FROM PEG 3 TO PEG 2
```

(d) The second of the three instructions in (a) is executed, resulting in the output

```
MOVE DISK FROM PEG 1 TO PEG 3
```

(e) The third instruction in (a), MOVES$(2, 2, 3, 1)$, is now executed. Since $2 \neq 1$, the following statements are executed:

```
MOVES(1,2,1,3);
WRITELN('MOVE DISK FROM PEG', 2, 'TO PEG', 3);
MOVES(1,1,3,2);
```

(f) The three instructions in (e) result in the following output:

```
MOVE DISK FROM PEG 2 TO PEG 1
MOVE DISK FROM PEG 2 TO PEG 3
MOVE DISK FROM PEG 1 TO PEG 3
```

Notice that the actual output consists of the seven lines given earlier.

Example 6.37 Write a Pascal procedure CONVERT that accepts two positive integers N and B and computes and outputs the base B representation of N. Assume $2 \leq B \leq 9$. The procedure CONVERT is to be recursive.

Solution The solution is derived from the fact that a positive integer n is uniquely expressible in the form

$$n = a_0 + a_1 b + a_2 b^2 + \cdots + a_t b^t$$

where $0 \leq a_i < b$ for $i = 0, 1, \ldots, t$ and $a_t \neq 0$. The desired output string is then $a_t a_{t-1} \ldots a_1 a_0$. In building the recursive procedure CONVERT, it is important to note two things:

1. If $n < b$, then (the digit) n should be output. (This is the anchor case.)
2. If $n \geq b$, then $n \operatorname{DIV} b = a_1 + a_2 b + \cdots + a_t b^{t-1}$ and $n \operatorname{MOD} b = a_0$.

We shall output the $t + 1$ base b digits of n using WRITE statements, writing a_t first, a_{t-1} second, and so on. For example, with $n = 59$ and $b = 3$, the following steps indicate how CONVERT should run:

$$59 \operatorname{MOD} 3 = 2 = a_0 \qquad n := 59 \operatorname{DIV} 3 = 19$$
$$19 \operatorname{MOD} 3 = 1 = a_1 \qquad n := 19 \operatorname{DIV} 3 = 6$$
$$6 \operatorname{MOD} 3 = 0 = a_2 \qquad n := 6 \operatorname{DIV} 3 = 2 \ (2 < 3)$$
$$2 \operatorname{MOD} 3 = 2 = a_3$$

Thus 2012 is the base 3 representation of 59. The recursion descends from $n = 59$ to $n = 2$, the anchor case, and the value 2 is output. Thereafter, the values 0, 1, and 2 are output in succession. In effect, this tells us how to code the body of CONVERT. The procedure follows.

```
PROCEDURE CONVERT(N, B: INTEGER);
 (*This procedure takes a positive integer N and an integer
 B, 2 <= B <= 9, and outputs the base B representation of N.*)

       BEGIN
        IF N < B THEN WRITE(N:1)
        ELSE
         BEGIN
          CONVERT(NDIVB,B);
          WRITE(NMODB:1)
         END;
         WRITELN;
       END;
```

The student is encouraged to trace this procedure for various values of N and B in order to gain a proper understanding of the recursion.   ■

Exercises 6.7

1. Write a recursive function DIV that takes integers a and b with $a \neq 0$ and returns b DIV a. (If $a < 0$, how is b DIV a related to b DIV $(-a)$? Now suppose $a > 0$. If $0 \leq b < a$, then b DIV $a = 0$. If $b < 0$, how is b DIV a related to $(b + a)$ DIV a? If $b \geq a$, how is b DIV a related to $(b - a)$ DIV a?)

2. Write a recursive function MOD that takes integers a and b with $a \neq 0$ and returns b MOD a.

3. Implement the Euclidean algorithm as a recursive function GCD that takes integers a and b (not both 0) and returns their greatest common divisor.

4. Implement the extended Euclidean algorithm as a recursive procedure EUCLID that takes integers a and b (not both 0) and returns $d = \gcd(a, b)$ and integers s and t such that $d = a \cdot s + b \cdot t$. (Suppose $1 \leq a < b$; let $q = b$ DIV a and $r = b$ MOD a. Then $d = \gcd(r, a)$; suppose $d = r \cdot s' + a \cdot t'$. How are s and t related to s' and t'? Hint: Use the identity $r = b - a \cdot q$.)

5. Write a recursive function MOD_EXPO that, given integers a, m, and n with $0 \leq a < n$ and $m \geq 0$, returns a^m MOD n. (See Chapter Problem 37 in Chapter 3.)

6. Write an iterative version of the function FIB to find the nth Fibonacci number. Compare its execution time to that of the recursive version given in Example 6.35. For instance, try computing the 25th Fibonacci number.

7. Write a Pascal procedure PRIME_FACTORS that takes an integer $n \geq 2$ and uses a recursive procedure HELPER to find and output the prime factors of n.
8. Write an iterative version of the function FACTORIAL, and compare its execution time with that of the recursive version given in the text.
9. Write a recursive function to compute values of the function $h: \mathbb{N} \to \mathbb{N}$ defined recursively by $h(1) = 1$ and $h(n) = 2h(n-1) + 1$, $n \geq 2$.

6.8   SOLVING RECURRENCE RELATIONS

In this section we describe a method for solving a special type of recurrence relation. A recurrence relation of the form

$$h(n) = a_1 h(n-1) + a_2 h(n-2) + \cdots + a_r h(n-r) + g(n) \quad (1)$$

where $n \geq r + 1$ and $a_1, a_2, \ldots, a_r$ are constants, is called a *linear recurrence relation with constant coefficients*. It is called "linear" because no products of functional values (such as $h(n-t)h(n-s)$ or $[h(n-t)]^3$) appear. For example, the recurrence relation

$$h(n) = 3h(n-1) + 6h(n-2) - 4h(n-3) + n^2$$

for $n \geq 4$ is linear, with constant coefficients. If, in addition, $g(n) = 0$, then we say that (1) is *homogeneous*. For example, the Fibonacci sequence $f(1), f(2), \ldots$ satisfies the recurrence relation

$$f(n) = f(n-1) + f(n-2)$$

for $n \geq 3$, and this is a linear homogeneous recurrence relation with constant coefficients. (Recall that $f(1) = f(2) = 1$.) In general, a recurrence relation of the form

$$F(n) = F(n-1) + F(n-2)$$

for $n \geq 3$ is called a *Fibonacci recurrence relation*.

Although it is not readily apparent, the terms of the Fibonacci sequence $1, 1, 2, 3, 5, 8, 13, 21, 34, \ldots$ increase exponentially. Suppose we assume a solution of the form

$$f(n) = \alpha^n$$

for some $\alpha \neq 0$. Then, substituting into the recurrence relation, we obtain

$$\alpha^n = \alpha^{n-1} + \alpha^{n-2}$$

for $n \geq 3$. Thus, upon division by α^{n-2}, we get that $\alpha^2 = \alpha + 1$ or

$$\alpha^2 - \alpha - 1 = 0$$

The solutions of this equation are, by the quadratic formula,

$$\alpha_1 = \frac{1 + \sqrt{5}}{2} \quad \text{and} \quad \alpha_2 = \frac{1 - \sqrt{5}}{2}$$

Thus $F_1(n) = \alpha_1^n$ and $F_2(n) = \alpha_2^n$ are both solutions to the Fibonacci recurrence relation; in fact, if c_1 and c_2 are any real constants, then

$$F(n) = c_1 \alpha_1^n + c_2 \alpha_2^n$$

is also a solution. (See Exercise 2.) Can c_1 and c_2 be chosen so that $F(n) = c_1 \alpha_1^n + c_2 \alpha_2^n$ also satisfies $F(1) = F(2) = 1$? This requires that

$$c_1 \alpha_1 + c_2 \alpha_2 = F(1) = 1$$

and

$$c_1 \alpha_1^2 + c_2 \alpha_2^2 = F(2) = 1$$

Keep in mind that $\alpha_1^2 - \alpha_1 = 1 = \alpha_2^2 - \alpha_2$. Hence, subtracting the first equation from the second yields

$$c_1 \left(\alpha_1^2 - \alpha_1 \right) + c_2 \left(\alpha_2^2 - \alpha_2 \right) = 0$$

or simply $c_1 + c_2 = 0$. So $c_2 = -c_1$, and it is then easily determined from the first equation that $c_1 = 1/\sqrt{5}$. We have now established the following result.

THEOREM 6.20   The terms of the Fibonacci sequence satisfy the relation

$$f(n) = \frac{1}{\sqrt{5}} \left(\frac{1 + \sqrt{5}}{2} \right)^n - \frac{1}{\sqrt{5}} \left(\frac{1 - \sqrt{5}}{2} \right)^n$$

for all $n \geq 1$.

Let's return to the general case of a linear homogeneous recurrence relation with constant coefficients:

$$h(n) = a_1 h(n - 1) + a_2 h(n - 2) + \cdots + a_r h(n - r) \qquad \text{for } n \geq r + 1$$

or

$$h(n) - a_1 h(n - 1) - a_2 h(n - 2) - \cdots - a_r h(n - r) = 0 \quad (2)$$

Taking the suggested lead from our work with the Fibonacci recurrence relation, we define the *characteristic equation* of this relation to be

$$x^r - a_1 x^{r-1} - a_2 x^{r-2} - \cdots - a_{r-1} x - a_r = 0 \qquad (3)$$

(Note that if $a_r \neq 0$, then 0 is not a solution of the equation.) The roots of (3) are called the *characteristic roots* of the associated recurrence relation

(2). Suppose now that $h(n) = \alpha^n$ is a solution to (2) for some $\alpha \neq 0$. Then

$$\alpha^n - a_1\alpha^{n-1} - a_2\alpha^{n-2} - \cdots - a_r\alpha^{n-r} = 0$$

Dividing by α^{n-r}, we obtain

$$\alpha^r - a_1\alpha^{r-1} - a_2\alpha^{r-2} - \cdots - a_r = 0$$

so that α is a characteristic root of (2). Conversely, if $\alpha \neq 0$ is a characteristic root of (2), then $h(n) = \alpha^n$ is a solution to (2). Thus we have the following result.

THEOREM 6.21 Let α be a nonzero complex number. Then $h(n) = \alpha^n$ is a solution of (2) if and only if α is a characteristic root of (2).

Suppose that (3) has r distinct nonzero roots $\alpha_1, \alpha_2, \ldots, \alpha_r$. Then

$$h(n) = c_1\alpha_1^n + c_2\alpha_2^n + \cdots + c_r\alpha_r^n$$

is a solution of (2) for any choice of the constants $c_1, c_2, \ldots, c_r$. (See Exercise 4.) We call this the *general solution* of (2). Suppose we are given the r initial conditions

$$h(1) = b_1, \, h(2) = b_2, \ldots, h(r) = b_r$$

and we seek a particular solution of (2) satisfying these conditions. Making use of them, we obtain the following system of equations

$$c_1\alpha_1 + c_2\alpha_2 + \cdots + c_r\alpha_r = b_1$$
$$c_1\alpha_1^2 + c_2\alpha_2^2 + \cdots + c_r\alpha_r^2 = b_2$$
$$\vdots$$
$$c_1\alpha_1^r + c_2\alpha_2^r + \cdots + c_r\alpha_r^r = b_r$$

Since $\alpha_1, \alpha_2, \ldots, \alpha_r$ are distinct and nonzero, this system has a unique solution for $c_1, c_2, \ldots, c_r$. (This depends on the fact that the coefficient matrix of the system is nonsingular.)

Example 6.38 Define $h: \mathbb{N} \to \mathbb{N}$ recursively by

1. $h(1) = h(2) = 1$
2. $h(n) = 2h(n-1) + 3h(n-2)$ for $n \geq 3$

Find an explicit formula for $h(n)$.

Solution The characteristic equation for h is readily obtained from the recurrence relation for h, as follows:

$$x^n = 2x^{n-1} + 3x^{n-2}$$
$$x^n - 2x^{n-1} - 3x^{n-2} = 0$$
$$x^2 - 2x - 3 = 0$$
$$(x+1)(x-3) = 0$$

Hence the characteristic roots are -1 and 3, so the general solution is

$$h(n) = c_1(-1)^n + c_2 3^n$$

For the initial conditions $h(1) = h(2) = 1$, we obtain the system of equations

$$-c_1 + 3c_2 = 1$$
$$c_1 + 9c_2 = 1$$

and the solution is easily seen to be $c_1 = -1/2$ and $c_2 = 1/6$. Thus h is given by

$$h(n) = \frac{-1}{2}(-1)^n + \frac{1}{6} 3^n$$

$$= \frac{1}{2}\left[(-1)^{n+1} + 3^{n-1}\right] \qquad \blacksquare$$

Example 6.39 Find and solve a recurrence relation for the number of n-digit ternary (base 3) sequences with no two consecutive digits equal.

Solution Let $g(n)$ denote the number of n-digit ternary sequences with no two consecutive digits equal. Clearly $g(1) = 3$. Moreover, if $n \geq 2$ and $x = a_1 a_2 \dots a_{n-1}$ is any such sequence with $n - 1$ digits, then there are two ways that a digit a_n can be appended to x to obtain a sequence of the same type with n digits (since $a_n \in \{0, 1, 2\}$ and $a_n \neq a_{n-1}$). Hence g satisfies the recurrence relation

$$g(n) = 2 \cdot g(n - 1) \quad \text{for } n \geq 2$$

The characteristic equation for $g(n)$ is obtained as follows:

$$x^n = 2x^{n-1}$$
$$x^n - 2x^{n-1} = 0$$
$$x - 2 = 0$$

Thus $x = 2$ and we obtain $g(n) = c2^n$. From $g(1) = 3$, we get $c = 3/2$, so $g(n) = 3 \cdot 2^{n-1}$. $\qquad \blacksquare$

Example 6.40 Solve the recurrence relation

$$z(n) = 3z(n - 1) + z(n - 2) - 3z(n - 3)$$

subject to the initial conditions $z(1) = 3$, $z(2) = 15$, and $z(3) = 27$.

Solution From the given recurrence relation we obtain the characteristic equation

$$x^3 - 3x^2 - x + 3 = 0$$

Observe that

$$x^3 - 3x^2 - x + 3 = x^2(x - 3) - (x - 3)$$
$$= (x - 3)(x^2 - 1) = (x - 3)(x - 1)(x + 1)$$

So the characteristic roots are 3, 1, and -1, and the general solution of the recurrence relation is

$$z(n) = a \cdot 1^n + b(-1)^n + c \cdot 3^n$$
$$= a + b(-1)^n + c \cdot 3^n$$

From the initial conditions we obtain the system of equations

$$a - b + 3c = 3$$
$$a + b + 9c = 15$$
$$a - b + 27c = 27$$

Subtracting the first equation from the third yields $c = 1$. The rest of the way one easily obtains $a = 3$ and $b = 3$. Hence the solution is

$$z(n) = 3 + 3(-1)^n + 3^n \qquad \blacksquare$$

The foregoing examples pertain to the case where the characteristic roots of the recurrence relation (2) are distinct. Suppose now that (3) has repeated roots; say α is a root of multiplicity t, meaning that

$$x^r - a_1 x^{r-1} - a_2 x^{r-2} - \cdots - a_{r-1}x - a_r = (x - \alpha)^t q(x)$$

where $q(\alpha) \neq 0$. Then it turns out that

$$\alpha^n, n\alpha^n, n^2\alpha^n, \ldots, n^{t-1}\alpha^n$$

are all solutions of (2) and that

$$c_1\alpha^n + c_2 n\alpha^n + c_3 n^2\alpha^n + \cdots + c_t n^{t-1}\alpha^n$$
$$= \left(c_1 + c_2 n + c_3 n^2 + \cdots + c_t n^{t-1}\right)\alpha^n$$

is that portion of the general solution corresponding to α. (See Exercise 6.)

Example 6.41 Solve the recurrence relation

$$h(n) = 2h(n - 1) + 4h(n - 2) - 8h(n - 3)$$

subject to the initial conditions $h(1) = 2$, $h(2) = 4$, and $h(3) = 12$.

Solution The characteristic equation of the given recurrence relation is

$$x^3 - 2x^2 - 4x + 8 = 0$$

and we see that

$$x^3 - 2x^2 - 4x + 8 = (x - 2)^2(x + 2)$$

Thus the characteristic roots are 2, 2, and -2, with the root 2 having multiplicity 2. The general solution is then

$$h(n) = (a + bn) \cdot 2^n + c(-2)^n$$

Using the initial conditions, we obtain the system of equations

$$2a + 2b - 2c = 2$$
$$4a + 8b + 4c = 4$$
$$8a + 24b - 8c = 12$$

The solution to this system is $a = 5/8$, $b = 1/4$, and $c = -1/8$; hence

$$h(n) = \left(\frac{5}{8} + \frac{1}{4}n\right)2^n - \frac{1}{8}(-2)^n \qquad \blacksquare$$

Exercises 6.8

1. Each part recursively defines a function from $\mathbb{N}$ to $\mathbb{Z}$. Find an explicit formula for the function.
   (a) $h_1(1) = 1$, $h_1(2) = 3$, $h_1(n) = 3h_1(n - 1) + 4h_1(n - 2)$, $n \geq 3$
   (b) $h_2(1) = h_2(2) = 2$, $h_2(n) = 2h_2(n - 1) - h_2(n - 2)$, $n \geq 3$
   (c) $h_3(1) = 0$, $h_3(2) = 1$, $h_3(n) = h_3(n - 2)$, $n \geq 3$
   (d) $h_4(1) = 1$, $h_4(2) = 5$, $h_4(3) = 17$, $h_4(n) = 3h_4(n - 1) + 3h_4(n - 2) - h_4(n - 3)$, $n \geq 4$
   (e) $h_5(1) = 1$, $h_5(2) = 0$, $h_5(n) = 4h_5(n - 2)$, $n \geq 3$
   (f) $h_6(1) = 2$, $h_6(n) = (n + 1)h_6(n - 1)$, $n \geq 2$
   (g) $h_7(1) = 1$, $h_7(2) = 2$, $h_7(3) = 11$, $h_7(n) = h_7(n - 1) + 9h_7(n - 2) - 9h_7(n - 3)$, $n \geq 4$
   (h) $h_8(1) = -1$, $h_8(2) = 0$, $h_8(n) = 8h_8(n - 1) - 16h_8(n - 2)$, $n \geq 3$
   (i) $h_9(1) = h_9(2) = 0$, $h_9(3) = -2$, $h_9(n) = 3h_9(n - 2) - 2h_9(n - 3)$, $n \geq 4$
   (j) $h_0(1) = -3$, $h_0(2) = 5$, $h_0(3) = 39$, $h_0(4) = 177$, $h_0(n) = 5h_0(n - 1) - 6h_0(n - 2) - 4h_0(n - 3) + 8h_0(n - 4)$, $n \geq 5$

2. Consider the linear homogeneous recurrence relation with constant coefficients (2):

$$h(n) = a_1 h(n - 1) + a_2 h(n - 2) + \cdots + a_r h(n - r)$$

for $n \geq r + 1$. Suppose that $h_1(n)$ and $h_2(n)$ are both solutions of this recurrence relation; prove that $c_1 h_1(n) + c_2 h_2(n)$ is also a solution for any constants c_1 and c_2.

3. Consider the function $g: \mathbb{N} \to \mathbb{Z}$ defined recursively by $g(1) = -1$, $g(2) = 1$, and $g(n) = 5g(n - 1) - 6g(n - 2)$, $n \geq 3$.
   (a) Compute $g(3)$ and $g(4)$.
   (b) Find an explicit formula for $g(n)$.

4. Suppose that the characteristic equation (3) has r distinct nonzero roots $\alpha_1, \alpha_2, \ldots, \alpha_r$. Show that

$$h(n) = c_1\alpha_1^n + c_2\alpha_2^n + \cdots + c_r\alpha_r^n$$

   is a solution of (2) for any choice of the constants $c_1, c_2, \ldots, c_r$.

5. Consider the function: $h\colon \mathbb{N} \to \mathbb{Z}$ defined recursively by $h(1) = 1$, $h(2) = 0$, and $h(n) = -h(n - 2)$, $n \geq 3$.
   (a) Compute the first several values of $h(n)$ and guess an explicit formula for $h(n)$. (Hint: The value of $h(n)$ depends on the value of n MOD 4.)
   (b) Use the theory of linear homogeneous recurrence relations to find an explicit formula for $h(n)$.

6. Suppose that the characteristic equation (3) has a repeated root α with multiplicity $t \geq 2$. Show that

$$\alpha^n, n\alpha^n, n^2\alpha^n, \ldots, n^{t-1}\alpha^n$$

   are all solutions of (2) and that

$$c_1\alpha^n + c_2 n\alpha^n + c_3 n^2\alpha^n + \cdots + c_t n^{t-1}\alpha^n$$
$$= \left(c_1 + c_2 n + c_3 n^2 + \cdots + c_t n^{t-1}\right)\alpha^n$$

   is that portion of the general solution corresponding to α.

7. Each part recursively defines a function from $\mathbb{N} \cup \{0\}$ to $\mathbb{Z}$. Find an explicit formula for the function.
   (a) $a(0) = 0$, $a(1) = 3$, $a(2) = 21$, $a(n) = 3a(n - 1) - 4a(n - 3)$, $n \geq 3$
   (b) $a_2(0) = 1$, $a_2(1) = 7$, $a_2(2) = 65$,
   $a_2(n) = 5a_2(n - 1) - 3a_2(n - 2) - 9a_2(n - 3)$, $n \geq 3$

CHAPTER PROBLEMS

1. A city police department has ten detectives: seven males and three females. In how many ways can a team of three detectives be chosen to work on a case under these conditions?
   (a) There are no restrictions.
   (b) The team must have at least one male and at least one female.

2. A hand in a game of bridge consists of 13 of the 52 cards in a standard deck.
   (a) How many bridge hands are there?
   How many bridge hands contain the following?
   (b) exactly two aces
   (c) exactly three aces, two kings, one queen, and no jacks
   (d) at least one heart

(e) four spades, three hearts, three diamonds, and three clubs

(f) four cards in one suit and three cards in each of the other three suits

(g) at least one heart and at least one spade

(h) exactly six hearts or exactly seven diamonds

3. Given 11 different combinatorics books and seven different number theory books, how many ways are there for Curly to choose a combinatorics book, and then for Larry to choose a combinatorics or a number theory book, and then for Moe to choose both a combinatorics and a number theory book?

4. For each of the following words, determine the number of distinct ways in which the letters can be rearranged.

   (a) PIE   (b) APPLE   (c) QUEUE   (d) KALAMAZOO

   (e) In general, if $L_1, L_2, \ldots, L_m$ are distinct letters, how many different "words" can be formed using n_1 copies of L_1, n_2 copies of $L_2, \ldots$, and n_m copies of L_m, where each n_i is a positive integer and $n_1 + n_2 + \cdots + n_m = n$?

5. How many eight-letter sequences, constructed using the 26 (lowercase) letters of the alphabet, meet the following conditions?

   (a) contain exactly three a's

   (b) contain three or four vowels (a, e, i, o, u)

   (c) have no repeated letters   (d) contain an even number of e's

6. The registrar at a certain college has observed that, of 5000 students, 22% have an 8 A.M. class, 52% have a 9 A.M. class, 67% have an 11 A.M. class, 12% have an 8 A.M. and a 9 A.M. class, 9% have an 8 A.M. and an 11 A.M. class, 35% have a 9 A.M. and an 11 A.M. class, and 5% of the students have classes during all three of these hours. What percentage of students have a class at 8 A.M., 9 A.M., or 11 A.M.?

7. The mathematics club at a small college has six senior, five junior, four sophomore, and three freshmen members. Five members are to be chosen to represent the club on the mathematics department curriculum committee. In how many ways can these five students be chosen under these conditions?

   (a) There are no restrictions.

   (b) At least two seniors are chosen.

   (c) At most one freshman is chosen.

   (d) Exactly two sophomores are chosen.

   (e) At least two seniors or exactly two sophomores are chosen.

8. Let $P_{m,n}$ denote the set of polynomials of degree n with coefficients chosen from the set $\{0, 1, \ldots, m\}$. Find $|P_{m,n}|$.

9. Let $A = \{1, 2, 3, 4, 5\}$ and $B = \{1, 2, 3, 4\}$. Determine the cardinality of each of these sets.

   (a) the set of relations from A to B

   (b) the set of functions from A to B

   (c) the set of one-to-one functions from A to B

(d) the set of functions from A onto B

(e) the set of functions $f\colon A \to B$ with $f(1) = 1$ or $f(5) = 4$

10. How many relations are there from an m-element set to an n-element set?

11. An urn contains three red balls, four orange balls, five blue balls, and six green balls, numbered 1 through 18, respectively. A subset of four balls is to be selected from the urn at random. How many such subsets contain the following?

(a) exactly two orange balls and one green ball

(b) one ball of each color

(c) not all four balls the same color

(d) at least two balls that are not blue

12. How many relations are there on an n-element set? How many of these relations are:

(a) reflexive?   (b) irreflexive?

(c) symmetric?   (d) antisymmetric?

13. Determine the coefficient of $x^7 y^5$ in the expansion of $(2x - y/4)^{12}$.

14. Prove the identity

$$1 \cdot C(n,0) + 2 \cdot C(n,1) + \cdots + (n + 1) \cdot C(n, n) = (n + 2) \cdot 2^{n-1}$$

15. The row of Pascal's triangle corresponding to $n = 6$ is

$$1 \quad 6 \quad 15 \quad 20 \quad 15 \quad 6 \quad 1$$

Use Pascal's identity to find the rows of Pascal's triangle corresponding to:

(a) $n = 7$ (b) $n = 8$

16. Use binomial coefficients to give a formula for the sum

$$1 \cdot 2 + 2 \cdot 3 + \cdots + (n - 1) \cdot n$$

17. Compute $C(6, 3)$ using:

(a) Pascal's identity

(b) the identity $C(n, k) = (n + k + 1)C(n, k - 1)/k$

18. Let $n \in \mathbb{N}$. Use the binomial theorem to prove that

$$\frac{C(n,0)}{1} + \frac{C(n,1)}{2} + \cdots + \frac{C(n, n)}{n + 1} = \frac{2^{n+1} - 1}{n + 1}$$

19. Let m and n be integers with $0 \le m \le n$. Prove the identity

$$C(n,0)C(n, m) + C(n,1)C(n, m - 1) + \cdots + C(n, m)C(n,0) = C(2n, m)$$

20. Find a formula for the sum

$$\sum_{k=0}^{n} \frac{(-1)^k C(n, k)}{k + 1}$$

21. Generalize Problem 19; let m, n_1, and n_2 be integers with $0 \leq m \leq n_1 \leq n_2$. Prove the identity

$$\sum_{k=0}^{m} C(n_1, k)C(n_2, m-k) = C(n_1 + n_2, m)$$

(a) combinatorially

(b) using the binomial theorem and the identity $(1 + x)^{n_1}(1 + x)^{n_2} = (1 + x)^{n_1 + n_2}$

(c) by induction on m

22. Find a formula for the sum

$$\sum_{k=1}^{n} C(n, k-1)C(n, k)$$

23. In how many ways can the 26 letters A, B, ..., Z be placed

(a) in a row?

(b) in a row so that no two vowels are consecutive?

(c) in a circle?

(d) in a circle so that no two vowels are consecutive?

★24. Eight students and four faculty members must be seated at two (identical) six-person tables for lunch.

(a) In how many ways can this be done?

(b) In how many ways can this be done with at least one faculty member at each table?

25. The 12 face cards from a standard deck are arranged in a circle. Four cards are to be chosen.

(a) In how many ways can this be done?

(b) In how many ways can this be done if no two cards that are consecutive are chosen?

26. Let m and n be positive integers with $m < n$. We are interested in the following question: How many permutations of $\{1, 2, \ldots, n\}$ do not contain a permutation of $\{1, 2, \ldots, m\}$ as a "subpermutation"?

(a) First consider a specific case. How many permutations of $\{1, 2, 3, 4\}$ contain neither the permutation $(1, 2)$ nor the permutation $(2, 1)$?

(b) Try another specific case. How many permutations of $\{1, 2, 3, 4, 5, 6, 7\}$ contain none of the permutations $(1, 2, 3)$, $(1, 3, 2)$, $(2, 1, 3)$, $(2, 3, 1)$, $(3, 1, 2)$, or $(3, 2, 1)$?

(c) Formally define the term *subpermutation* and answer the general question.

27. To guarantee that there are ten diplomats from the same continent at a party, how many diplomats must be invited if they are chosen from these groups?

(a) 12 Australasian, 14 African, 15 Asian, 16 European, 18 South American, and 20 North American diplomats

(b) 7 Australasian, 14 African, 8 Asian, 16 European, 18 South American, and 20 North American diplomats

28. Show that, given any set of seven integers, there exist two integers in the set whose sum or difference is a multiple of ten.

29. How many arrangements are there of KALAMAZOO if it is required that the O's must appear before the K?

30. A sequence $(x_1, x_2, \ldots, x_m)$ is called *monotonic* if it is increasing or decreasing; that is, either $x_1 < x_2 < \cdots < x_m$ or $x_1 > x_2 > \cdots > x_m$. In this problem we explore the following theorem of Erdös and Szekeres: Given a sequence $(x_1, x_2, \ldots, x_t)$ of $t = n^2 + 1$ distinct integers, it contains a monotonic subsequence of $n + 1$ integers.

   (a) Give an example of a sequence (x_1, x_2, x_3, x_4) of four distinct integers that does not contain a monotonic subsequence of length three.

   (b) Give an example of a sequence $(x_1, x_2, \ldots, x_9)$ of nine distinct integers that does not contain a monotonic subsequence of length four.

   (c) For an arbitrary integer $n > 1$, give an example of a sequence $(x_1, x_2, \ldots, x_s)$ of $s = n^2$ distinct integers that does not contain a monotonic subsequence of length $n + 1$.

   (d) Prove the result of Erdös and Szekeres. (Hint: For $1 \leq i \leq n^2 + 1$, let k_i be the maximum length of an increasing subsequence that begins with x_i. If $k_i > n$ for some i, then the result is established, so suppose that $1 \leq k_i \leq n$ for each i. Apply the pigeonhole principle to show that

$$k_{i_1} = k_{i_2} = \cdots = k_{i_{n+1}}$$

for some $1 \leq i_1 < i_2 < \cdots < i_{n+1} \leq n^2 + 1$, and then argue that $(x_{i_1}, x_{i_2}, \ldots, x_{i_{n+1}})$ is a decreasing subsequence of the original sequence.)

31. How many ways are there to distribute 20 different homework problems to ten students under these conditions?

   (a) There are no restrictions

   (b) Each student gets two problems.

32. How many different positive integers can be formed using the nine digits given?

   (a) $1, 2, 3, 4, 5, 6, 7, 8, 9$ ★**(b)** $1, 1, 3, 3, 5, 5, 7, 7, 9$

   ★**(c)** $3, 3, 3, 6, 6, 6, 9, 9, 9$ ★**(d)** $8, 8, 9, 9, 9, 9, 9, 9, 9$

33. A local dairy offers 16 flavors of ice cream.

   (a) How many ways are there to purchase eight scoops of ice cream?

   (b) How many ways are there to purchase eight scoops of ice cream of eight different flavors?

   ★**(c)** How many ways are there to distribute the eight scoops of part (b) to four students so that each student gets two scoops?

   (d) How many ways are there to distribute eight scoops of chocolate ice cream to four students so that each student gets at least one scoop?

(e) If a professor wants to purchase a triple-scoop ice cream cone, how many choices does she have? (Does the order of flavors on the cone matter? Answer the question for both cases.)

34. Consider choosing three digits from $\{0, 1, \ldots, 9\}$.
    (a) In how many ways can this be done?
    (b) In how many ways can this be done if no two of the digits chosen are consecutive?

35. Given five calculus books, three linear algebra books, and two number theory books (all distinct), how many ways are there to do the following?
    (a) Select three books, one in each subject.
    (b) Make a row of three books.
    (c) Make a row of three books, with one book in each subject.
    (d) Make a row of three books, with exactly two of the subjects represented.
    (e) Make a row of three books all in the same subject.

36. A rumor is spread among m college presidents as follows: The person who starts the rumor telephones someone, and then that person telephones someone else, and so on. How many different paths through the group can the rumor follow in n calls?

37. Given the books in Problem 35, in how many ways can they be arranged on a shelf under these conditions?
    (a) There are no restrictions.
    (b) Books on the same subject must be placed together.

38. A mathematician has four colleagues. During a summer workshop, she had lunch alone six times; she lunched with each colleague seven times, with each subset of two colleagues four times, with each subset of three colleagues three times, and with all four colleagues on two occasions. How many days did the workshop last (assuming that the mathematician had lunch each day)?

39. For $1 \le k \le n$, consider the identity

$$kC(n, k) = (n - k + 1)C(n, k - 1).$$

    (a) Verify this identity using Theorem 6.9.
    (b) Give a combinatorial proof of this identity. (Hint: Argue that both sides of the identity count the number of ways to choose a k-combination of $\{1, 2, \ldots, n\}$, where one of the elements chosen is designated as "special.")

40. A simple coding scheme involves the replacement of each of the 36 alphanumeric characters by an alphanumeric character, such that no two characters are encoded the same. In how many ways can this be done? In how many ways can it be done if no character is encoded as itself?

41. In how many ways can 30 identical cans of beer be distributed to five fraternity brothers if Curly gets at least three beers, Larry gets at most four beers, Moe gets between two and six beers (inclusive), Joe gets

between four and seven beers, and Al gets between one and eight beers?

42. For $0 \le k \le m \le n$, give a combinatorial proof of Vandermonde's identity:

$$C(m + n, k) = C(m,0)C(n, k) + C(m,1)C(n, k - 1) + \cdots +$$
$$C(m, k)C(n,0)$$
$$= \sum_{r=0}^{k} C(m, r)C(n, k - r)$$

(Hint: Consider choosing a k-combination from $\{1, 2, \ldots, m, m + 1, \ldots, m + n\}$; each such subset is the union of an r-element subset of $\{1, 2, \ldots, m\}$ and a $(k - r)$-element subset of $\{m + 1, m + 2, \ldots, m + n\}$ for some r, $0 \le r \le k$; partition according to the value of r and apply the addition principle.)

43. Let $q(n)$ denote the number of permutations of $\{1, 2, \ldots, n\}$ such that 2 does not immediately follow 1, 3 does not immediately follow 2, ..., and n does not immediately follow $n - 1$.
(a) Find $q(1)$, $q(2)$, $q(3)$.   (b) Use a tree diagram to compute $q(4)$.
(c) Use the principle of inclusion–exclusion to find a general formula for $q(n)$.

44. Let $f(n)$ denote the nth Fibonacci number.
(a) Show that $f(1) + f(2) + \cdots + f(n) = f(n + 2) - 1$.
(b) Find and verify a formula for $f(1) + f(3) + \cdots + f(2n - 1)$.
(c) Find and verify a formula for $f(2) + f(4) + \cdots + f(2n)$.
(d) Find and verify a formula for $f(1) - f(2) + \cdots + (-1)^{n+1}f(n)$.

45. Let $f(n)$ denote the nth Fibonacci number. Show that

$$f(1)^2 + f(2)^2 + \cdots + f(n)^2 = f(n)f(n + 1)$$

46. Let $f(n)$ denote the nth Fibonacci number.
(a) Show that $f(n)f(n + 2) = [f(n + 1)]^2 + (-1)^{n+1}$ for all $n \ge 1$.
(b) Show that any two consecutive Fibonacci numbers are relatively prime.

47. A wizard must climb an infinite staircase whose steps are numbered with the positive integers. At each step, the wizard may go up one stair or three. Let $w(n)$ be the number of ways for the wizard to reach step number n.
(a) Find $w(1)$, $w(2)$, and $w(3)$.
(b) Find a recurrence relation for $w(n)$, where $n \ge 4$.

48. Let $g_3(n)$ be the number of ways to partition a set having cardinality $3n$ into n subsets with three elements in each subset. Find initial values and a recurrence relation for $g_3(n)$.

49. Generalize Problem 48; let $g_k(n)$ be the number of ways to partition a set having cardinality kn into n subsets with k elements in each subset. Find initial values and a recurrence relation for $g_k(n)$.

50. Each part recursively defines a function from $\mathbb{N} \cup \{0\}$ to $\mathbb{Z}$. Find an explicit formula for the function.

(a) $h_1(0) = 1$, $h_1(1) = -2$, $h_1(n) = 5h_1(n-1) - 6h_1(n-2)$, $n \geq 2$

(b) $h_2(0) = 0$, $h_2(1) = 1$, $h_2(n) = 9h_2(n-2)$, $n \geq 2$

(c) $h_3(0) = 0$, $h_3(1) = -3$, $h_3(2) = 15$,
$h_3(n) = 4h_3(n-1) + h_3(n-2) - 4h_3(n-3)$, $n \geq 3$

(d) $h_4(0) = -1$, $h_4(1) = 0$, $h_4(n) = 10h_4(n-1) - 25h_4(n-2)$,
$n \geq 2$

(e) $h_5(0) = 4$, $h_5(1) = 0$, $h_5(2) = 20$, $h_5(n) = 12h_5(n-2) + 16h_5(n-3)$, $n \geq 3$

(f) $h_6(0) = 1$, $h_6(1) = -3$, $h_6(2) = 25$,
$h_6(n) = -h_6(n-1) + 5h_6(n-2) - 3h_6(n-3)$, $n \geq 3$

(g) $h_7(0) = 1$, $h_7(n) = 2nh_7(n-1)$, $n \geq 1$

(h) $h_8(0) = -3$, $h_8(1) = -2$, $h_8(2) = 20$, $h_8(n) = 6h_8(n-1) - 12h_8(n-2) + 8h_8(n-3)$, $n \geq 3$

51. Each of the following parts recursively defines a function from $\mathbb{N} \cup \{0\}$ to $\mathbb{Z}$, where a, b and c are integers. Find an explicit formula for the function.

(a) $f_1(0) = b$, $f_1(n) = f_1(n-1) + a$, $n \geq 1$

(b) $f_2(0) = c$, $f_2(n) = f_2(n-1) + 2an - a$, $n \geq 1$

(c) $f_3(0) = c$, $f_3(n) = f_3(n-1) + 2an - a + b$, $n \geq 1$

(d) $f_4(0) = 1$, $f_4(n) = af_4(n-1)$, $n \geq 1$

(e) $f_5(0) = 1 + b$, $f_5(n) = af_5(n-1) + b(1-a)$, $n \geq 1$

(f) $f_6(0) = a$, $f_6(n) = f_6(n-1)$

52. Consider the general linear nonhomogeneous recurrence relation with constant coefficients:

$$h(n) - a_1h(n-1) - a_2h(n-2) - \cdots - a_rh(n-r) = g(n) \quad (*)$$

Suppose that $h_0(n)$ is the general solution of the corresponding homogeneous recurrence relation and $p(n)$ is any particular solution of the nonhomogeneous recurrence relation $(*)$. Show that $h_0(n) + p(n)$ is a solution of $(*)$; this is called the *general solution of* $(*)$.

53. Each part recursively defines a function from $\mathbb{N} \cup \{0\}$ to $\mathbb{Z}$ and gives a particular solution $p(n)$ to the recurrence relation for $n \geq 1$. Use the result of Problem 52 to find the general solution.

(a) $f_7(0) = 5$, $f_7(n) = 3f_7(n-1) - 8$; $p(n) = 4$

(b) $f_8(0) = 1$, $f_8(n) = 2f_8(n-1) - n$; $p(n) = n + 2$

(c) $f_9(0) = 9$, $f_9(n) = -2f_9(n-1) + 27(n^2 + n)$; $p(n) = 9n^2 + 21n + 8$

(d) $f_0(0) = 2$, $f_0(n) = -3f_0(n-1) + 4^n$; $p(n) = 4^{n+1}/7$

54. Let $a(n)$ denote the number of distinct ways to properly parenthesize the expression

$$x_1 - x_2 - \cdots - x_n$$

(where $x_1, x_2, \ldots, x_n \in \mathbb{Z}$). Let $a(1) = a(2) = 1$; for $n = 3$, we may parenthesize the expression in two ways: $(x_1 - x_2) - x_3$ and $x_1 - (x_2 - x_3)$, so $a(3) = 2$.

(a) Find $a(4)$.

(b) Show that $a(n)$ satisfies the recurrence relation

$$a(n) = \sum_{k=1}^{n-1} a(k)a(n-k)$$

for $n \geq 2$. (Hint: Partition the set of ways to parenthesize the expression according to which " $-$ " is the last to be applied, and apply the addition principle.)

★**(c)** Show that $a(n) = C(2n-2, n-1)/n$.

The numbers $a(n)$ are called the *Catalan numbers*.

55. Suppose that $f : \mathbb{N} \to \mathbb{Z}$ and $g : \mathbb{N} \to \mathbb{Z}$ are functions and, for all $n \geq 2$,

$$f(n) - f(n-1) = g(n) - g(n-1)$$

Prove that $f(n) = g(n) + c$ for some constant $c \in \mathbb{Z}$. (Hint: Consider $h(n) = f(n) - g(n)$.)

56. Eight scientists work at a secret research building. The building is to have D doors, each with four locks. Each scientist is to be issued a set of K keys. For reasons of security, it is required that at least four scientists be present together in order that, among them, they possess a subset of four keys that fit the four locks on one of the D doors. How many doors must the building have and how many keys must each scientist be issued?

57. If, in Problem 56, there is to be only one door with L locks, what should the values be for L and K if it is required that at least four scientists be present together in order to enter the building?

58. For a nonnegative integer n, consider the expansion of the trinomial

$$(x + y + z)^n$$

(a) Expand this trinomial for $n = 0, 1, 2, 3$.

(b) In general, prove that

$$(x + y + z)^n = \sum \frac{n!}{i!j!k!} x^i y^j z^k$$

where the sum is over all 3-tuples (i, j, k) of nonnegative integers such that $i + j + k = n$. For this reason, the numbers $n!/(i!j!k!)$ are often called *trinomial coefficients*.

59. Generalizing Problem 58, we let k and n be integers, with k positive and n nonnegative, and consider the expansion of the k-nomial

$$(x_1 + x_2 + \cdots + x_k)^n$$

Prove that

$$(x_1 + x_2 + \cdots + x_k)^n = \sum \frac{n!}{m_1! m_2! \cdots m_k!} x_1^{m_1} x_2^{m_2} \ldots x_k^{m_k}$$

where the sum is over all k-tuples $(m_1, m_2, \ldots, m_k)$ of nonnegative integers such that $m_1 + m_2 + \cdots + m_k = n$. This result is known as the *multinomial theorem* and, for this reason, the numbers $n!/(m_1! m_2! \cdots m_k!)$ are often called *multinomial coefficients*.

60. Use the result of Problem 58 to find the coefficient of:
   (a) $x^3 y^2 z$ in the expansion of $(x + y + z)^6$
   (b) $x^7 y^3 z^2$ in the expansion of $(3x - y + 2z)^{12}$

61. For $0 \le k \le 2n$, find the coefficient of x^k in the expansion of $(1 + x + x^2)^n$.

62. Use the result of Problem 59 to find the coefficient of:
   (a) $x^4 y^3$ in the expansion of $(w + x + y + z)^7$
   (b) $vw^2 x^3 y^4 z^5$ in the expansion of $(v - 2w - 3x + 2y + z)^{15}$

★63. Exercise 29 of Exercises 6.3 discusses an algorithm for listing the permutations of $\{1, 2, \ldots, n\}$ in lexicographic order. A different algorithm for listing these permutations is due to H. Trotter [*Communications of the ACM*, 5 (1962), 434–435] and S. Johnson [*Math. Comp.* 17 (1963), 282–285]. Their method has the advantage that the successor of a given permutation π is obtained from π merely by a transposition, or swapping, of two adjacent elements. For example, here are the lists for $n = 1, 2$, and 3:

$$(1)$$

$$(1,2), (2,1)$$

$$(1,2,3), (1,3,2), (3,1,2), (3,2,1), (2,3,1), (2,1,3)$$

   (a) To construct the list for $n = k + 1$ from the list for $n = k$, repeat each k-permutation $k + 1$ times; then insert the element $k + 1$ into each k-permutation so that its position follows the pattern

$$k + 1, k, \ldots, 1, 1, 2, \ldots, k, \ldots, k + 1$$

   Use this rule to construct the list for $n = 4$.

   ★(b) Modify the procedure PERMS(N) of Exercise 29 in Exercises 6.3 so that it lists the permutations of $\{1, 2, \ldots, N\}$ using the method of Trotter and Johnson.

64. If the numbers from 1 to 100,000 are listed, how many times does the digit 5 appear?

★65. Given a permutation $\pi = (a_1, a_2, \ldots, a_n)$ of $\{1, 2, \ldots, n\}$, the *Trotter–Johnson rank* of π is an integer between 0 and $n! - 1$ giving the number of predecessors of π in the list of all the permutations of $\{1, 2, \ldots, n\}$ in Trotter–Johnson order (see Problem 63).

(a) Modify the Pascal function PERM_TO_RANK of Exercise 30 in Exercises 6.3 so that it returns the Trotter–Johnson rank of a given permutation.

(c) Modify the Pascal function RANK_TO_PERM of Exercise 30 in Exercises 6.3 so that, given integers R and N with $0 \le R \le N! - 1$, it returns the permutation of $\{1, 2, \ldots, N\}$ with Trotter–Johnson rank R.

SEVEN

Graph Theory

INTRODUCTION

It is the objective of this chapter to introduce the student to some of the basic concepts and results of graph theory, and to indicate several of the more important applications of the subject.

Many people date the beginning of graph theory to the year 1736, when the great Swiss mathematician Leonhard Euler published an analysis of the *Königsberg bridges problem*. Although Euler did not use graph theory as we know it, it is clear that he was thinking in graphical terms. Another famous problem that came to involve graph theory is the *four-color problem*, which was first posed in 1852 in a letter from Augustus DeMorgan to Sir William Rowan Hamilton. We discuss the Königsberg bridges problem in the next section and return to the four-color problem later in the chapter. For now it suffices to point out that it took graph theory quite a while to develop into the popular and fertile area of mathematical research that it is today; for a long time its main use was to aid in the solution of mathematical puzzles and games.

Perhaps a good place to start our discussion is with the definition of "graph." One can find several meanings for this term in the mathematical literature; for now we consider two kinds of "graph" that are particularly natural and useful.

Figure 7.1 shows a *simple graph* G_1 and a *directed graph* G_2. Note that both G_1 and G_2 have four distinguished points labeled u, v, x, and y. These points are called *vertices* (the plural of *vertex*). Certain pairs of vertices of G_1 are joined by line segments or simple curves called *edges*, while in G_2 certain pairs of vertices are joined by directed line segments or simple curves called *directed edges* or *arcs*. There is also a special type of arc, called a *loop*, from the vertex u to itself. Basically the presence of an arc in a directed graph indicates that one vertex is related to another, while

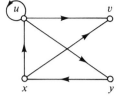

(a)   A simple graph G_1 (b)   A directed graph G_2

Figure 7.1

an edge in a simple graph indicates that two vertices are symmetrically related.

We now formally define the terms *simple graph* and *directed graph*.

DEFINITION 7.1

1. A (*simple*) *graph* G consists of a finite nonempty set V and a set E of two-element subsets of V. The set V is called the *vertex set* of G, E is called the *edge set* of G, and we write $G = (V, E)$ to denote the graph G with vertex set V and edge set E.
2. A *directed graph* or *digraph* G consists of a finite nonempty set V together with a subset A of the Cartesian product $V \times V$. We call V the *vertex set* of G and A the *arc set* of G, and write $G = (V, A)$ to denote the digraph G with vertex set V and arc set A.

For the remainder of this introduction we concern ourselves only with graphs, returning to digraphs in a later section. As a general remark, it can be noted that most of the terminology for graphs carries over to digraphs in a natural way.

Given a graph $G = (V, E)$, each element v of V is called a *vertex* of G (the elements of V are called *vertices*), while each element e of E is called an *edge* of G. If $e = \{u, v\}$ is an edge, we denote it simply by $e = uv$ (or $e = vu$). In this case, u and v are referred to as *adjacent vertices*, u and v are said to be *incident* with e, and e is *incident* with u and v. Similarly, if distinct edges e_1 and e_2 of G have a vertex in common, then e_1 and e_2 are called *adjacent edges*. Also, if several graphs are under discussion at the same time, then to avoid confusion we use $V(G)$ and $E(G)$ to denote the vertex set and edge set, respectively, of a particular graph G.

An alternate term for *vertex* is *node*. Also some authors use the terms *point* and *line* in place of *vertex* and *edge*, respectively.

As illustrated by Figure 7.1, a graph can be represented geometrically or "drawn" (in a plane) if each vertex is made to correspond to a point and an edge $e = uv$ is represented by joining the points corresponding to u and v

by a line segment or simple curve. For a graph $G = (V, E)$, note that the relation of "adjacency" is an irreflexive and symmetric relation on V. It is for this reason that an irreflexive and symmetric relation has a "graphical representation," as introduced in Chapter 4.

Example 7.1   For the graph G_1 of Figure 7.1, $G_1 = (V, E)$, where $V = \{u, v, x, y\}$ and $E = \{uv, ux, uy, vx, xy\}$. Thus the only nonadjacent vertices of G_1 are v and y. The edges uv and vx are adjacent, since both are incident with the vertex v. The edges uv and xy are nonadjacent. Two alternate ways of drawing the graph G_1 are shown in Figure 7.2. There is no unique way of drawing a graph; the relative positions of the points and curves have no special significance.   ∎

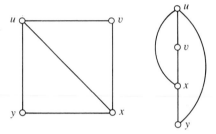

Figure 7.2 Two alternate representations of the graph G_1 of Figure 7.1

Two graphs G_1 and G_2 are called *equal*, denoted $G_1 = G_2$, provided $V(G_1) = V(G_2)$ and $E(G_1) = E(G_2)$. Keeping this in mind, consider the graphs G_1 and G_2 shown in Figure 7.3. It is clear that $G_1 \neq G_2$, since the vertex sets are different. However, suppose G_2 is drawn as in Figure 7.4. Then, except for the labeling of the vertices, the two graphs appear alike.

In fact, one can establish a one-to-one function $\phi: V(G_1) \to V(G_2)$ satisfying the condition

$$xy \in E(G_1) \leftrightarrow \phi(x)\phi(y) \in E(G_2)$$

for all $x, y \in V(G_1)$. In words, two vertices x and y are adjacent in G_1 if

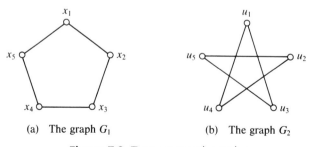

(a)   The graph G_1 (b)   The graph G_2

Figure 7.3 Two unequal graphs

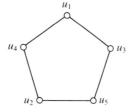

Figure 7.4 Another representation of the graph G_2

and only if their images $\phi(x)$ and $\phi(y)$ are adjacent in G_2. One such function is defined by

$$\phi(x_1) = u_1, \ \phi(x_2) = u_3, \ \phi(x_3) = u_5, \ \phi(x_4) = u_2, \ \phi(x_5) = u_4$$

A mapping such as ϕ establishes a very strong relationship between the graphs G_1 and G_2. Indeed, even though $G_1 \neq G_2$, the existence of ϕ implies that the graphs are "structurally equivalent."

DEFINITION 7.2

Two graphs G_1 and G_2 are called *isomorphic*, denoted $G_1 \simeq G_2$, provided there is a function $\phi: V(G_1) \rightarrow V(G_2)$ satisfying the following two conditions.

1. ϕ is one-to-one and onto.
2. For all $x, y \in V(G_1)$, $xy \in E(G_1) \leftrightarrow \phi(x)\phi(y) \in E(G_2)$.

Such a function ϕ is called an *isomorphism*.

In words, condition 2 of the definition states that the function ϕ *preserves adjacency*.

Example 7.2   Find an isomorphism between the two graphs G_1 and G_2 of Figure 7.5.

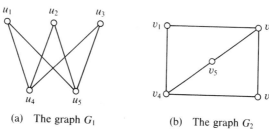

(a)   The graph G_1 (b)   The graph G_2

Figure 7.5 Two isomorphic graphs

Solution   Let $\phi: V(G_1) \to V(G_2)$ be defined by

$$\phi(u_1) = v_1, \ \phi(u_2) = v_3, \ \phi(u_3) = v_5, \ \phi(u_4) = v_2, \ \phi(u_5) = v_4$$

It is readily verified that ϕ is one-to-one and onto. It can also be checked that ϕ preserves adjacency. For example, $u_2u_5 \in E(G_1)$ and $\phi(u_2)\phi(u_5) = v_3v_4 \in E(G_2)$.

The student may have found a different isomorphism; if $U_1 = \{u_1, u_2, u_3\}$, $U_2 = \{u_4, u_5\}$, $V_1 = \{v_1, v_3, v_5\}$, and $V_2 = \{v_2, v_4\}$, then any one-to-one function $f: V(G_1) \to V(G_2)$ with $f(U_1) = V_1$ and $f(U_2) = V_2$ is an isomorphism.   ■

Definition 7.1 states explicitly that the vertex set of a graph is a nonempty finite set. If a graph G has p vertices and q edges, then G is said to have *order* p and *size* q.

Throughout this chapter we have occasion to use some special graphs, which are now identified. A graph in which every two vertices are adjacent is called a *complete graph*. Clearly, any two complete graphs of the same order are isomorphic, so it makes good sense to refer to *the* complete graph of order p; this graph is denoted K_p.

DEFINITION 7.3

A graph $G = (V, E)$ is called a *bipartite graph* if it is possible to partition V as $V_1 \cup V_2$ such that each edge $e \in E$ is incident with a vertex of V_1 and a vertex of V_2. The sets V_1 and V_2 are called *partite sets*, and we write $V = (V_1, V_2)$ to denote the associated partition of V.

The graph shown in Figure 7.6 is a bipartite graph with partite sets $V_1 = \{u_1, u_2, u_3, u_4\}$ and $V_2 = \{v_1, v_2, v_3\}$, and edge set $E = \{u_1v_1, u_2v_2, u_3v_1, u_3v_2, u_4v_2, u_4v_3\}$.

Let G be a bipartite graph with partite sets V_1 and V_2, where $|V_1| = r$ and $|V_2| = s$. If each $u \in V_1$ is adjacent to each $v \in V_2$, then G is called a *complete bipartite graph*. Let G_1 be another complete bipartite graph with $V(G_1) = (X_1, X_2)$, such that $|X_2| = r$ and $|X_2| = s$. Then it is not difficult to see that $G \simeq G_1$, and thus, up to isomorphism, there is a unique

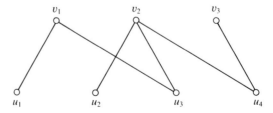

Figure 7.6 A bipartite graph

complete bipartite graph with partite sets of cardinalities r and s. This graph is denoted $K_{r,s}$ (and, by convention, the smaller of r and s is usually listed first). As an example, both of the graphs in Figure 7.5 are isomorphic to $K_{2,3}$.

Example 7.3   Find all nonisomorphic complete bipartite graphs of order 6.

Solution   They are $K_{1,5}$, $K_{2,4}$, and $K_{3,3}$.   ■

Example 7.4   Show that the graph G_1 of Figure 7.3 is not bipartite.

Solution   We proceed by contradiction. Suppose that G_1 is bipartite with partite sets X_1 and X_2. Without loss of generality, assume $x_1 \in X_1$. Since x_2 is adjacent to x_1, it must be that $x_2 \in X_2$. Similarly, we can argue that $x_3 \in X_1$ and $x_4 \in X_2$. But now consider x_5. Since x_5 is adjacent to x_1, we have $x_5 \notin X_1$, and, since x_5 is adjacent to x_4, we have $x_5 \notin X_2$. This contradicts the requirement that (X_1, X_2) be a partition of $V(G_1)$.   ■

In considering a given graph G, we are often interested in determining its structural properties. Of particular importance in this direction is the number of edges incident with a given vertex u. This number is called the *degree* of u and is denoted by $\deg(u)$. If several graphs are under consideration at the same time, then we use $\deg(u, G)$ to denote the degree of u in G. The maximum and minimum vertex degrees in G are denoted by $\Delta(G)$ and $\delta(G)$, respectively. The graph G of Figure 7.6, for example, has $\deg(u_1) = 1 = \delta(G)$, $\deg(v_1) = 2$, and $\deg(v_2) = 3 = \Delta(G)$.

If we sum the degrees of the vertices in a graph, then each edge is counted twice, once for each of its incident vertices. This observation yields the following basic and very useful result.

THEOREM 7.1   Let G be a graph of order p and size q with vertex set $V = \{v_1, v_2, \ldots, v_p\}$. Then

$$\deg(v_1) + \deg(v_2) + \cdots + \deg(v_p) = 2q$$

COROLLARY 7.2   The number of vertices of odd degree in any graph G is even.

Proof   Let $U = \{u_1, u_2, \ldots, u_r\}$ be the set of vertices of odd degree and let $W = \{w_1, w_2, \ldots, w_s\}$ be the set of vertices of even degree. Also let

$$m = \sum_{i=1}^{r} \deg(u_i) \quad \text{and} \quad n = \sum_{j=1}^{s} \deg(w_j)$$

If G has size q, then $m + n = 2q$ by Theorem 7.1. Since $\deg(w_j)$ is even for $1 \leq j \leq s$, we see that n is even. It follows that m must also be even. And since $\deg(u_i)$ is odd for $1 \leq i \leq r$, we have that r is even.   □

A graph G is called *k-regular* (or *regular of degree k*) provided $\deg(u) = k$ for all $u \in V(G)$. For example, the complete graph K_p is $(p - 1)$-regular and the complete bipartite graph $K_{r,r}$ is r-regular. Also notice that G is k-regular if and only if $\Delta(G) = \delta(G) = k$.

DEFINITION 7.4

Given a graph $G = (V, E)$, the *complement* of G is the graph $\overline{G} = (V, \overline{E})$, where

$$\overline{E} = \{ uv \mid uv \notin E \}$$

Notice that if G is a k-regular graph of order p, then its complement $\overline{G}$ is $(p - 1 - k)$-regular. This observation is useful for the next example.

Example 7.5   Give an example of a k-regular graph of order 6 for each k, where $0 \le k \le 5$.

Solution   Since K_6 is 5-regular, its complement $\overline{K}_6$ is 0-regular. Figure 7.7 shows a 1-regular graph G_1 of order 6 and its 4-regular complement $\overline{G}_1$. Since $K_{3,3}$ is a 3-regular graph of order 6, its complement $\overline{K}_{3,3}$ is 2-regular.   ∎

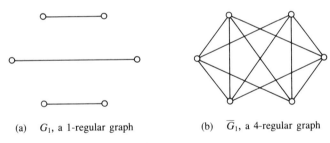

(a) G_1, a 1-regular graph   (b) $\overline{G}_1$, a 4-regular graph

Figure 7.7

Let $G = (V, E)$ be a graph, with $u, v \in V$. A *u-v walk* in G is a finite, alternating sequence of vertices and edges,

$$u, uu_1, u_1, u_1u_2, \ldots, u_{n-1}, u_{n-1}v, v$$

beginning with u and ending with v, such that each edge in the sequence is, as indicated, incident with the vertices that immediately precede and follow it. The *length* of a walk is the number of edges it contains, with repeated edges counted. A *u-v* walk is *closed* if $u = v$ and *open* if $u \neq v$. A *u-v trail* is a *u-v* walk in which no edge is repeated. A *path* is a trail with no repeated vertices, a *circuit* is a closed trail, and a *cycle* is a circuit in which the only repeated vertex is the first vertex, this being the same as the last vertex. A walk is called *trivial* if it contains no edges, that is, if it has length 0.

Note that it is superfluous to include the edges when listing the vertices and edges of a walk, for it is assumed that two consecutive vertices in a walk are adjacent. Thus, we agree to list only the vertices in a walk.

Example 7.6 In the country of Freedonia (a small, tropical island paradise) there are eight cities, a, b, c, d, u, v, x, y, with highways between certain pairs of cities. We can depict a road map of Freedonia as a graph by representing each city as a vertex and joining two vertices with an edge if the corresponding cities are (directly) linked by a highway. Assume that this process yields the graph H shown in Figure 7.8. Find examples of each of the following walks in the graph H:

(a) a u-v walk that is not a trail
(b) a u-v trail that is not a path
(c) a u-v path of length 5
(d) a u-u circuit that is not a cycle
(e) a u-u cycle of length 8

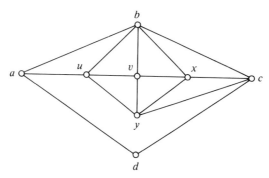

Figure 7.8 The graph of Freedonia

Solution
(a) The walk u, v, x, y, v, b, u, v is not a trail, since the edge uv is repeated. This walk has length 7.
(b) The trail u, b, x, y, c, x, v is not a path, since the vertex x is repeated. This trail has length 6.
(c) The path u, a, d, c, x, v has length 5.
(d) The circuit u, b, v, y, x, v, u is not a cycle, since the vertex v is repeated. This circuit has length 6.
(e) The cycle $u, v, x, y, c, d, a, b, u$ has length 8. ■

Let p be a positive integer and let $V = \{v_1, v_2, \ldots, v_p\}$. Define the graph $G_1 = (V, E_1)$, where

$$E_1 = \{v_i v_{i+1} \mid 1 \le i < p\}$$

and the graph $G_2 = (V, E_2)$, for $p \geq 3$, by

$$E_2 = E_1 \cup \{v_1v_p\}$$

Notice that the edges of G_1 form a path of length $p - 1$ from v_1 to v_p; for this reason, a graph isomorphic to G_1 is called a *path of order p* and is denoted by P_p. The edges of G_2 form a cycle of length p; any graph isomorphic to G_2 is called a *cycle of order p* or *p-cycle* and is denoted by C_p. For instance, the graphs in Figure 7.3 are 5-cycles. Note that removing an edge from C_p results in the graph P_p.

Graphs are often used to model communication or transportation networks. In the graph of a communications network, it is important that there be a way for any vertex to communicate with any other vertex. Similarly, in the graph of a transportation network, it is important that there be a way to travel from any vertex to any other. This leads to the next definition.

DEFINITION 7.5

A graph $G = (V, E)$ is termed *connected* provided, for any $u, v \in V$, there is a u-v walk in G. If G is not connected, then it is termed *disconnected*.

Example 7.7   One can observe that the graph H of Figure 7.8 is connected. Suppose that a hurricane hits Freedonia (one of the drawbacks of living in the tropics), and a resulting flood destroys several bridges, making highways *ab*, *bc*, *bu*, *cx*, *uv*, *vy*, and *xy* impassable. The graph of the resulting highway network H_1 is shown in Figure 7.9. We note that H_1 is disconnected; for example, there is no walk between *a* and *b*. ∎

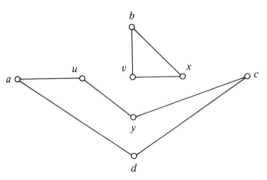

Figure 7.9 Freedonia after the flood

Exercises 7.1

1. Give an example of a graph G_1 of order 6 that has one vertex of degree 1, two vertices of degree 2, and three vertices of degree 3. What size must such a graph have?

2. If a graph G has no adjacent edges, what can be said about the degrees of its vertices?

3. Show the following:

    (a) A graph G of order 5 cannot have two vertices of degree 4 and one vertex of degree 1.

    (b) A graph G of order n cannot have two vertices of degree $n - 1$ and one vertex of degree 1.

4. Show that, in any set of two or more people, there are always two people with the same number of acquaintances in the set.

5. For an arbitrary positive integer n, give an example of a graph that has:

    (a) order n and size $n(n - 1)/2$

    (b) order $2n$ and size n

    (c) order n and size n $(n \geq 3)$

    (d) order $2n$ and size $3n$ $(n \geq 3)$

6. Let G_1 and G_2 be two isomorphic graphs and let $\phi: V(G_1) \to V(G_2)$ be an isomorphism. Prove each of these facts:

    (a) G_1 and G_2 have the same order.

    (b) For each $v \in V(G_1)$, $\deg(\phi(v)) = \deg(v)$.

    (c) G_1 and G_2 have the same size.

7. Each part gives three graphs. You are to (i) draw the graphs, (ii) find the two graphs that are isomorphic, (iii) exhibit an isomorphism, and (iv) give a reason why the third graph is not isomorphic to the other two.

    (a) $G_1 = (\{u_1, u_2, u_3, u_4, u_5, u_6\}, \{u_1u_2, u_1u_5, u_1u_6, u_2u_3, u_2u_5, u_3u_4, u_4u_5\})$
    $G_2 = (\{v_1, v_2, v_3, v_4, v_5, v_6\}, \{v_1v_5, v_1v_6, v_2v_4, v_2v_5, v_2v_6, v_3v_4, v_4v_6\})$
    $G_3 = (\{x_1, x_2, x_3, x_4, x_5, x_6\}, \{x_1x_2, x_1x_5, x_2x_3, x_3x_4, x_3x_6, x_4x_5\})$

    (b) $G_4 = (\{u_1, u_2, u_3, u_4, u_5\},$
    $\{u_1u_2, u_1u_3, u_1u_4, u_1u_5, u_2u_4, u_2u_5, u_3u_4, u_3u_5\})$
    $G_5 = (\{v_1, v_2, v_3, v_4, v_5\},$
    $\{v_1v_2, v_1v_5, v_2v_3, v_2v_4, v_2v_5, v_3v_4, v_3v_5, v_4v_5\})$
    $G_6 = (\{x_1, x_2, x_3, x_4, x_5\},$
    $\{x_1x_2, x_1x_3, x_1x_4, x_1x_5, x_2x_3, x_2x_5, x_3x_4, x_4x_5\})$

    (c) $G_7 = (\{u_1, u_2, u_3, u_4, u_5, u_6\},$
    $\{u_1u_2, u_1u_4, u_1u_5, u_2u_3, u_3u_4, u_3u_6, u_5u_6\})$
    $G_8 = (\{v_1, v_2, v_3, v_4, v_5, v_6\}, \{v_1v_2, v_1v_6, v_2v_3, v_2v_6, v_3v_4, v_3v_5, v_4v_5\})$
    $G_9 = (\{x_1, x_2, x_3, x_4, x_5, x_6\},$
    $\{x_1x_2, x_1x_5, x_2x_3, x_2x_6, x_3x_4, x_4x_5, x_4x_6\})$

8. Let G be a graph with vertex set $V = \{v_1, v_2, \ldots, v_p\}$. The list of numbers $(\deg(v_1), \deg(v_2), \ldots, \deg(v_p))$ is called a *degree sequence* of

G. Each part gives a list of nonnegative integers; give an example of a graph having that degree sequence or argue that no such graph exists.

(a) $(3, 2, 2, 2, 1)$ **(b)** $(3, 2, 2, 2, 1, 1)$ **(c)** $(4, 3, 2, 1, 0)$

(d) $(4, 4, 3, 3, 2, 2)$ **(e)** $(5, 4, 3, 3, 2, 1)$ **(f)** $(5, 5, 5, 4, 4, 3, 2)$

9. Give an example (preferably of the smallest possible order) of nonisomorphic graphs G_1 and G_2 such that the following conditions hold:
   (a) G_1 and G_2 have the same order.
   (b) G_1 and G_2 have the same order and the same size.
   (c) G_1 and G_2 have the same order and the same degree sequence.

10. Show that the relation "is isomorphic to" is an equivalence relation on the set of graphs.

11. Let $G_1 = (\{v_1, v_2, v_3, v_4, v_5, v_6, v_7\},$
    $\{v_1v_2, v_1v_4, v_2v_3, v_2v_5, v_3v_4, v_3v_6, v_4v_7, v_5v_6, v_6v_7\})$ and
    $G_2 = (\{v_1, v_2, v_3, v_4, v_5, v_6\}, \{v_1v_2, v_1v_4, v_1v_6, v_2v_3, v_2v_6, v_3v_4, v_4v_5, v_5v_6\})$.
    (a) Draw the graph G_1.   **(b)** Show that G_1 is bipartite.
    (c) Draw the graph G_2.   **(d)** Show that G_2 is not bipartite.

12. Given a graph G of order p, size q, minimum degree δ, and maximum degree Δ, prove that $p\delta \leq 2q \leq p\Delta$.

13. Let G be a k-regular graph of order p and size q.
    (a) Give an identity involving k, p, and q.
    (b) Show that k or p is even.

14. Let G be a bipartite graph of order p and size q.
    (a) Prove that $4q \leq p^2$.
    (b) If $4q = p^2$, to what graph is G isomorphic?

15. Let G be a k-regular bipartite graph with partite sets V_1 and V_2. Prove that $|V_1| = |V_2| \geq k$.

16. Give an example of each:
    (a) a 1-regular bipartite graph of order 4
    (b) a 2-regular bipartite graph of order 6
    (c) a 3-regular bipartite graph of order 8
    (d) an r-regular bipartite graph of order $2r + 2$, for an arbitrary $r \in \mathbb{N}$

17. Given that G is a k-regular graph of order p and size q, what can be said about G under these conditions?
    (a) $k = 3$ and $q = p + 3$ **(b)** $k = 4$ and $q = 3p - 5$

18. Prove: If G_1 and G_2 are isomorphic graphs and G_1 is bipartite, then G_2 is bipartite.

19. Determine, up to isomorphism, all graphs of order 5.

20. The intramural hockey league at a small college has seven teams: t, u, v, w, x, y, z. Each team is scheduled to play four games over a five-week season against four different opponents. During each of the first four weeks, six of the teams play a game while the remaining team has a bye. Then, for the last week, the three teams that have already played four games have a bye, while the remaining four teams play.
    (a) Design a schedule for the hockey league, using a graph to indicate which teams each team plays.

(b) Suppose the season is extended to six weeks. Is it possible for each of the seven teams to play five games against five different opponents?

21. Find $\overline{G}_1$, $\overline{G}_4$, and $\overline{G}_7$ for the graphs G_1, G_4, and G_7 of Exercise 7.

22. Let G be a graph of order p, size q, minimum degree δ, and maximum degree Δ.
    (a) Determine the size of $\overline{G}$.
    (b) If d is the degree of vertex v in G, determine the degree of v in $\overline{G}$.
    (c) Determine the minimum degree of $\overline{G}$.
    (d) Determine the maximum degree of $\overline{G}$.

23. Let $G = (\{t, u, v, w, x, y, z\},$
    $\{tu, tv, tw, ux, vw, vy, vz, wx, wz, xy, yz\})$. Give examples of the following walks in G:
    (a) a u-v walk that is not a trail
    (b) a u-v trail that is not a path
    (c) a u-v path of minimum length
    (d) a u-v path of maximum length
    (e) a u-u circuit that is not a cycle
    (f) a u-u cycle of minimum length
    (g) a u-u cycle of maximum length

24. Draw each of these graphs.
    (a) K_6 **(b)** $K_{3,5}$ **(c)** P_7 **(d)** C_7

25. Show that not both a graph G and its complement $\overline{G}$ can be disconnected.

26. Prove: If G_1 and G_2 are isomorphic graphs and G_1 is connected, then G_2 is connected.

7.2   EULERIAN GRAPHS

In this section we introduce the Königsberg bridges problem, which was mentioned at the outset of the chapter. Figure 7.10 shows a map of the town of Königsberg as it appeared in 1736. The river Pregel ran through the town and was spanned by seven bridges, which connected two islands in the river with each other and with the opposite banks.

The townsfolk had long amused themselves with the following problem: Is it possible to start at some point in the town and take a walk, crossing

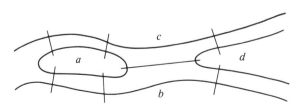

Figure 7.10 The bridges of Königsberg

each bridge exactly once and returning to the starting point? We can represent the situation as a kind of graph by representing each land mass as a vertex and joining two vertices with a number of edges equal to the number of bridges that join the corresponding land masses. The resulting "graph" is shown in Figure 7.11(a).

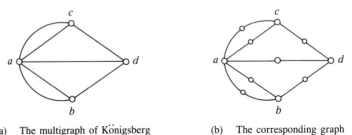

(a)   The multigraph of Königsberg   (b)   The corresponding graph

Figure 7.11

Of course, what we have in Figure 7.11 is not really a simple graph because some pairs of vertices are joined by more than one edge; such a graph is called a *multigraph*. However, one can obtain a graph from a multigraph by inserting a new vertex (of degree 2) into every edge of the multigraph. For example, the graph obtained from the multigraph that represents Königsberg is shown in Figure 7.11(b).

Note that, as far as the problem is concerned, the insertion of these extra vertices of degree 2 has no effect. This allows us to state the problem in terms of simple graphs. Let $G = (V, E)$ be a graph, with $u, v \in V$. Recall that a u-v trail is a u-v walk in which no edge is repeated, and that a circuit is a closed trail. With this terminology, we can state the Königsberg bridges problem in graphical terms:

Is there a circuit in the graph of Königsberg that includes every edge?

In general, a circuit in a graph G that includes every edge is called an *eulerian circuit*, and an open trail with this property is called an *eulerian trail*. Thus we ask the more general question:

When does a graph possess an eulerian circuit or trail?

Example 7.8   Recall the graph H of Freedonia from Example 7.6; it is repeated for convenience in Figure 7.12. The Freedonia Highway Patrol might be interested in finding an eulerian circuit or trail for this graph because a patrol car following such a trail would be able to cover each highway exactly once. As we will see shortly, the graph H does not contain an eulerian circuit; however, it does contain an eulerian trail, for instance,

$$a, b, c, d, a, u, b, x, c, y, u, v, x, y, v, b$$ ∎

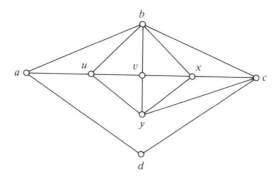

Figure 7.12

Let G be a graph, let v be a vertex of G, and suppose that G contains an eulerian circuit. Then each occurrence of v on the circuit contributes 2 to the degree of v, because the edges of the circuit that immediately precede and follow v are distinct and both are incident with v. Thus v has even degree in G and we have the following lemma.

LEMMA 7.3 Let G be a graph. If G contains an eulerian circuit, then every vertex of G has even degree.

In other words, the condition that the degree of every vertex is even is a necessary condition for a graph to possess an eulerian circuit. It is also sufficient, provided the graph is connected. In fact, if G contains an eulerian circuit C and G has no vertices of degree 0, then G is certainly connected. This is because, given any two distinct vertices x and y, they can be located on C, and then there is a part of C that is an x-y trail.

A connected graph that contains an eulerian circuit is called an *eulerian graph*. Thus we wish to prove that a connected graph G is eulerian if and only if every vertex of G has even degree. To do this, it is useful to have the following lemma.

LEMMA 7.4 Let G be a graph such that every vertex of G has even degree. If u and v are adjacent vertices of G, then there is a circuit of G that contains the edge uv.

Proof Consider the set S of all trails of G that begin

$$u, v, \ldots$$

Clearly S is nonempty. Also, since $E(G)$ is finite, so is S. Consider the set of lengths of trails in S; since this is a nonempty finite set of positive integers, it has a maximum element. Thus let W be a trail in S of maximum length, say

$$u, v, \ldots, x$$

We claim that W is a circuit, namely, that $x = u$. If not, then note that the number of edges of W incident with x is odd because each occurrence of x on W, except for the last occurrence, accounts for two edges of W incident with x, whereas the last occurrence of x accounts for one edge of W incident with it. However, x has even degree in G. Hence there is an edge xy of G that is not an edge of W. But then W can be extended (by one edge and one vertex) to the trail

$$u, v, \ldots, x, y$$

which has length 1 greater than that of W. This contradicts the choice of W as a trail in S of maximum length, thereby proving the result. □

THEOREM 7.5   Let G be a connected graph. Then G is eulerian if and only if every vertex of G has even degree.

Proof   The result is trivial if G is isomorphic to K_1 or K_2, so we may assume that G has at least three vertices. Moreover, in view of Lemma 7.3, it suffices to prove sufficiency. So let $G = (V, E)$ be a connected graph such that every vertex of G has (positive) even degree. We wish to prove that G contains an eulerian circuit.

Consider the set S of all nontrivial circuits of G; it follows from Lemma 7.4 that S is a nonempty finite set. Thus, as in the proof of Lemma 7.4, it makes sense to let W be an element of S of maximum length. We claim that W is an eulerian circuit of G. If not, then there is an edge uv of G that is not in W. Since G is connected, we may assume, without loss of generality, that u is on W. (If u is not on W, let w be a vertex on W. Since G is connected, there is a u-w walk in G; let $v_1 u_1$ be the first edge of this walk such that v_1 is not on W and u_1 is on W. Then replace u by u_1 and v by v_1 in the argument.) Now let $G_1 = (V, E_1)$ be the graph obtained from G by removing the edges of W; that is,

$$E_1 = E - \{\text{edges of } W\}$$

Then every vertex of G_1 has even degree, so by Lemma 7.4 there is a circuit W_1 of G_1 that includes the edge uv. It is then possible to describe a u-u circuit W' of G that extends W by "appending" W_1 to W at u; namely, proceed from u to u by following (all) the edges of W, and then proceed from u to u by following the edges of W_1. But W' has length greater than that of W, contradicting our choice of W. This proves the result. □

The preceding proof suggests an algorithm for finding an eulerian circuit in an eulerain graph. The formal description of such an algorithm is left as an exercise in a later section when, in fact, it can be considered for the case of directed graphs. For now, we present an example that illustrates the main idea of the algorithm.

Example 7.9   We consider the graph H of Figure 7.12. Since vertices a and b have odd degree, the graph H is not eulerian. However, let us add to H a new vertex

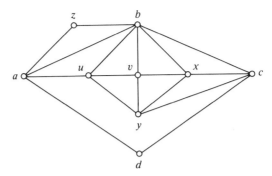

Figure 7.13 An eulerian graph

z, together with the edges az and bz. The resulting graph, call it G, is shown in Figure 7.13. Since G is connected and each vertex has even degree, G is eulerian, and we wish to construct an eulerian circuit. Choose a starting vertex, say a, and construct a circuit that includes every edge incident with that vertex. Then suppose we obtain the following circuit

$$W: a, u, v, x, c, d, a, b, z, a$$

Next look for a vertex of W that is incident with an edge of G not in W. If no such vertex were to exist, then W would be an eulerian circuit. However, in our case we find that vertex u is incident with edges of G that are not edges of W. We may then, as in the proof of Theorem 7.5, find a u-u circuit W_1 of G that has no edges in common with W. (In fact, we may construct W_1 so that, together, W and W_1 include every edge of G incident with u.) For example, suppose W_1 is the circuit

$$u, b, x, y, u$$

We may then replace W by the larger circuit obtained by appending W_1 to W at u; this yields the circuit

$$u, v, x, c, d, a, b, z, a, u, b, x, y, u$$

We next repeat the step of searching W for a vertex that is incident with an edge of G not in W. Note that v is such a vertex and that

$$W_2: v, b, c, y, v$$

is a v-v circuit of G having no edges in common with W. Again we replace W by the larger circuit obtained by appending W_2 to W at v, yielding the circuit

$$v, x, c, d, a, b, z, a, u, b, x, y, u, v, b, c, y, v$$

This time we find that no vertex of W is incident to an edge of G not in W, so that W is an eulerian circuit.

Note that if we delete from the eulerian circuit W the edges az and bz and the vertex z, we obtain an a-b eulerian trail

$$a, u, b, x, y, u, v, b, c, y, v, x, c, d, a, b$$

of the graph H representing the Freedonia highway network. This is a different eulerian trail than the one found in Example 7.8.   ■

Exercises 7.2

1. For each of these graphs, (i) draw it, (ii) determine whether it is eulerian, (iii) if it is, find an eulerian circuit, preferably using the algorithm suggested by Example 7.9.
   (a) $G_1 = (\{u_1, u_2, u_3, u_4, u_5\},$
   $\{u_1u_2, u_1u_3, u_1u_4, u_1u_5, u_2u_4, u_2u_5, u_3u_4, u_3u_5\})$
   (b) $G_2 = (\{v_1, v_2, v_3, v_4, v_5\}, \{v_1v_2, v_1v_5, v_2v_3, v_2v_4, v_2v_5, v_3v_4, v_3v_5, v_4v_5\})$
   (c) $G_3 = (\{u_1, u_2, u_3, u_4, u_5, u_6\},$
   $\{u_1u_2, u_1u_3, u_1u_4, u_1u_5, u_2u_3, u_3u_4, u_3u_6, u_5u_6\})$
   (d) $G_4 = (\{v_1, v_2, v_3, v_4, v_5, v_6\}, \{v_1v_2, v_1v_6, v_2v_3, v_2v_6, v_3v_4, v_3v_5, v_4v_5\})$
   (e) $G_5 = (\{v_1, v_2, v_3, v_4, v_5, v_6, v_7\},$
   $\{v_1v_2, v_1v_4, v_2v_3, v_3v_4, v_5v_6, v_5v_7, v_6v_7\})$
   (f) $G_6 = (\{v_1 v_2, v_3, v_4, v_5, v_6\},$
   $\{v_1v_2, v_1v_3, v_1v_4, v_1v_6, v_2v_3, v_2v_4, v_2v_6, v_3v_4, v_3v_6, v_4v_5, v_5v_6\})$
2. Let G be a connected graph of order at least 2. Prove: G contains an eulerian trail if and only if G has exactly two vertices of odd degree.
3. Determine which of the graphs of Exercise 1 contain an eulerian trail.
4. **(a)** Show that the graph of Königsberg (Figure 7.11(b)) does not contain an eulerian trail or an eulerian circuit.
   (b) In present-day Königsberg (Kaliningrad), there are two additional bridges joining regions b and c and regions b and d. Is it possible now to take a walk through the town that crosses each bridge exactly once? If so, devise such a walk.
5. Prove: If G_1 and G_2 are isomorphic graphs and G_1 is eulerian, then G_2 is eulerian.
6. Give an example of a multigraph G' of order 5 that has two vertices of degree 4 and one vertex of degree 1. Compare with Exercise 3 of Exercises 7.1. Is the result of Corollary 7.2 true for multigraphs?

7.3

DIRECTED GRAPHS

Consider the problem of traffic flow in a large city, where some streets are one-way and others are two-way. As part of modelling such a situation, a directed graph may be constructed as follows. Associate a vertex with each intersection of streets, and then place an arc from vertex a to vertex b if it is possible to travel directly from a to b without passing through a third intersection. Note that it is possible for both the arc from a to b and the arc from b to a to be included. (When does this happen?)

In this section we focus on some of the basic terminology associated with directed graphs. For convenience, we repeat the definition of a directed graph.

DEFINITION 7.6

A *directed graph* (or *digraph*) G consists of a finite nonempty set V together with a subset A of the Cartesian product $V \times V$. We call V the *vertex set* of G and A the *arc set* of G, and we write $G = (V, A)$.

Given a digraph $G = (V, A)$, the elements of V are called *vertices* (the plural of *vertex*) and the elements of A are called *arcs* or *directed edges*. An arc of the form (u, u), where u is a vertex, is called a *loop*. Also, for distinct vertices u and v, the arcs (u, v) and (v, u) are called *symmetric arcs*.

Example 7.10 A representation, or "drawing," of the digraph $G = (V, A)$, with $V = \{u, v, x, y, z\}$ and $A = \{(v, u), (v, x), (x, y), (y, x), (y, u), (z, u), (z, z)\}$ is shown in Figure 7.14. This digraph has a loop, (z, z), and one pair of symmetric arcs, (x, y) and (y, x). ∎

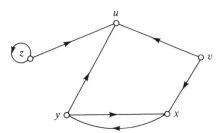

Figure 7.14 The digraph G

From the definition, notice that a digraph $G = (V, A)$ determines a relation on the set V, namely, A. It is for this reason that a relation on a finite set has a "digraph representation," as discussed in Chapter 4. If the relation A is irreflexive, then G has no loops and is called *loopless*. Moreover, if A is symmetric, then $(v, u) \in A$ whenever $(u, v) \in A$, and we may as well replace the symmetric pair of arcs (u, v) and (v, u) by the simple edge uv. Doing this for each pair of symmetric arcs results in a simple graph.

Suppose now that $G = (V, A)$ is a digraph and $(x, y) \in A$. We then say that x *is adjacent to* y, that y *is adjacent from* x, and that both x and y are *incident* with the arc (x, y). If both x is not adjacent to y and x is not adjacent from y, then x and y are called *nonadjacent* vertices.

For vertices x and y, an x-y *walk* in G is a finite alternating sequence W of vertices and arcs of the form:

$$ x = x_0, (x_0, x_1), x_1, (x_1, x_2), x_2, \ldots, (x_{n-1}, x_n), x_n = y $$

beginning with x and ending with y. We call n, the number of arcs in W, the *length* of W. Note that each arc in W joins the vertex immediately preceding it to the vertex immediately following it. For this reason, we may conveniently exhibit a walk by listing only its vertices. An x-y walk is *open* or *closed* according to whether $x \neq y$ or $x = y$, respectively. A walk with no repeated arcs is called a *trail*, and a trail with no repeated vertices is called a *path*. A *circuit* is a closed trail, and a *cycle* is a circuit in which the only repeated vertex is the first vertex, this being the same as the last vertex.

Example 7.11   For the digraph G_1 shown in Figure 7.15, find:

(a) a u-v walk that is not a trail
(b) a u-v trail that is not a path
(c) a u-v path of maximum length
(d) a v-v circuit that is not a cycle
(e) a v-v cycle of maximum length

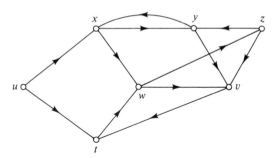

Figure 7.15 The digraph G_1

Solution   (a) The u-v walk u, x, w, z, y, x, w, v is not a trail since the arc (x, w) is repeated. The length of this walk is 7.
(b) The u-v trail u, t, w, z, y, x, w, v is not a path since the vertex w is repeated. The length of this trail is 7.
(c) The u-v path u, t, w, z, y, v has length 5; there are several other u-v paths having length 5, but no u-v paths of length 6, so 5 is the maximum length of a u-v path.
(d) The v-v circuit v, t, w, z, y, x, w, v is not a cycle since the vertex w is repeated.
(e) The v-v cycle v, t, w, z, y, v, with length 5, is a v-v cycle of maximum length. ∎

> **DEFINITION 7.7**
>
> Let $G = (V, A)$ be a digraph and let $x, y \in V$. If G contains an x-y walk, then we say that y is *reachable* from x. In this case, the *distance* from x to y in G is defined as the minimum length among all x-y walks in G and is denoted by $d(x, y, G)$, or simply $d(x, y)$ if the digraph under consideration is understood. If y is not reachable from x, then we write $d(x, y) = \infty$. If G contains an x-y walk for every ordered pair of vertices x and y, then G is called *strongly connected* or *strong*.

Example 7.12   Consider the digraph $G_1 = (V_1, A_1)$ of Figure 7.15.

(a) Find $d(u, a)$ for all $a \in V_1$.
(b) Find $d(a, z)$ for all $a \in V_1$.
(c) Show that G_1 is not strong.
(d) Let $G_2 = (V_1, A_2)$, where $A_2 = A_1 \cup \{(z, u)\}$. Show that G_2 is strong.

Solution

(a) We find that $d(u, u) = 0$, $d(u, t) = d(u, x) = 1$, $d(u, w) = d(u, y) = 2$, and $d(u, v) = d(u, z) = 3$.
(b) We find that $d(u, z) = d(v, z) = d(y, z) = 3$, $d(x, z) = d(t, z) = 2$, $d(w, z) = 1$, and $d(z, z) = 0$.
(c) Note that u is not reachable from any other vertex of G_1, showing that G_1 is not strong.
(d) In G_1 every vertex is reachable from u and every vertex can reach z, so this same property holds in G_2. Thus, to show that G_2 is strong, it suffices to observe that u is reachable from z. (Note that the relation "is reachable from" on the vertex set of a digraph is reflexive and transitive, but is not necessarily symmetric.) ∎

Next we state two basic but important results concerning reachability, paths, and distance in directed graphs.

THEOREM 7.6   Let $G = (V, A)$ be a digraph and let $x, y \in V$. If y is reachable from x, then G contains an x-y path.

Proof   Let $x, y \in V$ and assume y is reachable from x. The result of the theorem is trivial if $x = y$, so we assume $x \neq y$. Let S be the set of x-y walks in G; note that S is nonempty. Considering the lengths of the walks in S as a nonempty set of positive integers, we may apply the principle of well-ordering to obtain a walk W in S of minimum length. We claim that W is an x-y path. If not, then some vertex of W is repeated. If y is repeated, then W has the form

$$x, \ldots, y, z, \ldots, y$$

and that part of W from x to the first occurrence of y is an x-y walk whose length is less than that of W. This contradicts our choice of W. So we may assume that W has the form

$$x = u_0, u_1, \ldots, u_i, u_{i+1}, \ldots, u_j, u_{j+1}, \ldots, u_n = y$$

where $u_i = u_j$, $0 \leq i < j < n$. Then the walk

$$x = u_0, u_1, \ldots, u_i, u_{j+1}, \ldots, u_n = y$$

is an x-y walk whose length is less than that of W, contradicting our choice of W. This completes the proof.   □

Let x and y be vertices of a digraph G such that y is reachable from x. As a consequence of Theorem 7.6, $d(x, y)$ may be computed as the minimum length among all x-y paths in G.

THEOREM 7.7   Let $G = (V, A)$ be a digraph and let $x, y, z \in V$. If y is reachable from x and z is reachable from y, then z is reachable from x and

$$d(x, z) \leq d(x, y) + d(y, z)$$

Proof   Let $x, y, z \in V$ and assume y is reachable from x and z is reachable from y. Let P_1 be an x-y path of length $d(x, y)$ and P_2 be a y-z path of length $d(y, z)$. Define the x-z walk W by appending P_2 to P_1, that is, proceed from x to y along P_1 and then from y to z along P_2. Then the length of W is $d(x, y) + d(y, z)$. Hence, since $d(x, z)$ is defined as the minimum length among all x-z walks, we have

$$d(x, z) \leq d(x, y) + d(y, z)$$ □

What about other properties of the distance function? Let $G = (V, A)$ be a digraph with $x, y \in V$ and y reachable from x. It is clear that $d(x, y) \geq 0$. However, as illustrated by the digraph G_1 of Figure 7.15, the distance function is not symmetric in general. Notice in this digraph that $d(x, w) = 1$, whereas $d(w, x) = 3$.

Example 7.13   Show that the digraph $G_2 = (V, A)$ of Figure 7.16 is not strong, but that it is possible to obtain a strong digraph G_3 by reversing the direction of exactly one arc of G_2.

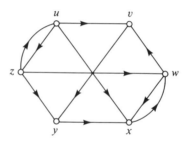

Figure 7.16 The digraph G_2

Solution   Let $U = \{u, z\}$. Note that each arc incident with a vertex of U and a vertex of $V - U$ is directed from U to $V - U$. Hence, neither of the two vertices in U is reachable from any of the four vertices in $V - U$, so that G_2 is not strong. Observe that, to obtain a strong digraph G_3, it suffices to reverse the direction of any one of the arcs (u, v), (u, x), (z, w), (z, y). For example, let $G_3 = (V, A_3)$, where

$$A_3 = \left(A - \{(u, v)\} \right) \cup \{(v, u)\}$$

We claim that G_3 is strong. An easy way to see this is to observe that

$$W: u, x, w, v, y, x, w, v, u, z, u$$

is a closed walk of G_3 that includes every vertex. Thus, given any two vertices a and b of G_3, we may locate a and b on W, and then that portion of W from a to b is an a-b walk of G_3.   ∎

Taking our lead from the preceding example, we present the following useful characterization of strongly connected directed graphs.

THEOREM 7.8   A digraph G is strongly connected if and only if G has a closed walk that includes every vertex.

Proof   Assume G is a strong digraph and let W be a closed walk of G that includes a maximum number of distinct vertices. We claim that W includes every vertex of G. If not, suppose that y is a vertex of G that is not on W. Let x be a vertex of G on W; then we may consider W as an x-x walk. Since G is strong, it contains an x-y walk W_1 and a y-x walk W_2. But then we may define an x-x walk W' by following W from x to x, then W_1 from x to y, and then W_2 from y to x, and W' includes more distinct vertices of G than does W. This contradicts our choice of W and thus proves necessity.

For sufficiency, let W be a closed walk of G that includes every vertex. Given arbitrary vertices u and v of G, we may locate u and v on W, and then that part of W from u to v is a u-v walk of G. This shows that G is strong.   □

Recall that, given a simple graph G and a vertex v of G, the degree of v is the number of vertices of G with which v is adjacent. Given a digraph $G = (V, A)$ and $v \in V$, then possibly there are vertices of G adjacent to v and vertices adjacent from v. Thus it makes sense to assign to v two different degree values. The *indegree* of v, denoted id(v), is the number of vertices of G adjacent to v, while the *outdegree* of v, denoted od(v), is the number of vertices of G adjacent from v. For example, the vertices of the digraph G_1 of Figure 7.15 have the following indegrees and outdegrees:

$$\text{id}(t) = 2, \text{id}(u) = 0, \text{id}(v) = 3, \text{id}(w) = \text{id}(x) = \text{id}(y) = 2, \text{id}(z) = 1$$

$$\text{od}(t) = 1, \text{od}(u) = 2, \text{od}(v) = 1, \text{od}(w) = \text{od}(x) = \text{od}(y) = \text{od}(z) = 2$$

Just as with graphs, the number of vertices and the number of arcs in a digraph are referred to as its *order* and *size*, respectively. The last result of this section gives the digraph analogue of Theorem 7.1; its proof is left to Exercise 10.

THEOREM 7.9 Let $G = (V, A)$ be a directed graph of order p and size q and let $V = \{v_1, v_2, \ldots, v_p\}$. Then

$$\sum_{i=1}^{p} \mathrm{id}(v_i) = q = \sum_{i=1}^{p} \mathrm{od}(v_i)$$

Exercises 7.3

1. For the digraph G_3 of Example 7.13, exhibit each of the following:
   (a) a cycle of length 4
   (b) a path of length 5 whose first vertex is adjacent to its last vertex
   (c) a trail of maximum length
   (d) a walk that includes every arc
2. Find all distances in the digraph $G_3 = (V, A)$, where $V = \{u, v, x, y, z\}$ and $A = \{(u, v), (v, z), (x, v), (y, x), (y, y), (z, u), (z, y)\}$.
3. Find all distances in the digraph G_3 of Example 7.13.
4. Give an example of a strong digraph G that does not have a circuit that includes every vertex.
5. Find the indegree and outdegree of each vertex in the digraph G_2 of Example 7.13.
6. Prove or disprove: In any digraph, the number of vertices with odd outdegree is even. (Compare with Corollary 7.2.)
7. Give an example of a digraph $G = (V, A)$ (of order p), with $V = \{v_1, v_2, \ldots, v_p\}$, such that $\mathrm{od}(v_i) = i - 1$, $1 \leq i \leq p$. (Compare with the result of Exercise 4 in Exercises 7.1.)
8. Prove or disprove: If G is a strong digraph, then there is a closed walk of G that includes every arc of G.
9. A digraph $G = (V, A)$ is called *r-regular* provided $\mathrm{id}(v) = \mathrm{od}(v) = r$ for all $v \in V$.
   (a) Given an example of a 2-regular digraph of order 5.
   (b) Prove: If r and p are integers with $0 \leq r < p$, then there exists an r-regular digraph of order p.
10. Prove Theorem 7.9
11. Given a digraph $G = (V, A)$, the *complement* of G is the digraph $\overline{G} = (V, \overline{A})$, where $\overline{A} = (V \times V) - A$. Find the complement of:
    (a) the digraph G of Figure 7.14
    (b) the digraph G_2 of Figure 7.16
12. Let G be a digraph of order p. Prove: If $\mathrm{od}(u) + \mathrm{id}(v) \geq p - 1$ whenever u and v are distinct vertices and u is not adjacent to v, then G is strong. Also show that this result is the best possible in the sense that the bound $p - 1$ cannot be lowered to $p - 2$.

13. Let $G_1 = (V_1, A_1)$ and $G_2 = (V_2, A_2)$ be two digraphs. We say that G_1 and G_2 is *isomorphic* to G_2 provided there is a bijection $\phi\colon V_1 \to V_2$ such that, for all x, $y \in V_1$, $(x, y) \in A_1 \leftrightarrow (\phi(x), \phi(y)) \in A_2$. Such a function ϕ is called an *isomorphism*.

(a) Show that G_1 is isomorphic to G_2, where $G_1 = (\{a, b, c, d, e\}$, $\{(a,b), (a, d), (a, e), (b, c), (c, d), (c, e), (d, e), (e, b)\})$ and $G_2 = (\{u, v, x, y, z\}$, $\{(u, v), (v, x), (x, u), (x, z), (y, u), (y, v), (y, z), (z, u)\})$.

(b) Determine whether the digraph $G_3 = (\{a, b, c, d, e\}, \{(a, a), (a, e), (b, c), (c, b), (d, c), (d, e), (e, b)\})$ is isomorphic to the digraph G of Figure 7.14.

14. Let digraph $G_1 = (V_1, A_1)$ be isomorphic to digraph $G_2 = (V_2, A_2)$, and let $\phi\colon V_1 \to V_2$ be an isomorphism. Prove:

(a) G_1 and G_2 have the same order.

(b) For each $v \in V_1$, $\mathrm{id}(\phi(v)) = \mathrm{id}(v)$.

(c) For each $v \in V_1$, $\mathrm{od}(\phi(v)) = \mathrm{od}(v)$.

(d) G_1 and G_2 have the same size.

(e) If $v_0, v_1, \ldots, v_n$ is a walk (trail, path, circuit, cycle) in G_1, then $\phi(v_0), \phi(v_1), \ldots, \phi(v_n)$ is a walk (trail, path, circuit, cycle) in G_2.

15. Give an example (preferably of smallest possible order) of nonisomorphic digraphs $G_1 = (V_1, A_1)$ and $G_2 = (V_2, A_2)$ such that:

(a) G_1 and G_2 have the same order.

(b) G_1 and G_2 have the same order and the same size.

(c) G_1 and G_2 have the same order, p, $V_1 = \{u_1, u_2, \ldots, u_p\}$, $V_2 = \{v_1, v_2, \ldots, v_p\}$, $\mathrm{id}(u_i) = \mathrm{id}(v_i)$ and $\mathrm{od}(u_i) = \mathrm{od}(v_i)$ for each i, $1 \le i \le p$.

16. Show that the relation "is isomorphic to" is an equivalence relation on the set of directed graphs.

17. Let $G = (V, A)$ be a loopless digraph such that, for some $k \in \mathbb{N}$, $\mathrm{od}(v) \ge k$ for all $v \in V$. Prove that G contains:

(a) a path whose length is at least k

(b) a cycle whose length is at least $k + 1$

18. Let G be a digraph. An *eulerian circuit* of G is a circuit of G that includes every arc of G. A strongly connected digraph that contains an eulerian circuit is called *eulerian*. Let $G = (V, A)$ be a strong digraph. Prove that G is eulerian if and only if $\mathrm{id}(v) = \mathrm{od}(v)$ for all $v \in V$.

19. Develop an algorithm that inputs a digraph $G = (V, A)$, determines whether G is eulerian, and, if so, outputs an eulerian circuit of G.

20. Illustrate the algorithm developed in Exercise 19 for each of the following digraphs.

(a) $G_1 = (\{t, u, v, w, x, y, z\}, \{(t, u), (t, w), (u, t), (u, x), (v, t), (v, u), (v, z), (w, v), (w, z), (x, w), (x, y), (y, v), (y, x), (z, v), (z, y)\})$

(b) $G_2 = (\{u, v, w, x, y, z\}, \{(u, v), (u, z), (v, x), (v, y), (w, v), (w, x), (x, w), (x, y), (y, u), (y, z), (z, u), (z, w)\})$

21. Determine, up to isomorphism, all loopless directed graphs of order 3.

22. Let $G_1 = (V_1, A_1)$ be the digraph of Figure 7.15, and let $G_1' = (V_1, A_1')$, where $A_1' = A_1 \cup \{(t, z), (v, u), (v, z), (z, u)\}$. Exhibit an eulerian circuit of G_1'.

23. Given a digraph $G = (V, A)$, the *converse* of G is the digraph $G^c = (V, A^c)$, where $A^c = \{(v, u) \mid (u, v) \in A\}$ (see Chapter 4 Problem 20). Find the converse of each:
    (a) the digraph G of Figure 7.14
    (b) the digraph G_2 of Figure 7.16

24. Let $G = (V, A)$ be a digraph such that, for some $k \in \mathbb{N}$, $\mathrm{id}(v) \geq k$ for all $v \in V$. Use the result of Exercise 17 and the idea of the converse (Exercise 23) to show that G contains:
    (a) a path whose length is at least k
    (b) a cycle whose length is at least $k + 1$

7.4

SUBDIGRAPHS, REACHABILITY, AND DISTANCE

Given a digraph $G = (V, A)$ and $u \in V$, we may be interested in knowing which vertices are reachable from u, or in finding $d(u, v)$ for all $v \in V$. More generally, we may be interested in knowing for which pairs of vertices (u, v) it is true that v is reachable from u, or in finding $d(u, v)$ for all $u, v \in V$. In this section we consider two algorithms for answering questions of this sort, namely, Dijkstra's algorithm and Warshall's algorithm. It is convenient to consider these algorithms in terms of various matrices that can be associated with digraphs.

DEFINITION 7.8

Let $G = (V, A)$ be a digraph of order p with $V = \{v_1, v_2, \ldots, v_p\}$. The *adjacency matrix* of G is the $p \times p$, $(0, 1)$-matrix $M = [m_{ij}]$ defined by

$$m_{ij} = \begin{cases} 0 & \text{if } (v_i, v_j) \notin A \\ 1 & \text{if } (v_i, v_j) \in A \end{cases}$$

(Note: A matrix, each of whose entries is 0 or 1, is called a $(0, 1)$-*matrix*.)

Example 7.14   Let $G = (V, A)$, where $V = \{v_1, v_2, v_3, v_4, v_5\}$ and $A = \{(v_1, v_4), (v_2, v_1), (v_2, v_3), (v_3, v_4), (v_4, v_3), (v_5, v_1), (v_5, v_5)\}$. The adjacency matrix of G is

$$M = \begin{bmatrix} 0 & 0 & 0 & 1 & 0 \\ 1 & 0 & 1 & 0 & 0 \\ 0 & 0 & 0 & 1 & 0 \\ 0 & 0 & 1 & 0 & 0 \\ 1 & 0 & 0 & 0 & 1 \end{bmatrix}$$

∎

> **DEFINITION 7.9**
>
> Let $G = (V, A)$ be a digraph of order p with $V = \{v_1, v_2, \ldots, v_p\}$. The *reachability matrix* of G is the $p \times p$, $(0, 1)$-matrix $R = [r_{ij}]$ defined by
>
> $$r_{ij} = \begin{cases} 0 & \text{if} \quad v_j \text{ is not reachable from } v_i \\ 1 & \text{if} \quad v_j \text{ is reachable from } v_i \end{cases}$$
>
> The *distance matrix* of G is the $p \times p$, $(0, 1)$-matrix $D = [d_{ij}]$, where $d_{ij} = d(v_i, v_j)$. (Recall that $d_{ij} = \infty$ if v_j is not reachable from v_i.)

Example 7.15   For the digraph G of Example 7.14, its reachability matrix R and distance matrix D are given as follows:

$$R = \begin{bmatrix} 1 & 0 & 1 & 1 & 0 \\ 1 & 1 & 1 & 1 & 0 \\ 0 & 0 & 1 & 1 & 0 \\ 0 & 0 & 1 & 1 & 0 \\ 1 & 0 & 1 & 1 & 1 \end{bmatrix}$$

$$D = \begin{bmatrix} 0 & \infty & 2 & 1 & \infty \\ 1 & 0 & 1 & 2 & \infty \\ \infty & \infty & 0 & 1 & \infty \\ \infty & \infty & 1 & 0 & \infty \\ 1 & \infty & 3 & 2 & 0 \end{bmatrix}$$ ∎

We can make several observations concerning the reachability matrix R and distance matrix D of a digraph $G = (V, A)$, with $V = \{v_1, v_2, \ldots, v_p\}$. First, since $d(v_i, v_i) = 0$, we see that $r_{ii} = 1$ and $d_{ii} = 0$ for each i, $1 \leq i \leq p$; that is, each main diagonal entry of R is 1, while each main diagonal entry of D is 0. Second, note that $r_{ij} = 0 \leftrightarrow d_{ij} = \infty$. Finally, suppose for some subscripts i, j, and k that $r_{ij} = 1 = r_{jk}$. Then, since the relation "is reachable from" on V is transitive, we know that $r_{ik} = 1$ and that $d_{ik} \leq d_{ij} + d_{jk}$.

This last observation suggests a method for computing the reachability matrix R from the adjacency matrix M for a digraph $G = (V, A)$, with $V = \{v_1, v_2, \ldots, v_p\}$. We begin by initializing R to the matrix M_1, obtained by replacing each 0 on the main diagonal of M by 1. The matrix M_1 reflects the fact that v_j is reachable from v_i if either $i = j$ or $(v_i, v_j) \in A$. We then search for subscripts i, j, k such that $r_{ij} = 1 = r_{jk}$ and $r_{ik} = 0$. If no such subscripts are found, then R is the correct reachability matrix; otherwise, set $r_{ik} = 1$ and repeat this last step. This is the essential idea of Warshall's algorithm for computing the reachability matrix R, except that

Warshall's algorithm does the "subscript search" in a rather efficient manner. Before describing this algorithm, we consider an example.

Example 7.16 Let $G = (\{v_1, v_2, v_3, v_4\}, \{(v_1, v_2), (v_2, v_2), (v_2, v_3), (v_3, v_1), (v_3, v_4)\})$. Use the method described in the preceding paragraph to compute the reachability matrix R of G.

Solution We begin by initializing R to

$$R = \begin{bmatrix} 1 & 1 & 0 & 0 \\ 0 & 1 & 1 & 0 \\ 1 & 0 & 1 & 1 \\ 0 & 0 & 0 & 1 \end{bmatrix}$$

This is the adjacency matrix for G, with each main diagonal entry set equal to 1. We then search for subscripts i, j, k such that $r_{ij} = 1 = r_{jk}$ and $r_{ik} = 0$. Note that this happens when $i = 3$, $j = 1$, and $k = 2$. We thus set $r_{32} = 1$, yielding

$$R = \begin{bmatrix} 1 & 1 & 0 & 0 \\ 0 & 1 & 1 & 0 \\ 1 & 1 & 1 & 1 \\ 0 & 0 & 0 & 1 \end{bmatrix}$$

Again we search for subscripts i, j, k such that $r_{ij} = 1 = r_{jk}$ and $r_{ik} = 0$. This time we find $i = 1$, $j = 2$, and $k = 3$. Hence, we set $r_{13} = 1$, and so

$$R = \begin{bmatrix} 1 & 1 & 1 & 0 \\ 0 & 1 & 1 & 0 \\ 1 & 1 & 1 & 1 \\ 0 & 0 & 0 & 1 \end{bmatrix}$$

Next we find $i = 1$, $j = 3$, and $k = 4$. It is important to note that here we are using r_{13}, which was changed to 1 in a previous step. Hence we set $r_{14} = 1$. Continuing in this way, we find:

$$i = 2, \ j = 3, \ k = 1, \text{ so that } r_{21} = 1$$

$$i = 2, \ j = 3, \ k = 4, \text{ so that } r_{24} = 1$$

At this point,

$$R = \begin{bmatrix} 1 & 1 & 1 & 1 \\ 1 & 1 & 1 & 1 \\ 1 & 1 & 1 & 1 \\ 0 & 0 & 0 & 1 \end{bmatrix}$$

One can check that there are no more subscripts i, j, k such that $r_{ij} = r_{jk} = 1$ and $r_{ik} = 0$, so this last matrix is the reachability matrix for G. ∎

We now present Warshall's algorithm.

ALGORITHM 7.1   (Warshall)

Let $G = (V, A)$ be a digraph, with $V = \{v_1, v_2, \ldots, v_p\}$, and let M be the adjacency matrix of G. This algorithm computes the reachability matrix R of G. ∎

> Step 0. Initially, $R := M_1$, where M_1 is obtained from M by replacing each 0 on the main diagonal by 1.
> Step 1. Repeat the following steps for each j, $1 \le j \le p$.
> Step 2. Repeat the following steps for each i, $1 \le i \le p$.
> Step 3. Repeat the following step for each k, $1 \le k \le p$.
> Step 4. If $r_{ij} = 1$ and $r_{jk} = 1$, then $r_{ik} := 1$.

Example 7.16 essentially employs Warshall's algorithm to compute the reachability matrix, so the student may wish to review that example at this point.

It is clear that if Algorithm 7.1 sets $r_{ik} = 1$ at some point, then v_k is reachable from v_i. What is not clearly true is the converse assertion that if v_t is reachable from v_m, then Algorithm 7.1 sets $r_{mt} = 1$ at some point. This is certainly true if either $m = t$ or $(v_m, v_t) \in A$, but what if $d(v_m, v_t) \ge 2$? In this case, let P be a v_m-v_t path of length 2 or more, and let v_n be the intermediate vertex on P with the largest subscript. We claim that $r_{mt} = 1$ after Step 4 of the algorithm is executed for $i = m$, $j = n$, and $k = t$. We now prove this claim by induction (the strong form) on n.

If $n = 1$, then necessarily P has length 2. Thus, $(v_m, v_n), (v_n, v_t) \in A$, so that $r_{mn} = r_{nt} = 1$ initially. Hence, when Step 4 is executed with $i = m$, $j = n$, and $k = t$, r_{mt} is set equal to 1.

Assume $n > 1$. The induction hypothesis is that, for all $v_a, v_b \in V$, if G contains a v_a-v_b path of length 2 or more and v_c is the intermediate vertex on this path with the largest subscript, where $1 \le c < n$, then $r_{ab} = 1$ after Step 4 of the algorithm is executed with $i = a$, $j = c$, and $k = b$.

Now let P' and P'' be those parts of P from v_m to v_n and from v_n to v_t, respectively, so that P' is a v_m-v_n path and P'' is a v_n-v_t path. If P' has length 1, then $(v_m, v_n) \in A$, and so $r_{mn} = 1$ initially. On the other hand, if the length of P' is greater than 1, let v_{n_1} be the intermediate vertex on P' with the largest subscript. Since v_{n_1} is also an intermediate vertex of P, $n_1 < n$ and so, by the induction hypothesis, $r_{mn} = 1$ after Step 4 is executed with $i = m$, $j = n_1$, and $k = n$. In a similar manner, working with P'', we can argue that $r_{nt} = 1$ either initially or after Step 4 is executed with $i = n$, $j = n_2$, and $k = t$, where $n_2 < n$. Hence, when Step 4 is executed with

$i = m$, $j = n$, and $k = t$, we have $r_{mn} = r_{nt} = 1$, and so r_{mt} is set to 1. This completes the proof of the correctness of Warshall's algorithm.

Warshall's algorithm can be easily modified so that it computes the distance matrix of a given directed graph; this modification is left to Exercise 10.

As is usually the case when a particular kind of mathematical structure is studied, one encounters the important notion of substructure. In this regard, graphs and digraphs are no exception.

DEFINITION 7.10

Let $G_1 = (V_1, A_1)$ and $G_2 = (V_2, A_2)$ be two digraphs. We call G_1 a *subdigraph* of G_2 provided $V_1 \subseteq V_2$ and $A_1 \subseteq A_2$. In this case we also call G_2 a *superdigraph* of G_1. The fact that G_1 is a subdigraph of G_2 is denoted by writing $G_1 \subseteq G_2$.

Let $G = (V, A)$ be a digraph. For $v \in V$, the notation $G - v$ is used to denote the subdigraph G_1 of G obtained by deleting v and all arcs of G incident with v; that is, $G_1 = (V_1, A_1)$, where $V_1 = V - \{v\}$ and $A_1 = A - \{(x, y) \mid x = v \text{ or } y = v\}$. Similarly, if $(u, v) \in A$, then $G - (u, v)$ denotes the subdigraph of G with vertex set V and arc set $A - \{(u, v)\}$. On the other hand, if $x, y \in V$ and $(x, y) \notin A$, then $G + (x, y)$ denotes the superdigraph of G with vertex set V and arc set $A \cup \{(x, y)\}$. In general, if $H = (V', A')$ is a subdigraph of G with $V' = V$, then H is called a *spanning subdigraph* of G. Thus, for example, if $u, v, x, y \in V$, $(u, v) \in A$, and $(x, y) \notin A$, then $G - (u, v)$ is a spanning subdigraph of G, which in turn is a spanning subdigraph of $G + (x, y)$.

Example 7.17   Recall the digraph $G_1 = (\{t, u, v, w, x, y, z\}, \{(t, w), (u, t), (u, x), (v, t), (w, v), (w, z), (x, w), (x, y), (y, v), (y, x), (z, v), (z, y)\})$ shown in Figure 7.15. The subdigraphs $H_1 = G_1 - w$, $H_2 = G_1 - (y, x)$, and the superdigraph $H_3 = G_1 + (w, u)$ are shown in Figure 7.17(a), (b), and (c), respectively. ∎

Again let $G = (V, A)$ be a digraph and let U be a nonempty subset of V. The *subdigraph* of G *induced by* U, denoted by $\langle U \rangle$, has vertex set U and arc set consisting of those arcs of G joining vertices of U; that is, $\langle U \rangle = (U, A')$, where

$$A' = \{(x, y) \in A \mid x \in U \text{ and } y \in U\}$$

A subdigraph H of G is called an *induced subdigraph* provided $H = \langle U \rangle$ for some subset U of V. Similarly, for a nonempty subset B of A, the *subdigraph of G induced by B*, denoted $\langle B \rangle$, has arc set B and vertex set consisting of those vertices of V incident with some arc of B. A subdigraph H of G is called *arc-induced* if $H = \langle B \rangle$ for some subset B of A.

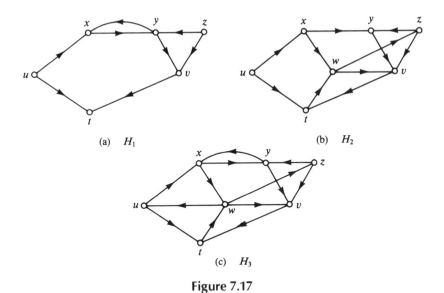

(a) H_1 (b) H_2

(c) H_3

Figure 7.17

Example 7.18   Again let G_1 be the digraph of Figure 7.15. The subdigraphs $H_4 = \langle \{v, w, y, z\} \rangle$ and $H_5 = \langle \{(x, y), (y, v), (z, y)\} \rangle$ are shown in Figure 7.18(a) and (b), respectively.   ■

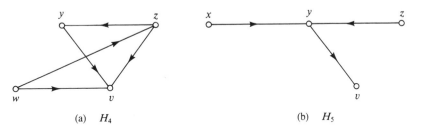

(a) H_4 (b) H_5

Figure 7.18

Let $G = (V, A)$ be a digraph with $V = \{v_1, v_2, \ldots, v_p\}$, let M be the adjacency matrix, and let R be the reachability matrix of G. We can interpret R as the adjacency matrix M_1 of a digraph $G_1 = (V, A_1)$, such that G is a spanning subdigraph of G_1. Note that G_1 has a loop at every vertex and, given any $x, y, z \in V$, if $(x, y) \in A_1$ and $(y, z) \in A_1$, then $(x, z) \in A_1$. (Why?) Thus the relation A_1 on V is both reflexive and transitive. In fact, Warshall's algorithm is sometimes presented as a method to find the reflexive, transitive closure of a given relation A on a finite set V. (See Chapter 4 Problems 21 and 25.)

We now consider another question mentioned at the beginning of this section, that of finding the distances from a fixed vertex of a digraph G to the other vertices of G. However, we wish to consider this question in the more general context of finding "shortest paths" from a fixed vertex to the other vertices of a weighted directed graph. A *weighted directed graph* is a directed graph $G = (V, A)$, together with a "weight function" w from A to the set of nonnegative real numbers. In many applications, the weight of an arc represents some "cost" associated with its incident vertices. For instance, in the traffic flow problem mentioned at the beginning of Section 7.3, the weight of an arc (u, v) might be the time it takes, on average, to travel from intersection u to intersection v during a weekday morning rush hour. A weighted directed graph G with vertex set V, arc set A, and weight function w is denoted by (V, A, w).

Given a weighted directed graph $G = (V, A, w)$ and a walk W of G, the (*weighted*) *length* of W is defined to be the sum of the weights of the arcs of W. For $u, v \in V$, the (*weighted*) *distance* from u to v is denoted $d(u, v)$ and is defined as the minimum length among all u-v paths in G; a u-v path P whose length is $d(u, v)$ is called a *shortest u-v path*. (If v is not reachable from u, then $d(u, v) = \infty$.)

Suppose $V = \{u_0, u_1, \ldots, u_{p-1}\}$. For $1 \le i \le p - 1$, we wish to find a shortest u_0-u_i path in G. We present an algorithm for doing this that is due to E. W. Dijkstra and was discovered in 1959.

As an example of this type of problem, suppose that a company has manufacturing plants in cities $c_1, c_2, c_3, c_4, c_5, c_6$ and its headquarters in city c_0. Frequently the company must send executives from c_0 to one of the cities c_i, where $1 \le i \le 6$. For this problem we could form a weighted directed graph G with vertex set $\{c_0, c_1, \ldots, c_6\}$, where an arc from c_i to c_j indicates the existence of a direct flight from c_i to c_j and the weight of such an arc is the cost of this flight. For the specific example we are going to consider, the arc set is symmetric and so we represent G as the weighted graph shown in Figure 7.19 (each edge represents a pair of symmetric arcs, and the number near it is the weight of each arc). The company would very

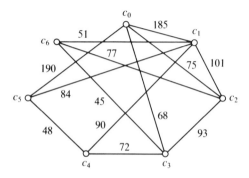

Figure 7.19 A weighted graph

much like to determine the cheapest routes from c_0 to each of the other cities. Notice that such routes correspond to the shortest c_0-c_i paths in the weighted graph G, where $1 \le i \le 6$.

We now discuss Dijkstra's algorithm for finding shortest u_0-u_i paths in a weighted directed graph $G = (V, A, w)$, where $V = \{u_0, u_1, \ldots, u_{p-1}\}$. In order to comprehend this algorithm, it is necessary to understand an important property of shortest paths. Suppose that U is a proper subset of V with $u_0 \in U$. Define a shortest path from u_0 to $V - U$ to be a path of minimum length among all u_0-v paths with $v \in V - U$, and let $d(u_0, V - U)$ be the weight of such a path. Suppose that

$$P: \quad u_0 = v_0, v_1, \ldots, v_{n-1}, v_n = y$$

is such a path in G. It is then not difficult to argue the following (see Exercise 12):

1. $v_0, v_1, \ldots, v_{n-1} \in U$ and $y \in V - U$
2. The path

$$P': u_0 = v_0, v_1, \ldots, v_{n-1}$$

representing that portion of P from u_0 to v_{n-1} is a shortest u_0-v_{n-1} path in G.

It follows that P is a shortest u_0-y path and that

$$d(u_0, V - U) = d(u_0, y) = \min\{d(u_0, v) + w((v, y))\}$$

where the minimum is taken over all vertices v and y with $v \in U$ and $y \in V - U$.

The ideas of the preceding paragraph form the gist of Dijkstra's algorithm. During its application, U denotes the set of vertices for which shortest paths from u_0 have been found; initially $U := \{u_0\}$. At each iteration, another vertex x is added to U; in fact, x is that vertex such that

$$d(u_0, x) = d(u_0, V - U)$$

The algorithm continues until $U = V$. The algorithm is to compute $d(u_0, v)$ for each vertex v. We denote this distance simply by $d(v)$; initially $d(u_0) := 0$ and $d(v) := \infty$ for $v \ne u_0$. (In practice, "∞" is replaced by some appropriately large positive number.) In addition, we want the algorithm to compute shortest u_0-v paths. A slick way to handle this problem is to have the algorithm compute, for each vertex v, the predecessor $p(v)$ of v on a shortest u_0-v path. Then, given a vertex v, one may find a shortest u_0-v path by working backward from v to $p(v)$, to $p(p(v))$, and so on, until u_0 is reached. For convenience, the algorithm initially sets $p(v) = v$ for each vertex v.

We now present Dijkstra's algorithm and then illustrate its application.

ALGORITHM 7.2 (Dijkstra)

Let $G = (V, A, w)$ be a weighted directed graph and let $u_0 \in V$. This algorithm computes $d(v) = d(u_0, v)$ and $p(v) =$ the predecessor of v on a shortest u_0-v path for all $v \in V$.

Step 0. Initially $U := \{u_0\}$, $d(u_0) := 0$, $d(v) := \infty$ for all $v \in V - \{u_0\}$, $p(v) := v$ for all $v \in V$, and $x := u_0$.

Step 1. Repeat the following steps until $U = V$.

Step 2. Repeat the following step for each $y \in V - U$.

Step 3. If $d(x) + w((x, y)) < d(y)$, then $d(y) := d(x) + w((x, y))$ and $p(y) := x$.

Step 4. Let x be a vertex in $V - U$ such that $d(x)$ is a minimum; $U := U \cup \{x\}$. □

Example 7.19 Apply Algorithm 7.2 to find $d(c_0, c_i)$ and shortest c_0-c_i paths for the weighted graph G of Figure 7.19.

Solution In Step 0, $U := \{c_0\}$, $d(c_0) := 0$, $d(c_i) := \infty$, $1 \le i \le 6$, $p(c_i) := c_i$, $0 \le i \le 6$, and $x := c_0$. We then trace the steps as follows.

Steps 1 and 2. Since $U \ne V$, Step 3 is executed for each $y \in V - U = \{c_1, c_2, c_3, c_4, c_5, c_6\}$.

Step 3. Those vertices y adjacent from c_0 have $d(y)$ and $p(y)$ updated. Thus $d(c_1) := 185$, $p(c_1) := c_0$, $d(c_2) := 75$, $p(c_2) := c_0$, $d(c_3) := 68$, $p(c_3) := c_0$, $d(c_5) := 190$, and $p(c_5) := c_0$.

Step 4. The vertex in $V - U$ with the minimum distance from c_0 is c_3, so $x := c_3$ and $U := U \cup \{x\} = \{c_0, c_3\}$.

Steps 1 and 2. Since $U \ne V$, Step 3 is executed for each $y \in V - U = \{c_1, c_2, c_4, c_5, c_6\}$.

Steps 3. Those vertices $y \in V - U$ adjacent from $x = c_3$ may have $d(y)$ and $p(y)$ updated. For $y = c_2$, $d(y) = 75$, whereas $d(x) + w((x, y)) = 68 + 93 = 163$. So $d(c_2)$ and $p(c_2)$ are not changed. On the other hand, for $y = c_4$, $d(y) = \infty$, whereas $d(x) + w((x, y)) = 68 + 72 = 140$. So $d(c_4) := 140$ and $p(c_4) := c_3$. Similarly, $d(c_6) := 68 + 45 = 113$ and $p(c_6) := c_3$.

Step 4. The vertex in $V - U$ with minimum distance from c_0 is c_2, so $x := c_2$ and $U := U \cup \{x\} = \{c_0, c_2, c_3\}$.

Steps 1 and 2. Since $U \ne V$, Step 3 is executed for each $y \in V - U = \{c_1, c_4, c_5, c_6\}$.

Step 3. The values updated are $d(c_1) := 75 + 101 = 176$ and $p(c_1) := c_2$.

Step 4. The vertex in $V - U$ with minimum distance from c_0 is c_6, so $x := c_6$ and $U := U \cup \{x\} = \{c_0, c_2, c_3, c_6\}$.

Steps 1 and 2. Since $U \ne V$, Step 3 is executed for each $y \in \{c_1, c_4, c_5\}$.

Step 3. The values updated are $d(c_1) := 113 + 51 = 164$ and $p(c_1) := c_6$.

Step 4. The vertex in $V - U$ with minimum distance from c_0 is c_4, so $x := c_4$ and $U := U \cup \{x\} = \{c_0, c_2, c_3, c_4, c_6\}$.

Steps 1 and 2. Since $U \neq V$, Step 3 is executed for each $y \in \{c_1, c_5\}$.
Step 3. The values updated are $d(c_5) := 140 + 48 = 188$ and $p(c_5) := c_4$.
Step 4. The vertex in $V - U$ with minimum distance from c_0 is c_1, so $x := c_1$ and $U := U \cup \{x\} = \{c_0, c_1, c_2, c_3, c_4, c_6\}$.
Steps 1 and 2. Since $U \neq V$, Step 3 is executed for $y = c_5$.
Step 3. No values are updated.
Step 4. Here $x := c_5$ and $U := U \cup \{x\} = V$.
Step 1. Since $U = V$, the algorithm terminates. The final computed values are:

$$d(c_1) = 164, \ d(c_2) = 75, \ d(c_3) = 68,$$
$$d(c_4) = 140, \ d(c_5) = 188, \ d(c_6) = 113$$
$$p(c_1) = c_6, \ p(c_2) = c_0, \ p(c_3) = c_0,$$
$$p(c_4) = c_3, \ p(c_5) = c_4, \ p(c_6) = c_3$$

As pointed out before, the predecessor values may be used to find shortest c_0-c_i paths, where $1 \leq i \leq 6$. To illustrate, suppose we desire a shortest c_0-c_1 path. Since $p(c_1) = c_6$, $p(c_6) = c_3$, and $p(c_3) = c_0$, one such path is c_0, c_3, c_6, c_1. We leave it as Exercise 2 to find shortest paths for the other vertices of G. ∎

Exercises 7.4

1. Apply Algorithm 7.1 to the directed graph of Example 7.14.
2. Complete Example 7.19 by finding shortest c_0-c_i paths for $2 \leq i \leq 6$.
3. Apply Algorithm 7.1 to each of the directed graphs in Figure 7.20. (Ignore the weights, and index the rows and columns of the adjacency and reachability matrices using 0, 1, 2, and so on, to correspond to the subscripts on the vertex labels.)
4. Let $G = (V, A, w)$, where $V = \{u_0, u_1, u_2, u_3, u_4, u_5\}$, $A = \{(u_0, u_1), (u_0, u_3), (u_0, u_4), (u_1, u_2), (u_2, u_3), (u_3, u_1), (u_3, u_2), (u_3, u_4), (u_4, u_3), (u_4, u_5), (u_5, u_0)\}$, and $w((u_0, u_1)) = 12$, $w((u_0, u_3)) = 6$, $w((u_0, u_4)) = 1$, $w((u_1, u_2)) = 6$, $w((u_2, u_3)) = 10 = w((u_3, u_2))$, $w((u_3, u_1)) = 5$, $w((u_3, u_4)) = 9 = w((u_4, u_3))$, $w((u_4, u_5)) = 3$, $w((u_5, u_0)) = 5$.
   (a) Apply Algorithm 7.2 to this weighted directed graph.
   (b) Find shortest u_0-u_i paths for $1 \leq i \leq 5$.
5. Apply Algorithm 7.2 to each of the weighted directed graphs in Figure 7.20.
6. Let $G = (V, A)$, where $V = \{v_1, v_2, \ldots, v_p\}$, and let M be the adjacency matrix for G. What information about G does the sum of the elements in the ith row of M give us? What about the sum of the elements in the jth column?

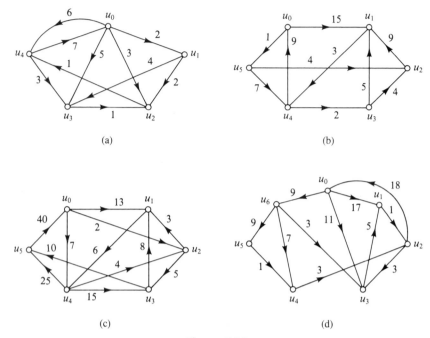

Figure 7.20

7. Exhibit each of the following subdigraphs of the digraph G_2 of Figure 7.16.
   (a) $H_1 = G_2 - z$ (b) $H_2 = (G_2 - z) - w$
   (c) $H_3 = \langle\{u, v, x, y\}\rangle$ (d) $H_4 = G_2 - (w, x)$
   (e) $H_5 = \langle\{(u, v), (u, z), (v, y), (y, x)\}\rangle$

8. Let $G = (V, A)$ be a digraph. Identify each of these statements as true or false.
   (a) Any induced subdigraph of G can be obtained by successively deleting vertices.
   (b) Any arc-induced subdigraph of G can be obtained by successively deleting arcs.

9. Let $G = (V, A, w)$ be a weighted directed graph with $V = \{u_0, u_1, \ldots, u_{p-1}\}$ such that no two arcs of G have the same weight.
   (a) Show that it is not always the case that the arc of smallest weight is used in some shortest u_0-u_i path.
   (b) Give a sufficient condition under which the arc of smallest weight must be used in some shortest u_0-u_i path.
   (c) For each $i \in \{1, 2, \ldots, p - 1\}$, is it true that G contains a unique shortest u_0-u_i path?

10. Modify Algorithm 7.1 so that it computes the distance matrix D of G. (Optional: Allow G to be a weighted directed graph.)

11. Apply the algorithm of Exercise 10 to each of the directed graphs in Figure 7.20. (Ignore the weights, unless you did the optional part of Exercise 10.)

12. Let $G = (V, A, w)$ be a weighted directed graph and let U be a proper subset of V with $u_0 \in U$. Suppose that

$$P: u_0 = v_0, v_1, \ldots, v_{n-1}, v_n = y$$

is a shortest path from u_0 to $V - U$ in G. Show the following:

(a) $v_0, v_1, \ldots, v_{n-1} \in U$ and $y \in V - U$

(b) The path P': $u_0 = v_0, v_1, \ldots, v_{n-1}$, representing that portion of P from u_0 to v_{n-1}, is a shortest u_0-v_{n-1} path in G.

(c) $d(u_0, V - U) = \min\{d(u_0, v) + w((v, y))\}$, where the minimum is taken over all vertices v and y with $v \in U$ and $y \in V - U$.

13. Let M be the adjacency matrix of the digraph $G = (\{v_1, v_2, v_3, v_4\}, \{(v_1, v_2), (v_1, v_3), (v_2, v_1), (v_2, v_2), (v_2, v_3), (v_3, v_4), (v_4, v_1)\})$.

(a) Verify that the $(2, 3)$-entry of M^2 is the number of distinct walks of length 2 from v_2 to v_3.

(b) Verify that the $(2, 3)$-entry of M^3 is the number of distinct walks of length 3 from v_2 to v_3.

(c) Verify that the $(2, 3)$-entry of M^4 is the number of distinct walks of length 4 from v_2 to v_3.

14. Let $G = (V, A)$ be a directed graph with $V = \{v_1, v_2, \ldots, v_p\}$, and let M be the adjacency matrix of G.

(a) Show that the (i, j)-entry of M^2 is the number of distinct v_i-v_j walks of length 2 in G.

(b) Generalizing the result of part (a), show that, for $n \in \mathbb{N}$, the (i, j)-entry of M^n is the number of v_i-v_j walks of length n in G. (Hint: Use induction on n.)

(c) Let M^0 denote the $p \times p$ identity matrix. Prove: G is strong if and only if each entry of $M^0 + M^1 + \cdots + M^{p-1}$ is positive.

15. Implement Algorithm 7.1 as (part of) a Pascal program.

16. Implement Algorithm 7.2 as (part of) a Pascal program.

7.5 ACYCLIC DIGRAPHS AND ROOTED TREES

An *acyclic* digraph is one having no cycles. In this section we discuss acyclic digraphs, including an important special class of such digraphs, called *rooted trees*.

Let $G = (V, A)$ be a digraph. To begin, we know that the relation "is reachable from" on V is both reflexive and transitive. If G is acyclic, then this relation is also antisymmetric. For suppose that u and v are distinct vertices of G and both v is reachable from u and u is reachable from v. Then G contains a u-v walk W_1 and a v-u walk W_2. If W denotes the u-u walk obtained by appending W_2 to W_1, then W is a nontrivial closed walk.

Now, in a manner similar to the proof of Theorem 7.6, we may show that G contains a cycle, contradicting the fact that G is acyclic. Thus, if $G = (V, A)$ is an acyclic digraph, then the relation "is reachable from" on V is a partial ordering of V. It is easy to see that the converse is also true.

THEOREM 7.10   Let $G = (V, A)$ be a digraph. Then G is acyclic if and only if the relation "is reachable from" on V is antisymmetric.

An application where acyclic directed graphs arise concerns the decomposition of a large, complicated problem into smaller, more manageable tasks. For example, we often decompose a complicated algorithm into smaller steps, some of which might conceivably be executed "in parallel." To introduce some consistent terminology into our discussion, we shall call such a smaller step, which is always executed as a unit, a *task*. We form a directed graph G whose vertex set is the set of tasks and which contains an arc from task u to task v provided u must be executed before v. This might be required, for instance, because u calculates some results that are needed by v's computation, or because u and v require independent access to the same external device. Notice that such a digraph must be acyclic, for otherwise there would exist tasks u and v such that u must be executed prior to v and v must be executed prior to u.

Suppose now that a schedule is desired for executing the tasks; that is, we desire a list

$$(u_1, u_2, \ldots, u_p)$$

of the tasks, indicating the order in which they are to be executed, such that, whenever u_i is required to be executed prior to u_j, then $i < j$. More generally, let $G = (V, A)$ be an acyclic digraph of order p. We desire a list $(u_1, u_2, \ldots, u_p)$ of the vertices such that, whenever u_j is reachable from u_i, then $i \le j$. One method of finding such a list is called the *topological sort*, and it is presented in a slightly different context as Chapter 4 Problem 32. The topological sort is based on the following lemma, which is an immediate consequence of Exercise 24 of Exercises 7.3.

LEMMA 7.11   If G is an acyclic digraph, then G contains a vertex u such that $\mathrm{id}(u) = 0$.

Proof   We prove the contrapositive. Suppose $\mathrm{id}(u) > 0$ for every vertex u of G. Then, by Exercise 24 of Exercises 7.3, G contains a cycle.   □

Given an acyclic digraph $G = (V, A)$, let $u_1 \in V$ such that $\mathrm{id}(u_1) = 0$. Clearly u_1 is not reachable from any other vertex of G. Next consider the subdigraph of G induced by $V - \{u_1\}$. Note that this digraph is also acyclic. Thus we may choose $u_2 \in V - \{u_1\}$ such that $\mathrm{id}(u_2) = 0$ in this digraph, and then u_2 is not reachable from any vertex of $V - \{u_1, u_2\}$. Continuing in this fashion gives the list $(u_1, u_2, \ldots, u_p)$ of the vertices of G such that u_j is reachable from u_i only if $i \le j$.

ALGORITHM 7.3   Let $G = (V, A)$ be an acyclic digraph of order p. This algorithm finds a list $(u_1, u_2, \ldots, u_p)$ of the vertices of G such that u_j is reachable from u_i only if $i \leq j$.

Step 0.   Initially let $U := V$.
Step 1.   Repeat Step 2 for each i, $1 \leq i \leq p$.
Step 2.   Choose u_i such that $\text{id}(u_i) = 0$ in $\langle U \rangle$. Let $U := U - \{u_i\}$.   □

Example 7.20   Illustrate Algorithm 7.3 for the digraph $G = (\{u, v, w, x, y, z\}, \{(u, v), (u, x), (v, w), (v, y), (w, z), (x, y), (y, z)\})$.

Solution   Let $U := \{u, v, w, x, y, z\}$. Step 2 is now repeated six times, for $i := 1, 2, 3, 4, 5, 6$. For $i := 1$, we find $\text{id}(u) = 0$, so $u_1 := u$ and $U := \{v, w, x, y, z\}$. For $i := 2$, note that $\text{id}(v) = \text{id}(x) = 0$, so we have a choice for u_2; suppose we let $u_2 := v$, so that $U := \{w, x, y, z\}$. For $i := 3$, $\text{id}(w) = \text{id}(x) = 0$; let $u_3 := x$ and $U := \{w, y, z\}$. For $i := 4$, $\text{id}(w) = \text{id}(y) = 0$; let $u_4 := w$ and $U := \{y, z\}$. For $i := 5$, $\text{id}(y) = 0$, so that $u_5 := y$ and $U := \{z\}$. Finally $u_6 := z$. Thus the list obtained is (u, v, x, w, y, z). Note that other possibilities are (u, v, w, x, y, z), (u, v, x, y, w, z), (u, x, v, w, y, z), and (u, x, v, y, w, z).   ∎

One class of acyclic digraphs that is particularly important in the study of computer science is rooted trees.

DEFINITION 7.11

An acyclic digraph $T = (V, A)$ is called a *rooted tree* if T contains a distinguished vertex u, called the *root*, such that, for every vertex $v \in V$, the digraph T contains a unique u-v path.

An example of a rooted tree is the digraph $T = (V, A)$, where $V = \{a, b, c, d, e, f, g, h, i, j\}$ and $A = \{(a, b), (a, c), (a, d), (b, e), (b, f), (d, g), (f, h), (g, i), (g, j)\}$, shown in Figure 7.21. The vertex a is the root.

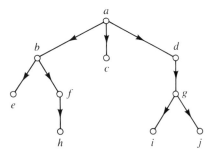

Figure 7.21 A rooted tree

Note, for instance, that T contains a unique path from a to h, namely, a, b, f, h.

We remark that a rooted tree necessarily has a unique root.

There is quite a bit of standard terminology associated with rooted trees. Let $T = (V, A)$ be a rooted tree and let $v \in V$. Those vertices adjacent from v are called *children* of v; if v has no children, then v is called a *leaf* of T. If v is not the root of T, then it follows from the definition that there is a unique vertex of T from which v is adjacent; we call this vertex the *parent* of v. Vertices of T that have the same parent are called *siblings*. The notions of "child" and "parent" extend naturally to the notions of "descendant" and "ancestor."

As mentioned, there is a unique path in T from the root to v. The length of this path is the *level* of v in T. Thus the root of T has level 0, the children of the root have level 1, and so on. The *height* of T is the maximum level among its vertices.

Example 7.21   For the rooted tree T of Figure 7.21, find (a) the children of b, (b) the ancestors of i, (c) the vertices having level 2, and (d) the height of T.

Solution   (a) The children of b are e and f.
(b) The ancestors of i are g, d, and a.
(c) The vertices at level 2 are e, f, and g.
(d) The height of T is 3.   ∎

Let m be a positive integer. A rooted tree T is said to be an *m-ary rooted tree* (or simply an *m-ary tree*) provided each vertex of T has at most m children. If $m = 1$, then a 1-ary tree is simply a path (from the root to the leaf). A 2-ary tree is also called a *binary tree*. Note that the tree of Figure 7.21 is a 3-ary tree.

One of the many uses made of rooted trees in the study of computer science is as a convenient method of representing and working with expressions. For instance, consider the expression

$$(a + b) * (c/\text{SQRT}(d))$$

where SQRT denotes the square root function. This expression can be represented by the rooted tree of Figure 7.22. Each operation corresponds to a nonleaf vertex in such a way that its subtrees (the subtrees of which its children are the roots) represent the operands for that operator. Note that a rooted tree represents an expression in an unambiguous manner, whereas the expression itself may require parentheses or the use of implicit precedence rules to make clear the order in which the operations are to be applied.

Let $G = (V, A)$ be a digraph and let $u \in V$. Then there is a subdigraph T_u of G that is a rooted tree with root u such that the vertex set of T_u is the set of vertices of G that are reachable from u. One method for finding such a rooted tree is called a *depth-first search*, and a rooted tree so found is

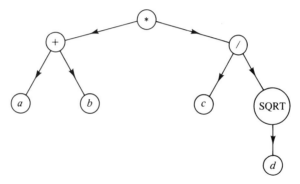

Figure 7.22 Representing an expression using a rooted tree

called a *depth-first search tree* of G. If every vertex of G is reachable from u —in particular, if G is strong, then T_u is a spanning subdigraph of G, called a *depth-first search spanning tree* of G.

Perhaps it is easier to illustrate how to find a depth-first search tree of a given directed graph than to define what one is. Thus, before giving a definition, we consider an example.

Example 7.22 Consider the directed graph $G = (V, A)$ of Figure 7.23. We find a depth-first search tree $T_a = (V_0, A_0)$ of G having root a. Initially $V_0 := \{a\}$ and $A_0 := \emptyset$. Since $(a, b) \in A$, we add b to V_0 and (a, b) to A_0. Next, since $(b, d) \in A$, we add d to V_0 and (b, d) to A_0. Here we can see one property of a depth-first search; it always prefers to move to the next level of the rooted tree being constructed, rather than to stay at the same level. That is, suppose the depth-first search has just added the vertex v and the arc (u, v) to the tree; it next prefers to add a vertex x that is adjacent from v, rather than adding another vertex y adjacent from u. Continuing with G, since $(d, t) \in A$, we add t to V_0 and (d, t) to A_0. At this point, $V_0 = \{a, b, d, t\}$ and $A_0 = \{(a, b), (b, d), (d, t)\}$. Now t is adjacent to no vertex of $V - V_0$, so the depth-first search "backs up" the tree to the parent of

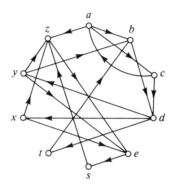

Figure 7.23

t, namely, d. From d we may proceed to x, then from x to e, from e to s, and from s to z, adding the vertices x, e, s, and z to V_0 and the arcs (d, x), (x, e), (e, s), and (s, z) to A_0. At this point, z is adjacent to no vertex of $V - V_0$, so we back up to s, the parent of z. However, s also is not adjacent to any vertex of $V - V_0$, so we back up to its parent, e. Similarly, we must back up from e to x and then from x to d. Again, from d we may proceed, this time to y, so that y is added to V_0 and (d, y) to A_0. Then we must back up from y to d, then from d to b, and then from b to a. Finally the vertex c is added to V_0 and the arc (a, c) is added to A_0, so that $V_0 = V$ and $A_0 = \{(a, b), (a, c), (b, d), (d, t), (d, x), (d, y), (e, s), (s, z), (x, e)\}$. Note that T_a is a spanning subdigraph of G. ∎

DEFINITION 7.12

Let $G = (V, A)$ and let $u \in V$. A *depth-first search tree of G with root u* is a rooted tree $T = (V_0, A_0)$ defined recursively as follows.

0. Let V_0 be the subset of V consisting of those vertices reachable from u. If $V_0 = \{u\}$, then $A_0 = \emptyset$.
1. Otherwise let $u_1 \in V_0 - \{u\}$, such that $(u, u_1) \in A$. Let V_1 be the subset of $V_0 - \{u\}$ consisting of those vertices reachable from u_1, and let $T_1 = (V_1, A_1)$ be a depth-first search tree of $\langle V_1 \rangle$ with root u_1. If $V_1 \cup \{u\} = V_0$, then let $A_0 = A_1 \cup \{(u, u_1)\}$.
2. Otherwise let $u_2 \in V_0 - (V_1 \cup \{u\})$ such that $(u, u_2) \in A$. Let V_2 be the subset of $V_0 - (V_1 \cup \{u\})$ consisting of those vertices reachable from u_2, and let $T_2 = (V_2, A_2)$ be a depth-first search tree of $\langle V_2 \rangle$ with root u_2. If $V_1 \cup V_2 \cup \{u\} = V_0$, then $A_0 = A_1 \cup A_2 \cup \{(u, u_1), (u, u_2)\}$.
3. Otherwise continue this process, finding depth-first search trees $T_1, T_2, \ldots, T_m$, where $T_i = (V_i, A_i)$ and has root u_i, until $V_1 \cup V_2 \cup \cdots \cup V_m \cup \{u\} = V_0$. Then $A_0 = A_1 \cup A_2 \cup \cdots \cup A_m \cup \{(u, u_i) \mid 1 \leq i \leq m\}$.

Example 7.23 Let us apply Definition 7.12 to the digraph $G = (V, A)$ of Figure 7.23 to find a depth-first search tree with root $u = a$. In Step 0, we find that $V_0 \neq \{a\}$, since, for instance, $(a, b) \in A$. Thus, in Step 1, let $u_1 = b$. Without going into the details of the recursion, suppose we find that $T_1 = (V_1, A_1)$ is a depth-first search tree of $\langle V_1 \rangle$, where $V_1 = \{b, d, e, s, t, x, y, z\}$ and $A_1 = \{(b, d), (d, t), (d, x), (d, y), (e, s), (s, z), (x, e)\}$. Note that $V_1 \cup \{a\} \neq V_0$, since $(a, c) \in A$ and $c \notin V_1 \cup \{a\}$. Thus, in Step 2, let $u_2 = c$. Applying the definition recursively, we find that $T_2 = (V_2, A_2)$ is a depth-first search tree of $\langle V_2 \rangle$, where $V_2 = \{c\}$ and $A_2 = \emptyset$. Since there is no vertex u_3 such that $u_3 \notin V_1 \cup V_2 \cup \{a\}$ and $(u, u_3) \in A$, we have that $V_1 \cup V_2 \cup \{a\} = V = V_0$, so that

$A_0 = A_1 \cup A_2 \cup \{(a, b), (a, c)\}$. This gives the tree $T = (V_0, A_0)$ found in Example 7.22. ∎

We next present an algorithm for finding a depth-first search tree with given root u in a given digraph G. One can regard such an algorithm as "searching" for all the vertices of G that are reachable from u, and this is the reason for the word *search* in the name "depth-first search."

ALGORITHM 7.4   Given a digraph $G = (V, A)$ and $u \in V$, this algorithm finds a depth-first search tree $T = (V_0, A_0)$ of G with root u.

Step 0. Initially $V_0 := \{u\}$, $A_0 := \emptyset$, and $v := u$.
Step 1. Repeat Step 2 until it is not possible to find w as indicated. At that point, execute Step 3.
Step 2. Choose $w \in V - V_0$ such that $(v, w) \in A$. Let $V_0 := V_0 \cup \{w\}$, $A_0 := A_0 \cup \{(v, w)\}$, and $v := w$.
Step 3. If $v = u$, then the algorithm terminates; otherwise let $v :=$ the parent of v in T, and return to Step 1.   □

Example 7.24   Illustrate Algorithm 7.4 for the digraph $G = (V, A)$ of Figure 7.23, with $u = a$.

Solution   In Step 0, $V_0 := \{a\}$, $A_0 := \emptyset$, and $v := a$. We now proceed to Step 1, and then repeat Step 2 as follows.
Step 2.   Choose $w := b$, so that $V_0 := \{a, b\}$, $A_0 := \{(a, b)\}$, and $v := b$.
Step 2.   Choose $w := d$, so that $V_0 := \{a, b, d\}$, $A_0 := \{(a, b), (b, d)\}$, and $v := d$.
Step 2.   Choose $w := t$, so that $V_0 := \{a, b, d, t\}$, $A_0 := \{(a, b), (b, d), (d, t)\}$, and $v := t$.
At this point, there is no $w \in V - V_0$ such that $(t, w) \in A$, so we execute Step 3. This gives $v :=$ the parent of t in $T = d$, and we return to Step 1. Step 2 is then repeated as follows.
Step 2.   Choose $w := x$, so that $V_0 := \{a, b, d, t, x\}$, $A_0 := \{(a, b), (b, d), (d, t), (d, x)\}$, and $v := x$.
Step 2.   Choose $w := e$, so that $V_0 := \{a, b, d, e, t, x\}$, $A_0 := \{(a, b), (b, d), (d, t), (d, x), (x, e)\}$, and $v := e$.
Step 2.   Choose $w := s$, so that $V_0 := \{a, b, d, e, s, t, x\}$, $A_0 := \{(a, b), (b, d), (d, t), (d, x), (e, s), (x, e)\}$, and $v := s$.
Step 2.   Choose $w := z$, so that $V_0 := \{a, b, d, e, s, t, x, z\}$, $A_0 := \{(a, b), (b, d), (d, t), (d, x), (e, s), (s, z), (x, e)\}$, and $v := z$.
At this point, there is no $w \in V - V_0 = \{c, y\}$ such that $(z, w) \in A$, so we execute Step 3. This gives $v :=$ the parent of z in $T = s$, and we return to Step 1. However, again there is no $w \in V - V_0$ such that $(s, w) \in A$, so Step 3 is executed. This gives $v :=$ the parent of s in $T = e$, and control is returned to Step 1. Again, there is no $w \in V - V_0$ such that $(e, w) \in A$, so Step 3 is executed, giving $v := x$. One more time in Step 1, there is no $w \in V - V_0$ such that $(x, w) \in A$, so Step 3 yields $v := d$. We then return

to Step 2.

Step 2. Choose $w := y$, so that $V_0 := \{a, b, d, e, s, t, x, y, z\}$, $A_0 :=$ $\{(a, b), (b, d), (d, t), (d, x), (d, y), (e, s), (s, z), (x, e)\}$, and $v := y$.

Since there is no $w \in V - V_0 = \{c\}$ such that $(y, w) \in A$, we execute Step 3. This gives $v := d$, and we return to Step 1. Again, there is no $w \in V - V_0$ such that $(d, w) \in A$, so we execute Step 3. This gives $v := b$, and we return to Step 1. One final time, there is no $w \in V - V_0$ such that $(b, w) \in A$, so we execute Step 3, giving $v := a$. We can now return to Step 2.

Step 2. Choose $w := c$ so that $V_0 := V$, $A_0 := \{(a, b), (a, c), (b, d), (d, t), (d, x), (d, y), (e, s), (s, z), (x, e)\}$, and $v := c$.

In Step 1, there is no $w \in V - V_0 = \emptyset$, so Step 3 is executed with the result that $v := a$. We return to Step 1 a final time, and then come back to Step 3. At this point $v = u = a$, so the algorithm terminates. ∎

Exercises 7.5

1. For the digraph $G = (V, A)$ of Figure 7.23, apply Algorithm 7.4 to find a depth-first search tree with root:
   (a) b **(b)** e **(c)** x **(d)** y
2. Apply algorithm 7.4 to each of these directed graphs to find a depth-first search tree with root u_1.
   (a) $G_1 = (\{u_1, u_2, u_3, u_4, u_5, u_6, u_7, u_8\}, \{(u_1, u_2), (u_2, u_5), (u_3, u_4), (u_4, u_7), (u_5, u_1), (u_5, u_4), (u_5, u_6), (u_6, u_1), (u_7, u_3), (u_7, u_8), (u_8, u_1), (u_8, u_3)\})$
   (b) $G_2 = (\{u_1, u_2, u_3, u_4, u_5, u_6, u_7, u_8, u_9\}, \{(u_1, u_2), (u_1, u_7), (u_2, u_3), (u_3, u_1), (u_3, u_4), (u_3, u_8), (u_4, u_6), (u_5, u_2), (u_5, u_4), (u_6, u_5), (u_7, u_8), (u_8, u_9), (u_9, u_1), (u_9, u_3), (u_9, u_4), (u_9, u_8)\})$
3. Let G be a digraph. Prove: If G contains a nontrivial closed walk, then G contains a cycle.
4. Let G be a digraph and let H be a subdigraph of G. Prove: If G is acyclic, then H is acyclic.
5. Let $T = (V_0, A_0)$ be a rooted tree of order p (that is, $|V_0| = p$).
   (a) Show that the size of T is $p - 1$ (that is, $|A_0| = p - 1$).
   (b) Prove or disprove: If H is an acyclic digraph of order p and size $p - 1$, then H is a rooted tree.
6. Let T be an m-ary tree with height h, and let v be a vertex of T at level k, where $k \geq 1$.
   (a) How many ancestors does v have?
   (b) Give an upper bound on the number of descendants of v.
7. An m-ary tree is called *full* provided every vertex that is not a leaf has m children.
   (a) Give an example of a full binary tree with height 3 that has as many vertices as possible.
   (b) Give an example of a full 3-ary tree with height 3 that has as few vertices as possible.
   (c) Give a lower bound on the order of a full m-ary tree with height h.

8. A *complete m-ary tree with height h* is a full *m*-ary tree such that every leaf is at level h; thus each vertex at level k, where $k < h$, has m children. What is the order of such a rooted tree?

9. A rooted tree with height h is called *balanced* provided every leaf is at level h or level $h - 1$. Let T be a full, balanced *m*-ary tree with height $h \geq 1$ such that T has exactly k leaves. Show that

$$m^{h-1} + m - 1 \leq k \leq m^h.$$

10. Apply Algorithm 7.3 to the digraph $G = (V, A)$, where $V = \{a, b, c, d, e, f, g, h, i\}$ and $A = \{(i, h), (i, g), (i, f), (h, e), (h, d), (g, e), (g, d), (g, c), (f, c), (e, b), (d, a), (c, a), (b, a)\}$.

11. What happens if the digraph G input to Algorithm 7.3 contains a cycle? For instance, let G be the digraph of Exercise 10, and try applying Algorithm 7.3 to the digraph $G + (a, h)$.

7.6 TREES

Rooted trees are defined in the previous section as acyclic digraphs that are "weakly connected" in the sense that every vertex of a rooted tree is reachable from the root by a unique path. In this section we investigate the analogous concept for simple graphs.

> ### DEFINITION 7.13
>
> An *acyclic graph* is one with no cycles. Such a graph is called a *forest* and a connected forest is called a *tree*.

Several examples of trees are shown in Figure 7.24. It is not surprising that the term *tree* is used to describe such graphs. Notice that T_1 is the complete bipartite graph $K_{1,6}$ and T_4 is a path of order 5. In general, a graph isomorphic to $K_{1,n-1}$ is called a *star* of order n; also, since there is a unique path of order n, up to isomorphism, it is denoted by P_n.

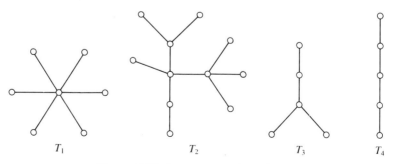

Figure 7.24 Some examples of trees

In many ways, the graphs P_n and $K_{1,n-1}$ are extreme examples of trees of order n. For example, consider the maximum degree among trees of order $n \geq 4$. The path P_n is the unique tree with maximum degree 2, while $K_{1,n-1}$ is the only such tree having maximum degree $n - 1$ (see Exercise 8).

For $n = 1, 2, 3$, there is a unique tree of order n (up to isomorphism), namely, P_n. The nonisomorphic trees of order 4 are P_4 and $K_{1,3}$.

Example 7.25   Find (a) three nonisomorphic trees of order 5 and (b) three nonisomorphic trees of order 6 having maximum degree 3.

Solution (a) The nonisomorphic trees of order 5 are P_5, $K_{1,4}$, and the tree T_3 of Figure 7.24.
  (b) The trees shown in Figure 7.25 are three nonisomorphic trees of order 6 having maximum degree 3.   ∎

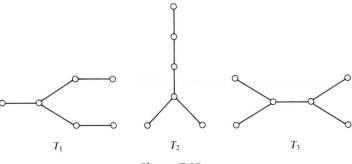

Figure 7.25

An important characterization of a tree is that it is a graph in which every two vertices are joined by a unique path. (Notice how this fact conforms with the definition of rooted trees.) In order to prove this result, we use the following lemma.

LEMMA 7.12   Let $G = (V, E)$ be a graph. Then G contains a cycle if and only if, for some distinct vertices $u, v \in V$, G contains two different u-v paths.

Proof Let

$$P: \quad u = x_0, x_1, \ldots, x_m = v$$
$$Q: \quad u = y_0, y_1, \ldots, y_n = v$$

be two distinct u-v paths in G. Without loss of generality, we may assume that $x_1 \neq y_1$. Then there is a least subscript $i > 0$ such that $x_i = y_j$ for some j, where $1 \leq j \leq n$. Let $w = x_i$. Then that portion of P from u to w, followed by that part of Q from w back to u, that is,

$$u = x_0, x_1, \ldots, x_i = y_j, y_{j-1}, \ldots, y_0 = u$$

is a cycle in G.

Conversely, suppose that

$$u = u_0, u_1, u_2, \ldots, u_n = u$$

is a cycle (of length $n \geq 3$). Then P: u, u_1 and Q: $u, u_{n-1}, \ldots, u_2, u_1$ are two distinct u-u_1 paths.   □

THEOREM 7.13   A graph $T = (V, E)$ is a tree if and only if, for all $u, v \in V$, there is a unique u-v path in T.

Proof   Necessity follows from Lemma 7.12 and the fact that a tree is acyclic. Conversely, assume that, for all $u, v \in V$, T contains a unique u-v path. Then T is clearly connected and it follows, again from Lemma 7.12, that T is acyclic. Therefore T is a tree.   □

As a consequence of Theorem 7.13, if T is a tree and e is any edge of T, then the graph $T - e$ is disconnected. (Why?) In fact, if $e = uv$, then $T - e$ contains exactly two connected parts, one containing u and the other containing v. To make this last statement more precise and to introduce several ideas needed in subsequent sections, we need the concepts of "subgraph" and "component."

DEFINITION 7.14

Let $G_1 = (V_1, E_1)$ and $G_2 = (V_2, E_2)$ be two graphs. We call G_1 a *subgraph* of G_2 provided $V_1 \subseteq V_2$ and $E_1 \subseteq E_2$. The fact that G_1 is a subgraph of G_2 is denoted by writing $G_1 \subseteq G_2$.

Much of the terminology and notation for subgraphs are analogous to those for subdigraphs, as introduced in Section 7.4. In particular, for a graph $G = (V, E)$, $v \in V$, and $e \in E$, we use $G - v$ to denote the subgraph of G obtained by deleting the vertex v (and all edges incident with v) and $G - e$ to denote the subgraph obtained by deleting the edge e. Also, for $U \subseteq V$, the notation $\langle U \rangle$ denotes the *subgraph of G induced by U*, that is, the subgraph with vertex set U and edge set consisting of those edges of G joining vertices of U.

Now, given a graph $G = (V, E)$ and $v \in V$, it is clear that the subgraph $\langle \{v\} \rangle$ consisting of only the vertex v is connected. It thus makes sense to consider those subgraphs of G that are maximal with respect to the property of being connected. It turns out that such subgraphs are always induced.

DEFINITION 7.15

Let $G = (V, E)$ be a graph and let $U \subseteq V$. The induced subgraph $\langle U \rangle$ is called a *component* of G provided it is maximal with respect to the property of being connected; that is, $\langle U \rangle$ is connected and $\langle U_1 \rangle$ is disconnected for any subset U_1 of V such that $U \subset U_1$.

Example 7.26   Let $G = (V, E)$, where $V = \{m, n, r, s, t, u, v, w, x, y, z\}$ and $E = \{mn, mu, rs, st, su, uv, uw, vx, vy, vz\}$. Then G is isomorphic to the tree T_2 of Figure 7.24. Since G is connected, G itself is its only component. Let $G_1 = G - uv$. As mentioned above, this subgraph of G is disconnected and consists of two components, each of which is a tree. The component containing u is $\langle \{m, n, r, s, t, u, w\} \rangle$, which has edge set $\{mn, mu, rs, st, su, uw\}$, and the component containing v is $\langle \{v, x, y, z\} \rangle$, which has edge set $\{vx, vy, vz\}$. Let $G_2 = G - u$; this subgraph is a forest with four components: $\langle \{m, n\} \rangle$, $\langle \{r, s, t\} \rangle$, $\langle \{w\} \rangle$, and $\langle \{v, x, y, z\} \rangle$. In general, one can prove that if u is a vertex of a tree T, then $T - u$ is a forest with $\deg(u)$ components.   ∎

In general, an edge e of a connected graph G is called a *bridge* provided $G - e$ is disconnected. Thus every edge of a tree is a bridge; it turns out that trees are partly characterized by this property.

THEOREM 7.14   A graph $T = (V, E)$ is a tree if and only if both T is connected and every edge of T is a bridge.

Proof   Necessity follows from the comments made preceding the theorem. Conversely, it is clear that a tree is connected; so let $T = (V, E)$ be a connected graph with the property that every edge is a bridge. We wish to show that T is acyclic. If not, then T contains a cycle, say

$$u = u_0, v = u_1, u_2, \ldots, u_n = u$$

Now consider the graph $T - e$, where $e = uv$. Note that this graph contains the u-v path

$$u = u_n, u_{n-1}, \ldots, u_2, u_1 = v$$

as in the proof of Theorem 7.13. Therefore $T - e$ is connected, showing that e is not a bridge. This gives a contradiction, thus proving the result.   □

As an important corollary to Theorem 7.14, we can show that every connected graph G contains a *spanning tree*. This is a subgraph T of G such that T is a tree and $V(T) = V(G)$.

COROLLARY 7.15   If $G = (V, E)$ is a connected graph, then G contains a spanning tree.

Proof   Since G is connected, G is a connected spanning subgraph of itself. Thus it makes sense to let $T = (V, E_0)$ be a connected spanning subgraph of G of minimum size (that is, $|E_0|$ is a minimum). We claim that T is a tree. If not, then, by Theorem 7.14, T contains an edge e that is not a bridge. This means that the graph $T_1 = (V, E_0 - \{e\})$ is connected, contradicting the choice of T as the connected spanning subgraph of G of minimum size. This proves the result.   □

Example 7.27 Consider the graph $G = (V, E)$, where $V = \{u, v, x, y, z\}$ and $E = \{uv, ux, vy, vz, xy, xz\}$.

(a) Find a spanning tree T_1 isomorphic to P_5.
(b) Find a spanning tree T_2 isomorphic to the tree T_3 of Figure 7.24.
(c) Give a reason why G cannot have a spanning tree isomorphic to $K_{1,4}$.

Solution

(a) There are six possibilities; for example, let $T_1 = (V, E_1)$, where $E_1 = \{uv, vz, xy, xz\}$.
(b) Again there are six possibilities, one being $T_2 = (V, E_2)$, where $E_2 = \{uv, ux, vy, vz\}$.
(c) Since $\Delta(K_{1,4}) = 4$ and $\Delta(G) = 3$, G cannot have a spanning tree isomorphic to $K_{1,4}$.   ■

Let us summarize the characterizations of trees that we have so far; namely, a graph G is a tree if and only if G satisfies one of the following conditions:

1. G is connected and acyclic (Definition 7.13).
2. There is a unique path joining any two vertices of G (Theorem 7.13).
3. G is connected and every edge is a bridge (Theorem 7.14).

Two additional characterizations are given in the next theorem, with the proofs left to Exercises 11 and 12.

THEOREM 7.16 A graph G of order p and size q is a tree if and only if G satisfies one of the following conditions:

4. G is connected and $q = p - 1$.
5. G is acyclic and $q = p - 1$.

At this point in our discussion of trees, we describe an application to transportation and communication networks. For this we need the concept of a *weighted graph*, which is analogous to the concept of a weighted directed graph introduced in Section 7.4.

Given a graph $G = (V, E)$, suppose there is associated with each $e \in E$ a positive number $w(e)$, called the *weight* of the edge e. The graph G, together with the function $w: E \to (0, \infty)$, is called a *weighted graph*. Given a subgraph $H = (V_0, E_0)$ of G, we define its *weight* to be the sum of the weights of the edges in E_0, and we denote this weight by $w(H)$.

Suppose we are given the problem of connecting p computers with a network of communications links, so that, perhaps indirectly, any computer may transmit data to any other computer. If a vertex is associated with each computer and if it is feasible to install a direct communications link between vertices c_1 and c_2 at a cost $w(c_1 c_2)$, then this number becomes the weight of the edge $c_1 c_2$ in our weighted graph. On the other hand, if it is determined not to be feasible to directly link c_1 and c_2, then c_1 and c_2 are nonadjacent in our weighted graph. Our goal may then be to build such a network so that these conditions are met:

1. It is possible (perhaps indirectly) to send data from any computer to any other computer in the network.
2. The cost of installing the communications links is a minimum.

This means that we desire a connected spanning subgraph of our weighted graph of minimum weight. Clearly such a subgraph is a spanning tree, and it is called a *minimum spanning tree* of the weighted graph.

Example 7.28 Find a minimum spanning tree of the graph G of Example 7.27 if the weights of the edges are $w(uv) = 2$, $w(ux) = 4$, $w(vy) = 4$, $w(vz) = 6$, $w(xy) = 3$, and $w(xz) = 5$.

Solution The graph G has order 5 and size 6 so, to obtain a spanning tree, two edges of G must be excluded. We would like to exclude the two edges of largest weight, namely, vz and xz; however, the resulting subgraph is not connected. The best we can do is exclude vz and one of the edges with weight 4, say ux. The resulting spanning tree T is a minimum spanning tree of G having $w(T) = w(uv) + w(vy) + w(xy) + w(xz) = 14$.   ■

In working with spanning trees, we use a particularly helpful operation on graphs. Let $G = (V, E)$ be a graph and let $u, v \in V$ be nonadjacent vertices. Then $G + uv$ denotes the graph obtained by adding the edge uv to G. Note that G is a proper spanning subgraph of $G + uv$. Now let $T = (V, E_0)$ be a spanning tree of G and let $e \in E - E_0$. It can be shown (see Exercise 13) that $T + e$ contains a unique cycle. If f is an edge of this cycle and $f \neq e$, then $T_1 = (T + e) - f$ is another spanning tree of G. By repeating this operation of "edge replacement" a finite number of times, it is possible to transform a given spanning tree T_1 into any other spanning tree T_2 of G (see Exercise 14).

Example 7.29 Find all minimum spanning trees of the weighted graph $G = (V, E)$ shown in Figure 7.26.

Solution Note that G has seven edges, while a spanning tree of G has five edges. Thus, to obtain a spanning tree, two edges of G must be excluded. If we exclude the two edges of largest weight, namely, uv and vw, the result is not a tree. Next let us try $T_0 = (G - uw) - yz$. This is a spanning tree of G and it is not hard to see that $T_0 = (V, E_0)$ is a minimum spanning tree, with weight 22. Now we ask, is it possible to replace an edge e_0 of T_0 with an edge e_1 of $E - E_0$ such that $w(e_0) = w(e_1)$ and the resulting graph

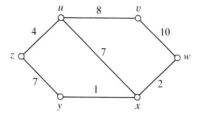

Figure 7.26

$T_1 = (T_0 + e_1) - e_0$ is acyclic? If so, then T_1 is another minimum spanning tree of G. Note that $T_0 + yz$ contains the cycle u, x, y, z, u. On this cycle, the edge ux has the same weight as the edge yz. Thus $T_1 = (T + yz) - ux$ is also a minimum spanning tree of G. Finally it can be argued that T_0 and T_1 are the only minimum spanning trees of G. ■

Given a connected weighted graph G, we desire an algorithm that produces a minimum spanning tree of G. One well-known algorithm for this problem was discovered by J. B. Kruskal, Jr., in 1956. Another algorithm is due to R. C. Prim and was discovered in 1957. Since Prim's algorithm is similar to Dijkstra's algorithm for finding shortest paths, we leave it to Exercise 22 and present Kruskal's algorithm next.

Let us first describe the algorithm informally by applying it to the graph $G = (V, E)$ of Figure 7.26. Our goal is to find a minimum spanning tree $T = (V, E_0)$ of G. We begin by sorting the edges of G into a list that is in nondecreasing order by weight; this gives the list $L = (xy, wx, uz, ux, yz, uv, vw)$. Initially $E_0 = \emptyset$; the main idea of the algorithm is to add to E_0, at each iteration, an edge of L of smallest weight such that (V, E_0) remains acyclic. In this sense, Kruskal's algorithm is an example of an important class of algorithms known as "greedy" algorithms. In our example, the first four edges of L may be added to E_0; this yields the forest $(V, \{ux, uz, wx, xy\})$. At this point, the next edge of L is yz; however, the graph $(V, \{ux, uz, wx, xy, yz\})$ contains the cycle u, x, y, z, u. Thus the edge yz is excluded from consideration. The next edge of L, namely, uv, causes no such problem, so that this edge may be added to E_0. This yields the minimum spanning tree $T = T_0 = (V, E_0)$, where $E_0 = \{uv, ux, uz, wx, xy\}$.

Several pertinent questions arise in connection with Kruskal's algorithm as just described. For one, how does one check that the subgraph (V, E_0) is acyclic at each stage? Suppose that $F = (V, E_0)$ is acyclic and $e = uv$ is an edge in L of smallest weight. We wish to know whether $(V, E_0 \cup \{e\})$ is acyclic. Bear in mind that, since F is acyclic, each of its components is a tree. Thus, if u and v belong to the same component of F, then $F + e$ contains a cycle, while if u and v belong to the different components of F, then $F + e$ remains acyclic. So the problem of checking whether $F + e$ is acyclic reduces to the problem of checking whether u and v belong to different components of F. This is handled by associating with each vertex v a number $c(v)$ (between 1 and $|V|$), such that $c(u) = c(v)$ if and only if u and v belong to the same component of (V, E_0). Initially the values $c(v)$, where $v \in V$, are distinct, representing each of the values $1, 2, \ldots, |V|$ exactly once. As the algorithm progresses and an edge uv is added to E_0, with $c(u) < c(v)$, then $c(x)$ is assigned the number $c(u)$ for every vertex x such that $c(x) = c(v)$. To this end, it is convenient to let V_i be the set of vertices v with $c(v) = i$, $1 \le i \le |V|$.

ALGORITHM 7.5 **(Kruskal)**

Given a connected weighted graph $G = (V, E, w)$, this algorithm produces a minimum spanning tree $T = (V, E_0)$ of G. The algorithm uses variables $c(v)$ and sets V_i as described in the preceding paragraph.

Step 0. Initially $E_0 := \emptyset$; $n := 0$. Repeat Step 1 once for each $v \in V$.
Step 1. $n := n + 1$; $c(v) := n$; $V_n := \{v\}$.
Step 2. $L := (e_1, e_2, \ldots, e_q)$ (L is a list of the edges of G in nondecreasing order by weight); $n := 1$.
Step 3. Repeat the following steps until $|E_0| = |V| - 1$. At that point, the algorithm terminates.
Step 4. Let u and v be the incident vertices of e_n, where $c(u) \le c(v)$. If $c(u) \ne c(v)$, then execute Step 5; otherwise proceed directly to Step 6.
Step 5. $k := c(u)$ and $m := c(v)$; for each $x \in V_m$, let $c(x) := k$; $V_k := V_k \cup V_m$; $V_m := \emptyset$; $E_0 := E_0 \cup \{uv\}$.
Step 6. $n := n + 1$.   □

Example 7.30 Illustrate Kruskal's algorithm using the weighted graph $G = (V, E, w)$ of Figure 7.27.

Solution In Step 0, $E_0 := \emptyset$. In Step 1, suppose $c(r) := 1$, $c(s) := 2$, $c(t) := 3$, $c(w) := 4$, $c(x) := 5$, $c(y) := 6$, and $c(z) := 7$. This means that $V_1 := \{r\}$, $V_2 := \{s\}$, $V_3 := \{t\}$, $V_4 := \{w\}$, $V_5 := \{x\}$, $V_6 := \{y\}$, and $V_7 := \{z\}$.

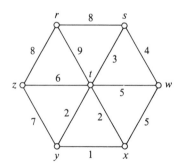

Figure 7.27

Next, in Step 2, we sort the edges of G in nondecreasing order by weight; suppose this yields the list $L := (e_1, e_2, \ldots, e_{12})$, where $e_1 = xy$, $e_2 = tx$, $e_3 = ty$, $e_4 = st$, $e_5 = sw$, $e_6 = tw$, $e_7 = wx$, $e_8 = tz$, $e_9 = yz$, $e_{10} = rs$, $e_{11} = rz$, and $e_{12} = rt$. Also $n := 1$.

We now repeat Steps 4, 5, and/or 6 until $|E_0| = |V| - 1 = 6$.

Step 4. We have $u := x$, $v := y$. Since $c(u) = 5$ and $c(v) = 6$, we execute Steps 5 and 6, giving $c(y) := 5$, $V_5 := \{x, y\}$, $V_6 := \emptyset$, $E_0 := \{xy\}$, and $n := 2$.

Step 4. We have $u := t$, $v := x$. Since $c(u) = 3$ and $c(v) = 5$, we execute Steps 5 and 6, giving $c(x) := 3$, $c(y) := 3$, $V_3 := \{t, x, y\}$, $V_5 := \emptyset$, $E_0 := \{tx, xy\}$, and $n := 3$.

Step 4. Here we have $u := t$, $v := y$. Since $c(u) = c(v) = 3$, we execute Step 6, so that $n := 4$.

Step 4. We have $u := s$, $v := t$. Since $c(u) = 2$ and $c(v) = 3$, we execute Steps 5 and 6, giving $c(t) := 2$, $c(x) := 2$, $c(y) := 2$, $V_2 := \{s, t, x, y\}$, $V_3 := \emptyset$, $E_0 := \{st, tx, xy\}$, and $n := 5$.

Step 4. We have $u := s$, $v := w$. Since $c(u) = 2$ and $c(v) = 4$, we execute Steps 5 and 6, giving $c(w) := 2$, $V_2 := \{s, t, w, x, y\}$, $V_4 := \emptyset$, $E_0 := \{st, sw, tx, xy\}$, and $n := 6$.

Step 4. Here we have $u := t$, $v := w$. Since $c(u) = c(v) = 2$, we proceed to Step 6, so that $n := 7$.

Step 4. Here we have $u := w$, $v := x$. Since $c(u) = c(v) = 2$, we again proceed to Step 6, giving $n := 8$.

Step 4. We have $u := t$, $v := z$. Since $c(u) = 2$ and $c(v) = 7$, we execute Steps 5 and 6, giving $c(z) := 2$, $V_2 := \{s, t, w, x, y, z\}$, $V_7 := \emptyset$, $E_0 := \{st, sw, tw, tz, xy\}$, and $n := 9$.

Step 4. Here $u := y$, $v := z$. Note that $c(u) = c(v) = 2$, so that $n := 10$ in Step 6.

Step 4. Here $u := r$, $v := s$. Note that $c(u) = 1$ and $c(v) = 2$, so that Steps 5 and 6 are executed. This yields $c(s) := 1$, $c(t) := 1$, $c(w) := 1$, $c(x) := 1$, $c(y) := 1$, $c(z) := 1$, $V_1 := \{r, s, t, w, x, y, z\}$, $V_2 := \emptyset$, $E_0 := \{rs, st, sw, tw, tz, xy\}$, and $n := 11$. However, we now have $|E_0| = 6$, so the algorithm terminates. The minimum spanning tree $T = (V, E_0)$, where $E_0 = \{rs, st, sw, tw, tz, xy\}$, has been produced. ∎

That Algorithm 7.5 produces the edge set of a spanning tree of G is clear, since $|E_0| = |V| - 1$ and (V, E_0) is acyclic. The next result states that the spanning tree (V, E_0) is, in fact, a minimum spanning tree.

THEOREM 7.17   Let $G = (V, E, w)$ be a connected weighted graph and let $T = (V, E_0)$ be a spanning tree of G that is produced by an application of Algorithm 7.5. Then T is a minimum spanning tree of G.

Proof   Since the number of spanning trees of G is finite, we may let $T_1 = (V, E_1)$ be a minimum spanning tree of G having a maximum number of edges in common with T. If $E_0 = E_1$, then $T = T_1$ and the proof is done, so assume $E_0 \neq E_1$. Let $E_0 = \{e_1, e_2, \ldots, e_{p-1}\}$ such that $w(e_1) \leq w(e_2) \leq \cdots \leq w(e_{p-1})$. Choose i as the smallest subscript such that $e_i \in E_0 - E_1$. It follows that the graph $T_1 + e_i$ contains a unique cycle, say C. Since T is a tree, there must be an edge f of C such that $f \notin E_0$. Now consider the graph $T_2 = (T_1 + e_i) - f$. By our earlier remarks concerning the operation of edge replacement, T_2 is also a spanning tree of G and its weight is

$$w(T_2) = w(T_1) + w(e_i) - w(f)$$

Since T_1 is a minimum spanning tree, it must be that $w(e_i) \geq w(f)$. In addition, $F = \langle \{e_1, \ldots, e_{i-1}, f\} \rangle$ is an acyclic subgraph of G, since the edges that induce F all come from T_1. If $w(f) < w(e_i)$, then f occurs before e_i in any list L of the edges of G sorted by weight, and so f would have been chosen, instead of e_i, as an edge of T, a contradiction to our assumption that T is produced by an application of Algorithm 7.5. Hence we must conclude that $w(e_i) = w(f)$, so that $w(T_2) = w(T_1)$. Therefore T_2 is a minimum spanning tree of G. But T_2 has one more edge in common with T than T_1 does, and this contradicts our choice of T_1. This completes the proof.   □

Exercises 7.6

1. Find all nonisomorphic trees of order 6 with maximum degree 4.
2. Find all nonisomorphic trees of order 7.
3. Find all nonisomorphic spanning trees of the graph $G = (V, E)$, where $V = \{u, v, w, x, y, z\}$ and $E = \{uv, ux, uy, uz, vy, wx, wz, xy, yz\}$.
4. For each weighted graph of Figure 7.28, apply Algorithm 7.5 to find a minimum spanning tree.
5. Apply Algorithm 7.5 to find a minimum spanning tree of the weighted graph $G = (V, E, w)$, where $V = \{r, s, t, x, y, z\}$, $E = \{rs, rx, ry, rz, sy, sx, tx, xy, yz\}$, and $w(rs) = 12$, $w(rx) = 6$, $w(ry) = 1$, $w(rz) = 5$, $w(sy) = 6$, $w(sx) = 5$, $w(tx) = 10$, $w(xy) = 9$, and $w(yz) = 3$.
6. Prove or disprove: If the edges of the connected weighted graph $G = (V, E, w)$ have distinct weights, then G has a unique minimum spanning tree.
7. Find a minimum spanning tree of the graph of Figure 7.13, assuming that $w(bv) = w(uv) = w(vx) = w(vy) < w(au) = w(az) = w(bz) = w(cx) = w(uy) = w(xy) < w(bu) = w(bx) < w(cy) < w(ab) = w(bc) < w(ad) = w(cd)$
8. Let T be a tree of order $n > 3$ and maximum degree Δ.
   (a) Show that $T \simeq P_n$ if and only if $\Delta = 2$.
   (b) Show that $T \simeq K_{1, n-1}$ if and only if $\Delta = n - 1$.

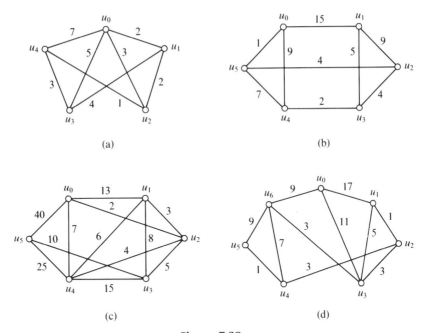

Figure 7.28

9. For $n > 4$, find the unique tree of order n, up to isomorphism, with maximum degree $n - 2$.

10. For $n > 5$, find the three nonisomorphic trees of order n with maximum degree $n - 3$.

11. Prove characterization 4 given in Theorem 7.16: a graph G of order p and size q is a tree if and only if G is connected and $q = p - 1$. (Hint: To prove necessity, proceed by induction on p; to prove sufficiency, let T be a spanning tree of G and show that $G = T$.)

12. Prove characterization 5 given in Theorem 7.16: a graph G of order p and size q is a tree if and only if G is acyclic and $q = p - 1$.

13. Let $G = (V, E)$ be a connected graph and let $T = (V, E_0)$ be a spanning tree of G. Let $e \in E - E_0$.
    (a) Show that $T + e$ contains a unique cycle; call it C.
    (b) Let f be an edge of C, where $f \neq e$. Show that $T_1 = (T + e) - f$ is a spanning tree of G.
    (c) If G has order p, how many edges do the trees T and T_1 have in common?

14. Let $G = (V, E)$ be a connected graph and let $T_1 = (V, E_1)$ and $T_2 = (V, E_2)$ be spanning trees of G. Show that T_2 may be obtained from T_1 by performing a finite sequence of edge replacement operations.

15. Every tree T of order $p > 1$ has at least two vertices of degree 1. Prove this fact:
   (a) using the fact that T has $p - 1$ edges
   (b) by considering the first and last vertices of a longest path

16. With regard to the result of Exercise 15, suppose further that T has maximum degree Δ.
   (a) Show that T has at least Δ vertices of degree 1.
   (b) What can be said about T if it has exactly Δ vertices of degree 1?

17. Let $T_1 = (V_1, E_1)$ and $T_2 = (V_2, E_2)$ be trees, where $V_1 = \{u_1, u_2, \ldots, u_n\}$, $V_2 = \{v_1, v_2, \ldots, v_n\}$, and $\deg(u_i) = \deg(v_i)$, $1 \le i \le n$.
   (a) If $n \le 5$, show that $T_1 \simeq T_2$.
   (b) For $n \ge 6$, give an example of two nonisomorphic such trees.

18. Prove that every tree is a bipartite graph.

19. Let F be a forest and let u and v be distinct nonadjacent vertices of F. Show that $F + uv$ is a forest if and only if u and v belong to different components of F.

20. Given that a forest F has order p and m components, what is its size q?

21. Let G be a graph. Prove that G is a forest if and only if every connected subgraph is an induced subgraph.

22. Given a weighted graph $G = (V, E, w)$, Prim's algorithm produces a minimum spanning tree $T = (V, E_0)$ of G. Here it is.

   Step 0. Initially $E_0 := \emptyset$; choose any vertex v of V and let $U := V - \{v\}$.

   Step 1. Repeat the following steps until $U = \emptyset$. At that point, the algorithm terminates.

   Step 2. Let $e = uv$ be an edge of smallest weight such that $u \in U$ and $v \in V - U$.

   Step 3. $U := U - \{u\}$; $E_0 := E_0 \cup \{uv\}$.

   (a) Show that Prim's algorithm produces a spanning tree.
   ★(b) Show that Prim's algorithm produces a minimum spanning tree.
   (c) Use Prim's algorithm to do Exercise 4.
   (d) Use Prim's algorithm to do Exercise 5.

23. Let T be a tree of order at least 4 that is not a star. Show that T has a vertex v such that $\deg(v) \ge 2$ and every vertex u of T adjacent to v, with one exception, has $\deg(u) = 1$.

24. Let $G = (V, E, w)$ be a connected weighted graph and let $T = (V, E_0)$ be a minimum spanning tree of G. Is it true, for some choice of the list L in Step 2, that Algorithm 7.5 produces T? In other words, is every minimum spanning tree of G obtainable from Algorithm 7.5?

25. Discuss what happens if Algorithm 7.5 is applied to a disconnected weighted graph $G = (V, E, w)$. Does the algorithm produce, or can it be modified to produce, a minimum spanning forest of G of maximum size?

HAMILTONIAN GRAPHS

7.7

In a previous section we considered the problem of determining whether a given graph (or digraph) contains a circuit that includes each of its edges, that is, an eulerian circuit. Another famous problem in graph theory is to determine whether a given graph (or digraph) contains a cycle through each of its vertices. This question, considered for graphs only, is the focus of this section.

DEFINITION 7.16

Let G be a graph. A path of G that includes every vertex is called a *hamiltonian path*, and a cycle that includes every vertex is called a *hamiltonian cycle*. A graph that has a hamiltonian path is called *traceable*, while a graph that has a hamiltonian cycle is called *hamiltonian*.

Hamiltonian graphs are named after the famous Irish mathematician, Sir William Rowan Hamilton. Hamilton had an idea for a puzzle in 1857. It made use of a solid wooden dodecahedron, with its 20 vertices labeled with important cities of the day. The object was to find a route along the edges of the dodecahedron that passed through each city exactly once, starting and ending at a given city. To keep track of which cities had been visited, the player was given 20 pegs and a long piece of string; the pegs were inserted at the vertices of the dodecahedron and the string was used to connect the pegs. Hamilton's puzzle was not commercially successful, but it was a forerunner of such modern-day puzzles as "Instant Insanity" and "Rubik's Cube." Perhaps Hamilton's puzzle was too easy to solve; the reader may attempt to find a hamiltonian cycle in the graph of the dodecahedron shown in Figure 7.29(a).

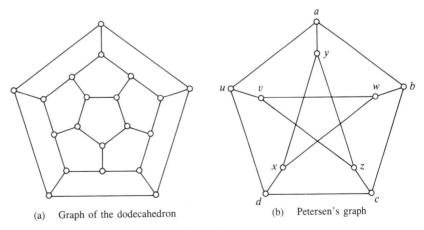

(a)   Graph of the dodecahedron

(b)   Petersen's graph

Figure 7.29

Of course, not all graphs are hamiltonian. The graph shown in Figure 7.29(b) is known as *Petersen's graph*. This graph is traceable, with one hamiltonian path being $a, b, c, d, u, v, w, x, y, z$. However, the Petersen graph is not hamiltonian, although this fact is surprisingly difficult to verify. The reader is encouraged to experiment with it until he or she is convinced that no hamiltonian cycle exists.

Unlike the situation for eulerian graphs, there is no useful necessary and sufficient condition known for a graph to be hamiltonian. We can develop some necessary conditions and some sufficient conditions but, in general, the problem of determining whether a given graph is hamiltonian is very difficult. In fact, it is a special case of a more general problem known as the "traveling salesman problem," that of finding a "shortest" hamiltonian cycle in a weighted graph. The hamiltonian problem is a member of an important class of problems in computer science known as "NP-complete" problems. These are very interesting for several reasons, one being the following: no (deterministic) polynomial-time algorithms are known for any of them; yet, if a polynomial-time algorithm is found for any one of the problems in the class, then polynomial-time algorithms can be constructed for all of them, or, if it is proven that no polynomial-time algorithm exists for some problem in the class, then no polynomial-time algorithms exist for any of them.

As far as necessary conditions for hamiltonian graphs are concerned, it is an immediate consequence of the definition that a hamiltonian graph is connected and has order at least 3. In fact, the removal of any vertex from a hamiltonian graph results in a graph that still contains a hamiltonian path and thus is still connected. In general, a graph $G = (V, E)$ of order at least 3 is called *2-connected* provided $G - v$ is connected for each $v \in V$. Thus we may remark that if a graph G is hamiltonian, then G is 2-connected.

Let $G = (V, E)$ be a hamiltonian graph and let $H = (V, E_1)$ be a spanning subgraph of G induced by the edges of some hamiltonian cycle of G. Suppose $|V| = n$ and $v_1, v_2, \ldots, v_m$ are distinct vertices, where $m < n$. Note that the graph $H - v_1$ is connected and that $(H - v_1) - v_2$ has at most two components. Similarly, $((H - v_1) - v_2) - v_3$ has at most three components, and so on. In general, if m vertices are removed from H, the resulting graph has at most m components. This observation leads to the following necessary condition for a graph G to be hamiltonian.

THEOREM 7.18   Let $G = (V, E)$ be a graph of order at least 3. If G is hamiltonian, then, for every proper subset U of V, the subgraph of G induced by $V - U$ has at most $|U|$ components.

Proof   Let $G = (V, E)$ be a hamiltonian graph, and let $H = (V, E_1)$ be a spanning subgraph of G induced by the edges of some hamiltonian cycle of G. Let U

be a proper subset of V, let m_1 denote the number of components in the subgraph of G induced by $V - U$, and let m_2 denote the number of components in the subgraph of H induced by $V - U$. Since H is a spanning subgraph of G, it is clear that $m_1 \leq m_2$. Also, by the remarks preceding the theorem, we have that $m_2 \leq |U|$. Hence $m_1 \leq |U|$, as was to be shown. $\square$

Example 7.31   Show that the graph $G = (V, E)$ of Figure 7.30 is not hamiltonian.

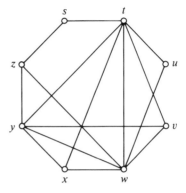

Figure 7.30

Solution   Let $U = \{t, w, y\}$ and apply Theorem 7.18. Note that $|U| = 3$, whereas the subgraph of G induced by $V - U$ has four components: $\langle \{s, z\} \rangle$, $\langle \{u\} \rangle$, $\langle \{v\} \rangle$, and $\langle \{x\} \rangle$. Thus G is not hamiltonian.   ∎

We now turn to sufficient conditions for hamiltonian cycles. Here one basic approach is to attempt to force the graph to have a relatively large number of edges, the idea being that a sufficiently "dense" graph is certain to possess a hamiltonian cycle. Of such results, one of the simplest to apply goes back to O. Ore and was published in 1960.

THEOREM 7.19   (Ore, 1960)

Let G be a graph of order $n \geq 3$ such that, for every pair of distinct nonadjacent vertices u and v,

$$\deg(u) + \deg(v) \geq n$$

Then G is hamiltonian.

In looking at the proof of Ore's theorem, J. A. Bondy and V. Chvátal observed that, in fact, a much stronger statement could be proved. (Recall that, if u and v are nonadjacent vertices of a graph G, then $G + uv$ denotes the graph obtained from G by adding the edge uv.)

THEOREM 7.20   **(Bondy and Chvátal, 1976)**

Let G be a graph of order $n \geq 3$ and suppose that u and v are distinct nonadjacent vertices of G such that

$$\deg(u) + \deg(v) \geq n$$

Then G is hamiltonian if and only if $G + uv$ is hamiltonian.

Proof   Let $G = (V, E)$, u, and v be as in the statement of the theorem. It is clear that, if G is hamiltonian, then $G + uv$ is hamiltonian.

To show the converse, we proceed by contradiction; assume that $G + uv$ is hamiltonian but G is not hamiltonian. Then G is traceable and contains a spanning u-v path, say,

$$u = u_1, u_2, \ldots, u_n = v$$

We now claim that there is some vertex u_k, $1 < k < n - 1$, such that $u_1 u_{k+1} \in E$ and $u_k u_n \in E$. Suppose this is not the case. Then for each of the $\deg(u_1)$ values of k ($1 \leq k < n - 1$) for which $u_1 u_{k+1} \in E$, we have $u_k u_n \notin E$. Thus

$$\deg(u) + \deg(v) = \deg(u_1) + \deg(u_n)$$
$$\leq \deg(u_1) + (p - 1 - \deg(u_1))$$
$$= p - 1$$

contradicting the hypothesis of the theorem. Now notice that the cycle

$$u_1, u_{k+1}, u_{k+2}, \ldots, u_n, u_k, u_{k-1}, \ldots, u_2, u_1$$

is a hamiltonian cycle of G.   □

Example 7.32   Consider the graph $G = G_0 = (V, E_0)$ shown in Figure 7.31. Note that G_0 has order 8 and that w and z are nonadjacent vertices such that

$$\deg(w) + \deg(z) = 8$$

Hence, by Theorem 7.20, G_0 is hamiltonian if $G_1 = G_0 + wz$ is hamil-

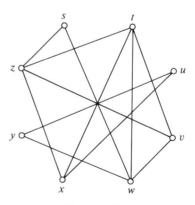

Figure 7.31

tonian. Now consider G_1; in this graph, vertices w and x are nonadjacent with $\deg(w) + \deg(x) = 8$. Again, using Theorem 7.20, we see that G_1 is hamiltonian if $G_2 = G_1 + wx$ is hamiltonian. We may continue to use this procedure, obtaining a sequence of graphs $G_0, G_1, G_2, \ldots, G_{16}$, such that G_{16} is isomorphic to K_{16}. Each of $G_1, G_2, \ldots, G_{16}$ is obtained from its predecessor by the addition of a single edge, and each of $G_0, G_1, \ldots, G_{15}$ is hamiltonian provided its successor in the sequence is hamiltonian. Thus, since any complete graph is obviously hamiltonian, we may conclude that G_0 is hamiltonian. ∎

Let us generalize the process of the preceding example. Given a graph G of order n, let $G_0 = G$. If G_0 contains two nonadjacent vertices u_0 and v_0 such that $\deg(u_0, G_0) + \deg(v_0, G_0) \geq n$, let $G_1 = G_0 + u_0 v_0$. Next, if G_1 contains two nonadjacent vertices u_1 and v_1 such that $\deg(u_1, G_1) + \deg(v_1, G_1) \geq n$, then let $G_2 = G_1 + u_1 v_1$. Continue this process, obtaining a sequence of graphs $G = G_0, G_1, \ldots, G_k$, $k \geq 0$, such that G_k does not contain two nonadjacent vertices whose degree sum in G_k is at least n. Then G is hamiltonian if G_k is hamiltonian. In particular, if G_k is a complete graph, then G is hamiltonian.

Bondy and Chvátal called G_k the "closure" of G. To be a bit more precise, perhaps it is instructive to give a recursive definition of the closure operation.

DEFINITION 7.17

Given a graph G of order n, the *closure* of G is the graph $Cl(G)$ defined recursively as follows: If G contains two distinct nonadjacent vertices u and v such that $\deg(u) + \deg(v) \geq n$, then $Cl(G) = Cl(G + uv)$; otherwise, $Cl(G) = G$.

It can be shown that the closure operation is well-defined (see Exercise 17). Also note that G is always a spanning subgraph of $Cl(G)$.

The next result is an immediate corollary of Theorem 7.20.

COROLLARY 7.21 Let G be a graph of order $n \geq 3$. Then G is hamiltonian if and only if $Cl(G)$ is hamiltonian.

The next result presents a series of sufficient conditions for a graph to be hamiltonian. These are listed so that the corresponding theorems are in order of decreasing strength; that is, each condition from the second on implies the condition preceding it. Moreover, notice that the theorems are, with one exception, in reverse chronological order as to date of discovery.

COROLLARY 7.22 Let $G = (V, E)$ be a graph of order $n \geq 3$ and let $d_1, d_2, \ldots, d_n$ be the degrees of the vertices of G, listed so that $d_1 \leq d_2 \leq \cdots \leq d_n$. If G satisfies any of the following conditions, then G is hamiltonian.

(a) (Bondy and Chvátal, 1976) $Cl(G) \simeq K_n$.

(b) (M. Las Vergnas, 1971) There exists a bijection $f: V \to \{1, 2, \ldots, n\}$ such that, if u and v are distinct nonadjacent vertices with $f(u) < f(v)$, $f(u) + f(v) \geq n$, $\deg(u) \leq f(u)$, and $\deg(v) < f(v)$, then $\deg(u) + \deg(v) \geq n$.

(c) (Chvátal, 1972) For each i with $2 \leq 2i < n$, if $d_i \leq i$, then $d_{n-i} \geq n - i$.

(d) (Bondy, 1969) For all i, j with $1 \leq i < j \leq n$, if $d_i \leq i$ and $d_j \leq j$, then $d_i + d_j \geq n$.

(e) (Pósa, 1962) For each i with $2i < n$, $|\{v \mid \deg(v) \leq i\}| < i$.

(f) (Ore, 1960) For any two distinct nonadjacent vertices u and v, $\deg(u) + \deg(v) \geq n$.

(g) (G. A. Dirac, 1952) For any vertex v, $2\deg(v) \geq n$.

As mentioned, each of conditions (b) through (g) in Corollary 7.22 implies the conditions preceding it; that is, (g) $\to$ (f) $\to$ (e) $\to$ (d) $\to$ (c) $\to$ (b) $\to$ (a). It is a direct consequence of Corollary 7.21 that condition (a) is sufficient for a graph G to be hamiltonian, and it then follows that each of conditions (b) through (g) is sufficient as well.

It can also be verified that no two of the conditions (a) through (g) are equivalent. To say this another way, let S_a denote the set of graphs (of order $n \geq 3$) that satisfy condition (a), let S_b denote the set of graphs that satisfy condition (b), and define sets $S_c, S_d, \ldots, S_g$ in a similar manner. Then what we are saying is that

$$S_g \subset S_f \subset \cdots \subset S_b \subset S_a$$

(recall that $\subset$ denotes "is a proper subset of"). Verification of most of the above inclusions is left to the exercises, but we prove a few of them as examples to illustrate the general approach. (Also recall that, to show $S \subset T$, we need to show $S \subseteq T$ and to exhibit an element $x \in T - S$.)

Example 7.33 We show that $S_d \subseteq S_c$. To do this, let G be a graph of order n and let $d_1 \leq d_2 \leq \cdots \leq d_n$ be as in the statement of Corollary 7.22. We proceed by contrapositive, showing that if G does not satisfy Chvátal's condition (c), then G does not satisfy Bondy's condition (d). If G does not satisfy condition (c), then for some $i, 2 \leq 2i < n$, we have $d_i \leq i$ and $d_{n-i} < n - i$. Let $j = n - i$; then $1 \leq i < j \leq n$, $d_i \leq i$, $d_j \leq j$, but $d_i + d_j < i + (n - i) = n$. Thus G does not satisfy condition (d). ∎

Example 7.34 We give an example of a graph $G_a \in S_a - S_b$ and an example of a graph $G_b \in S_b - S_c$.

The graph $G_a = (V_a, E_a)$ is shown in Figure 7.32(a). It is easy to verify that $Cl(G_a) \simeq K_7$, so that $G_a \in S_a$. To show that $G_a \notin S_b$, let f be any

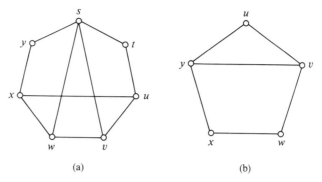

Figure 7.32

bijection from V_a to $\{1, 2, 3, 4, 5, 6, 7\}$. Of the two vertices of degree 2, let z_1 denote the one with the larger f-value. Among the four vertices not adjacent to z_1, let z_2 denote the one with the largest f-value. If $f(z_1) < f(z_2)$, then we have $f(z_1) \geq 2 = \deg(z_1)$, $f(z_2) - 1 \geq 3 \geq \deg(z_2)$, and $f(z_1) + f(z_2) \geq 7$. On the other hand, if $f(z_2) < f(z_1)$, then we have $f(z_2) \geq 4 > \deg(z_2)$, $f(z_1) - 1 \geq 4 > \deg(z_1)$, and $f(z_2) + f(z_1) \geq 7$. However, in either case, $\deg(z_1) + \deg(z_2) \leq 5$, so that G_a does not satisfy condition (b).

The graph $G_b = (V_b, E_b)$ is shown in Figure 7.32(b). Define $f: V_b \to \{1, 2, 3, 4, 5\}$ by $f(u) = 1$, $f(v) = 5$, $f(w) = 3$, $f(x) = 2$, $f(y) = 4$. Then it can be checked that f satisfies condition (b). For condition (c), we have $d_1 = d_2 = d_3 = 2$, $d_4 = d_5 = 3$. Let $i = 2$; note that $d_2 \leq 2$ but that $d_3 < 3$. Thus G_b does not satisfy condition (c), and so $G_b \in S_b - S_c$.   ∎

It should be mentioned that none of conditions (a) through (g) in Corollary 7.22 is a necessary condition for a graph to be hamiltonian. In fact, a result of C. St. J. A. Nash-Williams says that if G is a $2m$-regular graph of order $4m + 1$, then G is hamiltonian. Note that such a graph G does not satisfy Dirac's condition (g).

We remark that the Bondy–Chvátal condition, besides being easy to apply, can be developed into an algorithm for finding a hamiltonian cycle in any graph G of "Bondy–Chvátal type," that is, one for which $Cl(G) \simeq K_n$.

First of all, it is a simple matter to develop an algorithm for finding the closure $Cl(G)$ of a graph G of order n. Moreover, assuming $G = (V, E_0)$ has size q_0 and $Cl(G) = (V, E_1)$ has size q_1, we can have the closure-finding algorithm construct a "weight function" $w: E_1 \to \{1, 2, \ldots, q_0, \ldots, q_1\}$ so that w is a bijection and $w(E_0) = \{1, 2, \ldots, q_0\}$. In this way, we can tell from the weight of an edge whether it is an edge of G or not. Development of such an algorithm is left to Exercise 18.

We next find, by some manner, a hamiltonian cycle C_1 in $Cl(G)$. Of course, if $Cl(G)$ is complete, then finding such a cycle is trivial. What is needed at this point is a method for modifying the cycle C_1 into a hamiltonian cycle C_0 of the original graph G. For this, the idea of the proof of Theorem 7.20 is used.

ALGORITHM 7.6 Let $G = (V, E_0)$ be a graph of order n and size q_0, let $Cl(G) = (V, E_1)$ be the closure of G, of size q_1, let $w: E_1 \to \{1, 2, \ldots, q_1\}$ be a bijection with $w(E_0) = \{1, 2, \ldots, q_0\}$, and let C_1 be a hamiltonian cycle of $Cl(G)$. This algorithm finds a hamiltonian cycle C_0 of G.

> begin
>   loop
>   **Step 1.** Let uv be the edge of C_1 for which $w(uv)$ is a maximum. Exit from the loop when $w(uv) \le q_0$.
>   **Step 2.** Let
> $$u = u_1, u_2, \ldots, u_n = v$$
>   be the hamiltonian path obtained by deleting uv from C_1. As in the proof of Theorem 7.20, find a subscript k, $1 < k < n - 1$, such that $uu_{k+1} \in E_1$, $u_k v \in E_1$, $w(uu_{k+1}) < w(uv)$, and $w(u_k v) < w(uv)$.
>   **Step 3.** As in the proof of Theorem 7.20, replace C_1 by the cycle
> $$u, u_{k+1}, u_{k+2}, \ldots, v, u_k, u_{k-1}, \ldots, u_2, u$$
>   end loop
> end   □

It can be argued that Algorithm 7.6 is $O(n^3)$; this and a verification of correctness are left to Exercise 19.

Example 7.35 We apply Algorithm 7.6 to the graph $G = (V, E_0)$, where $V = \{r, s, t, x, y, z\}$ and $E_0 = \{rs, rt, ry, sx, sz, ty, xy, xz\}$. Here $q_0 = 8$ and $Cl(G)$ is complete; suppose that $w: E_0 \to \{1, 2, \ldots, 15\}$ is given by $w(rs) = 1$, $w(rt) = 2$, $w(ry) = 3$, $w(sx) = 4$, $w(sz) = 5$, $w(ty) = 6$, $w(xy) = 7$, $w(xz) = 8$, $w(rx) = 9$, $w(rz) = 10$, $w(st) = 11$, $w(sy) = 12$, $w(tx) = 13$, $w(tz) = 14$, $w(yz) = 15$. Also suppose that C_1 is the cycle

$$r, s, t, x, y, z, r$$

In Step 1, we find that yz is the edge of C_1 for which w is a maximum. Considering the path

$$y = u_1, x = u_2, t = u_3, s = u_4, r = u_5, z = u_6$$

note that $w(ty) < w(yz)$ and $w(xz) < w(yz)$, so that $k = 2$ in Step 2. In Step 3, C_1 becomes the cycle

$$y, t, s, r, z, x, y$$

Repeating Step 1, we find that $w(st) = 11$ is the maximum weight of an edge of C_1. Considering the path

$$s = u_1, r = u_2, z = u_3, x = u_4, y = u_5, t = u_6$$

we find that $w(sz) < w(st)$ and $w(rt) < w(st)$, so again $k = 2$ in Step 2. Thus, in Step 3, C_1 becomes the cycle

$$s, z, x, y, t, r, s$$

Now back to Step 1, where we find that $w(xz) = 8$ is the maximum weight of an edge of C_1. Since $q_0 = 8$, this means that each edge of C_1 is an edge of G. Hence the loop is exited, and the algorithm terminates after letting C_0 be the hamiltonian cycle s, z, x, y, t, r, s of G.   ■

Exercises 7.7

1. Show that the graphs of the five "regular polyhedra" are hamiltonian. These are the graphs of the dodecahedron, icosahedron, octahedron, cube, and tetrahedron shown in Figure 7.33.

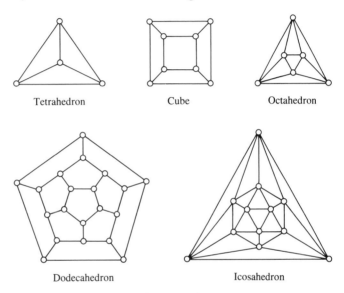

Tetrahedron   Cube   Octahedron

Dodecahedron   Icosahedron

Figure 7.33

2. Let m and n be integers with $1 \leq m \leq n$, and consider the complete bipartite graph $K_{m, n}$.
   (a) Show that $K_{3, 3}$ is hamiltonian.
   (b) Show that $K_{3, 4}$ is not hamiltonian.
   (c) Give a necessary and sufficient condition for $K_{m, n}$ to be hamiltonian.
3. Let G be a graph of order $n \geq 3$ and size q, where $2q \geq n^2 - 3n + 6$.
   (a) Show that G satisfies Ore's condition; thus G is hamiltonian.
   (b) Show that G need not necessarily satisfy Dirac's condition.
4. Find $Cl(G)$ for the graph G of:
   (a) Figure 7.30   (b) Figure 7.27   (c) Figure 7.8
5. Find the closure of each graph in Figure 7.28.
6. Each part gives a graph having vertex set $V = \{t, u, v, w, x, y, z\}$. You are to (i) find the closure of the graph and (ii) determine whether the graph is hamiltonian, giving a reason for your answer.
   (a) $G_1 = (V, E_1)$, where
   $$E_1 = \{tu, tz, uy, uz, vw, vx, vy, wx, wy, xy, yz\}$$

(b) $G_2 = (V, E_2)$, where $E_2 = \{tu, tw, tx, tz, ux, uy, vy, vz, wz\}$
(c) $G_3 = (V, E_3)$, where $E_3 =$
$\{tv, ty, uv, uw, uz, vw, vx, wy, xy, xz, yz\}$

7. Prove or disprove:
   (a) If $G = (V, E)$, where $V = \{v_1, v_2, \ldots, v_9\}$, $\deg(v_1) = 2$, $\deg(v_2) = \deg(v_3) = \deg(v_4) = 3$, $\deg(v_5) = 5$, $\deg(v_6) = \deg(v_7) = 6$, and $\deg(v_8) = \deg(v_9) = 7$, then G is hamiltonian.
   (b) If $G = (V, E)$, where $V = \{v_1, v_2, \ldots, v_7\}$, $\deg(v_1) = 2$, $\deg(v_2) = \deg(v_3) = \deg(v_4) = 3$, $\deg(v_5) = \deg(v_6) = 4$, and $\deg(v_7) = 5$, then G is hamiltonian.
   (c) If $G = (V, E)$, where $V = \{v_1, v_2, \ldots, v_8\}$, $\deg(v_1) = \deg(v_2) = \deg(v_3) = 3$, $\deg(v_4) = \deg(v_5) = \deg(v_6) = 4$, $\deg(v_7) = 5$, and $\deg(v_8) = 6$, then G is hamiltonian.

8. Let G be a graph of order $n \geq 3$. With reference to Corollary 7.22, prove:
   (a) If G satisfies Las Vergnas' condition (b), then G satisfies the Bondy–Chvátal condition (a).
   (b) If G satisfies Chvátal's condition (c), then G satisfies Las Vergnas' condition (b).
   (c) If G satisfies Pósa's condition (e), then G satisfies Bondy's condition (d).
   (d) If G satisfies Ore's condition (f), then G satisfies Pósa's condition (e).
   (e) If G satisfies Dirac's condition (g), then G satisfies Ore's condition (f).

9. With reference to Corollary 7.22, give examples of the following:
   (a) a graph G_c that satisfies Chvátal's condition (c) but not Bondy's condition (d)
   (b) a graph G_d that satisfies Bondy's condition (d) but not Pósa's condition (e)
   (c) a graph G_e that satisfies Pósa's condition (e) but not Ore's condition (f)
   (d) a graph G_f that satisfies Ore's condition (f) but not Dirac's condition (g)

10. Determine whether the digraph G of Figure 7.23 is hamiltonian or traceable.

11. For each digraph shown in Figure 7.20, determine whether it is hamiltonian or traceable.

12. Let G be a loopless digraph of order $n \geq 2$. Define the "closure" of G to be the digraph $Cl(G)$ defined recursively as follows: If G contains distinct vertices u and v such that u is not adjacent to v and $od(u) + id(v) \geq n$, then $Cl(G) = Cl(G + (u, v))$; otherwise $Cl(G) = G$. Prove or disprove: If $Cl(G)$ is a complete symmetric, loopless digraph, then G is hamiltonian.

13. State an analogue of Theorem 7.18 for digraphs. (Hint: For a digraph G, define a "traceable component" of G to be a subdigraph

that is maximal with respect to the property of being traceable.)

14. Let $G = (V, E)$ be a graph. Prove: If G is traceable, then, for every proper subset U of V, the number of components of the subgraph of G induced by $V - U$ is at most $|U| + 1$.

15. For a graph G of order $n \geq 2$, define the graph $Cl_1(G)$ recursively as follows: If G contains distinct nonadjacent vertices u and v for which $\deg(u) + \deg(v) \geq n - 1$, then $Cl_1(G) = Cl_1(G + uv)$; otherwise $Cl_1(G) = G$.

   (a) Prove: If $Cl_1(G) \simeq K_n$, then G is traceable.

   (b) For each $n \geq 3$, give an example of a graph G of order n for which $Cl_1(G) \simeq K_n$ and G is not hamiltonian.

16. Let $G = (V, E)$ be a graph of order $n \geq 4$ and let $v \in V$.

   (a) Prove: If $\deg(v) > (n - 1)/2$ and $G - v$ is hamiltonian, then G is hamiltonian.

   (b) Does the result of part (a) help in showing that the graph of Figure 7.31 is hamiltonian?

   (c) Does the result of part (a) help in showing that the graph of Figure 7.8 is hamiltonian?

17. Show that the operation of forming the closure of a graph is well-defined. (Hint: Let $G = (V, E_0)$ be a graph of order n, and suppose that $G_1 = (V, E_1)$ and $G_2 = (V, E_2)$ are two graphs formed by applying the closure process to G; we must show that $E_1 = E_2$. This is clearly the case if $E_0 = E_1 = E_2$, so suppose that $E_1 - E_0 = \{e_1, e_2, \ldots, e_k\}$ and $E_2 - E_0 = \{f_1, f_2, \ldots, f_m\}$, where the edges in the sets $E_1 - E_0$ and $E_2 - E_0$ are listed in the order in which they were added to G. We wish to show that $E_1 - E_0 = E_2 - E_0$; if this is not the case, then, without loss of generality, there is a smallest subscript t such that $e_t \notin E_2$. Consider the subgraph $H = (V, E_0 \cup \{e_i \mid 1 \leq i < t\})$ and obtain a contradiction.)

★18. Develop an algorithm for finding the closure $Cl(G)$ of a graph G of order n. Moreover, assuming $G = (V, E_0)$ has size q_0 and $Cl(G) = (V, E_1)$ has size q_1, have the closure-finding algorithm construct a "weight function" $w: E_1 \rightarrow \{1, 2, \ldots, q_0, \ldots, q_1\}$ so that w is a bijection and $w(E_0) = \{1, 2, \ldots, q_0\}$.

★19. With respect to Algorithm 7.6:

   (a) Argue that Algorithm 7.6 is $O(n^3)$.

   (b) Give a justification for its correctness.

7.8

VERTEX COLORING AND PLANAR GRAPHS

In this section we introduce a part of graph theory that arose from a very famous mathematics problem, the *four-color problem*. This problem seems to have first surfaced in a letter from Augustus DeMorgan to Sir William Rowan Hamilton, two mathematicians whose names have come up several times in this text. The letter is dated October 23, 1852, and part of it is reproduced in the book *Graph Theory* 1736–1936 by Norman L. Biggs, Keith Lloyd, and Robin J. Wilson. It is sufficiently interesting to bear repeating here.

> A student of mine asked me today to give him a reason for a fact which I did not know was a fact—and do not yet. He says that if a figure be anyhow divided and the compartments differently coloured so that figures with any portion of common boundary line are differently coloured—four colours may be wanted, but not more—the following is the case in which four are wanted [see Figure 7.34(a)]. Query cannot a necessity for five or more be invented. As far as I see at this moment if four ultimate compartments have each boundary line in common with one of the others, three of them inclose the fourth and prevent any fifth from connexion with it. If this be true, four colours will colour any possible map without any necessity for colour meeting colour except at a point.
>
> "Now it does seem that drawing three compartments with common boundary ABC two and two you cannot make a fourth take boundary from all, except by enclosing one [see Figure 7.34(b) and (c)]. But it is tricky work, and I am not sure of the convolutions—what do you say? And has it, if true, been noticed? My pupil says he guessed it in colouring a map of England. The more I think of it, the more evident it seems. If you retort with some very simple case which makes me out a stupid animal, I think I must do as the Sphynx did . . . .

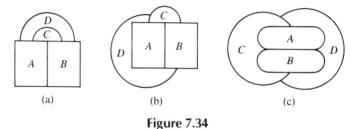

| (a) | (b) | (c) |

Figure 7.34

In ordinary language, the four-color problem can be stated as follows:

> A map consisting of *n* countries with well-defined boundaries is given in a plane. Two countries are said to be adjacent if their common boundary contains a nontrivial segment (not the empty set or a set of isolated points). Can this map be colored with four (or fewer) different colors so that no two adjacent countries are of the same color?

For more than 100 years, many attempts were made to solve the problem, and several fallacious solutions were submitted to professional journals. In some cases, such alleged solutions were first judged to be valid and were actually published. As is often the case in mathematics, this work, though unsuccessful in solving the four-color problem, introduced a great deal of interesting and useful mathematics. Finally, in 1977, the four-color problem became the *four-color theorem* (in other words, the answer to the question posed above is yes) with the publication of the celebrated paper, "Every Planar Map is Four-Colorable," by Kenneth Appel, Wolfgang Haken, and John Koch. The article appeared in the *Illinois Journal of Mathematics* and, besides generating much excitement, was quite controversial. The reason for this was that the solution of the four-color problem was an important instance in which a computer was used to help do a mathematical proof. Moreover, this use was truly significant, as over 1200 hours of computing time were required to analyze a large number of special cases! Those readers interested in learning more about the solution of the four-color problem may wish to consult an article that Appel and Haken wrote for *Scientific American* in 1977 entitled "The Solution of the Four-Color Map Problem."

As we will see shortly, it is possible to associate a graph with a plane map in such a way that each vertex of the graph corresponds to a country of the map, and so that two vertices are adjacent if and only if the two corresponding countries are adjacent. Such a graph is a "planar graph," which is a term defined later in this section, and we may then consider the problem of whether it is possible to assign one of four given colors to each vertex in such a way that any two adjacent vertices receive different colors.

More generally, given a graph $G = (V, E)$ and a set $P = \{1, 2, \ldots, k\}$ of k "colors," does there exist a "coloring function" $c: V \to P$ such that, whenever u and v are distinct vertices and $uv \in E$, $c(u) \neq c(v)$? If so, then we speak more informally of "coloring the vertices" of G with k (or fewer) colors, and the assignment c of colors to vertices is called a *k-coloring* of G. We also say that G is *k-colorable*.

The idea of coloring the vertices of a graph arises naturally in certain kinds of scheduling problems. For example, suppose that a department of mathematics at some university is concerned with scheduling its courses for the next semester. There are several (sections of) courses to be offered and, no doubt, a smaller number of time periods during which courses may be given. Two different courses may be intended for roughly the same audience of students and thus should not be given during the same time period. This suggests modeling the situation with a graph G whose vertices correspond to the courses to be offered, such that two vertices are adjacent provided the two corresponding courses should not be offered during the same time period. Suppose now that a k-coloring of G exists. Since all the vertices of a given color are mutually nonadjacent, the corresponding courses could all be offered during the same time period; hence all the courses offered by the department can be accommodated using k time periods.

DEFINITION 7.18

The minimum positive integer k for which a graph G is k-colorable is called the *chromatic number* of G and is denoted by $\chi(G)$.

Note that if G has order n, then $1 \leq \chi(G) \leq n$. Moreover $\chi(G) = 1$ if and only if G is empty ($G \simeq \overline{K_n}$), and $\chi(G) = n$ if and only if G is complete ($G \simeq K_n$).

The chromatic number of a cycle is also easily determined. If C_p denotes the cycle of order p, then for $n \in \mathbb{N}$, we call C_{2n} an *even cycle* and C_{2n+1} an *odd cycle*. It is readily verified that $\chi(C_{2n}) = 2$ and $\chi(C_{2n+1}) = 3$.

Example 7.36   Consider the graph G shown in Figure 7.35. Such a graph is called a *wheel of order 6*. Find $\chi(G)$.

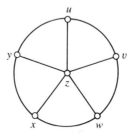

Figure 7.35 A wheel of order 6

Solution   Notice that $G - z$ is isomorphic to an odd cycle C_5. Also z is adjacent to each of the five vertices on this cycle. Thus, since $\chi(C_5) = 3$ and z requires its own fourth color, $\chi(G) = 4$.

In general, let W_p denote the wheel of order p. Then it can be shown, for $n \geq 2$, that $\chi(W_{2n}) = 4$ and $\chi(W_{2n+1}) = 3$. (See Exercise 11.)   ∎

A standard technique for showing that $\chi(G) = k$ is to exhibit a k-coloring of G and to argue that G is not $(k - 1)$-colorable. This technique is illustrated in the following example.

Example 7.37   Find $\chi(G)$ for the graph G shown in Figure 7.36.

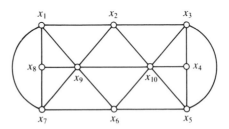

Figure 7.36

Solution   We first note that $\langle \{x_1, x_7, x_8, x_9\} \rangle$ is an induced subgraph of G isomorphic to K_4. This shows that $\chi(G) \geq 4$. To show that $\chi(G) = 4$, we must exhibit a 4-coloring of G. One possibility is to color x_1 and x_{10} with one color, say red, x_2, x_4, x_6, and x_8 with blue, green for x_3 and x_9, and yellow for x_5 and x_7.   ∎

Suppose that a graph $G = (V, E)$ is k-colorable and a particular k-coloring of G is given. Define a relation on V as follows: For $x, y \in V$, say x is related to y provided x and y are the same color. This relation is an equivalence relation on V and, as such, partitions V into equivalence classes, with each class consisting of all vertices of G of a particular color. These classes are called the *color classes* of the k-coloring of G. If the color classes are $V_1, V_2, \ldots, V_k$, then each V_i, $1 \le i \le k$, has the property that no two of its vertices are adjacent.

DEFINITION 7.19

Let $G = (V, E)$ be a graph. A nonempty subset U of V is called an *independent set* (of vertices) provided no two of its vertices are adjacent in G. An independent set U is called a *maximal independent set* provided U is not a proper subset of some other independent set of G.

Thus the color classes for a given k-coloring of G are independent sets in G. It should be noted that the chromatic number of G is the minimum number of independent sets into which the vertex set of G may be partitioned.

Example 7.38   The sets $V_1 = \{x_1, x_{10}\}$, $V_2 = \{x_2, x_4, x_6, x_8\}$, $V_3 = \{x_3, x_9\}$, and $V_4 = \{x_5, x_7\}$, which are the color classes for the 4-coloring of Example 7.37, are independent sets of the graph $G = (V, E)$ of Figure 7.36. In fact, V_1, V_2, and V_3 are maximal independent sets, but V_4 is not maximal, since it is properly contained in the independent set $V_4 \cup \{x_2\}$. Notice that if we let $U_1 = V_1$, $U_2 = V_2 - \{x_2\}$, $U_3 = V_3$, and $U_4 = V_4 \cup \{x_2\}$, then U_1, U_2, U_3, U_4 are independent sets in G whose union is V.

As was mentioned earlier, a graph has chromatic number 1 if and only if the graph is empty. The following theorem characterizes those graphs having chromatic number 2.

THEOREM 7.23   A graph $G = (V, E)$ has chromatic number 2 if and only if G is a nonempty bipartite graph.

Proof   If G is a nonempty bipartite graph, then clearly $\chi(G) \ge 2$. Moreover, if V_1 and V_2 are partite sets of G, then V_1 and V_2 are independent sets in G whose union is V. Thus $\chi(G) = 2$.
　　Conversely, if $\chi(G) = 2$, then E is nonempty. Let G be 2-colored, with B_1 and B_2 the resulting color classes. Then each of B_1 and B_2 is an independent set in G, and so every edge of G must join a vertex of B_1 with a vertex of B_2. It follows that G is bipartite.   □

In Exercise 6 the student is asked to show that a graph G is bipartite if and only if G has no odd cycles. With this result, we can restate Theorem 7.23 as follows:

> A graph G has chromatic number 2 if and only if G is nonempty and has no odd cycles.

Thus, if G has an odd cycle, then $\chi(G) \geq 3$; conversely, if $\chi(G) \geq 3$, then G must contain an odd cycle. Unfortunately, for $k \geq 3$, no nice characterization is known of those graphs with chromatic number k.

There is a simple and efficient algorithm for deciding whether an arbitrary given graph is bipartite (see Exercise 6). However, it is curious and interesting that the problem of deciding whether an arbitrary given graph is even 3-colorable is NP-complete, as described in the last section. Thus at present we have no algorithm for finding the chromatic number of a given input graph that runs in polynomial time for all graphs.

We next survey a few of the known results concerning chromatic number. For these, the following definition is helpful.

DEFINITION 7.20

A graph $G = (V, E)$ is called *critical* (with respect to χ) provided $\chi(G - v) < \chi(G)$ for every $v \in V$. Specifically, G is called *k-critical*, where $k \geq 2$, provided G is critical and $\chi(G) = k$.

As a consequence of this definition, we can readily show that if G is k-critical, then $\chi(G - v) = k - 1$ for all $v \in V$.

It is not difficult to verify that K_2 is the only 2-critical graph, while the only 3-critical graphs are the odd cycles (see Exercise 12). The wheels W_{2n}, where $n \geq 2$, are 4-critical graphs; we illustrate the verification of this fact using the wheel W_6, an instance of which is the graph G of Figure 7.35. In Example 7.36 we showed that $\chi(G) = 4$. Since $G - z$ is the odd cycle C_5, we have that $\chi(G - z) = 3$. Now consider any vertex on the "rim" of the wheel, such as u. Note that $(G - u) - z$ is a path and hence is bipartite, so that $\chi((G - u) - z) = 2$. Since z is adjacent to each of v, w, x, and y, a third color is required for z; hence $\chi(G - u) = 3$. It follows that G is 4-critical.

For $k \geq 4$, the k-critical graphs have not as yet been determined. However, we can obtain the following useful result. (Recall that $\delta(G)$ denotes the minimum degree among the vertices of G.)

THEOREM 7.24   If G is a k-critical graph, then $\delta(G) \geq k - 1$.

Proof   We proceed by contradiction; let G be a k-critical graph and suppose that u is a vertex of G such that $\deg(u) < k - 1$. We know that $\chi(G - u) =$

$k - 1$, so let a $(k - 1)$-coloring of $G - u$ be given. Since $\deg(u) < k - 1$, one of the $k - 1$ colors used, say green, is such that none of the vertices adjacent to u in G is colored green. We may then color u with green to obtain a $(k - 1)$-coloring of G. This contradicts the fact that $\chi(G) = k$ and thus establishes the result.

If $\chi(G) = k$, then G may or may not be k-critical, but G does contain a k-critical subgraph. For if G is not k-critical, then there is some vertex u of G such that $\chi(G - u) = k$. If $G - u$ is k-critical, then we have the sought-after subgraph; if not, then the foregoing process may be repeated and, after a finite number of steps, we arrive at a k-critical subgraph of G. (In fact, notice that we actually obtain a k-critical induced subgraph.)

By applying Theorem 7.24 to a $\chi(G)$-critical subgraph of G, one obtains the following upper bound on $\chi(G)$. (Recall that $\Delta(G)$ denotes the maximum degree among the vertices of G.)

COROLLARY 7.25   Let G be a graph. Then $\chi(G) \leq \Delta(G) + 1$.

Proof   Let $\chi(G) = k$ and let H be a k-critical subgraph of G. By Theorem 7.24, $\delta(H) \geq k - 1$ and so $\Delta(H) \geq k - 1$. Certainly $\Delta(H) \leq \Delta(G)$, and so we have that $k \leq \Delta(H) + 1 \leq \Delta(G) + 1$. $\square$

There is a stronger result than Corollary 7.25, proved by R. L. Brooks in 1941.

THEOREM 7.26   **(Brooks)**

Let G be a connected graph that is not isomorphic to either a complete graph or an odd cycle. Then $\chi(G) \leq \Delta(G)$.

It should be mentioned that the bound provided by Theorem 7.26 is not particularly satisfying for certain classes of graphs. For example, $\chi(K_{1,n}) = 2$, while $\Delta(K_{1,n}) = n$.

Example 7.39   Applying Brooks' theorem to the graph G of Figure 7.36, we find that $\chi(G) \leq \Delta(G) = 6$. For a lower bound, we observe that G contains a subgraph isomorphic to K_4; thus $4 \leq \chi(G)$. Another useful observation is the following: if v is a vertex of degree less than k in G, then G is k-colorable if and only if $G - v$ is k-colorable. Let us apply this observation to the graph G using $k = 4$. Note that x_4 has degree 3 in G, so G is 4-colorable if and only if $G_1 = G - x_4$ is 4-colorable. Next, x_8 has degree 3 in G_1, so G_1 is 4-colorable if and only if $G_2 = G_1 - x_8$ is 4-colorable. Applying Brooks' theorem to G_2, we find that $\chi(G_2) \leq \Delta(G_2) = 5$, so let's continue removing vertices of "small" degree. Note that both x_1 and x_3 have degree 3 in G_2, so let $G_3 = G_2 - x_1$ and $G_4 = G_3 - x_3$. Then G_2 is 4-colorable if and only if G_4 is 4-colorable. Now $\chi(G_4) \leq \Delta(G_4) = 4$, which shows that G_4 is 4-colorable. Therefore $\chi(G) = 4$.   ∎

At this point it is appropriate to return to a discussion of the four-color problem with which we began this section. In particular, we attempt to explain how the map-coloring problem translates into a vertex-coloring problem for a particular class of graphs. A rigorous and formal treatment would take considerable space and would, perhaps, simply cloud the issue. Since it is our intent to provide only an intuitive description, we are necessarily imprecise, but clear we hope.

DEFINITION 7.21

A graph G is called a *planar graph* provided it can be represented in the plane so that edges intersect only at incident vertices. Such a representation of a planar graph G is called a *plane embedding* of G. A graph that is not planar is called *nonplanar*.

Example 7.40   Figure 7.36 shows a plane embedding of a planar graph G. Likewise Figure 7.35 shows a plane embedding of the wheel W_6; thus W_6 is planar. The graph H shown in Figure 7.37 is isomorphic to K_5 minus an edge. Show that H is planar by giving a plane embedding of H.

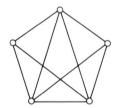

Figure 7.37 A planar graph H

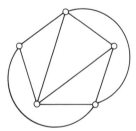

Figure 7.38 A plane embedding of H

Solution   A plane embedding of H is shown in Figure 7.38.   ∎

Two noteworthy and interesting examples of nonplanar graphs are K_5 and $K_{3,3}$; the student is asked to show that these graphs are nonplanar in Exercise 23. Indeed, by a famous result of K. Kuratowski (1930), nonplanar graphs can be characterized in terms of K_5 and $K_{3,3}$. In order to state his result, we require some additional terminology. An *elementary subdivision* of a nonempty graph $G = (V, E)$ is a graph $(V \cup \{w\}, (E - uv) \cup \{uw, vw\})$ obtained from G by removing an edge $e = uv$ of G and then adding a new vertex $w \notin V$, along with the edges uw and vw. For example, in Figure 7.39(b) is shown an elementary subdivision of the graph in Figure 7.39(a). A *subdivision* of G is a graph obtained from G by applying the following recursive definition.

1. G is a subdivision of G.
2. If H_1 is a subdivision of G and H_2 is an elementary subdivision of H_1, then H_2 is a subdivision of G.

Figure 7.39(c) shows a subdivision of the graph in Figure 7.39(a).

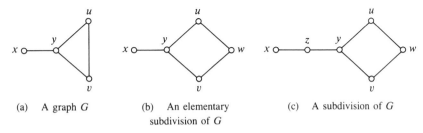

(a) A graph G

(b) An elementary
subdivision of G

(c) A subdivision of G

Figure 7.39

We now state Kuratowski's result.

THEOREM 7.27 **(Kuratowski)**

A graph G is planar if and only if G does not contain a subgraph H that is isomorphic to a subdivision of K_5 or $K_{3,3}$.

Example 7.41 The graph G shown in Figure 7.40 is known as "Petersen's graph" and was introduced in the last section. Show that G is nonplanar by finding a subgraph H of G that is isomorphic to a subdivision of $K_{3,3}$.

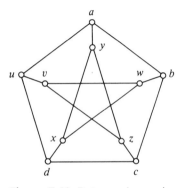

Figure 7.40 Petersen's graph

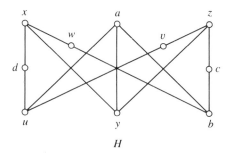

H

Figure 7.41 A subgraph of Petersen's graph

Solution   One possibility is the subgraph H shown in Figure 7.41, which is easily seen to be a subdivision of $K_{3,3}$.   ■

Given a planar graph G, a plane embedding of G partitions the plane into a finite number of regions, where each region is bounded by a closed walk of G. This is shown, for example, by the plane embedding of the graph G in Figure 7.42. Here there are four regions, numbered 1 through 4 as shown. Note that region 2 is bounded by the cycle t, y, z, t, while region 4 is bounded by the closed walk $s, t, y, x, u, v, w, x, y, z, s$. In general, we refer to the regions of a (plane embedding of a) planar graph as its *faces*. The closed walk that bounds a given face f is called the *boundary* of f.

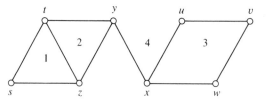

Figure 7.42

We now describe how the four-color problem for plane maps is equivalent to the problem of finding the maximum chromatic number among all planar graphs. Let a plane map M, such as the one in Figure 7.43(a), be given. We construct a planar graph G from M as follows. In each region of M place one vertex of G and make two vertices adjacent provided the associated regions have a nontrivial segment of their boundaries in common. Figure 7.43(b) shows the planar graph G corresponding to the map M of Figure 7.43(a).

Our problem is to color the "countries" of a given plane map M so that any two countries sharing a common boundary receive different colors. Note that this problem is equivalent to that of coloring the vertices of the associated planar graph so that adjacent vertices are colored differently. Thus the four-color problem, which asks whether four colors suffice to "properly color" the countries of any plane map, is equivalent to asking

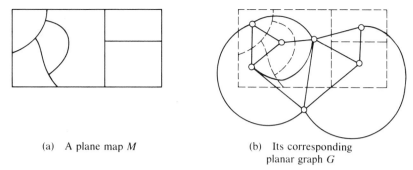

(a)   A plane map M (b)   Its corresponding
planar graph G

Figure 7.43

whether any planar graph has chromatic number at most 4. We can now restate the celebrated result of Appel, Haken, and Koch.

THEOREM 7.28 **(The four-color theorem)**

For any planar graph G, we have $\chi(G) \le 4$.

Exercises 7.8

1. Draw a plane map with six countries such that each country shares a boundary with exactly four others.
2. Consider plane maps such as the one shown in Figure 7.44(a), which are formed by some finite number of line segments joining points on different sides (or opposite corners) of a given rectangle.
   (a) Color the map of Figure 7.44(a) using just two colors.
   (b) Use induction (on the number of line segments) to show that any such map may be colored using two colors.
3. Consider the map of Figure 7.44(b); note that each country is adjacent to exactly five others.
   (a) Color the map using four colors.
   (b) Give the planar graph G corresponding to this map.
   (c) Show that it is not possible to color the map (or to color the vertices of G) with only three colors.

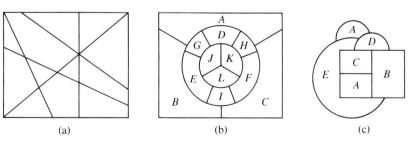

(a)   (b)   (c)

Figure 7.44

4. Consider maps such as the one in Figure 7.44(c); such maps are allowed to contain "disconnected" countries, like A, whose territory consists of two (or more) disconnected regions.
   (a) Show that such maps must be excluded in order that the four-color theorem hold.
   (b) Give an example of a map with six countries, two of which are disconnected, that requires six colors.

5. Find the chromatic number of each graph shown in Figure 7.45.

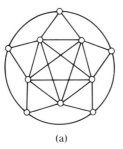
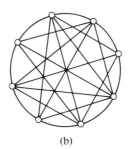
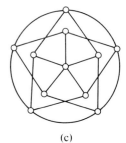

(a)   (b)   (c)

Figure 7.45

6. Let $G = (V, E)$ be a graph.
   (a) Show that G is bipartite if and only if G has no odd cycles.
   (b) Develop a "fast" algorithm to determine whether a given input graph is bipartite.

7. Find $\chi(G)$ for the graph G of Figure 7.40 (Petersen's graph).

8. Suppose that each of n students $s_1, s_2, \ldots, s_n$ is taking some subset of the courses $c_1, c_2, \ldots, c_p$. We wish to schedule a final exam for each course in such a way that no student has the conflict of two finals scheduled for the same time period. What is the minimum number of time periods required?
   (a) Formulate this problem as determining the chromatic number of a graph.
   (b) Suppose the set of students is {Al, Dick, Fred, Joe, Nancy}; Al is taking MA 332, MA 337, and MA 351; Dick is taking MA 332 and MA 457; Fred is taking MA 332, MA 351, and MA 423; Joe is taking MA 423 and MA 457; and Nancy is taking MA 332, MA 351, and MA 423. Construct the graph and solve the problem.

9. Let T be a nonempty tree. Find $\chi(T)$.

10. Given two graphs $G_1 = (V_1, E_1)$ and $G_2 = (V_2, E_2)$, their *join* is the graph $G = (V, E)$ defined by $V = V_1 \cup V_2$ and $E = E_1 \cup E_2 \cup \{v_1 v_2 \mid v_1 \in V_1 \wedge v_2 \in V_2\}$. We write $G = G_1 + G_2$ to denote that G is the join of G_1 and G_2.
   (a) Given $G_1 = (\{u\}, \emptyset)$ and $G_2 = (\{w, x, y, z\}, \{wx, wz, xy, yz\})$, find $G = G_1 + G_2$.
   (If $G_1 \simeq K_1$ and $G_2 \simeq C_{p-1}$, then $G_1 + G_2 \simeq W_p$, the wheel of order p. In part (a), for example, $G \simeq W_5$.)

(b) Given $G_1 = (\{u, v\}, \{uv\})$ and $G_2 = (\{w, x, y, z\}, \{wx, xy, yz\})$, find $G = G_1 + G_2$.

(c) Prove: If $G = G_1 + G_2$, then $\chi(G) = \chi(G_1) + \chi(G_2)$.

11. Let W_p denote the wheel of order p. Show for $n \geq 2$ that $\chi(W_{2n}) = 4$ and $\chi(W_{2n+1}) = 3$.

12. Concerning critical graphs:

   (a) Show that the 3-critical graphs are precisely the odd cycles.

   (b) Give an example of a 4-critical graph that is not isomorphic to any wheel.

   (c) Give an example of a 5-critical graph (other than K_5).

13. Let $U = \{0, 1, 2, \ldots, 9\}$ and consider the following subsets of U:

$$A = \{0, 1, 8\}, \; B = \{2, 5\}, \; C = \{0, 3, 7\}, \; D = \{1, 4, 9\},$$

$$E = \{2, 6, 8\}, \; F = \{5, 7, 9\}, \; G = \{3, 4, 6\}$$

   Form a graph H whose vertex set is $\{A, B, C, D, E, F, G\}$ such that two vertices of H are adjacent provided they are not disjoint.

   (a) Find $\chi(H)$.

   (b) Given a $\chi(H)$-coloring of H, what property do the color classes have?

   (c) Is there a subset of $\{A, B, C, D, E, F, G\}$ that is a partition of U?

14. Show that if G is k-critical, then $K_1 + G$ is $(k + 1)$-critical. (See Exercise 10.)

15. Show that if H is a subgraph of G, then $\chi(H) \leq \chi(G)$.

16. Prove or disprove the following statements about a graph G:

   (a) If G contains a unique odd cycle, then $\chi(G) = 3$.

   (b) If any two odd cycles of G have a vertex in common, then $\chi(G) \leq 5$.

17. Show that $\chi(G) \leq 1 + \max \delta(H)$, where the maximum is taken over all induced subgraphs H of G. (Hint: Modify the proof of Corollary 7.25.)

18. Let G be a graph of order p and let $\beta(G)$ denote the maximum cardinality of an independent set of vertices in G. Show that

$$p/\beta(G) \leq \chi(G) \leq p + 1 - \beta(G)$$

19. Recall that a graph $G = (V, E)$ is 2-connected provided G is connected and, for every $v \in V$, $G - v$ is connected. Prove: If G is k-critical, where $k \geq 2$, then G is 2-connected.

20. Prove Euler's formula: Given a plane embedding of a connected planar graph G with p vertices, q edges, and r faces, then $p - q + r = 2$. (Hint: Proceed by induction on q. If $q = p - 1$, then G is a tree, so that $r = 1$ and the formula holds. If $q \geq p$, let e be a cycle edge of G and apply the induction hypothesis to $G - e$.)

21. Let G be a planar graph of order $p \geq 3$ and size q. Prove that $q \leq 3p - 6$. (Hint: Let r be the number of faces in some plane

embedding of G; count the number of edges on the boundary of each face and sum over all faces. Since each edge is on the boundary of exactly two faces and each face boundary contains at least three edges, we have that $3r \le 2q$. Now apply Euler's formula from Exercise 20.)

22. Modify the proof of the result in Exercise 21 to show that, if G is a planar graph of order $p \ge 3$ and size q, and G does not contain a subgraph isomorphic to K_3, then $q \le 2p - 4$.

23. Concerning the remarks preceding Theorem 7.27:
    (a) Use the result of Exercise 21 to show that K_5 is nonplanar.
    (b) Use the result of Exercise 22 to show that $K_{3,3}$ is nonplanar.

24. Prove: If G is a planar graph, then $\delta(G) \le 5$.

25. Give an example of a planar graph G with $\delta(G) = 5$.

26. Use the result of Exercise 24 to prove that if G is a planar graph, then $\chi(G) \le 6$.

27. Determine whether each of the following graphs is planar or nonplanar.
    (a) $K_{2,4}$ (b) $G - ax$, where G is the graph of Figure 7.40
    (c) a graph isomorphic to the "Grötzsch graph" of Figure 7.45(c)
    (d) a graph isomorphic to the 4-regular subgraph of K_6

28. Find the maximal independent sets in the graph G of Figure 7.46. Try to do this in a systematic way.

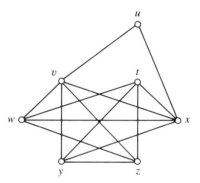

Figure 7.46

7.9   FINITE-STATE AUTOMATA

The process of compiling a computer program written in a high-level language, like Pascal or Ada, can be broken down into several phases. In the first phase, known as "lexical analysis," the source program is viewed as an input string of characters and the goal is to output the program as a sequence of "tokens," such as identifiers, keywords, operators, and special symbols like parentheses and semicolons. (Also extraneous characters such as nonsignificant blanks and comments may be removed.) In this section our goal is to indicate how the mathematics of "regular languages" and "deterministic finite-state automata" is applicable to the problem of lexical analysis. It is seen that a deterministic finite-state automata, or DFA, can be represented as a kind of labeled directed graph.

To begin, we need an *alphabet*, which is simply a nonempty set of characters. This might be the ASCII character set or the set $\{0, 1\}$ of binary digits. Next, given an alphabet Σ, a *string* over Σ is any finite sequence $\alpha = (x_1, x_2, \ldots, x_n)$ of characters from Σ. We use lowercase Greek letters to denote strings, and we agree to denote the string $\alpha = (x_1, x_2, \ldots, x_n)$ by the more familiar notation $\alpha = \text{“}x_1 x_2 \cdots x_n\text{”}$ or more simply by $x_1 x_2 \cdots x_n$ when there is no confusion. The *length* of a string is the number of characters it contains, so that the string $\alpha = x_1 x_2 \cdots x_n$ has length n. There is a unique string of length 0, the *empty string*, and it is denoted by ε.

DEFINITION 7.22

Given an alphabet Σ, a *language* over Σ is a set of strings over Σ.

Three special languages over an alphabet Σ are the empty language $\emptyset$, the language Σ^* of all strings over Σ, and the language $\Sigma^+ = \Sigma^* - \{\varepsilon\}$ of all nonempty strings over Σ.

Example 7.42   (a) If $\Sigma_1 = \{0\}$, then $\Sigma_1^* = \{\varepsilon, 0, 00, 000, \ldots\}$.

(b) If $\Sigma_2 = \{0, 1\}$, then Σ_2^+ is the language of all nonempty binary strings.

(c) For any alphabet Σ, the set $\{\varepsilon\}$, consisting of only the empty string, is a language over Σ.

(d) In the Ada programming language, an identifier is composed of letters, digits, and the underline character _. An identifier must begin with a letter, may not end with the underline character, and may not contain two consecutive underline characters. Consider the language of all valid Ada identifiers; we can represent this language, in an abstract way, by replacing each letter by 0, each digit by 1, and each underline character by 2. We obtain a language over $\Sigma_3 = \{0, 1, 2\}$ consisting of those strings that begin with a 0, do not end with a 2, and do not contain two consecutive 2's. We denote this language by L_7 and return to it later.

(e) Another language over $\Sigma_2 = \{0, 1\}$ is the language L_5 of all "palindromes" over Σ_2;

$$L_5 = \{\varepsilon, 0, 1, 00, 11, 000, 010, 101, 111, \ldots\}$$

(Given a string $\alpha = x_1 x_2 \cdots x_n$, the *reverse* of α is the string $r(\alpha) = x_n \cdots x_2 x_1$; α is a *palindrome* provided $\alpha = r(\alpha)$.)

(f) Note that if L_1 and L_2 are two languages over alphabet Σ, then their *union* $L_1 \cup L_2$ is also a language over Σ.   ∎

Given an alphabet Σ and strings α, $\beta \in \Sigma^*$, say $\alpha = x_1 x_2 \cdots x_m$ has length m and $\beta = y_1 y_2 \cdots y_n$ has length n, the *concatenation of α with β* is denoted $\alpha \& \beta$ and is the string in Σ^* of length $m + n$ defined by

$$\alpha \& \beta = x_1 x_2 \cdots x_m y_1 y_2 \cdots y_n$$

For example, with $\Sigma_2 = \{0, 1\}$, $\alpha = 010$, and $\beta = 1101$, we have $\alpha \& \beta = 0101101$. Note that $\alpha \& \varepsilon = \alpha = \varepsilon \& \alpha$ for any string α over Σ. The idea of concatenating strings over Σ can be used to define a concatenation operation on languages over Σ.

DEFINITION 7.23

For two languages L_1 and L_2 over an alphabet Σ, the *concatenation of L_1 with L_2* is the language $L_1 L_2$ defined by

$$L_1 L_2 = \{\alpha_1 \& \alpha_2 \mid \alpha_1 \in L_1 \wedge \alpha_2 \in L_2\}$$

Moreover, given a nonnegative integer n, the *nth power* of the language L over Σ is the language L^n defined recursively as follows:

$$L^0 = \{\varepsilon\}$$

$$L^n = L L^{n-1} \qquad n \geq 1$$

and the language L^* is defined by

$$L^* = \bigcup_{n=0}^{\infty} L^n$$

In the preceding definition note that L^n consists of those strings over Σ that are obtained by concatenating together n strings from L, while L^* consists of those strings over Σ that are obtained by concatenating together a finite number of strings from L.

Example 7.43   Let $L_1 = \{00, 01\}$ and $L_2 = \{01, 1, 111\}$ be languages over $\Sigma_2 = \{0, 1\}$. Find or describe each of the following languages:
  (a) $L_1 \cup L_2$ (b) $L_1 L_2$ (c) L_1^3 (d) L_1^*

Solution
  (a) $L_1 \cup L_2 = \{00, 01, 1, 111\}$
  (b) $L_1 L_2 = \{0001, 001, 00111, 0101, 011, 01111\}$
  (c) Note that $L_1^3 = \{(\alpha \& \beta) \& \gamma \mid \alpha, \beta, \gamma \in L_1\}$. Thus $L_1^3 = \{000000, 000001, 000100, 000101, 010000, 010001, 010100, 010101\}$.
  (d) A little thought should convince the student that L_1^* consists precisely of those strings over $\{0, 1\}$ either having an even number of 0's

and no 1's or having an odd number of 0's before the first 1 and an odd number of 0's between any two 1's.   ■

An important class of languages over an alphabet Σ is the class of regular languages over Σ.

DEFINITION 7.24

Let Σ be an alphabet. A *regular language* over Σ is any subset of Σ^* that can be constructed from the following recursive definition:

0. The empty language $\emptyset$ is a regular language.
1. The language $\{\varepsilon\}$ is a regular language.
2. For any $x \in \Sigma$, the language $\{x\}$ is a regular language.
3. If L_1 and L_2 are regular languages, then so is $L_1 \cup L_2$.
4. If L_1 and L_2 are regular languages, then so is $L_1 L_2$.
5. If L is a regular language, then so is L^*.

Example 7.44   Let $\Sigma_2 = \{0, 1\}$, $L_1 = \{0\}$, and $L_2 = \{1\}$. Then L_1 and L_2 are regular languages by part 2 of Definition 7.24.

(a) Note that Σ_2^* is regular, since $\Sigma_2^* = (L_1 \cup L_2)^*$. Here we are applying parts 3 and 5 of Definition 7.24.

(b) Let $L_3 = \{00, 01\}$. Then L_3 is regular, since $L_3 = L_1(L_1 \cup L_2)$. Here we are applying parts 3 and 4 of the definition.

(c) Let $L_4 = \{\varepsilon, 0, 1, 01, 10, 010, 101, \ldots\}$ be the language of those strings over Σ_2 having no two consecutive characters the same. To see that L_4 is regular, we make the following observation. Let $\alpha \in L_4$. If the first character of α is a 1, delete it, and if the last character of α is a 0, delete it. Let α' be the resulting string. Then observe that α' belongs to the regular language $(L_1 L_2)^*$. Thus

$$L_4 = (L_2 \cup \{\varepsilon\})(L_1 L_2)^*(L_1 \cup \{\varepsilon\})$$

and we see that L_4 is regular.

(d) Let L_5 be the language of palindromes over Σ_2. It can be shown that L_5 is not a regular language; we sketch the proof later in this section.

(e) Similarly, it can be shown that $L_6 = \{\varepsilon, 01, 0011, 000111, \ldots\}$, the language of those strings consisting of some number of 0's followed by the same number of 1's, is not regular.   ■

Given an alphabet Σ, a language L, and a string α over Σ, it is useful to have a method for deciding whether $\alpha \in L$. For instance, in compiling an Ada program, some method is needed for recognizing identifiers. Finite-state automata are useful mechanisms in this regard.

> **DEFINITION 7.25**
>
> A *deterministic finite-state automaton*, or DFA, is a quintuple (S, Σ, t, s_0, A), where S is a finite set of *states*, Σ is an alphabet, $t: S \times \Sigma \to S$ is a *transition function*, $s_0 \in S$ is a special state called the *initial state*, and A is a subset of S called the *set of accepting states*.

Let $G = (S, \Sigma, t, s_0, A)$ be a DFA; think of G as being "in" exactly one of its states at any moment and being allowed to change the state s it is currently in by reading a character $x \in \Sigma$ as input and then moving to the state $t((s, x))$. (This is why t is called the *transition function* of G.) Now, given $\alpha = x_1 x_2 \cdots x_n \in \Sigma^*$, the DFA G can "process" α as follows: G begins in the initial state s_0 and then inputs the characters of α one by one, each time changing the state it is in according to the value of the transition function for the current state and the character of α being input. More precisely, G moves through a series of $n + 1$ (not necessarily distinct) states $s_0, s_1, s_2, \ldots, s_n$, where $s_1 = t((s_0, x_1))$, $s_2 = t((s_1, x_2))$, and so on, until finally $s_n = t((s_{n-1}, x_n))$. If $s_n \in A$, then we say that G *accepts* α; otherwise we say that G *rejects* α. (This is the reason for calling A the "set of accepting states.")

Extending the notion of G accepting or rejecting a string α, we can consider the language L_G of all strings in Σ^* accepted by G; L_G is called the *language accepted by* G. It turns out that such a language is always regular. Moreover we have the following important and interesting result.

THEOREM 7.29   A language L over the alphabet Σ is a regular language if and only if there is a DFA $G = (S, \Sigma, t, s_0, A)$ such that $L = L_G$.

The proof of Theorem 7.29 is beyond the scope of this text, but we do want to indicate some of its implications.

Example 7.45   We use Theorem 7.29 to sketch a proof that the language L_5 of all palindromes over $\Sigma_2 = \{0, 1\}$ is not regular. Suppose, to the contrary, that L_5 is regular; then there is a DFA $G = (S, \Sigma_2, t, s_0, A)$ such that $L_5 = L_G$. Let $\alpha_1 = 01$, $\alpha_2 = 001$, $\alpha_3 = 0001$, and so on, and let s_i be the state that G is in after having processed α_i. We claim that the states $s_1, s_2, s_3, \ldots$ are distinct. To see this, let $\beta_1 = 0$, $\beta_2 = 00$, $\beta_3 = 000$, and so on. Suppose $s_i = s_j$ for some $i \neq j$. Then G either accepts both $\alpha_i \,\&\, \beta_i$ and $\alpha_j \,\&\, \beta_i$ or rejects both of them; however, G should accept $\alpha_i \,\&\, \beta_i$ and reject $\alpha_j \,\&\, \beta_i$. This establishes our claim and hence G has an infinite number of states. But this contradicts the requirement that S is finite. ∎

A DFA $G = (S, \Sigma, t, s_0, A)$ has a representation as a "labeled directed graph." This digraph, which is also called G, has a vertex for each state; if there is an $x \in \Sigma$ and states $s_i, s_j \in S$ such that $t((s_i, x)) = s_j$, then the digraph has an arc from s_i to s_j labeled x. Alternately we could label an arc from s_i to s_j with the set $\{x \mid t((s_i, x)) = s_j\}$. Such representations are illustrated in the next example.

Example 7.46   Let $\Sigma_2 = \{0, 1\}$. For each DFA in this example and the next, we follow the convention of letting the set of states be $\{-1, 0, 1, \ldots, n\}$ for some $n \in \mathbb{N}$, where 0 is the initial state and -1 is a special state called the (*certain*) *rejection state*.

(a) We design a DFA $G_3 = (S_3, \Sigma_2, t_3, 0, A_3)$ that accepts the regular language $L_3 = \{00, 01\}$. For $\alpha \in \Sigma_2^*$, we have that $\alpha \in L_3$ only if the first character of α is 0. Thus, starting from the initial state 0, we need two other states 1 and -1 in the DFA G_3, and we let $t_3((0, 0)) = 1$ and $t_3((0, 1)) = -1$. State 1 represents the state of having read 0 as the first character of α, in which case it is possible that $\alpha \in L_3$; state -1 represents the state of having read 1 as the first character of α, in which case G_3 should reject α. Note that state 1 is not an accepting state of G_3 because "0" $\notin L_3$. Next $\alpha \in L_3$ only if α has a second character, and it may be either 0 or 1. Thus we add another state 2 to the DFA, with transitions $t_3((1, 0)) = t_3((1, 1)) = 2$. Now state 2 is an accepting state; moreover, if α has length greater than 2, then α should be rejected. Hence we add the transitions $t_3((2, 0)) = t_3((2, 1)) = -1$ and $t_3((-1, 0)) = t_3((-1, 1)) = -1$. These last two transitions represent the fact that the rejection state should have no arcs to any other state because, once it has been decided to reject α, no additional input characters can alter this decision. The DFA G_3 is shown in Figure 7.47(a); the special form of vertex 2 is used to indicate that it is an accepting state.

(b) Let $L_4 = \{\varepsilon, 0, 1, 01, 10, 010, 101, \ldots\}$ be the regular language of those strings over Σ_2 having no two consecutive characters the same. Let's

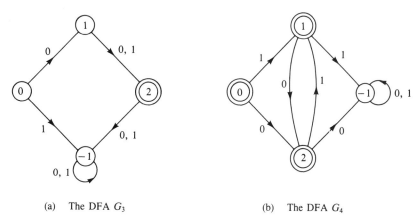

(a)   The DFA G_3 (b)   The DFA G_4

Figure 7.47

design a DFA $G_4 = (S_4, \Sigma_2, t_4, 0, A_4)$ that accepts L_4. The main idea is to have accepting states 1 and 2, where the DFA is in state 1 if the last character read is 1 and in state 2 if the last character read is 0. Thus $t_4((0,0)) = 2$, $t_4((0,1)) = 1$, $t_4((1,0)) = 2$, and $t_4((2,1)) = 1$. Since $\varepsilon \in L_4$, the state 0 is an accepting state; also, reflecting the condition that a string in L_4 cannot have two consecutive characters the same, we add the transitions $t_4((1,1)) = t_4((2,0)) = -1$ to the rejection state. The complete DFA G_4 is shown in Figure 7.47(b). ■

Example 7.47   In Example 7.42, part (d), we introduced a language L_7 over $\Sigma_3 = \{0, 1, 2\}$ that represents, abstractly, the language of Ada identifiers; L_7 contains those strings that begin with a 0, do not end with a 2, and do not contain two consecutive 2's. Design a DFA $G_7 = (S_7, \Sigma_3, t_7, 0, A_7)$ that accepts L_7, thereby showing that L_7 is a regular language.

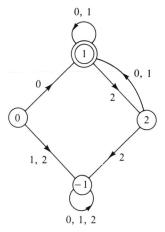

Figure 7.48 The DFA G_7

Solution   The DFA G_7 is shown in Figure 7.48. ■

A finite-state automaton G like those discussed above is "deterministic" in the sense that, given the current state s of G and the next input character x, the next state $t((s, x))$ is well-determined by the transition function t of G. There is another type of finite-state automaton, a "nondeterministic" variety, in which, given the current state s and the next input character x, there may be a choice of next states. Like a DFA, a *nondeterministic finite-state automaton* (NFA) may be defined formally as a quintuple (S, Σ, t, s_0, A), with the only difference being that, for an NFA, the transition function t is from $S \times \Sigma$ to $\mathscr{P}(S) - \{\emptyset\}$. Thus, given the current state s of an NFA G and an input character x, there is a nonempty subset $t((s, x))$ of S of possible next states. The NFA G *accepts* a string $\alpha = x_1 x_2 \cdots x_n$ of length n if there is a sequence of $n + 1$ states

$s_0, s_1, \ldots, s_n$ such that s_0 is the initial state, s_n is an accepting state, and for each i, $0 \le i < n$, $s_{i+1} \in t((s_i, x_i))$; if no such sequence of states exists, then G *rejects* α. The *language* L_G *accepted by* G is the set of all strings in Σ^* that G accepts.

Like a DFA, an NFA G has a representation as a labeled digraph. For instance, consider the NFA $G = (\{0, 1, 2, 3, 4\}, \{0, 1\}, 0, t, \{2, 4\})$ shown in Figure 7.49. Notice that there are two possible transitions from state 0 on input 0 and two possible transitions from state 1 on input 1. In fact, $t((0, 0)) = \{0, 1\}$ and $t((1, 1)) = \{2, 3\}$. It is not too hard to show that the language accepted by G is the regular language $(\{0\} \cup \{1\})^*((\{0\}\{1\}) \cup (\{0\}\{1\}\{0\}))$.

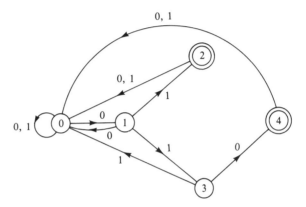

Figure 7.49 An NFA

Given a regular language L, there is a recursive method, based on the definition of a regular language, for constructing an NFA G_0 that accepts L. Unfortunately, the "nondeterminism" of an NFA makes it difficult to use as a tool for deciding if a given string α is in L. For instance, consider the NFA G of Figure 7.49 and the string $\alpha = 010$. Being in the initial state 0 and receiving the first character 0 of α as input, G can either stay in state 0 or go to state 1. How is this choice to be made? One way is to make the choice at random, say, by flipping a coin. Suppose it is decided to go to state 1. Then, on seeing the second character 1 of α, G must decide whether to go to state 2 or to state 3. Continuing this analysis, we see that the four possible paths through the NFA on input α are the following:

$$0, 0, 0, 0 \qquad 0, 0, 0, 1 \qquad 0, 1, 2, 0 \qquad 0, 1, 3, 4$$

Note that only the last of these possibilities ends up in an accepting state of G.

Fortunately, given a language L and an NFA G_0 that accepts L, there is an algorithm for converting G_0 to a DFA G_1 that accepts the same language. The bad news is that G_1 may have exponentially more states than does G_0; however, the good news is that this worst case rarely happens in practice. Moreover there is another algorithm, applied to G_1, that produces a DFA G_2 with the minimum number of states among all DFA's that accept L. Reluctantly, we must defer a more thorough discussion of these ideas to a course in the theory of computation.

Exercises 7.9

1. Let $X = \{1, 10\}$ and $Y = \{011, 1, 11\}$ be languages over $\{0, 1\}$. Find and/or describe precisely each of these languages.
   (a) $X \cup Y$ (b) $X \cap Y$ (c) XY (d) YX
   (e) X^3 (f) Y^2 (g) X^* (h) Y^*
2. For each of the following regular languages over $\{0, 1\}$, construct a DFA that accepts the language.
   (a) $\emptyset$ (b) $\{\varepsilon\}$ (c) $\{0\}$ (d) $\{1\}$
   (e) $\{0\}\{1\}^*\{0\}$ (f) $\{0\}^* \cup \{1\}^*$
   (g) $\{101, 000\}^*$ (h) $(\{11\} \cup \{00\}^*)^*$
3. For each of the following regular languages over $\{0, 1, 2\}$, construct a DFA that accepts the language.
   (a) $(\{0, 11\})^*\{012\}$ (b) $\{2\}(\{00\}^*\{1\})^*$
   (c) $((\{0\}\{1\}^*) \cup \{20\})^*$ (d) $\{0, 1\}\{2\}^*\{20, 12\}$
4. For each of the following languages over $\{0, 1\}$, show that the language is regular by constructing a DFA that accepts it.
   (a) the language X of all strings that begin with 0 and end with 1
   (b) the language Y of all strings that begin and end with the same character
   (c) the language Z of all strings that have an even number of 1's
   (d) the language W of all strings that have odd length and do not contain the substring 1001
5. Prove that the language L_6 of Example 7.44 is not regular.
6. Construct a DFA that accepts the same language as the NFA of Figure 7.49.
7. Explain why any finite language over $\{0, 1\}$ is regular.
8. If X is a regular language over $\{0, 1\}$, explain why $\{0, 1\}^* - X$ is also a regular language.
9. Explain in what sense a DFA is an NFA.
10. The purpose of this exercise is to develop a recursive method for constructing an NFA that accepts a given regular language over $\Sigma = \{0, 1\}$. Note that it is easy to give NFA's that accept $\emptyset$, $\{\varepsilon\}$, $\{0\}$, and $\{1\}$; see Exercise 2. Let X and Y be regular languages over $\Sigma = \{0, 1\}$, and let $G_X = (S_X, \Sigma, t_X, s_X, A_X)$ and $G_Y = (S_Y, \Sigma, t_Y, s_Y, A_Y)$ be NFA's that accept X and Y, respectively.
    (a) Explain how to construct an NFA (S, Σ, t, s, A) that accepts $X \cup Y$. (Hint: $s \notin S_X \cup S_Y$ is a new initial state, $S = S_X \cup S_Y \cup \{s\}$, and $A = A_X \cup A_Y$ or $A_X \cup A_Y \cup \{s\}$ depending on whether $\varepsilon \in X \cup Y$.)

(b) Explain how to construct an NFA (S, Σ, t, s, A) that accepts XY. (Hint: $S = S_X \cup S_Y$, $s = s_X$, and $A = A_Y$ or $A_Y \cup s$ depending on whether $\varepsilon \in X \cap Y$.)

(c) Explain how to construct an NFA (S, Σ, t, s, A) that accepts X^*. (Hint: $s \notin S_X$ is a new initial state, $S = S_X \cup \{s\}$, $A = A_X \cup \{s\}$.)

11. For each of the following regular languages over $\{0, 1\}$, use the method developed in Exercise 10 to construct an NFA that accepts the language.
   (a) $\{0, 1\}^*\{01, 010\}$ (see Figure 7.49)
   (b) $\{1, 101\}\{0\}^*\{01, 10\}$

12. Write a (Pascal) program that inputs a string $\alpha \in \{0, 1, 2\}^*$ and determines whether α belongs to the language L_7 of Example 7.42, part (d). The program should do this by simulating the DFA G_7 of Example 7.47.

CHAPTER PROBLEMS

1. Let $G = (V, E)$ be a graph. Prove that G is connected if and only if, for every partition of V as $V_1 \cup V_2$, there is $v_1 v_2 \in E$ with $v_1 \in V_1$ and $v_2 \in V_2$.

2. Each part gives two graphs G_1 and G_2. Draw the two graphs and then either exhibit an isomorphism between them or give a reason why they are not isomorphic.
   (a) $G_1 = (\{u_1, u_2, u_3, u_4, u_5, u_6\},$
   $\{u_1 u_4, u_1 u_5, u_1 u_6, u_2 u_4, u_2 u_5, u_2 u_6, u_3 u_4, u_3 u_5, u_3 u_6\});$
   $G_2 = (\{v_1, v_2, v_3, v_4, v_5, v_6\},$
   $\{v_1 v_2, v_1 v_3, v_1 v_6, v_2 v_3, v_2 v_5, v_3 v_4, v_4 v_5, v_4 v_6, v_5 v_6\})$
   (b) $G_1 = (\{u_1, u_2, u_3, u_4, u_5, u_6, u_7\}, \{u_1 u_2, u_1 u_3, u_1 u_4, u_1 u_5,$
   $u_1 u_6, u_1 u_7, u_2 u_5, u_2 u_7, u_3 u_4, u_3 u_5, u_4 u_5, u_4 u_6, u_4 u_7, u_5 u_6\});$
   $G_2 = (\{v_1, v_2, v_3, v_4, v_5, v_6, v_7\}, \{v_1 v_2, v_1 v_3, v_1 v_4, v_1 v_5,$
   $v_1 v_6, v_1 v_7, v_2 v_3, v_2 v_5, v_3 v_4, v_3 v_5, v_3 v_7, v_4 v_5, v_5 v_6, v_6 v_7\}$

3. A graph G is called *self-complementary* provided G is isomorphic to its complement $\overline{G}$.
   (a) Prove: If G is a self-complementary graph of order p, then p is congruent to 0 or 1 modulo 4.
   ★(b) Prove: If p is a positive integer and $p \equiv 0$ or 1 (mod 4), then there exists a self-complementary graph of order p.

4. Let m, n, and p be integers with $0 \le m \le n < p$. We are interested in the question, does there exist a graph G of order p such that $\delta(G) = m$ and $\Delta(G) = n$? The answer is obviously no if $m = 0$ and $n = p - 1$, or if $m = n$ and mp is odd, so we exclude these cases from consideration.
   (a) If $m = n$ and mp is even, then there exists an m-regular graph of order p. (See Section 8.2 of *Graphs and Digraphs*, 2nd ed., by Chartrand and Lesniak.) Give examples of (i) a 4-regular graph of order 9 and (ii) a 5-regular graph of order 10.

(b) Using the result stated in part (a), answer the question in the case $m = n - 1$.

(c) Answer the question in the remaining cases.

5. Let $d_1 \geq d_2 \geq \cdots \geq d_p$ be p positive integers. Show that the following two conditions are necessary for $(d_1, d_2, \ldots, d_p)$ to be the degree sequence of some graph G of order p (see Exercise 8 of Exercises 7.1):

(a) $d_1 + d_2 + \cdots + d_p$ is even.

(b) for each n, $1 \leq n \leq p - 1$,

$$\sum_{i=1}^{n} d_i \leq n(n - 1) + \sum_{j=n+1}^{p} \min\{n, d_j\}.$$

(In fact, Erdös and Gallai have shown that conditions (a) and (b) are also sufficient for $(d_1, d_2, \ldots, d_p)$ to be the degree sequence of some graph G of order p.)

6. Let $G = (V, A)$, where $V = \{s, t, u, v, w, x, y, z\}$ and $A = \{(s, w), (t, v), (t, z), (u, y), (v, x), (v, z), (w, t), (w, v), (x, t), (x, u), (y, w), (z, s), (z, x)\}$.

(a) Draw the digraph G.

(b) Find an eulerian circuit of G, preferably using the algorithm developed in Exercise 19 of Exercises 7.3.

7. Let G be a graph of order p and size q.

(a) Give a lower bound on q as a function of p if it is known that G is connected.

(b) Give an upper bound on q as a function of p if it is known that G is disconnected.

(c) Give examples to show that the bounds found in parts (a) and (b) are sharp (cannot be improved in general).

8. $G = (V, A)$, where $V = \{s, t, u, v, w, x, y, z\}$ and $A = \{(s, t), (t, u), (u, v), (v, w), (w, t), (w, x), (x, y), (y, w), (y, z), (z, x)\}$.

(a) Draw the digraph G.

(b) Give the adjacency matrix of G.

(c) Apply Algorithm 7.1 (Warshall) to find the reachability matrix of G.

(d) Find the distance matrix of G, preferably using the algorithm developed in Exercise 10 of Exercises 7.4.

9. A graph $G = (V, E)$ is called an *n-partite graph* provided it is possible to partition V as $V_1 \cup V_2 \cup \cdots \cup V_n$ such that each edge $e \in E$ is incident with a vertex of V_i and a vertex of V_j for some $i \neq j$. The sets $V_1, V_2, \ldots, V_n$ are called *partite sets* of G. For each graph shown in Figure 7.50, find the minimum n for which it is *n*-partite.

10. Let G be an *n*-partite graph with partite sets $V_1, V_2, \ldots, V_n$, where $|V_i| = p_i$, $1 \leq i \leq n$. If each vertex of V_i is adjacent with each vertex of V_j whenever $i \neq j$, then G is called a *complete n-partite graph*. Up to isomorphism, there is a unique complete *n*-partite graph with partite

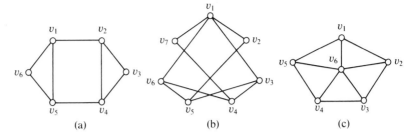

Figure 7.50

sets of cardinalities $p_1, p_2, \ldots, p_n$, and this graph is denoted $K_{p_1, p_2, \ldots, p_n}$. Draw each of the following graphs.

(a) $K_{1,2,3}$ (b) $K_{2,2,2}$ (c) $K_{1,2,2,3}$ (d) $K_{2,2,2,2}$

11. Find the order and size of the complete n-partite graph $K_{p_1, p_2, \ldots, p_n}$.

12. Let G be an n-partite graph of order p and size q. Give an upper bound on q in each case:

    (a) $n = 3$ and $p = 9$ (b) $n = 3$ and $p = 10$

    (c) $n = 4$ and $p = 10$ (d) $n = 4$ and $p = 12$

   ★(e) Generalize parts (a) through (d).

13. Let $G = (V, E)$, where $V = \{t, u, v, w, x, y, z\}$ and $E = \{tu, tv, tw, ux, vw, vy, vz, wx, wz, xy, yz\}$.

    (a) Find a subgraph $H_1 \simeq K_{2,3}$.

    (b) Find an induced subgraph $H_2 \simeq K_{1,3}$.

    (c) Find nonisomorphic spanning subgraphs H_3 and H_4 such that each is regular of degree 2.

    (d) Show that G does not contain a subgraph isomorphic to $K_{1,1,3}$.

14. Let $H = (V, E)$, where $V = \{u, v, x, y, z\}$ and $E = \{uv, ux, uz, vx, vz, yz\}$. Find, up to isomorphism, all connected spanning subgraphs of H.

15. Characterize those graphs with the property that every induced subgraph is connected.

16. Characterize those graphs with the property that every induced subgraph on three or more vertices is connected.

17. Generalize Problems 15 and 16. (Hint: See Problem 10.)

18. Let G be a graph with $\Delta(G) = \Delta$.

    (a) Prove: There exists a graph H such that G is an induced subgraph of H and H is Δ-regular.

    (b) Disprove: There exists a graph H such that G is a spanning subgraph of H and H is Δ-regular.

19. Apply Algorithm 7.2 (Dijkstra) to each of the weighted graphs shown in Figure 7.51 to find shortest paths from x. (Consider each edge to represent a pair of symmetric arcs.)

20. Apply Algorithm 7.5 (Kruskal) to each of the weighted graphs shown in Figure 7.51.

21. Adapt Algorithm 7.5 to solve the following problem: Given a connected

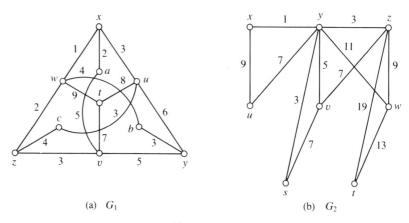

(a) G_1

(b) G_2

Figure 7.51

weighted graph $G = (V, E, w)$ and $E_1 \subseteq E$, find a connected spanning subgraph $H = (V, E_0)$ of minimum weight such that $E_1 \subseteq E_0$.

22. Consider the acyclic digraph $G = (V, A)$, where $V = \{a, b, c, d, e, x, y, z\}$ and $A = \{(a, d), (a, b), (a, c), (b, e), (b, x), (c, x), (c, y), (d, y), (e, y), (e, z), (x, z)\}$. Each part gives a list of the vertices of G; indicate, with justification, whether it is a valid result of applying Algorithm 7.3 (topological sort).
    (a) (a, b, c, d, e, x, y, z)
    (b) (a, d, c, b, x, e, z, y)
    (c) (a, d, b, e, y, c, x, z)

23. Consider the directed graph $G = (V, A)$, where $V = \{a, b, c, d, e, x, y, z\}$ and $A = \{(a, b), (a, c), (b, d), (b, e), (b, x), (c, e), (c, y), (d, e), (e, a), (x, y), (x, z), (y, z), (z, d)\}$. Apply Algorithm 7.4 to G to find a depth-first search tree with root a.

24. Let $T = (V, E)$ be a tree, with $V = \{v_1, v_2, \ldots, v_n\}$, and let m denote the number of vertices of degree 1 in T. Show that

$$m = 1 + \tfrac{1}{2} \sum_{i=1}^{n} |\deg(v_i) - 2|$$

25. Prove: Every tree T of order $p \geq 4$ that is not a star is isomorphic to a spanning tree of $\bar{T}$.

26. Let n be a positive integer and let G be a graph with $\delta(G) \geq n - 1$. Prove: If T is any tree of order n, then G contains a subgraph isomorphic to T.

27. Let G be a connected graph and let $e = uv$ be an edge of G. Prove:
    (a) The edge e is a bridge of G if and only if e is not on any cycle of G.
    (b) If e is a bridge, then G has at least two vertices of odd degree.
    (c) If e is a bridge, then G is not eulerian.

28. Prove: A graph G is a forest if and only if every induced subgraph of G contains a vertex of degree 0 or 1.

29. A graph G is said to be *unicyclic* provided G is connected and contains a unique cycle. Prove: A graph G of order p and size q is unicyclic if and only if G is connected and $q = p$.

30. A tree $T = (V, E)$ of order n is termed *graceful* provided there is a bijection $f: V \to \{1, 2, \ldots, n\}$ such that

$$\{|f(u) - f(v)| \mid uv \in E\} = \{1, 2, \ldots, n - 1\}$$

It is conjectured that all trees are graceful.
  (a) Show that the star $K_{1, n-1}$ is graceful.
  (b) Show that the path P_n is graceful.

31. When is a graph isomorphic to the graph of an equivalence relation?

32. Let G be a connected graph that is not a tree. The minimum length among the cycles of G is called the *girth* of G. Consider the following problem: For $r \geq 2$ and $n \geq 3$, determine those graphs of smallest order among all r-regular graphs with girth n; such graphs are called (r, n)-*cages* and are guaranteed to exist by a result of Erdös and Sachs.
  (a) Show that the cycle C_n is the unique $(2, n)$-cage.
  (b) Show that the complete graph K_{r+1} is the unique $(r, 3)$-cage.
  (c) Find the unique $(r, 4)$-cage.
  ★**(d)** Find a $(3, 5)$-cage.   ★**(e)** Find a $(3, 6)$-cage.

33. For positive integers m and n, the *ramsey number* $r(m, n)$ (named for the mathematician Frank Ramsey) is the least positive integer p such that every graph of order p either contains an induced subgraph isomorphic to K_m (m mutually adjacent vertices) or contains an induced subgraph isomorphic to $\overline{K}_n$ (an independent set of n vertices). It can be remarked that $r(m, n) = r(n, m)$, $r(1, n) = 1$, and $r(2, n) = n$.
  (a) Show that $r(3, 3) = 6$.
  (b) Prove, for $m \geq 2$ and $n \geq 2$, that $r(m, n) \leq r(m - 1, n) + r(m, n - 1)$. (This shows that the ramsey numbers exist.)
  (c) If follows from part (b) that $r(3, 4) \leq r(2, 4) + r(3, 3) = 4 + 6 = 10$. Show, in fact, that $r(3, 4) \leq 9$.
  (d) Show that $r(3, 4) \geq 9$ by exhibiting a graph of order 8 that contains neither K_3 nor $\overline{K}_4$ as an induced subgraph.

Thus $r(3, 4) = 9$. The only other known ramsey numbers are $r(3, 5) = 14$, $r(3, 6) = r(4, 4) = 18$, and $r(3, 7) = 23$.

34. Let $G = (V, A)$ be a digraph. For $n \geq 2$, the *nth power of* G is the digraph $G^n = (V, A_n)$, where $A_n = A \cup \{(u, v) \mid 2 \leq d(u, v) \leq n\}$.
  (a) Let $G = (V, A)$, where $V = \{s, t, u, v, w, x, y, z\}$ and $A = \{(s, t), (t, u), (u, v), (v, w), (w, t), (w, x), (x, y), (y, w), (y, z), (z, s), (z, x)\}$. Find G^2, G^3, and so on.
  (b) Define the analogous concept for graphs.
  (c) A result of H. Fleischner says that if G is any 2-connected graph, then G^2 is hamiltonian. Give an example of a tree T such that T^2 is not hamiltonian.
  (d) Prove: If T is any tree and u and v are vertices of T, then T^3 contains a spanning u-v path.

35. Consider expressions involving integer operands and the operators +
   (addition), * (multiplication), DIV, and MOD.
   (a) Find the expression represented by the binary tree of Figure 7.52.
   What is the value of the expression?
   ★**(b)** Given such a binary tree, develop an algorithm that finds the value
   of the expression.

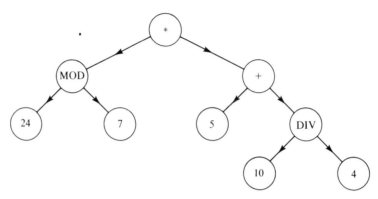

Figure 7.52

36. The 1989 NCAA basketball tournament began with 64 teams. Thirty-
   two games were played in the first round, eliminating 32 teams. In the
   second round, 16 games were played, eliminating 16 teams. The
   remaining 16 teams continued on to the next round of the tournament;
   from this point, half the remaining teams were eliminated at each
   round, until the winner of the tournament was determined.
   (a) Describe how such a single-elimination tournament can be
   represented as a binary tree.
   (b) What is the height of the binary tree representing the 1989 NCAA
   basketball tournament?
   (c) What was the total number of games played in the tournament?

37. An m-ary tree is called *ordered* provided each vertex x, other than the
   root, is assigned an integer $p(x)$ between 1 and m giving its "position"
   relative to its siblings, so that no two children of the same parent are
   assigned the same integer. We think of a child with a lower number as
   being "older" than a sibling with a higher number. For a binary tree, it
   is customary to assign the designations "left" and "right," rather than
   the integers 1 and 2, respectively. Let $T = (V, A)$ be an ordered
   complete m-ary tree, where $m > 2$. We may then define an ordered
   binary tree $\tilde{T} = (V, \tilde{A})$ as follows.

   (i) The root of $\tilde{T}$ is the root of T.
   (ii) If q is a nonleaf vertex of T, then the eldest child of q in T
   becomes the left child of q in $\tilde{T}$.
   (iii) If q is not the root and is not the youngest child of its parent
   in T, then its next youngest sibling becomes its right child in $\tilde{T}$.

   Using this method, it is possible to convert an m-ary tree T, $m > 2$,
   into a binary tree $\tilde{T}$ with the same vertex set.

Let $T = (V, A)$, where $V = \{a, b, c, d, e, f, g, h, i, j, k, m, n\}$, $A = \{(a, b), (a, c), (a, d), (b, e), (b, f), (c, g), (d, h), (d, i), (d, j), (e, k), (g, m), (g, n)\}$, ($a$ is the root) $p(b) = 1$, $p(c) = 2$, $p(d) = 3$, $p(e) = 1$, $p(f) = 2$, $p(g) = 1$, $p(h) = 1$, $p(i) = 2$, $p(j) = 3$, $p(k) = 1$, $p(m) = 1$, and $p(n) = 2$. Draw the 3-ary tree T, and then find and draw $\tilde{T}$.

38. With reference to the general ideas in Problem 37:
    (a) When is a vertex of T a leaf of $\tilde{T}$?
    (b) Given that T has height h, give an upper bound on the height of $\tilde{T}$ (in terms of m and h).
    (c) Given $\tilde{T}$, what information can be obtained about T?

39. A common operation on an ordered rooted tree is to *traverse* it—that is, to process or visit its vertices in some definite order. One method of traversing an ordered rooted tree is known as a *preorder traversal*. Let T be an ordered rooted tree with root r and let $T_1, T_2, \ldots, T_k$ be the subtrees of T whose roots are the children of r, in order from eldest to youngest. A preorder traversal of T is then defined recursively as follows: first visit r, and then perform preorder traversals of $T_1, T_2, \ldots, T_k$ in that order. For example, a preorder traversal of the tree T of Figure 7.21 visits the vertices in the order $a, b, e, f, h, c, d, g, i, j$.
    (a) Given the ordered rooted tree T of Problem 37, list the vertices of T in the order they are visited by a preorder traversal.
    (b) Informally describe an algorithm for performing a preorder traversal of a given ordered rooted tree T. Compare this algorithm with the depth-first search Algorithm 7.4.

40. Given integers k and p with $1 \le k \le p$, give an example of a graph G of order p such that $\chi(G) = k$ and $\chi(\overline{G}) = p + 1 - k$.

41. With reference to Problem 40, prove the following result of E. A. Nordhaus and J. W. Gaddum (1956): For any graph G of order p, $\chi(G) + \chi(\overline{G}) \le p + 1$.

42. A graph G is called *k-minimal* provided $\chi(G) = k$ and $\chi(G - e) = k - 1$ for every edge e of G.
    (a) Show that every connected k-minimal graph is k-critical.
    (b) Show that every 2-critical graph is 2-minimal.
    (c) Show that every 3-critical graph is 3-minimal.
    (d) Give an example of a 4-critical graph that is not 4-minimal.

43. Let $\omega(G)$ denote the largest order among the complete subgraphs of G. Then clearly $\chi(G) \ge \omega(G)$. Interestingly, there exists a graph G_k with $\omega(G_k) = 2$ and $\chi(G_k) = k$ for each $k \ge 2$. The following recursive method for constructing such graphs is due to J. Mycielski. Suppose $G_k = (V_k, E_k)$ has been constructed, with $V_k = \{v_1, v_2, \ldots, v_p\}$. We construct G_{k+1} by adding $p + 1$ new vertices $u_0, u_1, \ldots, u_p$ to G_k, making u_0 adjacent with each u_i, and making u_i adjacent with each v_j that is adjacent to v_i in G_k.
    (a) Starting with $G_2 \simeq K_2$, show that $G_3 \simeq C_5$.
    (b) Show that G_4 is isomorphic to the Grötzsch graph of Figure 7.45(c).

(c) Find the order of G_k.

(d) Verify that $\omega(G_k) = 2$ and $\chi(G_k) = k$.

44. Prove that for any graph G, $\chi(G)$ is at most 1 more than the maximum length among the paths of G.

45. Let G be a graph of order p with vertex degrees $d_1 \geq d_2 \geq \cdots \geq d_p$. Prove that

$$\chi(G) \leq \max\{\min\{i, d_i + 1\}\}$$

where the maximum is over all i, $1 \leq i \leq p$.

46. Given graphs $G_1 = (V_1, E_1)$ and $G_2 = (V_2, E_2)$, their *Cartesian product* is the graph $G_1 \times G_2$ with vertex set $V_1 \times V_2$ and edge set

$$\{(u_1, u_2)(v_1, v_2) \mid \text{either } u_1 = v_1 \text{ and } u_2 v_2 \in E_2 \text{ or } u_2 = v_2 \text{ and } u_1 v_1 \in E_1\}$$

Each part gives two graphs G_1 and G_2. Find $G_1 \times G_2$ and draw the graphs G_1, G_2, and $G_1 \times G_2$.

(a) $G_1 = (\{u, v\}, \{uv\})$, $G_2 = (\{x, y, z\}, \{xy, xz, yz\})$

(b) $G_1 = (\{t, u, v\}, \{tu, uv\})$, $G_2 = (\{w, x, y, z\}, \{wx, xy, yz\})$

(c) $G_1 = (\{t, u, v\}, \{tu, uv\})$, $G_2 = (\{w, x, y, z\}, \{wx, wz, xy, yz\})$

(d) $G_1 = (\{s, t, u, v\}, \{st, su, sv\})$, $G_2 = (\{w, x, y, z\}, \{wx, wz, xy, yz\})$

47. The *n-cube* is the graph Q_n defined recursively as follows: $Q_1 = (\{0, 1\}, \{01\})$; for $n \geq 1$, $Q_{n+1} = Q_1 \times Q_n$ (see Problem 46).

(a) Find and draw the graphs Q_2, Q_3, and Q_4.

(b) Find the order and size of Q_n.

(c) Show that Q_n is bipartite for each n.

(d) Show that Q_n is hamiltonian for each $n \geq 2$.

(e) Find the maximum distance between two vertices of Q_n.

(f) For which values of n is Q_n planar?

(The n-cubes are the underlying graphs for a certain class of computer networks known as "hypercube" networks.)

48. Show that the converse of Theorem 7.18 is false; that is, give an example of a nonhamiltonian graph $G = (V, E)$ having the property that, for each proper subset U of V, the subgraph of G induced by $V - U$ has at most $|U|$ components.

49. Give a necessary and sufficient condition for the complete tripartite graph $K_{m, n, p}$ to be hamiltonian.

50. Determine all values of n and $p_1 \leq p_2 \leq \cdots \leq p_n$ for which the complete n-partite graph $K_{p_1, p_2, \ldots, p_n}$ is planar.

51. Let G be a connected graph of order p and size q.

(a) Prove: If $q \leq p + 2$, then G is planar.

(b) For each $p \geq 6$, give an example of a nonplanar connected graph of order p and size $p + 3$.

52. Let G be a planar graph with girth g (see Problem 32).

(a) Generalize the results of Exercises 21 and 22 of Exercises 7.8.

(b) Prove: If $g \geq 4$, then $\chi(G) \leq 4$.

53. Describe an algorithm for finding a maximal independent set W in a given graph $G = (V, E)$.

54. Consider the following algorithm for purportedly finding $\chi(G)$ for a given graph $G = (V, E)$.

Step 0. Initially, $n := 0$; $U := V$.

Step 1. Repeat Step 2 until $U = \emptyset$; at that point the algorithm terminates and $\chi(G) = n$.

Step 2. Find a maximal independent set W of $\langle U \rangle$ (using the algorithm found in Problem 53); $n := n + 1$; $U := U - W$.

Give an example to show that this algorithm does not find $\chi(G)$ in general. What does the algorithm produce?

E I G H T

Algebraic Structures

BINARY OPERATIONS

Consider the operation of addition on the set $\mathbb{R}$. Associated with any two real numbers a and b there is a uniquely determined third real number $a + b$. This association can be formally defined as a function $f: \mathbb{R} \times \mathbb{R} \to \mathbb{R}$, where $f((a, b)) = a + b$ for all $(a, b) \in \mathbb{R} \times \mathbb{R}$. As another example, we can define a function $g: \mathbb{R} \times \mathbb{R} \to \mathbb{R}$ by $g((a, b)) = ab$. Each of these functions is an example of a "binary operation" defined on the set of real numbers.

DEFINITION 8.1

Given any nonempty set A, a function $f: A \times A \to A$ is called a *binary operation* on A.

As a general rule, we denote an abstract binary operation on a set by the symbol $*$. Instead of using the functional notation $*((a, b))$, we use the more standard notation $a * b$, which more closely parallels the way in which well-known binary operations (like $+$ and $\cdot$ on $\mathbb{R}$) are displayed.

Example 8.1 (a) The set operations of union and intersection on the set $\mathscr{P}(U)$ of all subsets of a universal set U are binary operations on $\mathscr{P}(U)$. Recall that if $A, B \in \mathscr{P}(U)$, then the set difference $A - B$ is defined by $A - B = \{x \in U \mid x \in A \text{ and } x \notin B\}$. Set difference is also a binary operation on $\mathscr{P}(U)$.

(b) Let A be a nonempty set and let $\mathscr{F}(A)$ denote the set of all functions from A to A. If f and g are two functions in $\mathscr{F}(A)$, then $f \circ g \in \mathscr{F}(A)$. (Recall that if $f, g \in \mathscr{F}(A)$, then $f \circ g$ is defined by $(f \circ g)(x) = f(g(x))$ for all $x \in A$.) Thus composition $\circ$ is a binary operation on $\mathscr{F}(A)$. More specifically, let $S(A)$ denote the set of permutations of A. It was previously shown that $f \circ g \in S(A)$ provided $f, g \in S(A)$ (see Corollary 5.8), so composition is also a binary operation on $S(A)$.

(c) Let C denote the set of all valid character strings in some programming language. Many languages (such as Ada and some versions of Pascal) define the binary operation of "concatenation" on C. For instance, the concatenation of the strings "PASC" and "AL" is the string "PASCAL".

(d) Matrix addition and multiplication are binary operations on the set $M_n(\mathbb{R})$ of all $n \times n$ matrices over $\mathbb{R}$.   ∎

Consider the set $\mathbb{Z}_m = \{[0], [1], [2], \ldots, [m-1]\}$ of residue classes modulo m. Is it perhaps possible to define, in a natural way, one or more binary operations on $\mathbb{Z}_m$? Indeed it is possible and well motivated by the natural binary operations of $+$ and $\cdot$ on $\mathbb{Z}$. For $[a], [b] \in \mathbb{Z}_m$, define $\oplus$ and $\odot$ as follows:

1. $[a] \oplus [b] = [a + b]$
2. $[a] \odot [b] = [ab]$

Bear in mind that if $\oplus$ and $\odot$ are to be binary operations on $\mathbb{Z}_m$, then both $[a] \oplus [b]$ and $[a] \odot [b]$ must be uniquely determined by $[a]$ and $[b]$; that is, $\oplus$ and $\odot$ must be well-defined functions from $\mathbb{Z}_m \times \mathbb{Z}_m$ to $\mathbb{Z}_m$. Perhaps the student can see a bit of a problem with this determination. The problem can be illustrated by the following example. In $\mathbb{Z}_7$ we have $[2] = [-5]$ and $[4] = [18]$. In order for $\oplus$ and $\odot$ to be binary operations, it is necessary that both $[2] \oplus [4] = [-5] \oplus [18]$ and $[2] \odot [4] = [-5] \odot [18]$. Both of these are easily seen to be true. In general, we must show that if $[a_1] = [a_2]$ and $[b_1] = [b_2]$ in $\mathbb{Z}_m$, then $[a_1] \oplus [b_1] = [a_2] \oplus [b_2]$ and $[a_1] \odot [b_1] = [a_2] \odot [b_2]$. If these can be shown to hold, then it follows that $\oplus$ and $\odot$ are binary operations on $\mathbb{Z}_m$.

THEOREM 8.1   For each $m \in \mathbb{N}$, $\oplus$ and $\odot$ are binary operations on $\mathbb{Z}_m$.

Proof   We present only the proof that $\odot$ is a binary operation on $\mathbb{Z}_m$, leaving the other part to Exercise 8. Suppose that $[a_1] = [a_2]$ and $[b_1] = [b_2]$ in $\mathbb{Z}_m$. We then have that $a_1 \equiv a_2 \pmod{m}$ and $b_1 \equiv b_2 \pmod{m}$. Hence $a_1 = a_2 + mk$ and $b_1 = b_2 + mt$ for some $k, t \in \mathbb{Z}$, so we obtain

$$a_1 b_1 = a_2 b_2 + m(a_2 t + b_2 k + mkt)$$

Thus $a_1 b_1 \equiv a_2 b_2 \pmod{m}$, and so $[a_1 b_1] = [a_2 b_2]$. Therefore

$$[a_1] \odot [b_1] = [a_2] \odot [b_2]$$

so that $\odot$ is a binary operation on $\mathbb{Z}_m$.   □

In general, if we are asked to prove that $*$ is a binary operation on a set A, then it should be emphasized that there are two properties to be verified:

1. If $a, b \in A$, then $a * b \in A$.
2. If $a, b \in A$, then $a * b$ is uniquely determined.

If property 1 holds, then we say that $*$ is *defined on A*; if property 2 holds, then we say that $*$ is *well-defined*. It is especially important to check property 2 when the elements of A are equivalence classes and can, as a consequence, be represented by different names, as in the case of $\mathbb{Z}_m$, for instance.

Example 8.2 Determine which of the following are binary operations:

(a) $a * b = a/(b^2 + 1)$ on $\mathbb{Z}$

(b) $a * b = \sqrt{(a + b)^2}$ on $\mathbb{Q}$ (c) $[a]*[b] = [a \text{ DIV } b]$ on $\mathbb{Z}_4$

Solution

(a) This is not a binary operation on $\mathbb{Z}$ since, for example, if $a = b = 1$, then $a * b = 1/2 \notin \mathbb{Z}$.

(b) This is a binary operation on $\mathbb{Q}$. Note that, if $a, b \in \mathbb{Q}$, then

$$a * b = \sqrt{(a + b)^2} = |a + b| \in \mathbb{Q}$$

(c) Here $*$ fails to be well-defined. For instance, $[2]*[2] = [2 \text{ DIV } 2] = [1]$, while $[6]*[2] = [6 \text{ DIV } 2] = [3]$. Hence we have $[2] = [6]$ but $[1] \neq [3]$. ∎

Given a binary operation $*$ on a finite set A, it is often helpful to construct an *operation table* for $*$, which is similar to the addition and multiplication tables for integers (with which the student is no doubt familiar). Such a table has a row and a column for each element of A; if $a, b \in A$, then the element $a * b$ is placed in the row corresponding to a and the column corresponding to b. The operation tables for $\mathbb{Z}_6$ under $\oplus$ and for $\mathbb{Z}_5^{\#}$ under $\odot$ are shown in Figure 8.1, where $\mathbb{Z}_5^{\#} = \mathbb{Z}_5 - \{[0]\}$.

$\oplus$	[0]	[1]	[2]	[3]	[4]	[5]
[0]	[0]	[1]	[2]	[3]	[4]	[5]
[1]	[1]	[2]	[3]	[4]	[5]	[0]
[2]	[2]	[3]	[4]	[5]	[0]	[1]
[3]	[3]	[4]	[5]	[0]	[1]	[2]
[4]	[4]	[5]	[0]	[1]	[2]	[3]
[5]	[5]	[0]	[1]	[2]	[3]	[4]

$\odot$	[1]	[2]	[3]	[4]
[1]	[1]	[2]	[3]	[4]
[2]	[2]	[4]	[1]	[3]
[3]	[3]	[1]	[4]	[2]
[4]	[4]	[3]	[2]	[1]

(a) $\mathbb{Z}_6$ under $\oplus$ (b) $\mathbb{Z}_5^{\#}$ under $\odot$

Figure 8.1 Two operation tables

The binary operations $\oplus$ and $\odot$ defined on $\mathbb{Z}_m$ satisfy some of the same properties that hold for $+$ and $\cdot$ on $\mathbb{Z}$. In general, given any binary operation $*$ on a set A, we are interested in what properties are satisfied by $*$. Two of the most common properties are investigated in the following definition.

DEFINITION 8.2

Let $*$ be a binary operation on a set A. We say that $*$ is *associative* provided $(a * b) * c = a * (b * c)$ for all $a, b, c \in A$. The operation $*$ is called *commutative* provided $a * b = b * a$ for all $a, b \in A$.

For example, it is clear that $+$ and $\cdot$ are associative and commutative binary operations on $\mathbb{R}$, and the student is asked to show in Exercise 14 that $\oplus$ and $\odot$ are associative and commutative binary operations on $\mathbb{Z}_m$. However, subtraction is a binary operation on $\mathbb{R}$ that is neither associative nor commutative. That subtraction is not associative can be seen, for example, from the inequality $(2 - 3) - 4 \neq 2 - (3 - 4)$.

In Example 8.1 we saw that composition is a binary operation on the set $\mathscr{F}(A)$ of all functions from a set A to itself. That composition is an associative binary operation follows from Theorem 5.10, and we restate it here.

THEOREM 5.10   Let A, B, C, and D be nonempty sets. Given $f: A \to B$, $g: B \to C$, and $h: C \to D$, then $(h \circ g) \circ f = h \circ (g \circ f)$.

If we simply apply Theorem 5.10 with $B = C = D = A$, then we obtain the following theorem as a corollary.

THEOREM 8.2   Let A be any nonempty set. Then composition is an associative binary operation on the set $\mathscr{F}(A)$ of all functions from A to A.

It can easily be seen that function composition is not, in general, a commutative binary operation. For example, define $f: \mathbb{Z} \to \mathbb{Z}$ by $f(m) = m^2$ and $g: \mathbb{Z} \to \mathbb{Z}$ by $g(m) = m + 1$. Then $(f \circ g)(m) = (m + 1)^2$, while $(g \circ f)(m) = m^2 + 1$. So $f \circ g \neq g \circ f$, and hence $\circ$ is not commutative on $\mathscr{F}(\mathbb{Z})$.

Given a set A, it is sometimes possible to define several different binary operations on A, although in a given discussion we may be interested in studying properties of one specific binary operation, say $*$, on A. To avoid any confusion in such cases, we shall refer to A together with $*$ by the notation $(A, *)$. For instance, we could consider any one of $(\mathbb{R}, +)$, $(\mathbb{R}, \cdot)$, or $(\mathbb{R}, -)$. We could also study the properties satisfied by a pair of binary operations defined on a set, such as $+$ and $\cdot$ on $\mathbb{R}$. In this case we use the notation $(\mathbb{R}, +, \cdot)$. Of course, this notation can be extended to three or more binary operations on a set. Systems like $(\mathbb{R}, +)$ and $(\mathbb{R}, +, \cdot)$ are examples of *algebraic structures*.

In addition to associativity and commutativity, there are some other interesting and natural properties that an operation might satisfy. For example, in $(\mathbb{R}, \cdot)$ there is a very special real number, namely 1, that satisfies the condition $1 \cdot a = a \cdot 1 = a$ for all $a \in \mathbb{R}$. Also, in $(\mathbb{R}, \cdot)$, for each $a \in \mathbb{R} - \{0\}$, there is a real number, namely $1/a$, such that $a \cdot (1/a) = 1$. These examples serve to motivate the following definition.

> **DEFINITION 8.3**
>
> Let $*$ be a binary operation on the set A.
>
> 1. An element $e \in A$ is called an *identity element* for $(A, *)$ provided
> $$a * e = e * a = a$$
>   for all $a \in A$.
>
> 2. Suppose $(A, *)$ has an identity element e. For $a \in A$, an element $a' \in A$ is called an *inverse* of a in $(A, *)$ provided
> $$a * a' = a' * a = e$$

Example 8.3   (a) In $(\mathbb{Z}, +)$, 0 is an identity element, and the inverse of the integer m is $-m$, since $m + (-m) = (-m) + m = 0$.

(b) In $(S(A), \circ)$, the function i_A is an identity element (Theorem 5.6) since $i_A \circ f = f \circ i_A = f$ for all $f \in S(A)$. Also, for any $f \in S(A)$, we know that f^{-1} exists, $f^{-1} \in S(A)$, and $f \circ f^{-1} = f^{-1} \circ f = i_A$. Thus f^{-1} is an inverse for f in $(S(A), \circ)$.

(c) An identity element for concatenation of character strings is the empty string. No nonempty string has an inverse under concatenation.

(d) Consider $(\mathscr{P}(U), \cap)$. Since for any $X \subseteq U$ it is true that $U \cap X = X \cap U = X$, the set U is an identity element in $(\mathscr{P}(U), \cap)$.

(e) The $n \times n$ zero matrix O_n (the $n \times n$ matrix each of whose entries is 0) is an identity element for $(M_n(\mathbb{R}), +)$, while the $n \times n$ identity matrix I_n (the $n \times n$ matrix each of whose main diagonal entries is 1, with all other entries 0) is an identity element for $(M_n(\mathbb{R}), \cdot)$.   ∎

Given an algebraic structure $(A, *)$, it may or may not be the case that $(A, *)$ has an identity element. Also, given that $(A, *)$ has an identity element, there is the question of the existence of inverses for the elements of A. For example, if E is the set of even integers, then $(E, \cdot)$ has no identity element. On the other hand, $(\mathbb{Z}, \cdot)$ has identity element 1, but inverses exist for only 1 and -1. Don't be fooled by this last statement; it states that if $a \in \mathbb{Z}$ and $a \neq \pm 1$, then there is no $b \in \mathbb{Z}$ such that $ab = 1$.

Suppose that $(A, *)$ is an algebraic structure with identity element e. Is e uniquely determined or can there possibly be more than one identity element for $(A, *)$? If $a \in A$ has an inverse a', is a' uniquely determined? These questions are treated in the following theorem.

THEOREM 8.3   Let $(A, *)$ be an algebraic structure with identity element e. Then e is the only identity element of $(A, *)$. If $a \in A$ has an inverse a' in $(A, *)$ and $*$ is associative, then a' is uniquely determined.

Proof   Suppose that both e and e_1 are identities for $(A, *)$. Since e is an identity element, we have $e * e_1 = e_1$. Also, since e_1 is an identity element, $e * e_1 = e$. Thus $e = e_1$, and $(A, *)$ has only one identity element.

Next assume $*$ is associative and let $a \in A$. If a' and a'' are inverses for a in $(A, *)$, then we have

$$a' = a' * e = a' * (a * a'') = (a' * a) * a'' = e * a'' = a'' \qquad \square$$

Notice how associativity was needed in Theorem 8.3 to prove that inverses are unique. In Exercise 6 the student is asked to provide an example of an algebraic structure in which some elements possess more than one inverse.

In what follows, the term *identity* is used interchangeably with the term *identity element*.

Exercises 8.1

1. Let $*$ be a binary operation on the set A and let B be a nonempty subset of A. Suppose that $b_1 * b_2 \in B$ for all $b_1, b_2 \in B$.
   (a) Show that $*$ can be considered a binary operation on B.
   (b) If $*$ is associative on A, does it follow that $*$ is associative on B?
   (c) If $*$ is commutative on A, does it follow that $*$ is commutative on B?
   (d) If $(A, *)$ has an identity, does it follow that $(B, *)$ has an identity?
   (e) Suppose that $(A, *)$ has identity e. If each $a \in A$ has an inverse $a' \in A$ under $*$, does it follow that each $b \in B$ has an inverse $b' \in B$ under $*$?

2. Show that if $A = \{a, b, c\}$, then $\circ$ is not a commutative binary operation on $S(A)$.

3. Let $\mathbb{C}^\#$ denote the set of nonzero complex numbers, $\mathbb{C}^\# = \{a + bi \mid a + bi \neq 0\}$. Recall that complex multiplication is defined by

   $$(a + bi)(c + di) = (ac - bd) + (ad + bc)i$$

   (We also use the familiar "$\cdot$" to denote complex multiplication.)
   (a) Show that complex multiplication is a binary operation on $\mathbb{C}^\#$.
   (b) Show that $(\mathbb{C}^\#, \cdot)$ has an identity element.
   (c) Show that each $z \in \mathbb{C}^\#$ has an inverse element in $(\mathbb{C}^\#, \cdot)$.

4. Let $\mathscr{R} = \{\cos \alpha + i \sin \alpha \mid \alpha \in \mathbb{R}\}$, the set of all complex numbers with magnitude 1. Recall that the magnitude of $a + ib$ is given by

   $$|a + ib| = \sqrt{a^2 + b^2}$$

   (a) Prove DeMoivre's theorem: For any $\alpha \in \mathbb{R}$ and any $n \in \mathbb{N}$,

   $$(\cos \alpha + i \sin \alpha)^n = \cos n\alpha + i \sin n\alpha$$

   (Hint: Use mathematical induction.)

(b) Show that $\cdot$ is a binary operation on $\mathscr{R}$.

(c) Why is $\cdot$ an associative operation on $\mathscr{R}$?

(d) Show that $(\mathscr{R}, \cdot\,)$ has an identity element.

(e) Show that each $z \in \mathscr{R}$ has an inverse element in $(\mathscr{R}, \cdot\,)$.

5. Let A be a nonempty set and let $a \in A$. Define the *stabilizer of a* to be the set

$$G_a = \{\, g \in S(A) \mid g(a) = a \,\}$$

(a) Show that $\circ$ is a binary operation on G_a.

(b) Why is $\circ$ an associative operation on G_a?

(c) Show that $(G_a, \circ)$ has an identity and that each $f \in G_a$ has an inverse in $(G_a, \circ)$.

6. Give an example of a binary operation $*$ on the set $\{\, a, b, c \,\}$ that has identity a, but in which the inverse of b is not unique.

7. In each of the following determine whether $*$ is a binary operation on $\mathbb{Z}$. If it is, then indicate whether $*$ is associative and whether it is commutative. Also discuss the existence of an identity and inverses.

(a) $m * n = mn + 1$ **(b)** $m * n = m + n - 1$

(c) $m * n = n$ **(d)** $m * n = 2m + n$

(e) $m * n = (m + n)/2$ **(f)** $m * n = 2^{mn}$

8. Show that $\oplus$ is a binary operation on $\mathbb{Z}_m$. (Hint: Parallel the proof of Theorem 8.1.)

9. Let $*$ be a binary operation on a nonempty set A. Complete each of the following statements.

(a) $*$ is not associative provided _____ .

(b) $*$ is not commutative provided _____ .

(c) $a \in A$ is not an identity for $*$ provided _____ .

(d) Given that $(A, *)$ has identity e, then $b \in A$ has no inverse provided _____ .

10. Give the operation table for each of the following algebraic structures.

(a) $(S(\{1, 2\}), \circ)$ **(b)** $(\mathbb{Z}_5, \oplus)$ **(c)** $(\mathbb{Z}_4, \odot)$

(d) $(\mathscr{P}(\{1, 2, 3\}), \cup)$ **(e)** $(\mathbb{Z}_6, \odot)$ **(f)** $(\mathbb{Z}_7^{\#}, \odot)$ **(g)** $(\mathbb{Z}_8^{\#}, \odot)$

11. This exercise concerns the following partial operation table for an algebraic structure $(\{\, a, b, c, d \,\}, *)$.

$*$	a	b	c	d
a	a		c	
b				a
c		d	a	
d			b	

Complete the table so as to make $*$

(a) associative   **(b)** commutative

(c) commutative with an identity

(Note: Part (a) has only one correct answer, but parts (b) and (c) have several correct answers.)

12. Consider the algebraic structure $(\mathscr{F}(A), \circ)$, where $A = \{1, 2\}$.
    (a) Construct the operation table for $(\mathscr{F}(A), \circ)$.
    (b) Is $\circ$ commutative in this case?
    (c) What is the identity of $(\mathscr{F}(A), \circ)$?
    (d) Find the inverse of each element that has one.

13. Let $T = \mathbb{Q} - \{0, 1\}$ and define the following six functions on T:

$$f_1(x) = x \qquad\qquad f_2(x) = 1 - x$$

$$f_3(x) = \frac{1}{x} \qquad\qquad f_4(x) = \frac{x - 1}{x}$$

$$f_5(x) = \frac{1}{1 - x} \qquad f_6(x) = \frac{x}{x - 1}$$

    Verify that $\circ$ is a binary operation on $F = \{f_1, f_2, f_3, f_4, f_5, f_6\}$ and construct the operation table for $(F, \circ)$.

14. Consider the algebraic structure $(\mathbb{Z}_m, \oplus, \odot)$.
    (a) Show that $\oplus$ is associative and commutative.
    (b) Show that $\odot$ is associative and commutative.

15. Division is a binary operation on the set $\mathbb{Q}^+$ of positive rational numbers.
    (a) Is division associative on $\mathbb{Q}^+$?
    (b) Is division commutative on $\mathbb{Q}^+$?
    (c) Note that $r/1 = r$ for every positive rational number r. Does this mean that 1 is an identity for division on $\mathbb{Q}^+$? Explain.

16. Each of the following defines a binary operation $*$ on $\mathbb{Z} \times \mathbb{N}$. Determine whether $*$ is associative and whether it is commutative. Also discuss the existence of an identity and inverses.
    (a) $(a, b)*(c, d) = (a + c, b + d)$
    (b) $(a, b)*(c, d) = (ac, bd)$
    (c) $(a, b)*(c, d) = (ad + bc, bd)$

17. Determine the smallest subset A of $\mathbb{Z}$ such that $2 \in A$ and addition is a binary operation on A.

18. Determine the smallest subset C of $\mathbb{Z}$ such that $\{2, 5\} \subseteq C$ and multiplication is a binary operation on C.

8.2

SEMIGROUPS AND GROUPS

In the last section we saw that it is often possible to define, in a natural way, one or more binary operations on a given set. Of particular interest is the case where these binary operations satisfy a specified set of properties. Such algebraic structures are the main objects of study in that branch of mathematics called abstract algebra. Many of the topics covered in this section will indeed seem very abstract to the student and will take some getting used to. Suffice it to say, however, that there are many interesting and significant applications of abstract algebra to such diverse areas as quantum mechanics, satellite communication, microwave transmission, electrical switching networks, crystallography, and computer science. In the rest of this chapter we survey several of the more important algebraic structures and then investigate a particular application of the theory of "Boolean algebras" to combinatorial circuits.

Two algebraic structures for which the theory is highly developed are *semigroups* and *groups*. In this section we introduce these and explore some of their elementary properties.

DEFINITION 8.4

Let $*$ be a binary operation on a nonempty set A. If $*$ is associative, then $(A, *)$ is called a *semigroup*.

Example 8.4   (a) Let A be any nonempty set and consider once again the set $\mathscr{F}(A)$ of all functions from A to A. According to Example 8.1(a), composition is a binary operation on the set $\mathscr{F}(A)$ and, by Theorem 8.2, we also have that composition is associative on $\mathscr{F}(A)$. Thus $(\mathscr{F}(A), \circ)$ is a semigroup.

(b) Let A be a set and consider the power set $\mathscr{P}(A)$ under the operation of intersection. We know that intersection is a binary operation on the set $\mathscr{P}(A)$ and, for any $B, C, D \in \mathscr{P}(A)$, we know that $(B \cap C) \cap D = B \cap (C \cap D)$. Thus, intersection is an associative binary operation on the set $\mathscr{P}(A)$ and hence $(\mathscr{P}(A), \cap)$ is a semigroup.

(c) Define $*$ as follows: for $a, b \in \mathbb{Z}$, $a * b = \sqrt{a^2 + b^2}$. Is $(\mathbb{Z}, *)$ a semigroup? We must first check that $*$ is a binary operation on $\mathbb{Z}$. In fact, we see that it is not since, for example,

$$2 * 3 = \sqrt{2^2 + 3^3} = \sqrt{13} \notin \mathbb{Z}$$

Thus $(\mathbb{Z}, *)$ is not a semigroup.

(d) Define $*$ as follows: for $a, b \in \mathbb{Z}$, $a * b = 3a + b$. Is $(\mathbb{Z}, *)$ a semigroup? Since, for any $a, b \in \mathbb{Z}$, $a * b = 3a + b$ is a uniquely determined element of $\mathbb{Z}$, we see that $*$ is a binary operation on $\mathbb{Z}$. Is $*$ associative? For $a, b, c \in \mathbb{Z}$, we have that $(a * b) * c = 3(a * b) + c = 3(3a + b) + c = 9a + 3b + c$, while $a * (b * c) = 3a + (b * c) = 3a + (3b + c) = 3a + 3b + c$. Thus, for example, if $a = b = c = 1$, then $(a * b) * c = 13$ and $a * (b * c) = 7$. Hence $*$ is not associative. Therefore $(\mathbb{Z}, *)$ is not a semigroup.

(e) Let C denote the set of all valid character strings in some programming language and let $*$ denote the operation of concatenation on C. It was shown in Example 8.1 that $*$ is a binary operation on C. If $a, b, c \in C$, then it is clear that $(a * b) * c = a * (b * c)$, so $*$ is associative. Therefore $(C, *)$ is a semigroup.

(f) According to Theorem 8.1 and Exercise 14 of Exercises 8.1, $\oplus$ and $\odot$ are associative binary operations on $\mathbb{Z}_m$. Hence both $(\mathbb{Z}_m, \oplus)$ and $(\mathbb{Z}_m, \odot)$ are semigroups.

(g) Remembering that matrix addition and multiplication are associative binary operations on $M_n(\mathbb{R})$, we see that both $(M_n(\mathbb{R}), +)$ and $(M_n(\mathbb{R}), \cdot)$ are semigroups. ∎

As illustrated in the preceding example, it is important to keep in mind that, in determining whether a given set A is a semigroup under an operation $*$, one must be certain, first of all, that $*$ is a binary operation on A.

It is easily checked that $(\mathbb{R}, +)$ is a semigroup with identity element 0, and that each $a \in \mathbb{R}$ has inverse element $-a \in \mathbb{R}$. If $\mathbb{R}^{\#} = \mathbb{R} - \{0\}$, then $(\mathbb{R}^{\#}, \cdot)$ is a semigroup with identity element 1, and each $a \in \mathbb{R}^{\#}$ has inverse element $1/a \in \mathbb{R}^{\#}$. In general, semigroups that have an identity element and in which each element has an inverse element belong to a very special class of algebraic structures.

DEFINITION 8.5

An algebraic structure $(G, *)$ is called a *group* provided the following conditions are satisfied:

1. The operation $*$ is associative.
2. $(G, *)$ has an identity element.
3. Each $g \in G$ has an inverse in $(G, *)$.

Example 8.5 (a) According to Example 8.4(f), $(\mathbb{Z}_6, \oplus)$ is a semigroup. Its operation table is shown in Figure 8.1(a). From this table we see that, for each $[a] \in \mathbb{Z}_6$, $[a] \oplus [0] = [0] \oplus [a] = [0 + a] = [a]$; thus $[0]$ is the identity element for $(\mathbb{Z}_6, \oplus)$. Furthermore, it is clear from the table that each element of $\mathbb{Z}_6$ has an inverse under $\oplus$. In particular, we see that

$[0] \oplus [0] = [0]$	(the inverse of $[0]$ is $[0]$)
$[1] \oplus [5] = [5] \oplus [1] = [0]$	(the inverse of $[1]$ is $[5]$ and that of $[5]$ is $[1]$)
$[3] \oplus [3] = [0]$	(the inverse of $[3]$ is $[3]$)
$[2] \oplus [4] = [4] \oplus [2] = [0]$	(the inverse of $[2]$ is $[4]$ and that of $[4]$ is $[2]$)

Therefore $(\mathbb{Z}_6, \oplus)$ is a group. (Note that $[-a] = [6 - a]$ in $\mathbb{Z}_6$; thus, for example, the inverse of $[2]$ is $[-2] = [6 - 2] = [4]$.)

(b) Consider $(\mathbb{Z}_5^{\#}, \odot)$, whose operation table is shown in Figure 8.1(b). We know that $\odot$ is an associative binary operation on $\mathbb{Z}_5$ so, according to Exercise 1(b) of Exercises 8.1, $\odot$ is an associative binary operation on $\mathbb{Z}_5^{\#}$. It is easily seen from the table that $[1]$ is the identity element of $(\mathbb{Z}_5^{\#}, \odot)$ and that the inverses of $[1]$, $[2]$, $[3]$, and $[4]$ in $(\mathbb{Z}_5^{\#}, \odot)$ are, respectively, $[1]$, $[3]$, $[2]$, and $[4]$. Thus $(\mathbb{Z}_5^{\#}, \odot)$ is a group.

(c) Is $(\mathbb{Z}_8, \odot)$ a group? Since it is a semigroup, we must determine if it has an identity element and, if so, whether its elements have inverses. Clearly [1] is the identity element for $(\mathbb{Z}_8, \odot)$. It is just as clear that [0] has no inverse in $(\mathbb{Z}_8, \odot)$, so $(\mathbb{Z}_8, \odot)$ is not a group. What about $\mathbb{Z}_8^{\#}$ under $\odot$? (As usual, $\mathbb{Z}_8^{\#} = \mathbb{Z}_8 - \{[0]\}$.) First of all, is $\odot$ a binary operation on $\mathbb{Z}_8^{\#}$? The answer is no, since $[2] \odot [4] = [0] \notin \mathbb{Z}_8^{\#}$. Hence $\mathbb{Z}_8^{\#}$ under $\odot$ is not even a semigroup.

(d) Let A be a nonempty set and consider the set $S(A)$ of all permutations of A. According to Example 8.1(b), composition is a binary operation on $S(A)$, and by Theorem 8.2, composition is also associative. In Example 8.3 we saw that i_A is the identity element for $(S(A), \circ)$ and that each element of $S(A)$ has an inverse in $S(A)$. Thus $(S(A), \circ)$ is a group. We call $(S(A), \circ)$ the *symmetric group on* A. In the special case that $A = \{1, 2, \ldots, n\}$, $(S(A), \circ)$ is called the *symmetric group of degree* n and is denoted by S_n. Recall that in Chapters 5 and 6 it was shown that $|S_n| = n!$.

(e) In Example 8.4(g) it was seen that $(M_n(\mathbb{R}), +)$ is a semigroup, and in Example 8.3(e) it was noted that O_n is the identity element for $(M_n(\mathbb{R}), +)$. Moreover, if $A \in M_n(\mathbb{R})$, say $A = [a_{ij}]$, then the $n \times n$ matrix $B = [-a_{ij}]$ satisfies the relation $A + B = B + A = O_n$. Thus $(M_n(\mathbb{R}), +)$ is a group.

(f) Define

$$\Delta_2(\mathbb{R}) = \left\{ \begin{bmatrix} a & 0 \\ 0 & b \end{bmatrix} \mid a, b \in \mathbb{R} \text{ and } ab \neq 0 \right\}$$

We show that $\Delta_2(\mathbb{R})$ is a group under matrix multiplication. Let $A, B \in \Delta_2(\mathbb{R})$, say

$$A = \begin{bmatrix} a & 0 \\ 0 & b \end{bmatrix} \quad \text{and} \quad B = \begin{bmatrix} c & 0 \\ 0 & d \end{bmatrix}$$

Then

$$A \cdot B = \begin{bmatrix} ac & 0 \\ 0 & bd \end{bmatrix}$$

so $A \cdot B \in \Delta_2(\mathbb{R})$. According to Exercise 1(b) of Exercises 8.1, it follows that matrix multiplication is associative on $\Delta_2(\mathbb{R})$. The 2×2 identity matrix I_2 is an element of $\Delta_2(\mathbb{R})$ and is the identity for $(\Delta_2(\mathbb{R}), \cdot)$. Moreover, the inverse of the matrix A given above is the matrix

$$A^{-1} = \begin{bmatrix} 1/a & 0 \\ 0 & 1/b \end{bmatrix}$$

and is clearly an element of $\Delta_2(\mathbb{R})$.   ∎

We hope that the student has gained some idea regarding how to show that a given set A is a group under a specified operation $*$. Keep in mind that one must first show that $*$ is a binary operation on A (note how this condition failed in Examples 8.4(c) and 8.5(c)). Then one must show that:

1. $*$ is associative.
2. There is an identity element in $(A, *)$.
3. Each $a \in A$ has an inverse element $a' \in A$ under $*$.

If, in the process of attempting to show that $(A, *)$ is a group, one of the above properties is found not to hold, then, of course, one can immediately conclude that $(A, *)$ is not a group.

In Example 8.5(a) we showed that $(\mathbb{Z}_6, \oplus)$ is a group. If m is any positive integer, then one can show, in almost the same fashion, that $(\mathbb{Z}_m, \oplus)$ is a group. The student is asked to provide the details in Exercise 3. We also saw that $(\mathbb{Z}_5^{\#}, \odot)$ is a group and that $\mathbb{Z}_8^{\#}$ under $\odot$ is not; in fact, $\odot$ is not even a binary operation on $\mathbb{Z}_8^{\#}$. In general, if $\mathbb{Z}_m^{\#} = \mathbb{Z}_m - \{[0]\}$, when is $\mathbb{Z}_m^{\#}$ a group under $\odot$? The next result provides the answer to this question.

THEOREM 8.4   Let $m \in \mathbb{N}$. Then $\mathbb{Z}_m^{\#}$ is a group under $\odot$ if and only if m is a prime.

Proof   We consider just the "if" portion of the theorem and leave the "only if" part to the student. So assume that m is a prime and set $m = p$.

We must first see that $\odot$ is a binary operation on $\mathbb{Z}_p^{\#}$. Let $[a], [b] \in \mathbb{Z}_p^{\#}$ and consider $[a] \odot [b] = [ab]$. The only way that $[ab]$ can fail to be an element of $\mathbb{Z}_p^{\#}$ is if $[ab] = [0]$. Suppose this were true. Then $ab \equiv 0 \pmod{p}$ and hence $p \mid ab$. But then, since p is a prime, we have either $p \mid a$ or $p \mid b$. Thus either $a \equiv 0 \pmod{p}$ or $b \equiv 0 \pmod{p}$; that is, either $[a] = [0]$ or $[b] = [0]$. But this contradicts the fact that $[a], [b] \in \mathbb{Z}_p^{\#}$. Hence $[ab] \in \mathbb{Z}_p^{\#}$.

Since $\odot$ is an associative binary operation on $\mathbb{Z}_p$, it follows that it is also associative on $\mathbb{Z}_p^{\#}$ (by Exercise 1 of Exercises 8.1). It is clear that $[1]$ is the identity of $(\mathbb{Z}_p^{\#}, \odot)$. To complete the proof, we must show that each element of $\mathbb{Z}_p^{\#}$ has an inverse in $(\mathbb{Z}_p^{\#}, \odot)$. Let $[a] \in \mathbb{Z}_p^{\#}$. Then we know that $\gcd(a, p) = 1$, so there exist integers s and t such that $as + pt = 1$. From this we have $as - 1 = pt$, or $as \equiv 1 \pmod{p}$. Thus $[as] = [1]$; that is, $[a] \odot [s] = [1]$. Clearly $[s] \neq [0]$, so that $[s] \in \mathbb{Z}_p^{\#}$, and therefore $[s]$ is the inverse for $[a]$ in $(\mathbb{Z}_p^{\#}, \odot)$. Hence $(\mathbb{Z}_p^{\#}, \odot)$ is a group. $\square$

As was pointed out in Example 8.5(c), $\odot$ is not a binary operation on $\mathbb{Z}_8^{\#}$ and hence $(\mathbb{Z}_8^{\#}, \odot)$ is not a group. Are there any subsets of $\mathbb{Z}_8^{\#}$ on which $\odot$ is a binary operation? Consider the subset $U_8 = \{[1], [3], [5], [7]\}$; the operation table for $(U_8, \odot)$ is provided in Figure 8.2. An inspection of the table shows that $(U_8, \odot)$ is, in fact, a group. What special properties do the elements of U_8 possess that make it a group under $\odot$? This question is answered in Theorem 8.5.

$\odot$	[1]	[3]	[5]	[7]
[1]	[1]	[3]	[5]	[7]
[3]	[3]	[1]	[7]	[5]
[5]	[5]	[7]	[1]	[3]
[7]	[7]	[5]	[3]	[1]

Figure 8.2 The operation table for $(U_8, \odot)$

THEOREM 8.5 Let $m \in \mathbb{N}$ and let $U_m = \{[a] \mid \gcd(a, m) = 1\}$. Then $\odot$ is a binary operation on U_m and $(U_m, \odot)$ is a group.

Proof We show that $\odot$ is a binary operation on U_m and leave the remaining details to Exercise 4. Let $[a]$ and $[b]$ be elements of U_m. Then, by the definition of U_m, $\gcd(a, m) = 1$ and $\gcd(b, m) = 1$. Hence there exist integers s, t, u, and v such that $as + mt = 1$ and $bu + mv = 1$. From these equalities it follows, upon multiplication, that $absu + m(asv + tbu + tmv) = 1$. Hence $\gcd(ab, m) = 1$ and we obtain $[ab] \in U_m$. Thus $\odot$ is a binary operation on U_m. $\qquad\square$

If $(G, *)$ is a group, then the results of the previous section imply directly that $(G, *)$ has a unique identity element and that each $a \in G$ has a unique inverse element in $(G, *)$. We shall denote the identity element of $(G, *)$ by e and denote the inverse of $a \in G$ by a^{-1}. Thus we have $a * e = e * a = a$ and $a * a^{-1} = a^{-1} * a = e$ for all $a \in G$. With regard to inverses in the group $(G, *)$, what can be said about the inverse of a^{-1}, or about $(a * b)^{-1}$? These questions are answered in the following theorem.

THEOREM 8.6 Let $(G, *)$ be a group and let a and b be any elements of G. Then the following hold:

1. $(a^{-1})^{-1} = a$
2. $(a * b)^{-1} = b^{-1} * a^{-1}$

Proof First let a be any element of G and consider $(a^{-1})^{-1}$. Note that this notation signifies "the inverse of a^{-1}." But it is also true that $a * a^{-1} = a^{-1} * a = e$, the identity of $(G, *)$. This string of equalities can also be interpreted as saying that a is the inverse of a^{-1}. And since the inverse of an element is unique, we conclude that $(a^{-1})^{-1} = a$.

Next consider $(a * b)^{-1}$, the inverse of the element $a * b$. As indicated in the preceding paragraph, to show that $(a * b)^{-1} = b^{-1} * a^{-1}$, it suffices to show that $(a * b) * (b^{-1} * a^{-1}) = (b^{-1} * a^{-1}) * (a * b) = e$. We proceed as follows:

$$(a * b) * (b^{-1} * a^{-1}) = \left[(a * b) * b^{-1}\right] * a^{-1}$$

$$= \left[a * (b * b^{-1})\right] * a^{-1}$$

$$= (a * e) * a^{-1}$$

$$= a * a^{-1}$$

$$= e$$

Similarly, it can be shown that $(b^{-1} * a^{-1}) * (a * b) = e$, and so it follows that $(a * b)^{-1} = (b^{-1} * a^{-1})$. $\qquad\square$

In the context of Theorem 8.6, it is very tempting to write $(a * b)^{-1} = a^{-1} * b^{-1}$ for all $a, b \in G$. However, this is true if and only if $*$ is commutative (see Exercise 6).

In general, if $(G, *)$ is a group in which $*$ is commutative, then $(G, *)$ is called an *abelian group*, after the famous Norwegian mathematician Niels Abel (1802–1829). He used ideas of group theory to prove that the general fifth-degree polynomial equation is not solvable in terms of radicals. A group that is not abelian is called *nonabelian*.

Example 8.6   Show that $(S_3, \circ)$ is a nonabelian group.

Solution   We display the six elements of S_3 as follows:

$$\iota: 1 \to 1 \qquad \beta: 1 \to 3 \qquad \delta: 1 \to 2$$
$$ 2 \to 2 \qquad 2 \to 2 \qquad 2 \to 3$$
$$ 3 \to 3 \qquad 3 \to 1 \qquad 3 \to 1$$

$$\alpha: 1 \to 1 \qquad \gamma: 1 \to 2 \qquad \rho: 1 \to 3$$
$$ 2 \to 3 \qquad 2 \to 1 \qquad 2 \to 1$$
$$ 3 \to 2 \qquad 3 \to 3 \qquad 3 \to 2$$

(We use this notation to describe the permutations of a small finite set. The notation indicates, for example, that $\beta(1) = 3$, $\beta(2) = 2$, and $\beta(3) = 1$.) To show that $(S_3, \circ)$ is nonabelian, it suffices to exhibit two of its elements that do not commute with each other. For example, consider α and β; we see that

$$(\alpha \circ \beta)(1) = \alpha(\beta(1)) = \alpha(3) = 2$$

whereas

$$(\beta \circ \alpha)(1) = \beta(\alpha(1)) = \beta(1) = 3$$

This shows that $\alpha \circ \beta \neq \beta \circ \alpha$. In fact, it can readily be verified that $\alpha \circ \beta = \delta$ and $\beta \circ \alpha = \rho$. In Exercise 5 the student is asked to complete the operation table for $(S_3, \circ)$. ∎

Exercises 8.2

1. Verify that each of the following algebraic structures is a semigroup. Discuss the existence of an identity element and inverses. Also note whether the operation is commutative. Which are groups?
   (a) $(\mathbb{Z}, +)$ **(b)** $(\mathbb{Q}^+, \cdot)$ **(c)** $(\mathcal{P}(A), \cup)$
   (d) $(\{2m \mid m \in \mathbb{Z}\}, +)$ **(e)** $(\{2m \mid m \in \mathbb{Z}\}, \cdot)$
   (f) $(\{2m + 1 \mid m \in \mathbb{Z}\}, \cdot)$

2. Show that $H = \{1, -1, i, -i\}$ is a group under complex multiplication.

3. Show that $(\mathbb{Z}_m, \oplus)$ is a group for all $m \in \mathbb{N}$.

4. Complete the details for the proof of Theorem 8.5 to show that $(U_m, \odot)$ is a group.

5. Complete the operation table for the group $(S_3, \circ)$.

6. Let $(G, *)$ be a group. Prove that $(G, *)$ is abelian if and only if $(a * b)^{-1} = a^{-1} * b^{-1}$ for all $a, b \in G$.

7. Use the Euclidean algorithm to find the inverse of $[37]$ in $(\mathbb{Z}_{73}^{\#}, \odot)$.

8. (Cancellation laws) Let $(G, *)$ be a group and let a, b, c be elements of G. Prove the following:
   (a) If $a * b = a * c$, then $b = c$.
   (b) If $b * a = c * a$, then $b = c$.

9. Prove: If x and y are elements of the group $(G, *)$ and $x * y = y$, then x is the identity.

10. Let $\mathcal{F}(\mathbb{R})$ denote the set of all functions from $\mathbb{R}$ to $\mathbb{R}$ and define the operation $+$ on $\mathcal{F}(\mathbb{R})$ as follows: For $f, g \in \mathcal{F}(\mathbb{R})$,

$$(f + g)(x) = f(x) + g(x)$$

for all $x \in \mathbb{R}$. Here the sum $f(x) + g(x)$ represents the ordinary sum of the real numbers $f(x)$ and $g(x)$. Show that $(\mathcal{F}(\mathbb{R}), +)$ is an abelian group.

11. Assume that $(\{x, y, z\}, *)$ is a group with identity x. Complete the operation table for the group.

12. Let S be the rectangle with vertices A, B, C, D and center O, as shown in Figure 8.3. Denote by G_S the following set of mappings from S to itself:

   i: the identity mapping

   r: the 180° rotation about O

   h: the reflection about the line d_1

   v: the reflection about the line d_2

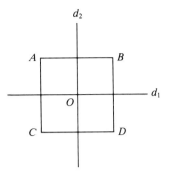

Figure 8.3

Show that $(G_S, \circ)$ is an abelian group by completing its operation table. In determining the product $\alpha \circ \beta$ of two given elements of G_S, consider first the effect of β on S and then apply α to the resulting configuration $\beta(S)$. For example, in determining $h \circ r$ notice that

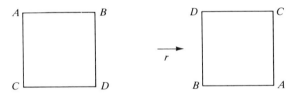

and then

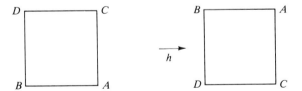

Thus the effect of $h \circ r$ is the same as that of v. Hence $h \circ r = v$. (This group is called the *Klein four-group*.) Since the elements of G_S leave S invariant, they are called *symmetries* of S.

13. A permutation $\sigma: \mathbb{R}^2 \to \mathbb{R}^2$ is called an *isometry* of $\mathbb{R}^2$ provided σ preserves distance. If we denote by $d(P, Q)$ the distance between the points P and Q, then σ is an isometry provided $d(\sigma(P), \sigma(Q)) = d(P, Q)$ for all $P, Q \in \mathbb{R}^2$. If we denote by E_2 the set of all isometries of $\mathbb{R}^2$, then prove that $(E_2, \circ)$ is a group.

14. Let S be a nonempty subset of the xy-plane $\mathbb{R}^2$. An isometry $\sigma: \mathbb{R}^2 \to \mathbb{R}^2$ is called a *symmetry* of S if $\sigma(S) = S$. Let G_S be the set of all symmetries of S; prove that $(G_S, \circ)$ is a group.

15. Let A be a nonempty set and let $a \in A$. In Exercise 5 of Exercises 8.1, the stabilizer of a, denoted G_a, was defined and parts (a)–(c) dealt with showing, in effect, that $(G_a, \circ)$ is a group. Prove that $(G_a, \circ)$ is abelian if and only if $|A| \leq 3$.

16. Recall that a 2×2 matrix $A \in M_2(\mathbb{R})$ is called *nonsingular* if it has an inverse under multiplication (that is, there is a 2×2 matrix B such that $AB = BA = I_2$). If $GL_2(\mathbb{R})$ denotes the set of all 2×2 nonsingular matrices, then show that $(GL_2(\mathbb{R}), \cdot)$ is a nonabelian group.

17. Let $(G, *)$ be a group. Prove that if $a * a = e$ for all $a \in G$, then $(G, *)$ is an abelian group.

18. If $\mathscr{R} = \{\cos \alpha + i \sin \alpha \,|\, \alpha \in \mathbb{R}\}$ (see Exercise 4 of Exercises 8.1), then show that $(\mathscr{R}, \cdot)$ is an abelian group.

19. (The *dihedral group* D_4) Let T be the square with vertices $A(1, -1)$, $B(1, 1)$, $C(-1, 1)$, and $D(-1, -1)$ and define the following mappings of the plane $\mathbb{R}^2$ to itself:

i: the identity mapping

ρ: 90° clockwise rotation about $O(0, 0)$

σ: the reflection in the x-axis

Let $D_4 = \{i, \rho, \rho^2, \rho^3, \sigma, \rho \circ \sigma, \rho^2 \circ \sigma, \rho^3 \circ \sigma\}$.

(a) Determine the geometric effect on T of each of the mappings ρ^2, ρ^3, $\rho \circ \sigma$, $\rho^2 \circ \sigma$, and $\rho^3 \circ \sigma$ (for example, show that ρ^2 is the 180° clockwise rotation about O).

(b) Prove that $(D_4, \circ)$ is a group by completing its operation table. Use the same technique that is suggested in Exercise 12. Give the inverse of each element.

(Note that $(D_4, \circ)$ is a group of symmetries of the square; in fact, it can be shown that D_4 consists of all such symmetries.)

20. Let $(G, *)$ be a group. Use induction and Theorem 8.6, part 2, to show that, if $a_1, a_2, \ldots, a_n$ are elements of G, then

$$\left(a_1 * a_2 * \cdots * a_n\right)^{-1} = a_n^{-1} * a_{n-1}^{-1} * \cdots * a_2^{-1} * a_1^{-1}$$

21. Consider Exercise 7 of Exercises 8.1. For each $*$ that is a binary operation on $\mathbb{Z}$, determine whether the algebraic structure $(\mathbb{Z}, *)$ is a group.

22. Verify that the algebraic structure $(F, \circ)$ of Exercise 13 in Exercises 8.1 is a group.

23. Which of the algebraic structures $(\mathbb{Z} \times \mathbb{N}, *)$ of Exercise 16 in Exercises 8.1 are (abelian) groups?

BASIC PROPERTIES OF GROUPS AND CYCLIC GROUPS

If $(G, *)$ is a group and $a \in G$, then we define integral powers of a inductively as follows:

$$a^0 = e$$
$$a^n = a * a^{n-1}$$
$$a^{-n} = (a^n)^{-1}$$

for all $n \in \mathbb{N}$. Thus a^m has meaning for any $m \in \mathbb{Z}$ and we refer to m as an *exponent*. The rules satisfied by exponents in a group $(G, *)$ are identical to the familiar rules for the real numbers under multiplication.

THEOREM 8.7   Let $(G, *)$ be a group and let $a \in G$. For any $m, n \in \mathbb{Z}$, the following hold:

1. $a^{-n} = (a^{-1})^n = (a^n)^{-1}$
2. $a^m * a^n = a^{m+n}$
3. $(a^m)^n = a^{mn}$

Proof   We prove 1 and 2 and leave 3 to Exercise 3.

To prove that $a^{-n} = (a^n)^{-1}$, it suffices, in view of the definition of a^{-n}, to prove it for $n \leq 0$. For $n = 0$, observe that $a^{-0} = a^0 = e = e^{-1} = (a^0)^{-1}$. If $n < 0$, let $n = -k$; then $k \in \mathbb{N}$, and we have that

$$a^{-n} = a^k$$
$$= \left((a^k)^{-1} \right)^{-1} \qquad \text{(by Theorem 8.6)}$$
$$= (a^{-k})^{-1} \qquad \text{(by definition)}$$
$$= (a^n)^{-1}$$

To prove that $(a^{-1})^n = (a^n)^{-1}$ for all integers n, we first proceed by induction to show the result for $n \geq 0$. Let S be the set of nonnegative integers for which the result holds. To see that $0 \in S$, simply note that

$$(a^0)^{-1} = e^{-1} = e = (a^{-1})^0$$

Assume $k \in S$ for an arbitrary $k \geq 0$; thus the induction hypothesis is that $(a^{-1})^k = (a^k)^{-1}$. To obtain $k + 1 \in S$, consider the following string of equalities:

$$(a^{k+1})^{-1} = (a * a^k)^{-1}$$
$$= (a^k)^{-1} * a^{-1} \qquad \text{(by Theorem 8.6)}$$
$$= (a^{-1})^k * a^{-1} \qquad \text{(by the induction hypothesis)}$$
$$= (a^{-1})^{k+1} \qquad \text{(see Exercise 18)}$$

Hence $k + 1 \in S$, and it follows that S is the set of all nonnegative integers. The result $(a^{-1})^n = (a^n)^{-1}$ is next established for $n < 0$; as before, let $n = -k$, with $k \in \mathbb{N}$. Then

$$(a^n)^{-1} = (a^{-k})^{-1}$$

$$= ((a^{-1})^k)^{-1} \qquad \text{(by the above proof)}$$

$$= (a^{-1})^{-k} \qquad \text{(by definition)}$$

$$= (a^{-1})^n$$

Thus we have shown that $(a^n)^{-1} = (a^{-1})^n$ for all $n \in \mathbb{Z}$.

To show that $a^m * a^n = a^{m+n}$ for all $m, n \in \mathbb{Z}$, we first consider the case where $n \geq 0$. The result is almost immediate for $n = 0$, so we proceed to the case for $n \geq 1$. We use mathematical induction on n and let S be the set of positive integers n for which the result is true. We must first show that $1 \in S$. To do so, we separate the cases $m \geq 0$ and $m < 0$. For $m \geq 0$, the relation $a^m * a^1 = a^{m+1}$ follows readily by induction on m and is left to Exercise 18. If $m < 0$, then set $m = -r$ and consider $a^{m+1} * a^{-1} = a^{-r+1} * a^{-1}$. We have

$$a^{-r+1} * a^{-1} = a^{-(r-1)} * a^{-1}$$

$$= (a * a^{r-1})^{-1}$$

$$= (a^r)^{-1} \qquad \text{(by definition)}$$

$$= a^{-r}$$

Hence $a^{-r+1} = a^{-r} * a$; that is, $a^{m+1} = a^m * a$. Thus $1 \in S$. Next assume that $k \in S$ for an arbitrary $k \geq 1$. The induction hypothesis then states that $a^m * a^k = a^{m+k}$ for all $m \in \mathbb{Z}$. We must now show that $k + 1 \in S$; that is, $a^m * a^{k+1} = a^{m+k+1}$. Proceed as follows:

$$a^m * a^{k+1} = a^m * (a^k * a^1) \qquad \text{(since } 1 \in S)$$

$$= (a^m * a^k) * a$$

$$= a^{m+k} * a \qquad \text{(by the induction hypothesis)}$$

$$= a^{m+k+1} \qquad \text{(since } 1 \in S)$$

Thus $k + 1 \in S$ and it follows that $S = \mathbb{N}$. $\qquad \square$

In many of the examples of abelian groups given, the operation is addition (for example, $(\mathbb{Z}, +), (\mathbb{R}, +), (\mathbb{Z}_n, \oplus)$). In many cases, it is convenient to adopt the "additive notation" in an abelian group $(G, *)$. When this happens, the following notational replacements are made (for $r \in \mathbb{N}$):

Multiplicative	Additive
$*$	$+$
e	0
a^{-1}	$-a$
a^r	$r \cdot a$

Here we understand that $a^r = a * a * \cdots * a$, with r factors, and that $r \cdot a = a + a + \cdots + a$, with r summands. With the above replacements, the following correspondence between relations is established:

Multiplicative	Additive
$a^0 = e$	$0 \cdot a = 0$
$a^{-n} = (a^{-1})^n = (a^n)^{-1}$	$(-n) \cdot a = n \cdot (-a) = -(n \cdot a)$
$a^m * a^n = a^{m+n}$	$(m + n) \cdot a = m \cdot a + n \cdot a$
$(a^m)^n = a^{mn}$	$n \cdot (m \cdot a) = (mn) \cdot a$

These hold for all integers m, n. In keeping with conventional language, one normally refers to $n \cdot a$ as a *multiple* of a.

Example 8.7   Determine each of the following:

(a) $5 \cdot [2]$ in $(\mathbb{Z}_7, \oplus)$ (b) $[2]^3$ in $(\mathbb{Z}_5^{\#}, \odot)$
(c) α^{-2} in $(S_3, \circ)$ (d) $[3]^{-2}$ in $(U_{10}, \odot)$

Solution   In keeping with the preceding conventions, we have the following:

(a) $5 \cdot [2] = [2] \oplus [2] \oplus [2] \oplus [2] \oplus [2] = [5 \cdot 2] = [10] = [3]$ in $\mathbb{Z}_7$
(b) $[2]^3 = [2] \odot [2] \odot [2] = [2 \cdot 2 \cdot 2] = [8] = [3]$ in $\mathbb{Z}_5^{\#}$
(c) Using the fact that $\alpha(1) = 1$, $\alpha(2) = 3$, and $\alpha(3) = 2$, we get $\alpha^{-2} = (\alpha^{-1})^2 = \alpha^2 = \iota$ in $(S_3, \circ)$
(d) $[3]^{-2} = ([3]^2)^{-1} = [9]^{-1} = [9]$ in $(U_{10}, \odot)$ ∎

If $(G, *)$ is a group and $a \in G$, then it is possible that $a^m = e$ for some positive integer m. If this happens, then the principle of well-ordering tells us that there is a smallest positive integer n such that $a^n = e$. For example, $[2]^6 = [1]$ in $(\mathbb{Z}_7^{\#}, \odot)$ and the smallest positive integer n for which $[2]^n = [1]$ is $n = 3$.

DEFINITION 8.6

Let $(G, *)$ be a group with identity element e and let $a \in G$. If there exists a positive integer m such that $a^m = e$, then the smallest positive integer n for which $a^n = e$ is called the *order of the element* a in $(G, *)$ and is denoted by $|a|$. If there is no positive integer m such that $a^m = e$, then we say that a has *infinite order* and write $|a| = \infty$. The number of elements in G is called the *order of the group* $(G, *)$.

If $(G, *)$ is a group, then we denote the order of $(G, *)$ by the conventional notation $|G|$. If $(G, *)$ has finite order, then we call $(G, *)$ a *finite group*; otherwise $(G, *)$ is called an *infinite group* and we write $|G| = \infty$. We remark that if $(G, *)$ is an abelian group, written additively, then the relation $a^n = e$ is replaced by $na = 0$.

Example 8.8   Find the order of each of the following group elements:

(a) $[2]$ in $(\mathbb{Z}_6, \oplus)$
(b) $[3]$ in $(\mathbb{Z}_5^\#, \odot)$
(c) 2 in $(\mathbb{Z}, +)$
(d) δ in $(S_3, \circ)$
(e) $g = \cos\dfrac{2\pi}{6} + i \sin\dfrac{2\pi}{6}$ in $(\mathbb{C}^\#, \cdot)$
(f) $A = \begin{bmatrix} -1 & 0 \\ 0 & 1 \end{bmatrix}$ in $\Delta_2(\mathbb{R})$

Solution

(a) Since $m \cdot [2] = [2m]$ for all $m \in \mathbb{N}$, the smallest such m for which $m \cdot [2] = [0]$ in $(\mathbb{Z}_6, \oplus)$ is $m = 3$. Thus $|[2]| = 3$.

(b) The smallest $m \in \mathbb{N}$ for which $[3]^m = [1]$ in $(\mathbb{Z}_5^\#, \odot)$ is $m = 4$, so $|[3]| = 4$.

(c) Since $2m \neq 0$ for all $m \in \mathbb{N}$, it follows that $|2| = \infty$.

(d) Recalling that $\delta(1) = 2$, $\delta(2) = 3$, $\delta(3) = 1$, we obtain $\delta^2(1) = \delta(\delta(1)) = \delta(2) = 3$, so $\delta^2 \neq \iota$. In fact, it is easily checked that $\delta^2 = \rho$. Moreover, $\delta^3(1) = \delta(\rho(1)) = \delta(3) = 1$, $\delta^3(2) = \delta(\rho(2)) = \delta(1) = 2$, and $\delta^3(3) = \delta(\rho(3)) = \delta(2) = 3$, so $\delta^3 = \iota$. Hence $|\delta| = 3$.

(e) By De Moivre's theorem, we have that

$$g^m = \left(\cos\frac{2\pi}{6} + i \sin\frac{2\pi}{6} \right)^m = \cos\frac{2\pi m}{6} + i \sin\frac{2\pi m}{6}$$

So $g^6 = \cos 2\pi + i \sin 2\pi = 1$ and $g^r \neq 1$ for $0 < r \leq 5$. Thus $|g| = 6$.

(f) Note that

$$A^2 = \begin{bmatrix} 1 & 0 \\ 0 & 1 \end{bmatrix}$$

is the identity element of $\Delta_2(\mathbb{R})$, so $|A| = 2$. ∎

THEOREM 8.8   Let $(G, *)$ be a group and let $x \in G$. Then the following hold:

1. $|x| = 1$ if and only if $x = e$.
2. $|x| = |x^{-1}|$
3. If $|x| = n$ and $x^m = e$, then $n \mid m$.

Proof

We prove part 3 and leave 1 and 2 to Exercise 5. For part 3 we first apply the division algorithm to m and n, thus obtaining integers q and r such that $m = nq + r$, where $0 \leq r < n$. If we show that $r = 0$, the proof is complete. To see this, observe that

$$e = x^m = x^{nq+r} = x^{nq} * x^r = \left(x^n \right)^q * x^r = e^q * x^r = e * x^r = x^r$$

So $x^r = e$. However, since $|x| = n$ and $0 \leq r < n$, it must be the case that $r = 0$. □

Example 8.9   (a) Show that each element of $(\mathbb{Z}_5^{\#}, \odot)$ is a power of [2].
(b) Show that each element of $(\mathbb{Z}_6, \oplus)$ is a multiple of [1].
(c) Recalling that $H = \{1, -1, i, -i\}$ is a group under complex multiplication, show that each element of $(H, \cdot)$ is a power of i.

Solution   (a) $\mathbb{Z}_5^{\#} = \{[1], [2], [3], [4]\}$ and $[2]^1 = [2]$, $[2]^2 = [4]$, $[2]^3 = [3]$, and $[2]^4 = [1]$. Hence each element of $\mathbb{Z}_5^{\#}$ is a power of [2].
(b) $\mathbb{Z}_6 = \{[0], [1], [2], [3], [4], [5]\}$ and we see that $k \cdot [1] = [k \cdot 1] = [k]$ for $k = 0, 1, \ldots, 5$. Hence each element of $\mathbb{Z}_6$ is a multiple of [1].
(c) Note that $i^1 = i$, $i^2 = -1$, $i^3 = -i$, and $i^4 = 1$, so each element of H is a power of i.   ∎

DEFINITION 8.7

A group $(G, *)$ is called *cyclic* provided there is an element $c \in G$ such that

$$G = \{c^m \mid m \in \mathbb{Z}\}$$

We call the element c a *generator* for $(G, *)$ and write $(G, *) = \langle c \rangle$.

From Example 8.9 we conclude that the groups $(\mathbb{Z}_5^{\#}, \odot)$, $(\mathbb{Z}_6, \oplus)$, and $(H, \cdot)$ are cyclic, with $(\mathbb{Z}_5^{\#}, \odot) = \langle [2] \rangle$, $(\mathbb{Z}_6, \oplus) = \langle [1] \rangle$, and $(H, \cdot) = \langle i \rangle$. Definition 8.7 makes no claim that generators for cyclic groups are unique; in fact, it can easily be shown that [3] is a generator for $(\mathbb{Z}_5^{\#}, \odot)$ and $-i$ is a generator for $(H, \cdot)$.

THEOREM 8.9   Let $(G, *)$ be a cyclic group with generator c.

1. If $|c| = \infty$, then $c^s = c^t$ implies $s = t$ and

$$G = \{c^r \mid r \in \mathbb{Z}\}$$
$$= \{\ldots, c^{-1}, c^{-2}, e, c, c^2, \ldots\}$$

2. If $|c| = n$, then $G = \{e, c, c^2, \ldots, c^{n-1}\}$.

Proof   We prove part 2 only, leaving 1 to Exercise 7.
Assume $|c| = n$ and let $m \in \mathbb{Z}$. By the division algorithm, there exist integers q and r such that $m = qn + r$, where $0 \le r < m$. So

$$c^m = c^{nq+r} = c^{nq} * c^r = (c^n)^q * c^r = e^q * c^r = c^r$$

Thus c^m is one of the elements $e, c, c^2, \ldots, c^{n-1}$. To see that $e, c, c^2, \ldots, c^{n-1}$ are distinct, suppose that $c^s = c^t$ for some integers s and t, where $0 \le t \le s \le n - 1$. Then $c^s * c^{-t} = c^t * c^{-t} = e$, which says that $c^{s-t} = e$. Since the order of c is n, it follows from Theorem 8.8, part 3, that $n \mid (s - t)$. But since $0 \le s - t < n$, we conclude that $s - t = 0$, or that $s = t$. This shows that $e, c, c^2, \ldots, c^{n-1}$ are distinct.   □

COROLLARY 8.10   If $(G, *)$ is a cyclic group with $(G, *) = \langle c \rangle$, then $|G| = |c|$.

Several other interesting properties of cyclic groups are left to the exercises.

Exercises 8.3

1. Determine the order of each of the elements of $(S_3, \circ)$.
2. Determine the order of each of the following group elements:
   (a) [2] in $(\mathbb{Z}_{11}, \oplus)$ **(b)** [2] in $(\mathbb{Z}_{11}^{\#}, \odot)$
   (c) $\cos(2\pi/5) + i\sin(2\pi/5)$ in the group $\mathscr{R}$ of Exercise 18 in Exercises 8.2
   (d) the matrices

$$A = \begin{bmatrix} -1 & 0 \\ 1 & 1 \end{bmatrix} \quad \text{and} \quad B = \begin{bmatrix} 1 & 1 \\ 0 & 1 \end{bmatrix}$$

   in the group $(GL_2(\mathbb{R}), \cdot)$
   (e) [3] in $(U_{16}, \odot)$
   (f) the permutation $\alpha \in S_6$, given by

$$\alpha(1) = 2, \ \alpha(2) = 4, \ \alpha(3) = 1, \ \alpha(4) = 3, \ \alpha(5) = 6, \ \alpha(6) = 5$$

   (g) c^2, c^3, c^5, c^{10} and c^{12} in $(G, *) = \langle c \rangle$, given that $|c| = 30$
3. Prove part 3 of Theorem 8.7.
4. Show that $(S_3, \circ)$ is not cyclic.
5. Prove parts 1 and 2 of Theorem 8.8.
6. Let $(G, *) = \langle c \rangle$, where $|c| = 24$. Find all elements of order:
   (a) 2   **(b)** 3   **(c)** 6   **(d)** 12
7. Prove part 1 of Theorem 8.9.
8. Find all generators of the group $(G, *)$ in Exercise 6.
9. Show that $(U_8, \odot)$ is not a cyclic group, but that $(U_9, \odot)$ is.
10. Let a be a nonidentity element of a group $(G, *)$. Show that $a = a^{-1}$ if and only if $|a| = 2$.
11. Show that every group of even order has an element of order 2. (Hint: Pair each nonidentity element with its inverse and apply the result of Exercise 10.)
12. Let $(G, *)$ be an abelian group and let $a, b \in G$.
    (a) Prove that $(a * b)^m = a^m * b^m$ for all $m \in \mathbb{Z}$.
    (b) If $|a| = m$ and $|b| = n$, then prove each of the following:
       (i) $|a * b|$ divides mn.
       (ii) If $\gcd(m, n) = 1$, then $|a * b| = mn$. (Hint: If $d = |a * b|$, then prove that $m \mid d$ and $n \mid d$.)
13. Let $(G, *)$ be a group and let a and b be nonidentity elements of G.
    (a) Find $|a|$ if $b^2 = e$ and $b * a * b^{-1} = a^2$.
    (b) Find $|a|$ if $|b| = 3$ and $a * b = b * a^2$.
14. For each $n \in \mathbb{N}$, give an example of a cyclic group of order n.

15. Determine the order of each element of the dihedral group $(D_4, \circ)$. (See Exercise 19 of Exercises 8.2.)

16. Prove that a group $(G, *)$ is abelian if and only if $(a * b)^2 = a^2 * b^2$ for all $a, b \in G$.

17. Let $(G, *) = \langle c \rangle$ be a cyclic group of order n. For $0 < m < n$, find the order of the element c^m.

18. Let $(G, *)$ be a group and let $a \in G$. Prove that $a^m * a = a^{m+1}$ for all $m \geq 0$.

8.4

SUBGROUPS

We have seen that $(\mathbb{Z}, +)$ and $(\mathbb{R}, +)$ are groups with the same binary operation. Thus $\mathbb{Z}$ is a subset of $\mathbb{R}$ that forms a group under the same binary operation defined in $(\mathbb{R}, +)$. The same relationship holds between the groups $(\mathbb{Q}^{\#}, \cdot)$ and $(\mathbb{R}^{\#}, \cdot)$.

DEFINITION 8.8

Let $(G, *)$ be a group and let H be a subset of G. If $*$ is a binary operation on H such that $(H, *)$ is itself a group, then $(H, *)$ is called a *subgroup* of $(G, *)$. We write $(H, *) \leq (G, *)$ to denote that $(H, *)$ is a subgroup of $(G, *)$.

It should be remarked that if $(G, *)$ is a group and $(H, *) \leq (G, *)$, then H must be a nonempty subset of G. This is because $(H, *)$, as a group, must possess an identity element. Is the identity element for $(H, *)$ the same as that for $(G, *)$? If $x \in H$, is the inverse of x in $(H, *)$ the same as the inverse of x in $(G, *)$? These questions are answered in the following theorem.

THEOREM 8.11   Let $(G, *)$ be a group with identity element e and let $(H, *) \leq (G, *)$. Then e is the identity element for $(H, *)$. Moreover, if $x \in H$, then the inverse of x in $(H, *)$ is x^{-1}.

Proof   If f denotes the identity of $(H, *)$, then $f * f = f$. Also, since e is the identity of $(G, *)$, we have $f = f * e$. Hence $f * f = f * e$ and, by the cancellation laws (see Exercise 8 of Exercises 8.2), we have $f = e$.

Next let $x \in H$ and suppose x' is the inverse of x in $(H, *)$. Then $x * x' = e$, so that x' is an inverse of x in $(G, *)$. Since inverses in $(G, *)$ are unique, we conclude that $x' = x^{-1}$.   □

Given a group $(G, *)$ and $H \subseteq G$, what basic steps must be taken to show that $(H, *) \leq (G, *)$? First one must show that $*$ is a binary operation on H; that is, one must show that $x, y \in H$ implies $x * y \in H$. Since $*$ is an associative binary operation on G, it must be associative on H; in other words, if $x * (y * z) = (x * y) * z$ for all $x, y, z \in G$, then (since $H \subseteq G$) we have $a * (b * c) = (a * b) * c$ for all $a, b, c \in H$. So one need not verify that $*$ is associative on H (this hereditary idea was actually addressed in Exercise 1(b) of Exercises 8.1). In order to show that $(H, *)$ is a group, it remains to verify that $e \in H$ and that $x^{-1} \in H$ for all $x \in H$. The next theorem provides a rather nice criterion to determine whether $(H, *) \le (G, *)$.

THEOREM 8.12   Let $(G, *)$ be a group and let $H \subseteq G$. Then $(H, *) \le (G, *)$ if and only if the following conditions hold:

1. $H \neq \emptyset$
2. If $x, y \in H$, then $x * y \in H$.
3. If $x \in H$, then $x^{-1} \in H$.

Proof   If $(H, *) \le (G, *)$, then properties 1, 2, and 3 follow readily. Conversely, assume that properties 1, 2, and 3 hold. Then, according to the discussion preceding the statement of the theorem, we need only show that $(H, *)$ contains the identity element e. Since $H \neq \emptyset$, there is some element $x \in H$. By 3, $x \in H$ implies $x^{-1} \in H$. And by 2, $x * x^{-1} \in H$. Hence $e \in H$ and therefore $(H, *) \le (G, *)$.   □

Example 8.10   Each part gives a group $(G, *)$ and a subset H of G. Determine whether $(H, *)$ is a subgroup of $(G, *)$.

(a) $(G, *) = (\mathbb{R}^{\#}, \cdot)$ and $H = \mathbb{Q}^{\#} = \mathbb{Q} - \{0\}$
(b) $(G, *) = (\mathbb{Z}, +)$ and $H = \mathbb{N}$
(c) $(G, *) = (\mathbb{C}^{\#}, \cdot)$ where $\mathbb{C}^{\#} = \mathbb{C} - \{0\}$ and $H = \{\cos \theta + i \sin \theta \mid \theta \in \mathbb{R}\}$
(d) $(G, *) = (\mathbb{Z}_{10}, \oplus)$ and $H = \{[0], [2], [4], [6], [8]\}$
(e) $(G, *) = (M_2(\mathbb{R}), +)$ and $H = \left\{ \begin{bmatrix} a & b \\ 0 & c \end{bmatrix} \mid a, b, c \in \mathbb{R} \right\}$

Solution   In each case we follow the criterion given in Theorem 8.12.
(a) Clearly $\mathbb{Q}^{\#} \neq \emptyset$. If $m/n, r/s \in \mathbb{Q}^{\#}$, then

$$\left(\frac{m}{n} \right) \cdot \left(\frac{r}{s} \right) = \frac{mr}{ns} \in \mathbb{Q}^{\#}$$

Finally, given $m/n \in \mathbb{Q}^{\#}$, we have

$$\left(\frac{m}{n} \right)^{-1} = \frac{n}{m} \in \mathbb{Q}^{\#}$$

Thus $(\mathbb{Q}^{\#}, \cdot) \le (\mathbb{R}^{\#}, \cdot)$.
(b) Again it is clear that $\mathbb{N} \neq \emptyset$. Also, if $m, n \in \mathbb{N}$, then $m + n \in \mathbb{N}$. However, for any $n \in \mathbb{N}$, observe that $-n \notin \mathbb{N}$. Therefore $(\mathbb{N}, +)$ is not a subgroup of $(\mathbb{Z}, +)$.

(c) Note that $1 = \cos 0 + i \sin 0 \in H$, so $H \neq \emptyset$. Let $u, v \in H$, say $u = \cos \theta + i \sin \theta$ and $v = \cos \alpha + i \sin \alpha$. Then, according to DeMoivre's theorem, we have

$$uv = \cos(\theta + \alpha) + i \sin(\theta + \alpha)$$

so $uv \in H$. Also $u^{-1} = \cos(-\theta) + i \sin(-\theta) \in H$. Hence $(H, \cdot) \leq (\mathbb{C}^{\#}, \cdot)$.

(d) Here $H = \{[2m] \mid m \in \mathbb{Z}\}$ and is clearly nonempty. For $m_1, m_2 \in \mathbb{Z}$, $[2m_1] \oplus [2m_2] = [2(m_1 + m_2)] \in H$ and $-[2m_1] = [-2m_1] = [2(-m_1)] \in H$. In particular, note that $-[2] = [8]$ and $-[4] = [6]$. Thus $(H, \oplus) \leq (\mathbb{Z}_{10}, \oplus)$.

(e) We see that $H \neq \emptyset$ since the zero matrix is in H. If $A, B \in H$, say

$$A = \begin{bmatrix} a & b \\ 0 & c \end{bmatrix} \quad \text{and} \quad B = \begin{bmatrix} r & s \\ 0 & t \end{bmatrix}$$

then

$$A + B = \begin{bmatrix} a+r & b+s \\ 0 & c+t \end{bmatrix}$$

which is again an element of H. Finally, for $A \in H$, described as above, the inverse of A is the matrix

$$B = \begin{bmatrix} -a & -b \\ 0 & -c \end{bmatrix}$$

and $B \in H$. Thus we have $(H, +) \leq (M_2(\mathbb{R}), +)$. ∎

Let $(G, *)$ be a group and let $a \in G$. Consider the subset

$$\langle a \rangle = \{a^m \mid m \in \mathbb{Z}\}$$

of G. Is $(\langle a \rangle, *) \leq (G, *)$? The subset $\langle a \rangle$ is clearly nonempty; for example, $a \in \langle a \rangle$. For arbitrary elements a^m and a^n in $\langle a \rangle$, we see from Theorem 8.7, part 2, that $a^m * a^n = a^{m+n} \in \langle a \rangle$. Also, for any $m \in \mathbb{Z}$, observe that $(a^m)^{-1} = a^{-m} \in \langle a \rangle$ since $-m \in \mathbb{Z}$. Hence by Theorem 8.12, $(\langle a \rangle, *) \leq (G, *)$. We call $(\langle a \rangle, *)$ the *cyclic subgroup* of $(G, *)$ *generated by a* and refer to a as a *generator* of the subgroup. In case $(G, *)$ is an abelian group written additively, the cyclic subgroup generated by the element a is given by

$$\langle a \rangle = \{ma \mid m \in \mathbb{Z}\}$$

Example 8.11 For each of the following, determine the indicated cyclic subgroup.

(a) $\langle [3] \rangle$ in $(\mathbb{Z}_7, \oplus)$ (b) $\langle [2] \rangle$ in $(\mathbb{Z}_7^{\#}, \odot)$

(c) $\langle \frac{1}{2} \rangle$ in $(\mathbb{Q}^{\#}, \cdot)$

(d) $\left\langle \begin{bmatrix} 1 & 1 \\ 0 & 1 \end{bmatrix} \right\rangle$ in $(M_2(\mathbb{R}), +)$

(e) $\langle \alpha \rangle$ in $(S_4, \circ)$, where α is the permutation given by

$$\alpha(1) = 2, \ \alpha(2) = 3, \ \alpha(3) = 4, \ \alpha(4) = 1$$

Solution   (a) In this case, the additive notation is used and we have $\langle [3] \rangle = \{m[3] \mid m \in \mathbb{Z}\} = \{[3m] \mid m \in \mathbb{Z}\}$. Thus $\{[0], [3], [6], [9], [12], [15], [18]\} \subseteq \mathbb{Z}_7$. But this set is precisely $\mathbb{Z}_7$. Hence $(\langle [3] \rangle, \oplus) = (\mathbb{Z}_7, \oplus)$.

(b) Using the multiplicative notation, we have $\langle [2] \rangle = \{[2]^m \mid m \in \mathbb{Z}\}$. Note that $[2]^0 = [1]$, $[2]^1 = [2]$, $[2]^2 = [4]$, and $[2]^3 = [1]$. Thus $\langle [2] \rangle = \{[1], [2], [4]\}$.

(c) By definition, $\langle \frac{1}{2} \rangle = \{(\frac{1}{2})^m \mid m \in \mathbb{Z}\} = \{2^m \mid m \in \mathbb{Z}\}$ in $(\mathbb{Q}^{\#}, \cdot)$.

(d) If we denote the given matrix by A, then we know that $\langle A \rangle = \{mA \mid m \in \mathbb{Z}\}$. And since

$$mA = \begin{bmatrix} m & m \\ 0 & m \end{bmatrix}$$

for all $m \in \mathbb{Z}$, we have

$$\langle A \rangle = \left\{ \begin{bmatrix} m & m \\ 0 & m \end{bmatrix} \mid m \in \mathbb{Z} \right\}$$

(e) We first determine α^2, α^3, and α^4. Note that

$$\alpha^2(1) = 3, \alpha^2(2) = 4, \alpha^2(3) = 1, \alpha^2(4) = 2$$

$$\alpha^3(1) = 4, \alpha^3(2) = 1, \alpha^3(3) = 2, \alpha^3(4) = 3$$

$$\alpha^4(1) = 1, \alpha^4(2) = 2, \alpha^4(3) = 3, \alpha^4(4) = 4$$

Hence $\alpha^4 = \iota$, the identity permutation. In fact, we claim that $\langle \alpha \rangle = \{\iota, \alpha, \alpha^2, \alpha^3\}$. Consider α^m, where $m \in \mathbb{Z}$. By the division algorithm applied to m and 4, we have $m = 4q + r$, where $0 \le r < 4$. Thus

$$\alpha^m = \alpha^{4q+r} = \alpha^{4q} \circ \alpha^r = (\alpha^4)^q \circ \alpha^r = \iota^q \circ \alpha^r = \alpha^r$$

So $\alpha^m = \alpha^r$, and we have $\alpha^m \in \{\iota, \alpha, \alpha^2, \alpha^3\}$. Therefore $\langle \alpha \rangle = \{\iota, \alpha, \alpha^2, \alpha^3\}$. Note that $|\langle \alpha \rangle| = |\alpha|$, the order of α. ∎

Given an arbitrary group $(G, *)$, one is interested in determining as much as possible about its structure. One can quite often derive important information about $(G, *)$ from its subgroups. Certain special subgroups turn out to be particularly useful in this regard. One of these is presented in the following definition.

DEFINITION 8.9

Let $(G, *)$ be a group. Define the *center* of $(G, *)$ to be the set

$$Z(G) = \{c \in G \mid c * g = g * c \text{ for all } g \in G\}$$

In other words, $Z(G)$ is the set of all elements in G that commute with every element of G.

It can be shown without too much difficulty that

$$Z(G) = \{ c \in G \mid c^{-1} * g * c = g \text{ for all } g \in G \}$$

This result is left to Exercise 10.

THEOREM 8.13 If $(G, *)$ is a group, then $(Z(G), *)$ is a subgroup of $(G, *)$.

Proof   We apply the criterion established in Theorem 8.12. To begin with, $Z(G) \neq \emptyset$ since $e \in Z(G)$. Next suppose $a, b \in Z(G)$. To show that $a * b \in Z(G)$, we must show that $(a * b) * g = g * (a * b)$ for all $g \in G$. Since $a, b \in Z(G)$, we have that $a * g = g * a$ and $b * g = g * b$; thus,

$$\begin{aligned}
(a * b) * g &= a * (b * g) \\
&= a * (g * b) \\
&= (a * g) * b \\
&= (g * a) * b \\
&= g * (a * b)
\end{aligned}$$

So $a * b \in Z(G)$. To complete the proof, it must be shown that if $a \in Z(G)$, then $a^{-1} \in Z(G)$. For any $g \in G$, we have $a^{-1} * g * a = g$; thus, $(a^{-1} * g * a) * a^{-1} = g * a^{-1}$. So $(a^{-1} * g) * (a * a^{-1}) = g * a^{-1}$. But this yields $(a^{-1} * g) * e = g * a^{-1}$ or $a^{-1} * g = g * a^{-1}$. Thus $a^{-1} \in Z(G)$ and this completes the proof. □

Other special types of subgroups are addressed in the exercises and Chapter Problems.

It should be emphasized that we have barely touched the surface of the subject of group theory. Suffice it to say it is a highly developed subject, rich in applications, and our treatment of it is quite elementary. Further study of the subject is better left to a first course in modern algebra.

Exercises 8.4

1. For each of the following subsets H, show that $(H, *)$ is a subgroup of the given group $(G, *)$.
   (a) $H = \{-1, 1\}$ in $(\mathbb{Q}^{\#}, \cdot)$
   (b) $H = \{ a + b\sqrt{5} \mid a, b \in \mathbb{Z} \}$ in $(\mathbb{R}, +)$
   (c) $H = \{ ai \mid a \in \mathbb{R} \}$ in $(\mathbb{C}, +)$

   (d) $H = \left\{ \begin{bmatrix} a & 0 \\ b & c \end{bmatrix} \mid a, b, c \in \mathbb{R} \right\}$ in $(M_2(\mathbb{R}), +)$

   (e) $H = \left\{ \begin{bmatrix} 1 & a \\ 0 & 1 \end{bmatrix} \mid a \in \mathbb{R} \right\}$ in $(GL_2(\mathbb{R}), \cdot)$

   (f) $H = \{ a + b\sqrt{3} \mid a, b \in \mathbb{Q} \text{ and } a + b\sqrt{3} \neq 0 \}$ in $(\mathbb{R}^{\#}, \cdot)$
   (g) $H = \{ \alpha \in S_n \mid \alpha(1) = 1 \}$ in $(S_n, \circ)$
   (h) $H = \{ f \in \mathscr{F}(\mathbb{R}) \mid f(m) = 0 \text{ for all } m \in \mathbb{Z} \}$ in $(\mathscr{F}(\mathbb{R}), +)$

2. In each part determine whether $(H, *)$ is a subgroup of the given group $(G, *)$.
   (a) $(\mathbb{R}^{\#}, \cdot)$, $H = \mathbb{Q}^{+}$ (b) $(\mathbb{Z}, +)$, $H = \{\ldots, -4, -2, 0, 2, 4, \ldots\}$
   (c) $(\mathbb{Q}^{+}, \cdot)$, $H = \mathbb{N}$
   (d) $(\mathbb{Q}^{+}, \cdot)$, $H = \{3^{m} \mid m \in \mathbb{Z}\}$ (Recall that $\mathbb{Q}^{+} = \{r \in \mathbb{Q} \mid r > 0\}$)
   (e) $(\mathbb{Z}, +)$, $H = \{-1, 0, 1\}$

3. Let $(G, *)$ be an abelian group. Show that each of the following subsets of G is a subgroup of $(G, *)$.
   (a) $H_{1} = \{a^{2} \mid a \in G\}$ (b) $H_{2} = \{a \in G \mid a^{2} = e\}$

4. Let $(G, *)$ be a group and let $(H, *)$ and $(K, *)$ be subgroups of $(G, *)$.
   (a) Prove that $(H \cap K, *)$ is also a subgroup of $(G, *)$.
   (b) Provide an example to show that, in general, $(H \cup K, *)$ is not a subgroup of $(G, *)$.

5. Let $(G, *)$ be a group and let $a \in G$. Prove that each of the following subsets is a subgroup of $(G, *)$.
   (a) $C(a) = \{c \in G \mid a * c = c * a\}$ $(C(a)$ is called the *centralizer of a.*)
   (b) $H = \{x^{-1} * a^{k} * x \mid k \in \mathbb{Z}\}$ where x is a fixed element of G.

6. Determine all cyclic subgroups of the dihedral group $(D_{4}, \circ)$.

7. Let D be the subset of S_{4} consisting of the following permutations:

ι:	$1 \to 1$	r_{1}:	$1 \to 2$	r_{2}:	$1 \to 3$	r_{3}:	$1 \to 4$
	$2 \to 2$		$2 \to 3$		$2 \to 4$		$2 \to 1$
	$3 \to 3$		$3 \to 4$		$3 \to 1$		$3 \to 2$
	$4 \to 4$		$4 \to 1$		$4 \to 2$		$4 \to 3$

h:	$1 \to 4$	v:	$1 \to 2$	d_{1}:	$1 \to 1$	d_{2}:	$1 \to 3$
	$2 \to 3$		$2 \to 1$		$2 \to 4$		$2 \to 2$
	$3 \to 2$		$3 \to 4$		$3 \to 3$		$3 \to 1$
	$4 \to 1$		$4 \to 3$		$4 \to 2$		$4 \to 4$

   (a) Show that $(D, \circ)$ is a subgroup of $(S_{4}, \circ)$ by completing its operation table.
   (b) Find the inverse and order of each element of D.
   (c) Verify that D is not cyclic and find a subgroup of D of order 4 that is not cyclic.
   (d) Determine all subgroups of D.
   (e) The elements of D can be viewed as permutations of the vertices of a square (think of the vertices with labels $1, 2, 3, 4$). With this convention, compare the operation tables for $(D, \circ)$ and the dihedral group $(D_{4}, \circ)$ and establish a meaningful one-to-one correspondence between D and D_{4}, in which the image of an element $\alpha \in D_{4}$ is determined by the effect of α on the vertices of the square.

8. Let $C(\mathbb{R}) = \left\{\begin{bmatrix} a & b \\ -b & a \end{bmatrix} \mid a, b \in \mathbb{R} \text{ and not both } a \text{ and } b \text{ are } 0\right\}$. Show that $(C(\mathbb{R}), \cdot) \le (GL_{2}(\mathbb{R}), \cdot)$. In fact, show that $(C(\mathbb{R}), \cdot)$ is abelian.

Does the product of two matrices in $C(\mathbb{R})$ bear any resemblance to a product in a well-known set of numbers? Explain.

9. Let $\mathscr{C}(\mathbb{R}) = \left\{ \begin{bmatrix} c & 0 \\ 0 & c \end{bmatrix} \mid c \in \mathbb{R}^{\#} \right\}.$

   Show that $(\mathscr{C}(\mathbb{R}), \cdot) \leq (Z(GL_2(\mathbb{R})), \cdot)$.
   (In fact, it can be shown that $(\mathscr{C}(\mathbb{R}), \cdot) = (Z(GL_2(\mathbb{R})), \cdot)$

10. For a group $(G, *)$, show that
    $$Z(G) = \{ c \in G \mid c^{-1} * g * c = g \text{ for all } g \in G \}.$$

11. Find the center of each of the following groups:
    (a) S_3 **(b)** $(D_4, \circ)$

12. Given an abelian group $(G, *)$, determine the center of $(G, *)$.

13. (The *alternating subgroup* A_4) Consider the following two elements of S_4:

$$\alpha: \begin{array}{l} 1 \to 2 \\ 2 \to 3 \\ 3 \to 1 \\ 4 \to 4 \end{array} \qquad \beta: \begin{array}{l} 1 \to 1 \\ 2 \to 3 \\ 3 \to 4 \\ 4 \to 2 \end{array}$$

   (a) Find the subgroup of minimum order that contains both α and β. This subgroup is called the *alternating subgroup* of S_4 and is denoted by A_4.
   (b) Show that A_4 has a unique subgroup of order 4.
   (c) Find all subgroups of A_4 of order 3.

14. If $(G, *)$ is a group and $(H, *)$ is a subgroup of $(G, *)$ such that $\{e\} \subset H \subset G$, then $(H, *)$ is called a *proper subgroup* of $(G, *)$.
    (a) Prove that if $(G, *)$ has no proper subgroups, then $(G, *)$ is cyclic.
    (b) Determine all proper subgroups of $(\mathbb{Z}_{12}, \oplus)$.
    (c) Determine all proper subgroups of $(S_3, \circ)$.
    (d) Show that $(\mathbb{Z}_m, \oplus)$ has no proper subgroups if and only if m is prime.

15. List all elements of each of the following cyclic subgroups:
    (a) $\langle [3] \rangle$ in $(\mathbb{Z}_{30}, \oplus)$ **(b)** $\langle (1 + i)/\sqrt{2} \rangle$ in $(\mathbb{C}^{\#}, \cdot)$
    (c) $\langle A \rangle$ in $(GL_2(\mathbb{R}), \cdot)$, where

$$A = \begin{bmatrix} 1 & 1 \\ 0 & 1 \end{bmatrix}$$

   (d) $\langle c^3 \rangle$ in $(G, *) = \langle c \rangle$, where $|c| = 14$

16. Let $(G, *) = \langle c \rangle$ be a cyclic group and let $(H, *) \leq (G, *)$. Show that $(H, *)$ is cyclic. (Hint: If $H \neq \{e\}$, then consider c^t, where t is the smallest positive integer such that $c^t \in H$.)

17. Use the result of Exercise 16 to find all subgroups of each of the following cyclic groups:
    (a) $(\mathbb{Z}, +)$ **(b)** $(\mathbb{Z}_{15}, \oplus)$ **(c)** $(\mathbb{Z}_{11}^{\#}, \odot)$ **(d)** $(\mathbb{Z}_7, \oplus)$

■■■■■■■	**RINGS, INTEGRAL DOMAINS, AND FIELDS**
8.5	

We know that a group $(G, *)$ is an algebraic structure composed of a nonempty set G, together with a single binary operation $*$. In this section we focus our attention on algebraic structures with two binary operations. In particular, we study *rings*, *integral domains*, and *fields*. Like groups, these structures have interested mathematicians for quite some time and continue to be very active areas of mathematical research. Applications exist to the solution of polynomial equations, coding theory, and switching networks, to mention just a few.

To begin, consider the structure $(\mathbb{Z}, +, \cdot)$; it is clear that $(\mathbb{Z}, +)$ is an abelian group and that $(\mathbb{Z}, \cdot)$ is a semigroup. Furthermore we can combine the operations $+$ and $\cdot$ to obtain properties such as

$$a \cdot (b + c) = a \cdot b + a \cdot c$$

which holds for all $a, b, c \in \mathbb{Z}$. This is the distributive law for ordinary arithmetic in $\mathbb{Z}$. (The expression $a \cdot b + a \cdot c$ is to be interpreted as $(a \cdot b) + (a \cdot c)$ in accordance with the usual precedence of multiplication over addition.)

As another example, consider $(\mathbb{Z}_m, \oplus, \odot)$, where m is any positive integer. It has already been established that $(\mathbb{Z}_m, \oplus)$ is a group; in fact, it is an abelian group, since $[a] \oplus [b] = [a + b] = [b + a] = [b] \oplus [a]$ for all $[a], [b] \in \mathbb{Z}_m$. We have also seen that $\odot$ is associative, so that $(\mathbb{Z}_m, \odot)$ is a semigroup. Moreover, it is easily checked that

$$[a] \odot ([b] \oplus [c]) = [a] \odot [b] \oplus [a] \odot [c]$$

holds for all $[a], [b], [c] \in \mathbb{Z}_m$.

Examples such as $(\mathbb{Z}, +, \cdot)$ and $(\mathbb{Z}_m, \oplus, \odot)$ serve to motivate the following definition.

DEFINITION 8.10

Let $+'$ and $\cdot'$ be binary operations on a nonempty set S. The algebraic structure $(S, +', \cdot')$ is called a *ring* provided the following conditions hold:

1. $(S, +')$ is an abelian group.
2. $(S, \cdot')$ is a semigroup.
3. The *distributive laws* hold for all $a, b, c \in S$:
   (a) $a \cdot'(b +' c) = a \cdot' b +' a \cdot' c$
   (b) $(b +' c) \cdot' a = b \cdot' a +' c \cdot' a$

The group $(S, +')$ is called the *additive group* of the ring. The identity element of $(S, +')$ is denoted by z and is called the *zero element* of the ring.

We refer to $+'$ as the *ring addition* and to $\cdot'$ as the *ring multiplication*. Consequently, in the group $(S, +')$, we denote the inverse of an element $a \in S$ by $-a$ and refer to $-a$ as the *additive inverse* of a.

As the student has probably noted, it is somewhat cumbersome to have to write $+'$ and $\cdot'$ to denote the operations in a general ring. We therefore agree simply to use $+$ and $\cdot$ to denote the operations, with the understanding that, in general, they are not the ordinary operations of addition and multiplication of real numbers. Also, if $(S, +, \cdot)$ is a ring and $a, b \in S$, then we agree to rewrite $a + (-b)$ as $a - b$ and refer to the operation $-$ as "subtraction."

Example 8.12   Let $\mathbb{Z}[\sqrt{2}\,] = \{a + b\sqrt{2} \mid a, b \in \mathbb{Z}\}$ and show that $(\mathbb{Z}[\sqrt{2}\,], +, \cdot)$ is a ring.

Solution   Notice that $+$ and $\cdot$ are the ordinary addition and multiplication for real numbers.

We must first show that $(\mathbb{Z}[\sqrt{2}\,], +)$ is an abelian group. For this it suffices to show that $(\mathbb{Z}[\sqrt{2}\,], +)$ is a subgroup of $(\mathbb{R}, +)$. We apply the criterion established in Theorem 8.12. Clearly $\mathbb{Z}[\sqrt{2}\,]$ is a nonempty set. Let $x = a + b\sqrt{2}$ and $y = c + d\sqrt{2}$ be elements of $\mathbb{Z}[\sqrt{2}\,]$. Then

$$x + y = (a + b\sqrt{2}) + (c + d\sqrt{2}) = (a + c) + (b + d)\sqrt{2}$$

so $x + y \in \mathbb{Z}[\sqrt{2}\,]$. Also

$$-x = -(a + b\sqrt{2}) = (-a) + (-b)\sqrt{2}$$

so $-x \in \mathbb{Z}[\sqrt{2}\,]$. Hence $(\mathbb{Z}[\sqrt{2}\,], +)$ is a subgroup of $(\mathbb{R}, +)$.

Next we must show that $(\mathbb{Z}[\sqrt{2}\,], \cdot)$ is a semigroup. Here it suffices to show that multiplication is a binary operation on $\mathbb{Z}[\sqrt{2}\,]$. Associativity then follows since $(\mathbb{R}, \cdot)$ is a semigroup. With x and y as given above, we have

$$xy = (a + b\sqrt{2})(c + d\sqrt{2}) = (ac + 2bd) + (ad + bc)\sqrt{2}$$

so $xy \in \mathbb{Z}[\sqrt{2}\,]$. Finally the distributive laws must be verified, but again these are inherited from $(\mathbb{R}, +, \cdot)$. Therefore $(\mathbb{Z}[\sqrt{2}\,], +, \cdot)$ is a ring.   ■

Example 8.13   Recall that $\mathscr{F}(\mathbb{R})$ denotes the set of all functions from $\mathbb{R}$ to $\mathbb{R}$. Define the operations $+$ and $\cdot$ on $\mathscr{F}(\mathbb{R})$ as follows: For $f, g \in \mathscr{F}(\mathbb{R})$,

(a) $(f + g)(x) = f(x) + g(x)$

(b) $(f \cdot g)(x) = f(x) \cdot g(x)$

for all $x \in \mathbb{R}$. In other words, the image of x under the sum $f + g$ is the ordinary real sum $f(x) + g(x)$, and the image of x under the product $f \cdot g$ is the ordinary real product $f(x) \cdot g(x)$. Show that $(\mathscr{F}(\mathbb{R}), +, \cdot)$ is a ring.

Solution   In Exercise 10 of Exercises 8.2 the student was asked to show that $(\mathscr{F}(\mathbb{R}), +)$ is an abelian group. The additive identity element for this group is the "zero function" $z: \mathbb{R} \to \mathbb{R}$, defined by $z(x) = 0$ for all $x \in \mathbb{R}$, and the additive inverse of a function $f \in \mathscr{F}(\mathbb{R})$ is the function $-f \colon \mathbb{R} \to \mathbb{R}$, defined by $(-f)(x) = -f(x)$ for all $x \in \mathbb{R}$. In $(\mathscr{F}(\mathbb{R}), \cdot)$ it must be shown that $\cdot$ is associative: For $f, g, h \in \mathscr{F}(\mathbb{R})$, we have

$$
\begin{aligned}
(f \cdot (g \cdot h))(x) &= f(x) \cdot (g \cdot h)(x) \\
&= f(x) \cdot (g(x) \cdot h(x)) \\
&= (f(x) \cdot g(x)) \cdot h(x) \\
&= ((f \cdot g)(x)) \cdot h(x) \\
&= ((f \cdot g) \cdot h)(x)
\end{aligned}
$$

for all $x \in \mathbb{R}$. Hence $(f \cdot g) \cdot h = f \cdot (g \cdot h)$. Finally it must be shown that the distributive laws hold. For $f, g, h \in \mathscr{F}(\mathbb{R})$, we have

$$
\begin{aligned}
(f \cdot (g + h))(x) &= f(x) \cdot (g + h)(x) \\
&= f(x) \cdot (g(x) + h(x)) \\
&= f(x) \cdot g(x) + f(x) \cdot h(x) \\
&= (f \cdot g)(x) + (f \cdot h)(x) \\
&= (f \cdot g + f \cdot h)(x)
\end{aligned}
$$

for all $x \in \mathbb{R}$. Thus $f \cdot (g + h) = f \cdot g + f \cdot h$. The remaining distributive law follows similarly. Hence $(\mathscr{F}(\mathbb{R}), +, \cdot)$ is a ring. ∎

Each of the rings mentioned so far is an example of a ring $(S, +, \cdot)$ in which the multiplication $\cdot$ is commutative; that is,

$$
x \cdot y = y \cdot x
$$

for all $x, y \in S$. Such rings are called *commutative rings*. Also note that each of the rings given possesses an identity element with respect to the ring multiplication. In general, a ring $(S, +, \cdot)$ is called a *ring with identity* provided it has an identity element with respect to the ring multiplication. In keeping with common practice, we denote this element by e and call it the *multiplicative identity* of $(S, +, \cdot)$. Thus

$$
a \cdot e = e \cdot a = a
$$

for all $a \in S$.

Example 8.14   Show that each of the rings $(\mathbb{Z}, +, \cdot)$, $(\mathbb{Z}_m, \oplus, \odot)$, and $(\mathscr{F}(\mathbb{R}), +, \cdot)$ is a commutative ring with identity.

Solution   We know that $a \cdot b = b \cdot a$ for all $a, b \in \mathbb{Z}$ and $a \cdot 1 = 1 \cdot a = a$ for all $a \in \mathbb{Z}$. So clearly $(\mathbb{Z}, +, \cdot)$ is a commutative ring with identity.

To see that $(\mathbb{Z}_m, \oplus, \odot)$ is the same, let $[a], [b] \in \mathbb{Z}_m$. Note that $[a] \odot [b] = [a \cdot b] = [b \cdot a] = [b] \odot [a]$ and $[a] \odot [1] = [a]$, implying that $(\mathbb{Z}_m, \oplus, \odot)$ is a commutative ring with identity $[1]$. Let $f, g \in \mathscr{F}(\mathbb{R})$ and define $f_1: \mathbb{R} \to \mathbb{R}$ by $f_1(x) = 1$ for all $x \in \mathbb{R}$. Then, for any $x \in \mathbb{R}$,

$$(f \cdot g)(x) = f(x) \cdot g(x) = g(x) \cdot f(x) = (g \cdot f)(x)$$

and

$$(f \cdot f_1)(x) = f(x) \cdot f_1(x) = f(x) \cdot 1 = f(x)$$

so that $f \cdot g = g \cdot f$ and $f \cdot f_1 = f$. Hence $(\mathscr{F}(\mathbb{R}), +, \cdot)$ is a commutative ring with identity f_1. ∎

Example 8.15 Consider $(M_2(\mathbb{R}), +, \cdot)$, where $+$ and $\cdot$ denote matrix addition and multiplication, respectively. Show that this structure is a noncommutative ring with identity.

Solution We already know that $(M_2(\mathbb{R}), +)$ is an abelian group and that matrix multiplication is associative. We leave the distributive laws to Exercise 18. Hence $(M_2(\mathbb{R}), +, \cdot)$ is a ring. The 2×2 identity matrix

$$I_2 = \begin{bmatrix} 1 & 0 \\ 0 & 1 \end{bmatrix}$$

satisfies the property $A \cdot I_2 = I_2 \cdot A = A$ for all $A \in M_2(\mathbb{R})$, so I_2 is the multiplicative identity for $(M_2(\mathbb{R}), +, \cdot)$. Finally note that if

$$A = \begin{bmatrix} 1 & 2 \\ 0 & 1 \end{bmatrix} \quad \text{and} \quad B = \begin{bmatrix} 0 & 0 \\ 0 & 1 \end{bmatrix}$$

then $A \cdot B \neq B \cdot A$. So $(M_2(\mathbb{R}), +, \cdot)$ is a noncommutative ring. ∎

THEOREM 8.14 Let $(S, +, \cdot)$ be a ring and let $a, b, c \in S$. Then the following are satisfied:

1. If $c + c = c$, then $c = z$.
2. $a \cdot z = z \cdot a = z$
3. $(-a) \cdot b = a \cdot (-b) = -(a \cdot b)$
4. $a \cdot (b - c) = (a \cdot b) - (a \cdot c)$
5. $(b - c) \cdot a = b \cdot a - c \cdot a$

Proof We prove parts 2 and 3 and leave the other parts to Exercise 4.
To establish part 2, note that

$$a \cdot z = a \cdot (z + z) = a \cdot z + a \cdot z$$

From $a \cdot z = a \cdot z + a \cdot z$ and part 1 it follows that $a \cdot z = z$. A similar proof shows that $z \cdot a = z$.
To show that $(-a) \cdot b = -(a \cdot b)$, first note that

$$(-a) \cdot b + a \cdot b = ((-a) + a) \cdot b = z \cdot b = z$$

so $(-a) \cdot b$ is an additive inverse of $a \cdot b$. Since additive inverses are unique, it must be that $(-a) \cdot b = -(a \cdot b)$. Similarly $a \cdot (-b) = -(a \cdot b)$. □

Recall that, in a group $(G, *)$, the expression a^n, $n \geq 0$, was defined by

$$a^0 = e \tag{1}$$
$$a^n = a * a^{n-1}, \qquad n > 0 \tag{2}$$

and the following properties of exponents were given:

$$a^{-n} = (a^n)^{-1} = (a^{-1})^n \tag{3}$$
$$a^m * a^n = a^{m+n} \tag{4}$$
$$(a^m)^n = a^{mn} \tag{5}$$

for all $m, n \in \mathbb{Z}$. If $(S, +, \cdot)$ is a ring, then $(S, +)$ is an abelian group. Since the additive notation is being used in this group, we state properties (1) through (5) with this notation. For example, in place of $a^2 = a * a$, we write $2a = a + a$ and, instead of $a^3 = a * a * a$, we write $3a = a + a + a$. In general, equations (1) and (2) become

$$0a = z \tag{1'}$$
$$na = a + (n - 1)a \tag{2'}$$

and we refer to na as a *multiple* of a. Properties (3) through (5) become

$$(-n)a = -(na) = n(-a) \tag{3'}$$
$$ma + na = (m + n)a \tag{4'}$$
$$n(ma) = (mn)a \tag{5'}$$

Similar properties involving ring multiplication are combined in the following theorem.

THEOREM 8.15   If $(S, +, \cdot)$ is a ring and $a, b \in S$, then for any integers m and n the following properties hold:

1. $m(a \cdot b) = (ma) \cdot b = a \cdot (mb)$
2. $(mn)(a \cdot b) = (ma) \cdot (nb)$

Proof   We prove part 1 and leave 2 to Exercise 5.
We first use induction to show that $m(a \cdot b) = (ma) \cdot b$ for all nonnegative integers m. If $m = 0$, then $0(a \cdot b) = z = z \cdot b = (0a) \cdot b$, and if $m = 1$, then $1(a \cdot b) = a \cdot b = (1a) \cdot b$, so the result holds in these two cases. Assume the result holds for an arbitrary integer $k \geq 1$. Our induction hypothesis is then

$$\text{IHOP:} \quad k(a \cdot b) = (ka) \cdot b$$

We then have

$$
\begin{array}{ll}
(k + 1)(a \cdot b) = k(a \cdot b) + a \cdot b & \text{(by definition)} \\
\qquad\qquad = (ka) \cdot b + a \cdot b & \text{(by IHOP)} \\
\qquad\qquad = [(ka) + a] \cdot b & \text{(by the distributive law)} \\
\qquad\qquad = [(k + 1)a] \cdot b & \text{(by definition)}
\end{array}
$$

This shows that the result holds for $k + 1$ and hence, by induction, $m(a \cdot b) = (ma) \cdot b$ for all $m \geq 0$.

Next we show that $m(a \cdot b) = (ma) \cdot b$ for all negative integers m. To do so, let $n \in \mathbb{N}$ and $m = -n$. Then

$$
\begin{aligned}
m(a \cdot b) &= (-n)(a \cdot b) \\
&= -(n(a \cdot b)) && \text{(by property (3'))} \\
&= -((na) \cdot b) && \text{(by proof for } m \geq 0) \\
&= (-(na)) \cdot b && \text{(by Theorem 8.14)} \\
&= ((-n)a) \cdot b && \text{(by property (3'))} \\
&= (ma) \cdot b
\end{aligned}
$$

In a like manner, it can be shown that $m(a \cdot b) = a \cdot (mb)$ for all $m \in \mathbb{Z}$. $\square$

Before proceeding further, there is one very simple ring that should be mentioned, namely, the ring consisting of a single element. The element is necessarily the zero element z. Notice that $z + z = z$ and $z \cdot z = z$ are the only additions and multiplications in the ring. Thus the ring has an identity element, namely, z. We call this ring the *trivial ring*.

THEOREM 8.16 If $(S, +, \cdot)$ is a ring with identity e and is not the trivial ring, then $e \neq z$.

Proof Proceed by contradiction and suppose that $e = z$. Since $S \neq \{z\}$, there is some element $a \in S$ such that $a \neq z$. But then

$$a = a \cdot e = a \cdot z = z$$

which is a contradiction. Hence $e \neq z$. $\square$

Suppose now that $(S, +, \cdot)$ is a ring with identity and is not the trivial ring. What can be said about the multiplicative structure $(S, \cdot)$ of the ring? For example, must each element of S possess a multiplicative inverse? The answer to this question is clearly no, since the element z possesses no multiplicative inverse in $(S, \cdot)$. Since $a \cdot z = z$ for all $a \in S$, it seems more reasonable to consider $(S^{\#}, \cdot)$, where $S^{\#} = S - \{z\}$. Again the answer is no, since the element $2 \in \mathbb{Z}$ possesses no inverse in the ring $(\mathbb{Z}, +, \cdot)$. As another example, consider the ring $(\mathbb{Z}_{12}, \oplus, \odot)$. The element $[2]$ satisfies the condition $[6] \odot [2] = [0]$, the zero element of the ring. If there were an element $[a] \in \mathbb{Z}_{12}$ for which $[2] \odot [a] = [1]$, then one would have

$$
\begin{aligned}
[6] &= [6] \odot [1] \\
&= [6] \odot ([2] \odot [a]) \\
&= ([6] \odot [2]) \odot [a] \\
&= [0] \odot [a] \\
&= [0]
\end{aligned}
$$

a contradiction. Thus the element [2] has no multiplicative inverse in $(\mathbb{Z}_{12}, \oplus, \odot)$.

Note that, in $(\mathbb{Z}_{12}, \oplus, \odot)$, [2] and [6] are nonzero elements whose product is the zero element [0]. Other rings have this interesting property of possessing nonzero elements whose product is the zero element of the ring.

DEFINITION 8.11

Let $(S, +, \cdot)$ be a ring. An element $a \in S$, $a \neq z$, is called a *zero divisor* of the ring provided there is an element $b \in S$, $b \neq z$, such that either $a \cdot b = z$ or $b \cdot a = z$.

Example 8.16 Show that each of the rings $(M_2(\mathbb{R}), +, \cdot)$ and $(\mathscr{F}(\mathbb{R}), +, \cdot)$ possesses zero divisors.

Solution First consider the ring $(M_2(\mathbb{R}), +, \cdot)$. The matrices

$$A = \begin{bmatrix} 1 & 0 \\ 0 & 0 \end{bmatrix} \quad \text{and} \quad B = \begin{bmatrix} 0 & 0 \\ 1 & 1 \end{bmatrix}$$

are nonzero and satisfy $A \cdot B = O_2$, the 2×2 zero matrix. Hence A is a zero divisor of $(M_2(\mathbb{R}), +, \cdot)$ (and so is B).

Next consider the ring $(\mathscr{F}(\mathbb{R}), +, \cdot)$. As mentioned in previous examples, the zero element of this ring is the function $z: \mathbb{R} \to \mathbb{R}$, given by $z(x) = 0$ for all $x \in \mathbb{R}$. To exhibit zero divisors in this ring, we must find functions $f, g \in \mathscr{F}(\mathbb{R})$ such that $f \neq z$, $g \neq z$, and $f \cdot g = z$; that is, $f(x) \cdot g(x) = 0$ for all $x \in \mathbb{R}$. Such examples are numerous and easy to construct. For example, define f and g by

$$f(x) = \begin{cases} 0 & \text{if } x < 0 \\ 1 & \text{if } x \geq 0 \end{cases} \quad \text{and} \quad g(x) = \begin{cases} -1 & \text{if } x < 0 \\ 0 & \text{if } x \geq 0 \end{cases} \qquad \blacksquare$$

Suppose next that $(S, +, \cdot)$ is a ring with identity and no zero divisors. Will it then be the case that each element $a \in S^{\#}$ has a multiplicative inverse? The answer is still no; in fact, we have already seen an example of a ring with identity and no zero divisors in which infinitely many nonzero elements do not have multiplicative inverses. This is the ring $(\mathbb{Z}, +, \cdot)$. Notice that each element different from ± 1 has no multiplicative inverse in $(\mathbb{Z}, +, \cdot)$. Yet rings like $(\mathbb{Z}, +, \cdot)$ do possess some added structure in that they have no zero divisors.

DEFINITION 8.12

Let $(S, +, \cdot)$ be a commutative ring with identity $e \neq z$. If $(S, +, \cdot)$ has no zero divisors, then $(S, +, \cdot)$ is called an *integral domain*.

Example 8.17   Show that each of the following rings is an integral domain.

(a) $(\mathbb{Z}[\sqrt{2}\,], +, \cdot)$ (b) $(\mathbb{Z}_p, \oplus, \odot)$, where p is a prime

Solution   First note that each of these rings is commutative with identity $(1 = 1 + 0\sqrt{2}$ is the identity for $(\mathbb{Z}[\sqrt{2}\,], +, \cdot)$ and [1] is the identity for $(\mathbb{Z}_p, \oplus, \odot))$. If $(S, +, \cdot)$ is a commutative ring with identity, then to show that it has no divisors of zero, it suffices to show that

$$xy = z \quad \text{implies} \quad x = z \text{ or } y = z$$

for all $x, y \in S$.

(a) Consider first the ring $(\mathbb{Z}[\sqrt{2}\,], +, \cdot)$, and let $x = a + b\sqrt{2}$ and $y = c + d\sqrt{2}$ be elements of this ring. Suppose that $xy = 0$. Then simply note that x and y are also real numbers and $xy = 0$ implies immediately that $x = 0$ or $y = 0$. Thus $(\mathbb{Z}[\sqrt{2}\,], +, \cdot)$ is an integral domain.

(b) For $(\mathbb{Z}_p, \oplus, \odot)$, suppose that $[x] \odot [y] = [0]$. Then $[xy] = [0]$, which implies that $p \mid xy$. Hence, by Euclid's lemma, either $p \mid x$ or $p \mid y$, so that either $[x] = [0]$ or $[y] = 0$. Therefore $(\mathbb{Z}_p, \oplus, \odot)$ is an integral domain when p is a prime. ∎

The definition of an integral domain is often given in terms of the "cancellation law," a condition that turns out to be equivalent to having no zero divisors.

DEFINITION 8.13

Let $(S, +, \cdot)$ be a commutative ring. We say that $(S, +, \cdot)$ satisfies the *cancellation law* provided, for all $a, b, c \in S$, if $a \cdot b = a \cdot c$ and $a \neq z$, then $b = c$.

THEOREM 8.17   Let $(S, +, \cdot)$ be a commutative ring. Then $(S, +, \cdot)$ has no zero divisors if and only if $(S, +, \cdot)$ satisfies the cancellation law.

Proof   Assume first that $(S, +, \cdot)$ has no zero divisors. Let $a, b, c \in S$, with $a \neq z$, and suppose that $a \cdot b = a \cdot c$. Then $a \cdot b - a \cdot c = z$, so by the distributive law, $a \cdot (b - c) = z$. Since $a \neq z$ and $(S, +, \cdot)$ has no zero divisors, it follows that $b - c = z$. So $b = c$. Thus $(S, +, \cdot)$ satisfies the cancellation law.

For the converse, assume that $(S, +, \cdot)$ satisfies the cancellation law and suppose that $a \cdot b = z$ for some $a, b \in S$. If $a = z$, then there is nothing to prove. Assume $a \neq z$. By Theorem 8.14, part 2, $a \cdot z = z$, so we have that $a \cdot b = a \cdot z$. Since $(S, +, \cdot)$ satisfies the cancellation law and $a \neq z$, we conclude that $b = z$. Hence $(S, +, \cdot)$ has no zero divisors. □

COROLLARY 8.18   Let $(S, +, \cdot)$ be a commutative ring with identity. If $(S, +, \cdot)$ satisfies the cancellation law, then $(S, +, \cdot)$ is an integral domain.

If $(S, +, \cdot)$ is a commutative ring with identity $e \neq z$, then the multiplicative structure $(S^{\#}, \cdot)$ is a commutative semigroup with identity. In general, $(S^{\#}, \cdot)$ need not be a group, as in the case of $(\mathbb{Z}^{\#}, \cdot)$. In fact, $(S^{\#}, \cdot)$ is an abelian group if and only if each element of $S^{\#}$ has a multiplicative inverse in $(S^{\#}, \cdot)$. The resulting ring $(S, +, \cdot)$ is then called a *field*.

DEFINITION 8.14

Let $(F, +, \cdot)$ be a commutative ring with identity. If $(F^{\#}, \cdot)$ is an abelian group, then $(F, +, \cdot)$ is called a *field*.

As already noted, we denote the multiplicative identity of a field $(F, +, \cdot)$ by e. In addition, if $a \in F^{\#}$, then the multiplicative inverse of a is denoted by the familiar notation a^{-1}.

Example 8.18   The multiplicative inverse of the real number a, where $a \neq 0$, is $1/a$. It follows that each of the rings $(\mathbb{R}, +, \cdot)$ and $(\mathbb{Q}, +, \cdot)$ is a field. We call $(\mathbb{R}, +, \cdot)$ the *field of real numbers* and $(\mathbb{Q}, +, \cdot)$ the *field of rational numbers*. ∎

Example 8.19   It can be shown that $(\mathbb{Z}[\sqrt{2}], +, \cdot)$ is not a field (see Exercise 6). However, consider the ring $(\mathbb{Q}[\sqrt{2}], +, \cdot)$, where

$$\mathbb{Q}[\sqrt{2}] = \{ a + b\sqrt{2} \mid a, b \in \mathbb{Q} \}$$

Just as with $(\mathbb{Z}[\sqrt{2}], +, \cdot)$, we can verify that $(\mathbb{Q}[\sqrt{2}], +, \cdot)$ is a commutative ring with identity. Show that $(\mathbb{Q}[\sqrt{2}], +, \cdot)$ is a field.

Solution   We must show that each nonzero element of $\mathbb{Q}[\sqrt{2}]$ has a multiplicative inverse in $(\mathbb{Q}[\sqrt{2}], +, \cdot)$. So consider $a + b\sqrt{2}$, where $a, b \in \mathbb{Q}$ and not both a and b are 0. We must find an element $c + d\sqrt{2} \in \mathbb{Q}[\sqrt{2}]$ such that

$$(a + b\sqrt{2})(c + d\sqrt{2}) = 1$$

From this equation, we obtain $ac + 2bd = 1$ and $ad + bc = 0$. These equations must be solved for c and d. We leave it to the student to check that the solution is $c = a/(a^2 - 2b^2)$ and $d = -b/(a^2 - 2b^2)$. Alternatively, we have

$$(a + b\sqrt{2})^{-1} = \frac{1}{a + b\sqrt{2}} = \frac{a - b\sqrt{2}}{(a + b\sqrt{2})(a - b\sqrt{2})}$$

$$= \frac{a}{a^2 - 2b^2} + \frac{-b}{a^2 - 2b^2}\sqrt{2}$$

Therefore $(\mathbb{Q}[\sqrt{2}], +, \cdot)$ is a field. ∎

THEOREM 8.19 If $(F, +, \cdot)$ is a field, then it is an integral domain.

Proof In order to show that $(F, +, \cdot)$ is an integral domain, it suffices to show, by Corollary 8.18, that $(F, +, \cdot)$ satisfies the cancellation law. Suppose that $a, b, c \in F$, with $a \neq z$ and $a \cdot b = a \cdot c$. Since $a \neq z$, we have that a^{-1} exists; thus we have the following:

$$a^{-1} \cdot (a \cdot b) = a^{-1} \cdot (a \cdot c)$$

$$(a^{-1} \cdot a) \cdot b = (a^{-1} \cdot a) \cdot c$$

$$e \cdot b = e \cdot c$$

$$b = c$$

Hence $(F, +, \cdot)$ satisfies the cancellation law. □

It has been pointed out that $(\mathbb{Z}, +, \cdot)$ is an integral domain that is not a field, and thus the collection of all fields is a proper subset of the set of all integral domains. Indeed, using the notations

$$\mathscr{R} = \text{set of all rings}$$

$$\mathscr{C} = \text{set of all commutative rings}$$

$$\mathscr{D} = \text{set of all integral domains}$$

$$\mathscr{F} = \text{set of all fields}$$

the following inclusion relations hold:

$$\mathscr{F} \subset \mathscr{D} \subset \mathscr{C} \subset \mathscr{R}$$

Under certain restrictions it happens that every integral domain is a field.

THEOREM 8.20 Every finite integral domain is a field.

Proof Let $(D, +, \cdot)$ be a finite integral domain, say $D = \{a_1, a_2, \ldots, a_m\}$. To show that $(D, +, \cdot)$ is a field, we must show that each nonzero element of D has a multiplicative inverse. Let $a \in D$, with $a \neq z$, and consider the elements

$$a \cdot a_1, a \cdot a_2, \ldots, a \cdot a_m$$

These elements are distinct; if $a \cdot a_i = a \cdot a_j$ for some i and j, then we can apply the cancellation law (since $a \neq z$) to conclude that $a_i = a_j$. Since $a \cdot a_1, a \cdot a_2, \ldots, a \cdot a_m$ are m distinct elements of D and $|D| = m$, we conclude that

$$D = \{a \cdot a_1, a \cdot a_2, \ldots, a \cdot a_m\}$$

Thus the identity element e must be one of the elements $a \cdot a_1$, $a \cdot a_2, \ldots, a \cdot a_m$, say $e = a \cdot a_i$. It follows that $a^{-1} = a_i$. Therefore $(D, +, \cdot)$ is a field. □

We know that $(\mathbb{Z}_m, \oplus, \odot)$ is an integral domain if and only if m is a prime, thus we have the following result.

COROLLARY 8.21   The ring $(\mathbb{Z}_m, \oplus, \odot)$ is a field if and only if m is a prime.

Exercises 8.5

1. Each of the following parts gives a set S and two binary operations $+'$ and $\cdot'$ on S. Determine whether $(S, +', \cdot')$ is a ring.
   (a) $S = \mathbb{R}$, $x +'y = x + y - 1$, $x \cdot'y = xy$
   (b) $S = \mathbb{Z}$, $x +'y = x + y - 1$, $x \cdot'y = x + y - xy$
   (c) $S = \mathbb{R}^+$ (the positive reals), $x +'y = xy$, $x \cdot'y = x + y$
   (d) $S = \mathbb{R}^+$, $x +'y = xy$, $x \cdot'y = x^y$
2. Consider the ring $(\mathbb{Z}, +', \cdot')$ defined in Exercise 1(b).
   (a) What is the zero element?
   (b) What is the multiplicative identity?
   (c) Is this ring commutative?
   (d) Is this ring an integral domain?
   (e) Is this ring a field?
3. Consider the set $\mathbb{R} \times \mathbb{R}$ of ordered pairs of real numbers. Define binary operations $+$ and $\cdot$ on $\mathbb{R} \times \mathbb{R}$ as follows:

$$(a, b) + (c, d) = (a + c, b + d)$$

$$(a, b) \cdot (c, d) = (ac - bd, ad + bc)$$

Note that $+$ and $\cdot$ are also binary operations on $\mathbb{Z} \times \mathbb{Z}$.
   (a) Show that $(\mathbb{Z} \times \mathbb{Z}, +, \cdot)$ is a commutative ring with identity.
   (b) Is $(\mathbb{Z} \times \mathbb{Z}, +, \cdot)$ an integral domain? a field?
   (c) Show that $(\mathbb{R} \times \mathbb{R}, +, \cdot)$ is a field.
4. Prove parts 1, 4, and 5 of Theorem 8.14.
5. Prove part 2 of Theorem 8.15.
6. Show that $(\mathbb{Z}[\sqrt{2}], +, \cdot)$ is not a field.
7. Consider the ring $(\mathbb{Q}, +', \cdot')$, where $+'$ and $\cdot'$ are defined by

$$x +'y = x + y - 1$$
$$x \cdot'y = x + y - xy$$

Is this ring a field?
8. For $k \in \mathbb{N}$, let M_k denote the set of integer multiples of k. Show that $(M_k, +, \cdot)$ (where $+$ and $\cdot$ denote ordinary addition and multiplication) is a commutative ring but has no identity for $k \geq 2$.
9. Let A be a nonempty set and consider $\mathscr{P}(A)$ under the operations of $*$ and $\cap$, where $B * C = (B - C) \cup (C - B)$, the symmetric difference of B and C. Show that $(\mathscr{P}(A), *, \cap)$ is a commutative ring with identity.

10. Is $(\mathcal{F}(\mathbb{R}), +, \circ)$ a ring? Explain.

11. Let $(S, +, \cdot)$ be a commutative ring with identity.
   (a) Prove: If $a \in S$ has a multiplicative inverse, then a is not a zero divisor of S.
   (b) What about the converse of the statement of part (a)?
   (c) Prove: If S is finite and $a \in S$ is not a zero divisor of S, then a has a multiplicative inverse in $(S, +, \cdot)$.

12. Consider the ring $(\mathbb{Z}_m, \oplus, \odot)$ and let $[a] \in \mathbb{Z}_m$, $[a] \neq [0]$.
   (a) Show that if $\gcd(a, m) > 1$, then $[a]$ is a zero divisor.
   (b) Show that $[a]$ has a multiplicative inverse if and only if $\gcd(a, m) = 1$.
   (c) Show that $(\mathbb{Z}_m, \oplus, \odot)$ is not an integral domain if m is composite.

13. Let E be the set of even integers. Define the operation $*$ on E as follows:

$$x * y = \frac{xy}{2}$$

Show that $(E, +, *)$ is an integral domain.

14. Define operations $+$ and $\cdot$ on $\mathbb{R} \times \mathbb{R}$ as follows:

$$(a, b) + (c, d) = (a + c, b + d)$$
$$(a, b) \cdot (c, d) = (ac, bd)$$

   (a) Show that $(\mathbb{R} \times \mathbb{R}, +, \cdot)$ is a commutative ring with identity.
   (b) Is $(\mathbb{R} \times \mathbb{R}, +, \cdot)$ an integral domain? (Compare with Exercise 3.)

15. We have noted that $(\mathbb{Z}_4, \oplus, \odot)$ is not a field. Show that the operation tables given here define a field $(F, +, \cdot)$ with four elements. (It can be shown that there exists a finite field of order m if and only if m is a power of a prime.)

+	z	e	a	b
z	z	e	a	b
e	e	z	b	a
a	a	b	z	e
b	b	a	e	z

$\cdot$	z	e	a	b
z	z	z	z	z
e	z	e	a	b
a	z	a	b	e
b	z	b	e	a

16. Let $(S, +, \cdot)$ be a ring with identity e. Show that $(-e) \cdot a = -a$ for all $a \in S$.

17. Let $(F, +, \cdot)$ be a field and let $a, b \in F$, with $a \neq z$. Show that the equation

$$a \cdot x + b = z$$

has a unique solution $x \in F$.

18. Recall that $M_n(\mathbb{R})$ denotes the set of all $n \times n$ matrices with real entries. Show that $(M_n(\mathbb{R}), +, \cdot)$ is a noncommutative ring with identity, where $+$ and $\cdot$ represent ordinary matrix addition and multiplication.

8.6 BOOLEAN ALGEBRAS

Recall that a lattice is a poset $(A, \preccurlyeq)$ in which every two elements possess both a least upper bound (lub) and a greatest lower bound (glb). For each $a, b \in A$, both $\mathrm{lub}(a, b)$ and $\mathrm{glb}(a, b)$ are uniquely determined elements of A; notationally, $\mathrm{lub}(a, b) = a \vee b$ and $\mathrm{glb}(a, b) = a \wedge b$. Also recall that $a \vee b$ is called the *join* of a and b, and $a \wedge b$ is called the *meet* of a and b. If we set $d = a \vee b$, then remember the two conditions satisfied by d:

1. d is an upper bound for the set $\{a, b\}$ (that is, $a \preccurlyeq d$ and $b \preccurlyeq d$).
2. If c is any upper bound for $\{a, b\}$, then $d \preccurlyeq c$.

Analogous statements can be made for $a \wedge b$. It is clear that $\vee$ and $\wedge$ are binary operations on A; we can then consider the algebraic structure $(A, \vee, \wedge)$.

THEOREM 8.22 If $(A, \preccurlyeq)$ is a lattice, then the following properties hold for all $a, b, c \in A$:

1. *Commutative laws:*

$$a \vee b = b \vee a \quad \text{and} \quad a \wedge b = b \wedge a$$

2. *Associative laws:*

$$(a \vee b) \vee c = a \vee (b \vee c) \quad \text{and} \quad (a \wedge b) \wedge c = a \wedge (b \wedge c)$$

3. *Idempotent laws:*

$$a \vee a = a \quad \text{and} \quad a \wedge a = a$$

4. *Absorption laws:*

$$(a \vee b) \wedge a = a \quad \text{and} \quad (a \wedge b) \vee a = a$$

Proof   We demonstrate the proof technique involved by considering the property $(a \vee b) \wedge a = a$. Let $d = (a \vee b) \wedge a$; then clearly $d \preccurlyeq a \vee b$ and $d \preccurlyeq a$. It suffices, by antisymmetry, to show that $a \preccurlyeq d$. Since $a \preccurlyeq a \vee b$ and $a \preccurlyeq a$, we may conclude that $a \preccurlyeq (a \vee b) \wedge a$. Hence $a \preccurlyeq d$. Thus $a = d$.

The proofs of the remaining properties are handled in a similar manner and are left to the exercises.   □

Beyond those listed in Theorem 8.22, there are some other important properties that certain lattices may satisfy. Recall that a lattice $(L, \preccurlyeq)$ is called a *distributive lattice* (see Chapter 4 Problem 30) provided

$$a \wedge (b \vee c) = (a \wedge b) \vee (a \wedge c)$$

and

$$a \vee (b \wedge c) = (a \vee c) \wedge (a \vee c)$$

hold for all $a, b, c \in L$.

The obvious example of a distributive lattice is $(\mathcal{P}(U), \subseteq)$, the lattice of all subsets of a set U. In this case, $A \vee B = A \cup B$ and $A \wedge B = A \cap B$, and we proved in Chapter 2 (Theorem 2.6) that

$$A \cap (B \cup C) = (A \cap B) \cup (A \cap C)$$

and

$$A \cup (B \cap C) = (A \cup B) \cap (A \cup C)$$

Let's examine $(\mathcal{P}(U), \subseteq)$ further to see if it satisfies any other interesting and notable properties. Right off we know that $\emptyset$ and U are elements of $\mathcal{P}(U)$ such that

$$A \cup \emptyset = A \quad \text{and} \quad A \cap U = A$$

for all $A \in \mathcal{P}(U)$. Thus it follows that $\emptyset$ is an identity element with respect to $\cup$ and U is an identity element with respect to $\cap$.

DEFINITION 8.15

Let $(L, \preccurlyeq)$ be a lattice. An element $1 \in L$ is called a *unit element* for L provided $a \wedge 1 = a$ for all $a \in L$. An element $0 \in L$ is called a *zero element* for L provided $a \vee 0 = a$ for all $a \in L$. If $(L, \preccurlyeq)$ has both a zero element and a unit element, then we say that $(L, \preccurlyeq)$ is a *lattice with zero and unity*. We sometimes write $(L, \preccurlyeq)$ *is a lattice with* 0 *and* 1.

In view of the above definition, we can say that $(\mathcal{P}(U), \subseteq)$ is a distributive lattice with unit element U and zero element $\emptyset$.

In addition to the above properties, the lattice $(\mathcal{P}(U), \subseteq)$ has the property that for each $A \in \mathcal{P}(U)$ there is an element $A' \in \mathcal{P}(U)$ such that

$$A \cup A' = U \quad \text{and} \quad A \cap A' = \emptyset$$

Recall that A' is called the *complement* of A in U.

DEFINITION 8.16

Let $(L, \preccurlyeq)$ be a lattice with 0 and 1. An element $a' \in L$ is called a *complement* of a in L provided

$$a \vee a' = 1 \quad \text{and} \quad a \wedge a' = 0$$

If each element of L has a complement in L, then we call $(L, \preccurlyeq)$ a *complemented lattice*.

In Definition 8.16 we do not insist that complements, should they exist, be uniquely determined. In Exercise 16 the student is asked to provide an example of a lattice that contains elements possessing more than one complement. In a distributive lattice, however, one can show that each element has at most one complement (see Exercise 17).

Example 8.20 Let L be the set of positive divisors of 30 and, for each $a, b \in L$, define $a \preccurlyeq b$ provided $a \mid b$. Show that $(L, \preccurlyeq)$ is a lattice with zero and unity that is both distributive and complemented.

Solution Note that $L = \{1, 2, 3, 5, 6, 10, 15, 30\}$. It is easily shown that $(L, \preccurlyeq)$ is a poset. In addition, for $a, b \in L$, the meet $a \wedge b$ is that element $d \in L$ satisfying

1. $d \mid a$ and $d \mid b$
2. If $c \mid a$ and $c \mid b$, then $c \mid d$.

It follows that $a \wedge b = \gcd(a, b)$. Similarly it can be shown that $a \vee b = \mathrm{lcm}(a, b)$. So $(L, \preccurlyeq)$ is a lattice.

For each $a \in L$, it is clear that $\mathrm{lcm}(a, 1) = a$. Hence $a \vee 1 = a$ for all $a \in L$, which shows that 1 is a zero element for $(L, \preccurlyeq)$. Also $\gcd(a, 30) = a$ for each $a \in L$, so 30 is a unit element for $(L, \preccurlyeq)$.

To see that $(L, \preccurlyeq)$ is complemented, we use the fact that 30 has no repeated prime factors. If $d \mid 30$, then $\gcd(d, 30/d) = 1$. Hence we define $a' = 30/a$ for each $a \in L$ and note that both $a \wedge a' = 1$ (the zero element) and $a \vee a' = 30$ (the unit element).

For the distributive laws, let $a, b, c \in L$ and let

$$a = 2^{r_1} \cdot 3^{s_1} \cdot 5^{t_1}$$
$$b = 2^{r_2} \cdot 3^{s_2} \cdot 5^{t_2}$$
$$c = 2^{r_3} \cdot 3^{s_3} \cdot 5^{t_3}$$

be their prime power factorizations. Since $30 = 2 \cdot 3 \cdot 5$, each of the exponents is either 0 or 1. To show the distributive law $a \wedge (b \vee c) = (a \wedge b) \vee (a \wedge c)$, we must verify that

$$\gcd(a, \mathrm{lcm}(b, c)) = \mathrm{lcm}(\gcd(a, b), \gcd(a, c))$$

Suppose that

$$\gcd(a, \mathrm{lcm}(b, c)) = 2^{m_1} \cdot 3^{m_2} \cdot 5^{m_3}$$

and

$$\mathrm{lcm}(\gcd(a, b), \gcd(a, c)) = 2^{q_1} \cdot 3^{q_2} \cdot 5^{q_3}$$

We must show that $q_1 = m_1$, $q_2 = m_2$, and $q_3 = m_3$. Note that

$$m_1 = \min(r_1, \max(r_2, r_3))$$

and

$$q_1 = \max(\min(r_1, r_2), \min(r_1, r_3))$$

With these formulations, it is not hard to show that $q_1 = m_1$. Similarly, it follows that $q_2 = m_2$ and $q_3 = m_3$. Thus $a \wedge (b \vee c) = (a \wedge b) \vee (a \wedge c)$. We leave it to the student to verify the other distributive law.   ■

It has already been established that $(\mathscr{P}(U), \subseteq)$ is a distributive, complemented lattice with zero and unity. It turns out that this special lattice is the prototype for a very important algebraic structure.

DEFINITION 8.17

A lattice with zero and unity that is both distributive and complemented is called a *Boolean algebra*.

In addition to $(\mathscr{P}(U), \subseteq)$ being a Boolean algebra, we see that the lattice $(L, \preceq)$ of Example 8.20 is also. In fact, if $n = p_1 p_2 \cdots p_r$, where $p_1, p_2, \ldots, p_r$ are distinct primes, and if L is the set of positive divisors of n, then $(L, |)$ is a Boolean algebra.

Example 8.21 Let L be the set of positive divisors of 72. Show that $(L, |)$ is not a Boolean algebra.

Solution It suffices to show that one of the properties for a Boolean algebra does not hold. Just as in Example 8.20, it can be shown that $(L, |)$ is a lattice with unit element 72 and zero element 1. However, $(L, |)$ is not complemented since there is no $a' \in L$ for which both $24 \wedge a' = 1$ and $24 \vee a' = 72$.   ■

Example 8.22 Determine whether each of the lattices whose Hasse diagrams are shown in Figure 8.4 is a Boolean algebra.

Solution The lattice of Figure 8.4(a) is a Boolean algebra. The unity and zero elements are shown as 1 and 0, respectively, and $a' = b$. Verifying the distributive laws takes a bit of effort in exhausting all cases.

The lattice in Figure 8.4(b) fails to satisfy the distributive laws since, for instance, $a \vee (b \wedge c) = a \vee 0 = a$ and $(a \vee b) \wedge (a \vee c) = 1 \wedge 1 = 1$. Thus this lattice is not a Boolean algebra.

An example of a lattice that is distributive but not complemented is shown in Figure 8.4(c). Neither a nor b has a complement and the lattice is clearly distributive.   ■

It should be emphasized at this point that the definition of a Boolean algebra originates with a lattice $(L, \preceq)$. The binary operations $\vee$ and $\wedge$ are defined by $a \vee b = \text{lub}(a, b)$ and $a \wedge b = \text{glb}(a, b)$. If $(B, \preceq)$ is a Boolean algebra, then the following properties hold for all $a, b, c \in B$:

1. *Commutative laws:*

$$a \vee b = b \vee a \quad \text{and} \quad a \wedge b = b \wedge a$$

2. *Associative laws:*

$$a \vee (b \vee c) = (a \vee b) \vee c \quad \text{and} \quad a \wedge (b \wedge c) = (a \wedge b) \wedge c$$

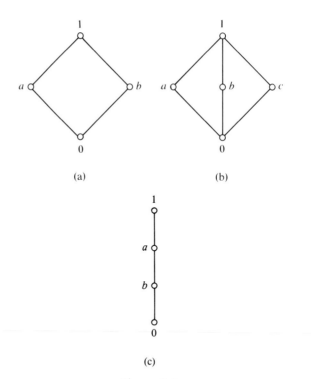

Figure 8.4

3. *Distributive laws:*

$$a \wedge (b \vee c) = (a \wedge b) \vee (a \wedge c)$$

and

$$a \vee (b \wedge c) = (a \vee b) \wedge (a \vee c)$$

4. *Existence of zero and unit elements:*   There exist elements 0 and 1 such that

$$a \vee 0 = a \quad \text{and} \quad a \wedge 1 = a$$

for all $a \in B$.

5. *Complements:*   For each $a \in B$ there is an element $a' \in B$ such that

$$a \vee a' = 1 \quad \text{and} \quad a \wedge a' = 0$$

If one consults the mathematical literature, one will frequently find that a Boolean algebra is defined to be an algebraic structure $(B, \vee, \wedge)$, where $\vee$ and $\wedge$ are binary operations on B that satisfy precisely properties 1 through 5. In fact, if $(A, \vee, \wedge)$ is such an algebraic structure, then $(A, \leqslant)$ is a Boolean algebra (in the sense of Definition 8.17) if we define

$$a \leqslant b \leftrightarrow a \vee b = b$$

for any $a, b \in A$.

If $(B, \preccurlyeq)$ is a Boolean algebra, then the zero element 0 is a *universal lower bound* in the sense that $0 \preccurlyeq x$ for all $x \in B$. What can be said about elements that are "immediate successors" of 0? In other words, do there exist elements $a \in B$ such that $0 \prec a$ and for which there is no $b \in B$ with $0 \prec b \prec a$. If so, then such an element is called an *atom*. In other words, an element $a \in B$ is called an *atom* provided, for any $b \in B$, the condition

$$0 \prec b \preccurlyeq a \rightarrow b = a \qquad\qquad (*)$$

holds. What are the atoms of the Boolean algebra $(L, |)$ given in Example 8.20? Recall that the zero element of this Boolean algebra is 1. If $p \in \{2, 3, 5\}$, then, for any $q \in L$, condition $(*)$ becomes

$$1 \neq q \quad \text{and} \quad q \mid p \rightarrow q = p$$

which certainly holds since p is a prime. Thus 2, 3, and 5 are atoms in $(L, |)$. In fact, it is easy to see that 6, 10, 15, and 30 are not atoms, so that 2, 3, and 5 are all the atoms of $(L, |)$.

Example 8.23   If $U = \{1, 2, \ldots, n\}$, then determine the atoms of the Boolean algebra $(\mathscr{P}(U), \subseteq)$. Also show that each nonempty subset of U is expressible as a join of atoms.

Solution   In this case, $\emptyset$ is the zero element of $(\mathscr{P}(U), \subseteq)$ and we seek those subsets A of U satisfying the condition

$$\emptyset \subset B \subseteq A \rightarrow B = A$$

for all $B \in \mathscr{P}(U)$. It is clear that the singleton subsets are the only subsets of U that satisfy this condition. Hence the atoms of $(\mathscr{P}(U), \subseteq)$ are $\{1\}, \{2\}, \ldots, \{n\}$. If B is a nonempty subset of U, say $B = \{a_1, a_2, \ldots, a_t\}$, then $B = \{a_1\} \cup \{a_2\} \cup \cdots \cup \{a_t\}$, a join of atoms.   ∎

A Boolean algebra $(B, \preccurlyeq)$ is called a *finite Boolean algebra* if B is a finite set. In Exercise 21 the student is asked to prove the following basic result concerning the atoms of a finite Boolean algebra.

THEOREM 8.23   If $(B, \preccurlyeq)$ is a finite Boolean algebra, then every nonzero element of B is expressible as a join of atoms. Moreover, this expression is unique up to the order in which the atoms appear.

Exercises 8.6

1. Each of the lattices in Figure 8.5 has a 0 and 1 as shown. First determine whether the lattice is complemented. If so, find a complement for each element. Next, if the lattice is complemented, determine whether it is distributive. Is the lattice a Boolean algebra?

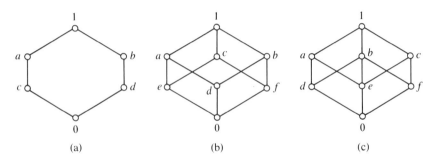

(a)   (b)   (c)

Figure 8.5

2. Construct the Hasse diagram and complete the operation tables for $\vee$ and $\wedge$ for the Boolean algebra of positive divisors of each of the following numbers:
   (a) 5   **(b)** 10   **(c)** 30

3. Show that any finite nontrivial Boolean algebra has an even number of elements.

4. Construct the Hasse diagram and complete the operation tables for $\vee$ and $\wedge$ for the Boolean algebra of positive divisors of n, where:
   (a) n is a prime.
   (b) n is a product of two distinct primes: $n = pq$.
   (c) n is a product of three distinct primes: $n = pqr$.

5. Consider the lattice of positive divisors of a given positive integer $n > 1$.
   (a) Show that this lattice is complemented if and only if n is a product of distinct primes. (Hint: If p^2 divides n, where p is a prime, show that p has no complement.)
   (b) Use part (a) to show that this lattice is a Boolean algebra if and only if n is the product of distinct primes, $n = p_1 p_2 \cdots p_m$.

6. If $n \in \mathbb{N}$ is a product of m distinct primes, $n = p_1 p_2 \cdots p_m$, and B is the Boolean algebra of positive divisors of n, how many elements does B have?

7. Show that $(\{0, 1\}, \preccurlyeq)$ is a Boolean algebra, where $0 \preccurlyeq 1$. Complete each of the following:
   (a) $0 \wedge 1 = $ _____   **(b)** $0 \vee 1 = $ _____   **(c)** $0' = $ _____
   (d) $1' = $ _____

8. For $n \in \mathbb{N}$, consider the set

$$B = \{(x_1, x_2, \ldots, x_n) \mid x_i = 0 \text{ or } 1, 1 \leq i \leq n\},$$

with $\wedge$ and $\vee$ defined as follows:

$$(x_1, x_2, \ldots, x_n) \wedge (y_1, y_2, \ldots, y_n) = (z_1, z_2, \ldots, z_n)$$

where $z_i = x_i y_i$, $1 \leq i \leq n$, and

$$(x_1, x_2, \ldots, x_n) \vee (y_1, y_2, \ldots, y_n) = (w_1, w_2, \ldots, w_n)$$

where $w_i = x_i + y_i - x_i y_i$.

(a) Show that $\wedge$ and $\vee$ satisfy properties 1 through 5 that are listed in the paragraphs following Example 8.22. This shows that $(B, \vee, \wedge)$ is a Boolean algebra.

(b) Discuss the properties of the partial ordering $\preccurlyeq$ on B defined by
$$(x_1, x_2, \ldots, x_n) \preccurlyeq (y_1, y_2, \ldots, y_n)$$
$$\leftrightarrow (x_1, x_2, \ldots, x_n) \vee (y_1, y_2, \ldots, y_n) =$$
$$(y_1, y_2, \ldots, y_n)$$

9. Consider the Boolean algebra of positive divisors of $n = pqr$, where p, q, and r are distinct primes. Simplify each of the following expressions:

   (a) $p \wedge q'$ (b) $p \vee q'$ (c) $(p \wedge q) \vee r$

   (d) $(p \vee q) \wedge r$ (e) $p' \wedge q'$ (f) $p' \vee q'$

   (g) $(p' \wedge q') \vee q$ (h) $(p \vee q) \wedge (q \vee r)$

10. For elements a and b in a Boolean algebra $(B, \preccurlyeq)$, show that the following conditions are equivalent:

    (i) $a \preccurlyeq b$ (ii) $a \wedge b' = 0$ (iii) $a' \vee b = 1$ (iv) $b' \preccurlyeq a'$

11. Show that every Boolean algebra $(B, \preccurlyeq)$ satisfies DeMorgan's laws for all $a, b \in B$:

    (a) $(a \vee b)' = a' \wedge b'$ (b) $(a \wedge b)' = a' \vee b'$

12. Prove: If $(B, \preccurlyeq)$ is a Boolean algebra and $a, b, c \in B$ satisfy $a \vee c = b \vee c$ and $a \wedge c = b \wedge c$, then $a = b$.

13. Prove: For all elements a, b, c in a Boolean algebra $(B, \preccurlyeq)$,
$$(a \preccurlyeq b \wedge c) \leftrightarrow (a \preccurlyeq b \text{ and } a \preccurlyeq c)$$

14. Complete the proof of Theorem 8.22 by showing that a lattice satisfies the commutative laws, the associative laws, the idempotent laws, and the absorption law $(a \wedge b) \vee a = a$.

★15. Does there exist a nondistributive lattice with five elements that has a 0 and 1, other than the one of Figure 8.4(b)?

16. Give an example of a lattice $(L, \preccurlyeq)$ in which some elements possess more than one complement.

17. Prove: If $(L, \preccurlyeq)$ is a distributive lattice, then each $a \in L$ has at most one complement. Thus every element of a Boolean algebra has a unique complement.

18. Show that the lattice whose Hasse diagram is given in Figure 8.6 is distributive and not complemented.

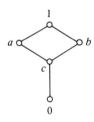

Figure 8.6

19. We defined a Boolean algebra as a special kind of lattice. As mentioned in the paragraph following Example 8.22, a different approach is to define a Boolean algebra as an algebraic structure $(B, \vee, \wedge)$ in which $\vee$ and $\wedge$ satisfy the properties 1 through 5 given. To show that these approaches are equivalent, define a relation $\preccurlyeq$ on B by $a \preccurlyeq b \leftrightarrow a \vee b = b$. Verify that $\preccurlyeq$ is a partial-order relation on B and that $(B, \preccurlyeq)$ is a lattice.

20. For the lattice of Example 8.20, verify the remaining distributive law: $a \vee (b \wedge c) = (a \vee b) \wedge (a \vee c)$ for all $a, b, c \in L$.

21. Prove Theorem 8.23 using the following outline: Let A be the set of all atoms of $(B, \preccurlyeq)$ and, for each $b \in B$, let $A_b = \{a \in A \mid a \preccurlyeq b\}$.
    (a) Show that A_b is nonempty if and only if $b \neq 0$.
    (b) For $b, c \in B$, show that $b \preccurlyeq c$ if and only if $A_b \subseteq A_c$.
    (c) For $b \in B$, $b \neq 0$, show that b is equal to the join of the atoms in A_b. (Hint: Let c be the join of the atoms in A_b. To show $c \preccurlyeq b$, show that $b \wedge c = c$. To show that $b \preccurlyeq c$, use part (b).)

22. If $(B, \vee, \wedge)$ is a finite Boolean algebra with m atoms, then prove that $|B| = 2^m$. (Hint: Let $C = \{c_1, c_2, \ldots, c_r\}$ and $D = \{d_1, d_2, \ldots, d_t\}$ be sets of atoms of B. It must be shown that, if $C \neq D$, then $c_1 \vee c_2 \vee \cdots \vee c_r \neq d_1 \vee d_2 \vee \cdots \vee d_t$. To do this assume, without loss of generality, that $c_1 \notin D$, and consider the elements $c_1 \wedge (c_1 \vee c_2 \vee \cdots \vee c_r)$ and $c_1 \wedge (d_1 \vee d_2 \vee \cdots \vee d_t)$.)

8.7

BOOLEAN FORMS AND SWITCHING FUNCTIONS

In this section we define the notion of an *n-variable switching function* and show that the set of all such functions forms a Boolean algebra. We also introduce the idea of a Boolean form and use it, together with Theorem 8.23, in developing the disjunctive normal form for switching functions. This turns out to be the representation of a switching function as a join of atoms. Throughout this section we assume that $B = \{0, 1\}$ and work with the Boolean algebra $(B, \vee, \wedge)$ (where 0 is the zero element and 1 is the unit element).

DEFINITION 8.18

A variable with values in B is called a *Boolean variable*. If $x_1, x_2, \ldots, x_n$ are Boolean variables, then a *Boolean form in the variables* $x_1, x_2, \ldots, x_n$ is defined recursively as follows:

1. 0 and 1 are Boolean forms.
2. $x_1, x_2, \ldots, x_n$ are Boolean forms.
3. If f and g are Boolean forms, then so are $f \vee g$, $f \wedge g$, and f'.

We denote the set of all Boolean forms in $x_1, x_2, \ldots, x_n$ by $\mathscr{B}_n$.

In other words, $\mathcal{B}_n$ consists of the elements $0, 1, x_1, x_2, \ldots, x_n$, together with all Boolean forms that can be obtained from these by applying any meaningful combination of the operations $\vee$, $\wedge$, and $'$.

Example 8.24 Each of the following is a Boolean form in the variables x_1, x_2, x_3, and x_4:

(a) $(x_1 \wedge x_2 \wedge x_3' \wedge x_4) \vee (x_1 \wedge x_2' \wedge x_3 \wedge x_4)$
(b) $(x_1 \vee x_2) \wedge x_3'$
(c) $(1 \wedge x_2 \wedge x_3) \vee (x_1' \wedge x_2 \wedge x_3) \vee x_4$
(d) $1 \vee x_2$

Notice, as in (b) and (d), that the expressions are still Boolean forms in the variables x_1, x_2, x_3, x_4, even though not all of these variables are present. ∎

If $p(x_1, x_2, \ldots, x_n)$ is any Boolean form, then for any assignment of values from B to the variables $x_1, x_2, \ldots, x_n$, the form can be evaluated and has a value in B. For example, if $p(x_1, x_2, x_3) = (x_1 \wedge x_2) \vee x_3'$, then $p(0, 1, 0) = (0 \wedge 1) \vee 0' = 0 \vee 1 = 1$. In fact, one can construct a table of values for a Boolean form. The table of values for $p(x_1, x_2, x_3)$ is given in Figure 8.7. Thus we can view $p(x_1, x_2, x_3) = (x_1 \wedge x_2) \vee x_3'$ as a function from $B \times B \times B$ to B.

x_1	x_2	x_3	$(x_1 \wedge x_2) \vee x_3'$
0	0	0	1
0	0	1	0
0	1	0	1
0	1	1	0
1	0	0	1
1	0	1	0
1	1	0	1
1	1	1	1

Figure 8.7 Table for $(x_1 \wedge x_2) \vee x_3'$

In general, a Boolean form in n variables $x_1, x_2, \ldots, x_n$ can be viewed as a function $f: B^n \to B$, where B^n denotes the Cartesian product $B \times B \times \cdots \times B$ of n copies of B (that is,

$$B^n = \{(a_1, a_2, \ldots, a_n) \mid a_i \in B \text{ for } 1 \leq i \leq n\}).$$

We call f an *n-variable switching function*. For example, $f: B^2 \to B$, given by $f(x_1, x_2) = x_1 \vee (x_2 \wedge x_1')$, is a 2-variable switching function. We denote the set of all *n-variable switching functions* by $\mathcal{F}_n$.

Example 8.25 List all the 2-variable switching functions.

Solution We know that $B^2 = \{(0, 0), (0, 1), (1, 0), (1, 1)\}$, so there are 2^4 2-variable switching functions in this case. We construct a table in which the left-hand column lists the elements of B^2 in their natural binary order. (Notice that, in Figure 8.7, the triples (x_1, x_2, x_3) are listed in their natural binary order.) Then, for $0 \leq m \leq 2^4 - 1$, the function f_m corresponds to the binary

	f_0	f_1	f_2	f_3	f_4	f_5	f_6	f_7	f_8	f_9	f_{10}	f_{11}	f_{12}	f_{13}	f_{14}	f_{15}
$(0,0)$	0	1	0	1	0	1	0	1	0	1	0	1	0	1	0	1
$(0,1)$	0	0	1	1	0	0	1	1	0	0	1	1	0	0	1	1
$(1,0)$	0	0	0	0	1	1	1	1	0	0	0	0	1	1	1	1
$(1,1)$	0	0	0	0	0	0	0	0	1	1	1	1	1	1	1	1

Figure 8.8 The 2-variable switching functions

representation of m. For example, consider f_{13}. We see that $13 = 1 + 2^2 + 2^3$, and so we define $f_{13}(0,0) = 1$, $f_{13}(0,1) = 0$, $f_{13}(1,0) = 1$, and $f_{13}(1,1) = 1$. In general, if $m = a_0 + a_1 \cdot 2^1 + a_2 \cdot 2^2 + a_3 \cdot 2^3$, then $f_m(0,0) = a_0$, $f_m(0,1) = a_1$, $f_m(1,0) = a_2$, and $f_m(1,1) = a_3$. These function values are then placed in the column headed by f_m. The complete table is shown in Figure 8.8.   ■

We now define operations $\vee$, $\wedge$, and $'$ on $\mathscr{F}_n$ and show that the resulting structure $(\mathscr{F}_n, \vee, \wedge)$ is a Boolean algebra. For $f_1, f_2 \in \mathscr{F}_n$, define $f_1 \vee f_2$, $f_1 \wedge f_2$, and f_1' as follows:

1. $(f_1 \vee f_2)(x_1, x_2, \ldots, x_n) = f_1(x_1, x_2, \ldots, x_n) \vee f_2(x_1, x_2, \ldots, x_n)$
2. $(f_1 \wedge f_2)(x_1, x_2, \ldots, x_n) = f_1(x_1, x_2, \ldots, x_n) \wedge f_2(x_1, x_2, \ldots, x_n)$
3. $f_1'(x_1, x_2, \ldots, x_n) = (f_1(x_1, x_2, \ldots, x_n))'$

for all $(x_1, x_2, \ldots, x_n) \in B^n$.

Example 8.26   Using the notation of Example 8.25, determine each of the following in $\mathscr{F}_2$:

(a) f_3' (b) $f_2 \wedge f_4$ (c) $f_2 \vee f_3$ (d) $(f_4 \wedge f_2) \vee f_3'$

Solution
(a) Using the table of Figure 8.8 and the definition of f', we find $f_3'(0,0) = 1' = 0$, $f_3'(0,1) = 1' = 0$, $f_3'(1,0) = 0' = 1$, and $f_3'(1,1) = 0' = 1$. Hence $f_3' = f_{12}$.
(b) Here it is easily checked that $(f_2 \wedge f_4)(x_1, x_2) = 0$ for all $(x_1, x_2) \in B^2$. Thus $f_2 \wedge f_4 = f_0$.
(c) Here we see that $(f_2 \vee f_3)(0,0) = 1$, $(f_2 \vee f_3)(0,1) = 1$, $(f_2 \vee f_3)(1,0) = 0$, and $(f_2 \vee f_3)(1,1) = 0$. Thus $f_2 \vee f_3 = f_3$.
(d) Set $g = (f_4 \wedge f_2) \vee f_3'$. Then we have $g(0,0) = 0$, $g(0,1) = 0$, $g(1,0) = 1$, and $g(1,1) = 1$, so $g = f_{12}$.   ■

THEOREM 8.24   The structure $(\mathscr{F}_n, \vee, \wedge)$ is a Boolean algebra.

Proof   We need only verify properties 1 through 5 listed in the paragraph following Example 8.22. Let $(x_1, x_2, \ldots, x_n) \in B^n$ and set $x = (x_1, x_2, \ldots, x_n)$. Then, for $f, g, h \in \mathscr{F}_n$, we have the following:

1. $(f \vee g)(x) = f(x) \vee g(x) = g(x) \vee f(x) = (g \vee f)(x)$. So $f \vee g = g \vee f$.
2. $((f \vee g) \vee h)(x) = (f \vee g)(x) \vee h(x) = (f(x) \vee g(x)) \vee h(x) = f(x) \vee (g(x) \vee h(x)) = f(x) \vee (g \vee h)(x) = (f \vee (g \vee h))(x)$. So $(f \vee g) \vee h = f \vee (g \vee h)$.

3. $(f \wedge (g \vee h))(x) = f(x) \wedge (g \vee h)(x) = f(x) \wedge (g(x) \vee h(x)) = (f(x) \wedge g(x)) \vee (f(x) \wedge h(x)) = (f \wedge g)(x) \vee (f \wedge h)(x) = ((f \wedge g) \vee (f \wedge h))(x)$. Thus $f \wedge (g \vee h) = (f \wedge g) \vee (f \wedge h)$.

In Exercise 3 the student is asked to show that $f \wedge g = g \wedge f$, $(f \wedge g) \wedge h = f \wedge (g \wedge h)$, and $f \vee (g \wedge h) = (f \vee g) \wedge (f \vee h)$.

If we define $e_0 \colon B^n \to B$ by $e_0(x) = 0$ and $e_1 \colon B^n \to B$ by $e_1(x) = 1$, for all $x \in B$, then the following properties are easily verified for all $f \in \mathscr{F}_n$:

$$f \vee e_0 = f$$
$$f \wedge e_1 = f$$
$$f \wedge f' = e_0$$
$$f \vee f' = e_1$$

Thus e_0 is a zero element and e_1 is a unit element. Hence $(\mathscr{F}_n, \vee, \wedge)$ is a Boolean algebra. □

How many elements are there in $\mathscr{F}_n$? That is, how many functions are there from B^n to B? Recall that, if we are given sets C and D with $|C| = r$ and $|D| = s$, then the number of functions from C to D is given by s^r. Since $|B^n| = 2^n$ and $|B| = 2$, we obtain the following result.

THEOREM 8.25 The number of elements in $\mathscr{F}_n$ is 2^{2^n}.

A function $h \in \mathscr{F}_n$ is an atom provided, for any $f \in \mathscr{F}_n$,

$$e_0 \prec f \preccurlyeq h \to f = h$$

Keep in mind that $f \preccurlyeq g$ if and only if $f \vee g = g$, and $f \prec g$ if and only if $f \preccurlyeq g$ and $f \neq g$. Moreover, by Theorem 8.23 and Exercise 22 of Exercises 8.6, there are 2^n atoms in $(\mathscr{F}_n, \vee, \wedge)$. Realizing this, let's see if we can determine the four atoms of $(\mathscr{F}_2, \vee, \wedge)$. Using the notational convention established in Example 8.25, consider f_4. Since $4 = 0 \cdot 1 + 0 \cdot 2^1 + 1 \cdot 2^2 + 0 \cdot 2^3$, we see that $f_4(1,0) = 1$ and $f_4(x) = 0$ for any other $x \in B^2$. We claim that f_4 is an atom. Suppose that $f_0 \prec g \preccurlyeq f_4$ for some $g \in \mathscr{F}_2$. Then $f_4 \vee g = f_4$; in particular, if $x \in B^2$, $x \neq (1,0)$, then

$$0 = f_4(x) = (f_4 \vee g)(x) = f_4(x) \vee g(x) = 0 \vee g(x) = g(x)$$

So $g(x) = 0$. But since $g \neq f_0$, we must have $g(1,0) = 1$. Thus $g = f_4$ and we conclude that f_4 is an atom. Similarly, f_1, f_2, and f_8 are atoms of $(\mathscr{F}_2, \vee, \wedge)$; thus f_1, f_2, f_4, and f_8 must be all the atoms of $(\mathscr{F}_2, \vee, \wedge)$. In fact, one can easily see how any other nonzero $f \in \mathscr{F}_2$ is a join of two or more of f_1, f_2, f_4, f_8. For example, notice that, since $6 = 2^1 + 2^2$, $f_6 = f_2 \vee f_4$.

In the general case, we extend the notation adopted in Example 8.25. If d_r is the binary representation of 2^r (expressed as an ordered n-tuple), then define $f_{2^r}: B^n \to B$ as follows: For $x \in B^n$,

$$f_{2^r}(x) = \begin{cases} 1 & \text{if} \quad x = d_r \\ 0 & \text{if} \quad x \neq d_r \end{cases}$$

If $1 \leq m \leq 2^n - 1$ and $m = 2^{r_1} + 2^{r_2} + \cdots + 2^{r_i}$ is the binary expansion of m, then $f_m: B^n \to B$ is defined by

$$f_m(x) = (f_{2^{r_1}} \vee f_{2^{r_2}} \vee \cdots \vee f_{2^{r_i}})(x)$$

Then the functions f_{2^k}, $0 \leq k \leq 2^n - 1$, are the 2^n atoms of $(\mathscr{F}_n, \vee, \wedge)$.

As pointed out earlier, any Boolean form in n variables gives rise to an n-variable switching function. According to Theorem 8.23, each element of $\mathscr{F}_n$ is expressible as a join of atoms in $\mathscr{F}_n$. We apply this result to show how every switching function is given by a Boolean form. In order to do this, we must first describe a special type of Boolean form called a *minterm*. For any Boolean variable y, we use the notation y^* to stand for either y or y'. Then any Boolean form of the type

$$x_1^* \wedge x_2^* \wedge \cdots \wedge x_n^*$$

is called a *minterm* of $\mathscr{B}_n$. For example, $x_1' \wedge x_2 \wedge x_3$ and $x_1 \wedge x_2 \wedge x_3$ are minterms in $\mathscr{B}_3$; however, $x_1 \vee x_2$ and $x_1 \wedge (x_2 \vee x_3)$ are not.

Example 8.27   Determine all the minterms in $\mathscr{B}_2$ and evaluate each.

Solution   The minterms in $\mathscr{B}_2$ have the general form $x_1^* \wedge x_2^*$. Since there are two choices for each of x_1^* and x_2^*, we see that there are four minterms in all; in fact, they are $x_1 \wedge x_2$, $x_1 \wedge x_2'$, $x_1' \wedge x_2$, and $x_1' \wedge x_2'$.

A table of their values is given in Figure 8.9.

x_1	x_2	$x_1 \wedge x_2$	$x_1 \wedge x_2'$	$x_1' \wedge x_2$	$x_1' \wedge x_2'$
0	0	0	0	0	1
0	1	0	0	1	0
1	0	0	1	0	0
1	1	1	0	0	0

Figure 8.9 The minterms in $\mathscr{B}_2$

Notice that each minterm in $\mathscr{B}_2$ has value 1 at exactly one element of B^2. Indeed, the switching functions they represent are precisely the atoms of $(\mathscr{F}_2, \vee, \wedge)$. We see that $f_1(x_1, x_2) = x_1' \wedge x_2'$, $f_2(x_1, x_2) = x_1' \wedge x_2$, $f_4(x_1, x_2) = x_1 \wedge x_2'$, and $f_8(x_1, x_2) = x_1 \wedge x_2$. In general, each minterm in $\mathscr{B}_n$ has value 1 at exactly one element of B^n and represents an atom of $(\mathscr{F}_n, \vee, \wedge)$.

We now demonstrate, by illustration, how a switching function can be expressed as a join of minterms in $\mathscr{B}_n$. Consider the switching function $f: B^3 \to B$ given in Figure 8.10.

x_1	x_2	x_3	$f(x_1, x_2, x_3)$
0	0	0	1
0	0	1	1
0	1	0	0
0	1	1	1
1	0	0	0
1	0	1	0
1	1	0	0
1	1	1	0

Figure 8.10

We focus our attention on those triples (x_1, x_2, x_3) at which f has value 1. In seeking a Boolean form for f, we look for Boolean forms $p_1(x_1, x_2, x_3)$, $p_2(x_1, x_2, x_3)$, and $p_3(x_1, x_2, x_3)$ that satisfy the following defining conditions:

$$p_1(x_1, x_2, x_3) = \begin{cases} 1 & \text{if } (x_1, x_2, x_3) = (0,0,0) \\ 0 & \text{otherwise} \end{cases}$$

$$p_2(x_1, x_2, x_3) = \begin{cases} 1 & \text{if } (x_1, x_2, x_3) = (0,0,1) \\ 0 & \text{otherwise} \end{cases}$$

$$p_3(x_1, x_2, x_3) = \begin{cases} 1 & \text{if } (x_1, x_2, x_3) = (0,1,1) \\ 0 & \text{otherwise} \end{cases}$$

It would then follow that $f = p_1 \vee p_2 \vee p_3$.

Can the student think of valid candidates for p_1, p_2, and p_3? We just introduced the notion of a minterm, so let's try to use it. Specifically, try to find a minterm in $\mathscr{B}_3$ that has value 1 only at $(0, 0, 0)$? Since there are only eight minterms in $\mathscr{B}_3$, we could simply list them and proceed by exhaustion. However, notice that $x_1 = x_2 = x_3 = 0$ if and only if $x_1' = x_2' = x_3' = 1$, so let's try $x_1' \wedge x_2' \wedge x_3'$ and see that it works. Hence we use

$$p_1(x_1, x_2, x_3) = x_1' \wedge x_2' \wedge x_3'$$

In the same way, we find that

$$p_2(x_1, x_2, x_3) = x_1' \wedge x_2' \wedge x_3 \quad \text{and} \quad p_3(x_1, x_2, x_3) = x_1' \wedge x_2 \wedge x_3$$

So f is a join of minterms in $\mathscr{B}_3$:

$$f = (x_1' \wedge x_2' \wedge x_3') \vee (x_1' \wedge x_2' \wedge x_3) \vee (x_1' \wedge x_2 \wedge x_3)$$

These facts are summarized in the following representation theorem for switching functions.

THEOREM 8.26   **(Disjunctive Normal Form)**

Every n-variable switching function is expressible as a join of distinct minterms in $\mathscr{B}_n$. Moreover, this representation is unique up to the order of the minterms.

If $f \in \mathcal{F}_n$, then the representation of f provided by Theorem 8.26 is called the *disjunctive normal form* (*normal form*, for short) of f. It has already been illustrated that, if we are given the table of values for an n-variable switching function f, then we can obtain the normal form for f. If f is expressed as a Boolean form, then another technique can be used to determine the normal form for f. This method uses the laws of a Boolean algebra and is illustrated in the next example.

Example 8.28   Find the normal form of each of the following switching functions in $(\mathcal{F}_3, \vee, \wedge)$.

(a) $(x_1 \wedge x_2) \vee (x_1 \wedge x_3)$ (b) $x_1 \vee x_2'$

Solution   (a) We proceed as follows:

$$(x_1 \wedge x_2) \vee (x_1 \wedge x_3) =$$
$$[(x_1 \wedge x_2) \wedge (x_3 \vee x_3')] \vee [(x_1 \wedge x_3) \wedge (x_2 \vee x_2')]$$
$$= (x_1 \wedge x_2 \wedge x_3) \vee (x_1 \wedge x_2 \wedge x_3') \vee (x_1 \wedge x_2 \wedge x_3) \vee (x_1 \wedge x_2' \wedge x_3)$$
$$= (x_1 \wedge x_2 \wedge x_3) \vee (x_1 \wedge x_2 \wedge x_3') \vee (x_1 \wedge x_2' \wedge x_3)$$

(b) For this switching function we have the following:

$$x_1 \vee x_2' = [x_1 \wedge (x_2 \vee x_2') \wedge (x_3 \vee x_3')] \vee [x_2' \wedge (x_1 \vee x_1') \wedge (x_3 \vee x_3')]$$
$$= (x_1 \wedge x_2 \wedge x_3) \vee (x_1 \wedge x_2 \wedge x_3') \vee (x_1 \wedge x_2' \wedge x_3) \vee$$
$$(x_1 \wedge x_2' \wedge x_3') \vee (x_1 \wedge x_2' \wedge x_3) \vee (x_1 \wedge x_2' \wedge x_3') \vee$$
$$(x_1' \wedge x_2' \wedge x_3) \vee (x_1' \wedge x_2' \wedge x_3')$$
$$= (x_1 \wedge x_2 \wedge x_3) \vee (x_1 \wedge x_2 \wedge x_3') \vee (x_1 \wedge x_2' \wedge x_3) \vee$$
$$(x_1 \wedge x_2' \wedge x_3') \vee (x_1' \wedge x_2' \wedge x_3) \vee (x_1' \wedge x_2' \wedge x_3') \qquad \blacksquare$$

Observe that the expression $x_1 \vee x_2'$ is much simpler than its normal form. Indeed, in applying the methods of this section to the design of switching networks, we are concerned with obtaining a more simplified form than that given by the normal form.

Given the table of values of a switching function f, one can use it to determine the normal form of f and then use the laws of Boolean algebra to simplify the normal form.

We have seen that there is associated with each n-variable switching function a Boolean form in n Boolean variables and, conversely, that each Boolean form in n Boolean variables gives rise to an n-variable switching function. Thus the two concepts are equivalent and we shall draw no essential distinction between them. For example, each has a uniquely determined normal form.

Exercises 8.7

1. For each of the following, find a minterm in $\mathscr{B}_4$ that has value 1 only for the combination of values given.
   (a) $x_1 = 0$, $x_2 = 1$, $x_3 = 1$, $x_4 = 0$
   (b) $x_1 = 1$, $x_2 = 0$, $x_3 = 0$, $x_4 = 1$
   (c) $x_1 = 1$, $x_2 = 0$, $x_3 = 1$, $x_4 = 1$
   (d) $x_1 = 0$, $x_2 = 0$, $x_3 = 1$, $x_4 = 1$

2. Exhibit the table of values for each of the following switching functions. In each case, use the table to express the function in normal form.
   (a) $f(x_1, x_2, x_3) = (x_1 \wedge x_2') \vee x_3$
   (b) $f(x_1, x_2, x_3) = (x_1 \vee x_3')' \vee [(x_1 \wedge x_2') \vee (x_1 \wedge x_3)]$
   (c) $f(x_1, x_2, x_3) = (x_1 \vee x_2)' \vee (x_3 \vee x_1) \wedge [(x_2 \wedge x_3) \vee (x_2' \wedge x_3')]$

3. If $f, g, h \in \mathscr{F}_n$, then prove the following:
   (a) $f \wedge g = g \wedge f$ (b) $f \wedge (g \wedge h) = (f \wedge g) \wedge h$
   (c) $f \vee (g \wedge h) = (f \vee g) \wedge (f \vee h)$

4. Use the technique of Example 8.28 to find the normal form of each of the following switching functions in $(\mathscr{F}_3, \vee, \wedge)$:
   (a) $(x_1 \wedge x_2) \vee (x_1' \wedge x_2 \wedge x_3')$
   (b) $(x_1 \vee x_2 \vee x_3) \wedge (x_1' \vee x_2 \vee x_3') \wedge (x_1' \vee x_2 \vee x_3)$
   (c) $(x_1 \vee x_3) \wedge x_2$ (d) $x_1 \wedge x_3'$ (e) x_2

5. Determine the normal form for each of the switching functions $f, g, h \in \mathscr{F}_3$ given by the table in Figure 8.11. Use the laws of Boolean algebra to simplify these forms.

x_1	x_2	x_3	f	g	h
0	0	0	0	0	0
0	0	1	1	0	1
0	1	0	1	0	1
0	1	1	0	1	1
1	0	0	1	0	1
1	0	1	0	1	1
1	1	0	0	1	1
1	1	1	1	1	0

Figure 8.11

6. Determine the normal form for each of the following Boolean forms in B_4.
   (a) $(x_1 \vee x_3) \wedge x_2 \wedge x_4'$
   (b) $(x_1 \wedge x_3' \wedge x_4) \vee (x_1 \wedge x_2 \wedge x_3 \wedge x_4)$

7. Prove that $f_m' = f_{15-m}$ for all $f_m \in \mathscr{F}_2$.

8. If $p(x_1, x_2, \ldots, x_n)$ and $q(x_1, x_2, \ldots, x_n)$ are distinct minterms in $\mathscr{B}_n$, prove that $p(x_1, x_2, \ldots, x_n) \wedge q(x_1, x_2, \ldots, x_n) = 0$.

9. How many distinct minterms are there in $\mathscr{B}_n$? What is the value of the meet of all the minterms in $\mathscr{B}_n$?

10. Prove: If $f, g \in \mathscr{F}_n$, then $f = g$ if and only if f and g have the same normal form.

8.8 SWITCHING CIRCUITS AND LOGIC GATES

Boolean algebra can be used in a very neat and clear way to analyze special types of electrical circuits. To explain the nature of such circuits, we must first agree on terminology. To begin with, a *switch* is a device having two possible states, *open* or *closed*. Current flows through the point at which a switch is located if and only if the switch is closed. Figure 8.12 shows a diagram of a circuit containing a single switch *a*. The points *s* and *t* are called *terminals* of the circuit, and it is clear that current flows from *s* to *t* if and only if switch *a* is closed.

Figure 8.12 A circuit with one switch.

In general, we consider circuits that contain only switches and two terminals *s* and *t* (and wire, of course); such circuits are called *switching circuits*. A switching circuit is said to be *closed* if current can pass from *s* to *t*, and it is *open* otherwise. Given a switching circuit, we are interested in those combinations of switching states for which the circuit is closed.

For example, consider the circuit of Figure 8.13. It seems evident that this circuit is closed if and only if either *a* or *b* is closed. This combination of switches *a* and *b* is denoted (strangely enough) by $a \vee b$. Moreover, switches *a* and *b* are said to be in *parallel*.

A contrasting situation is provided by the circuit of Figure 8.14. In this case, it is easy to see that the circuit is closed if and only if both *a* and *b* are closed. Here *a* and *b* are said to be in *series*, and the circuit is denoted by $a \wedge b$.

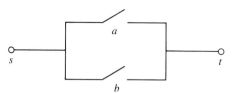

Figure 8.13 A parallel circuit.

Figure 8.14 A series circuit

Starting with a set *S* of switches, we can consider the set *C* of all switching circuits that can be constructed using combinations of series and parallel circuits. We call such circuits *series-parallel circuits*. (Keep in mind that $S \subseteq C$.) In effect, we have now defined two binary operations $\vee$ and

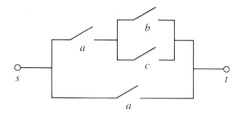

Figure 8.15 The circuit $[a \wedge (b \vee c)] \vee a$

$\wedge$ on C and, corresponding to any meaningful expression involving $\vee$, $\wedge$, and the elements of S, there is a uniquely determined series-parallel circuit. Conversely, each series-parallel circuit determines such an expression uniquely. Consequently, we shall draw no distinction between a given series-parallel circuit and its associated expression. Thus, for example, the circuit that is displayed in Figure 8.15 can also be referred to as $[a \wedge (b \vee c)] \vee a$.

For each switch $a \in S$, we assume there is a switch $a' \in S$ that is open when a is closed and closed when a is open. The switch a' is called the *complement* of a.

Consider the circuits $a \wedge a'$ and $a \vee a'$ displayed in Figure 8.16. The circuit in part (a) is always open, whereas the circuit in part (b) is always closed. Should we adopt special notations for these circuits? Indeed we denote a circuit that is always open by 0 and one that is always closed by 1.

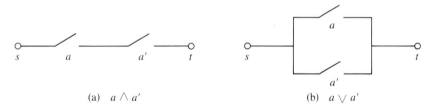

(a) $a \wedge a'$ (b) $a \vee a'$

Figure 8.16

To this end we write

$$a \wedge a' = 0 \quad \text{and} \quad a \vee a' = 1$$

In general, two circuits C_1 and C_2 are called *equal*, denoted $C_1 = C_2$, provided the switch positions that close C_1 are the same as those that close C_2. With this definition of equality, we can legitimately consider the algebraic structure $(C, \vee, \wedge)$. It turns out that $(C, \vee, \wedge)$ is a Boolean algebra. The zero and unit elements are the circuits 0 and 1, respectively, just defined. To see that the distributive law

$$a \vee (b \wedge c) = (a \vee b) \wedge (a \vee c)$$

holds, for instance, consider the circuits in Figure 8.17. A 4-minute perusal

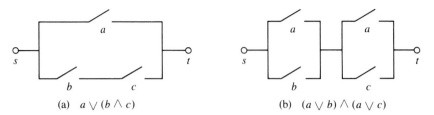

(a) $a \lor (b \land c)$ (b) $(a \lor b) \land (a \lor c)$

Figure 8.17 Circuits illustrating a distributive law

should be sufficient to determine that each circuit is closed if and only if either a is closed or both b and c are closed.

With the knowledge that $(C, \lor, \land)$ is a Boolean algebra, we can now apply the properties of a Boolean algebra to a given circuit, possibly simplifying it and saving considerably on switches and wire.

Example 8.29 Simplify the circuit of Figure 8.18.

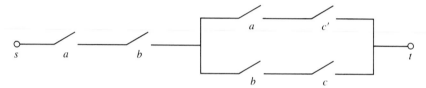

Figure 8.18

Solution   This circuit is given by the expression

$$a \land b \land [(a \land c') \lor (b \land c)]$$

Using the properties of a Boolean algebra, we obtain

$$a \land b \land [(a \land c') \lor (b \land c)]$$
$$= [(a \land b) \land (a \land c')] \lor [(a \land b) \land (b \land c)]$$
$$= [(a \land a) \land (b \land c')] \lor [(a \land (b \land b)) \land c]$$
$$= [a \land (b \land c')] \lor [(a \land b) \land c]$$
$$= [(a \land b) \land c'] \lor [(a \land b) \land c]$$
$$= (a \land b) \land (c' \lor c)$$
$$= (a \land b) \land 1$$
$$= a \land b$$

Thus we may replace the given circuit by the circuit $a \land b$.   ■

In current-day usage, electrical switches, as basic elements in the design of switching networks, have been replaced by other basic elements called *logic gates*. Switching circuits that employ logic gates are referred to as *logic gate circuits* or, more simply, *lg-circuits*. Logic gates are used to implement the Boolean operations $\vee$ (join), $\wedge$ (meet), and $'$ (complementation) and are presented in the next definition. In general, given a Boolean form $f(x_1, x_2, \ldots, x_n)$ for an lg-circuit, we refer to $x_1, x_2, \ldots, x_n$ as the *inputs* of the circuit and $f(x_1, x_2, \ldots, x_n)$ as the *output expression* (or simply, *output*) for the circuit.

DEFINITION 8.19

Let x and y be inputs.

1. The *OR gate*, symbolized in Figure 8.19(a), represents the lg-circuit with output $x \vee y$.
2. The *AND gate*, symbolized in Figure 8.19(b), represents the lg-circuit with output $x \wedge y$.
3. The *NOT gate* (or *inverter*), symbolized in Figure 8.19(c), represents the lg-circuit with output x'.

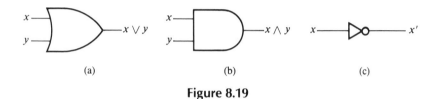

(a)   (b)   (c)

Figure 8.19

The AND gate and the OR gate can be extended to any finite number of inputs. Thus, for example, we can represent the lg-circuit with output expression $x_1 \vee x_2 \vee x_3 \vee x_4$ by the OR gate of Figure 8.20.

More generally, the AND and OR gates for n inputs are represented as shown in Figure 8.21, parts (a) and (b).

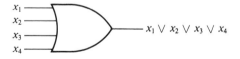

Figure 8.20 OR gate for $x_1 \vee x_2 \vee x_3 \vee x_4$

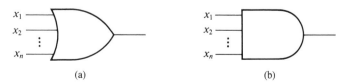

(a)   (b)

Figure 8.21 (a) OR gate for n inputs (b) AND gate for n inputs

Some very basic lg-circuits involving the NOT gate are exhibited in Figure 8.22, parts (a), (b), (c), and (d). In subsequent usage we adopt the

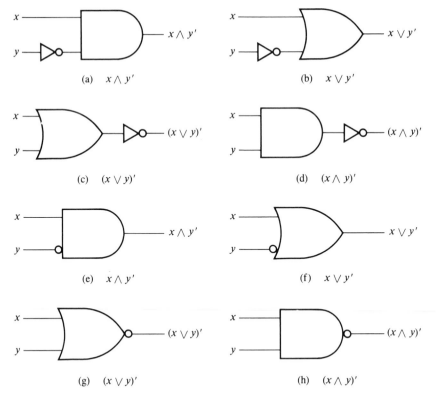

Figure 8.22

convenient shorthand notations for these depicted in Figure 8.22, parts (e), (f), (g), and (h), respectively.

The lg-circuit in part (c) is referred to as a *NOR gate*, while that in part (d) is called a *NAND gate*.

Consider the switching function $f(x, y, z) = [x \wedge (y \vee z)] \vee x$ that is shown in Figure 8.15. We use a combination of OR gates and AND gates to reproduce an associated lg-circuit. It is shown in Figure 8.23(a). In Figure 8.23(b) we describe the same lg-circuit without a repetition of the input x. Both an AND gate and an OR gate have the input x in common. The use of one form over the other is dictated strictly by convenience.

Example 8.30   Use a combination of AND, OR, and NOT gates to produce an lg-circuit for each of the following switching functions.

(a) $f(x, y, z) = (x' \wedge y)' \vee z$
(b) $g(x, y, z) = (x \wedge y') \vee (x' \wedge z) \vee (x \wedge y \wedge z)$

Solution   (a) The lg-circuit for $f(x, y, z)$ is shown in Figure 8.24(a).
(b) The lg-circuit for $g(x, y, z)$ is shown in Figure 8.24(b).   ∎

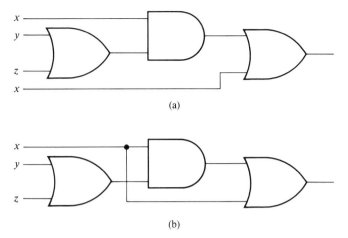

Figure 8.23 lg-circuit for $[x \wedge (y \vee z)] \vee x$

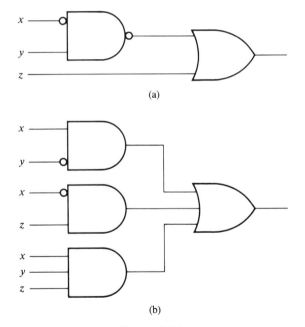

Figure 8.24

Example 8.31   Design an lg-circuit that allows four people to vote yes or no, and that has output 1 if and only if at least three of the four vote yes.

Solution   Let x, y, z, w denote inputs corresponding to the four votes cast. We associate the value 1 with a yes vote. Hence we must design an lg-circuit that has output 1 if and only if at least three of x, y, z, w have value 1. The normal form for the associated 4-variable switching function with these

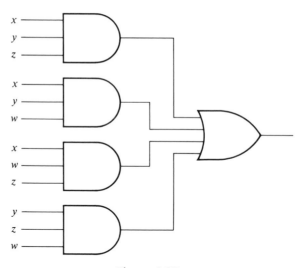

Figure 8.25

values is easily seen to be

$$f(x, y, z, w) = (x \wedge y \wedge z \wedge w') \vee (x \wedge y \wedge z' \wedge w) \vee$$
$$(x \wedge y' \wedge z \wedge w) \vee (x' \wedge y \wedge z \wedge w) \vee (x \wedge y \wedge z \wedge w)$$

Using the laws of Boolean algebra, the student can check (see Exercise 7) that, in fact,

$$f(x, y, z, w) = (x \wedge y \wedge z) \vee (x \wedge y \wedge w) \vee$$
$$(x \wedge w \wedge z) \vee (y \wedge w \wedge z)$$

The lg-circuit associated with this representation is given in Figure 8.25.   ∎

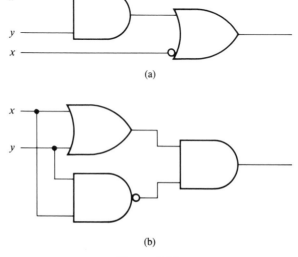

(a)

(b)

Figure 8.26

Example 8.32 Find the output of each of the lg-circuits exhibited in Figure 8.26.

Solution   (a) The output of the circuit in Figure 8.26(a) is

$$(x \wedge y) \vee x'$$

(b) The output of the circuit in Figure 8.26(b) is

$$(x \vee y) \wedge (x \wedge y)'$$

Exercises 8.8

1. Simplify each of the switching circuits of Figure 8.27.

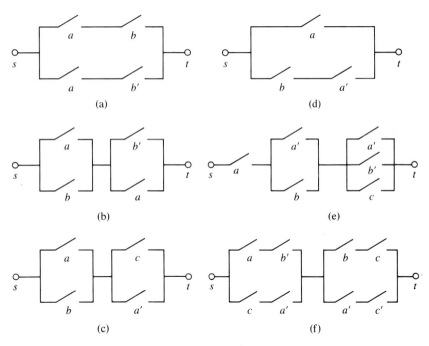

Figure 8.27

2. Draw and simplify the switching circuit

$$(a \vee b) \wedge (a \vee c) \wedge (b \vee c)$$

3. Consider the switching circuit $a' \vee (b \wedge c) \vee (b \wedge c')$.
     (a) Draw this switching circuit.
     (b) Use the properties of Boolean algebra to simplify the circuit.
4. Draw the lg-circuit for each of the following outputs.
     (a) $(x \vee y \vee z) \wedge (x' \vee y \vee z) \wedge (x' \vee y' \vee z)$
     (b) $(x \wedge y' \wedge z) \vee (x' \wedge y) \vee (x' \wedge z)$
     (c) $([(x \wedge y)' \wedge z \wedge w] \vee [(x \vee y)' \vee w])'$

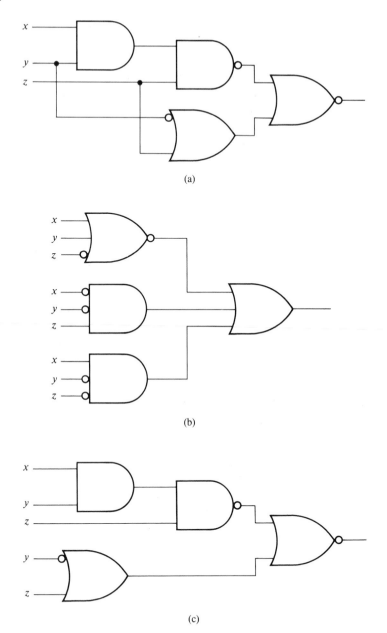

(a)

(b)

(c)

Figure 8.28

5. Find the output of each of the lg-circuits in Figure 8.28.
6. Design an lg-circuit whose output is 1 if and only if:
   (a) Exactly one of the inputs x and y has value 1.
   (b) At least one of x and y has value 1.
   (c) Exactly two of x, y, and z have value 1.
   (d) At most two of x, y, and z have value 1.

7. Prove that $f(x, y, z, w) = (x \wedge y \wedge z) \vee (x \wedge y \wedge w) \vee (x \wedge w \wedge z) \vee (y \wedge w \wedge z)$ for the switching function f of Example 8.31.

8. Design an lg-circuit for a garage light so that the light can be controlled from the house or the garage. In other words, this light will have two switches, one in the house and one in the garage, and it must be possible to turn the light on or off from either location.

9. Farmer Joe has a barn light that can be switched on or off from either the barn, the garage, or the house. Design an lg-circuit to represent this situation.

10. Suppose that in Exercise 9 the switch in the barn is a "master switch": when it is off the light is off, and when it is on the light may be controlled from either the house or the garage. Design an lg-circuit for this situation.

11. Design an lg-circuit that allows five people to vote yes or no and that has output 1 if and only if a majority vote yes.

12. Use only NAND gates to design lg-circuits with the following outputs:
    (a) $(x \wedge y)$ **(b)** $(x \vee y)$ **(c)** $(x \vee y) \wedge (x \vee z')$

13. Redesign the following lg-circuits so that each uses only NOR gates:
    (a) $(x \vee y)$ **(b)** $(x \wedge y)$

8.9 ISOMORPHISMS

Recall that the dihedral group $(D_4, \circ)$ consists of eight mappings of the square T with vertices $A(1, -1)$, $B(1, 1)$, $C(-1, 1)$, and $D(-1, -1)$. Recall that these mappings are called symmetries of T. Consider the cyclic subgroup $(\langle \rho \rangle, \circ)$ of $(D_4, \circ)$ generated by ρ, the clockwise rotation about $(0, 0)$ through $90°$. The operation table for this subgroup is displayed in Figure 8.29.

$\circ$	ι	ρ	ρ^2	ρ^3
ι	ι	ρ	ρ^2	ρ^3
ρ	ρ	ρ^2	ρ^3	ι
ρ^2	ρ^2	ρ^3	ι	ρ
ρ^3	ρ^3	ι	ρ	ρ^2

Figure 8.29 Operation table for $(\langle \rho \rangle, \circ)$

Next consider the group $(\mathbb{Z}_4, \oplus)$; its operation table is shown in Figure 8.30.

The two groups $(\langle \rho \rangle, \circ)$ and $(\mathbb{Z}_4, \oplus)$ are different, yet the student no doubt notices a similarity between the two operation tables. How can we make this resemblance more precise? It is tempting to say that ι corresponds to $[0]$, ρ to $[1]$, ρ^2 to $[2]$, and ρ^3 to $[3]$. Moreover, as is evident from the tables, such a correspondence is preserved by the operations $\circ$ and $\oplus$; that is, if $\rho^i \circ \rho^j = \rho^k$ in $\langle \rho \rangle$, then $[i] \oplus [j] = [k]$ in $\mathbb{Z}_4$. For instance, notice that $\rho^2 \circ \rho^3 = \rho$ and $[2] \oplus [3] = [1]$. This correspondence is just a

$\oplus$	[0]	[1]	[2]	[3]
[0]	[0]	[1]	[2]	[3]
[1]	[1]	[2]	[3]	[0]
[2]	[2]	[3]	[0]	[1]
[3]	[3]	[0]	[1]	[2]

Figure 8.30 Operation table for $(\mathbb{Z}_4, \oplus)$

function, so what we are saying is that we have a function $f: \langle \rho \rangle \to \mathbb{Z}_4$, defined by $f(\rho^i) = [i]$ for $i = 0, 1, 2, 3$. Moreover, the function f satisfies the interesting property

$$f(\rho^i \circ \rho^j) = f(\rho^i) \oplus f(\rho^j)$$

(At first glance this property may seem confusing, but keep in mind that $f(\rho^i)$ and $f(\rho^j)$ are elements of $\mathbb{Z}_4$, so the expression $f(\rho^i) \oplus f(\rho^j)$ has the proper meaning.)

Therefore, even though the two groups $(\langle \rho \rangle, \circ)$ and $(\mathbb{Z}_4, \oplus)$ are not equal, they are the same in the sense we have just demonstrated. We say that $(\langle \rho \rangle, \circ)$ is "isomorphic" to $(\mathbb{Z}_4, \oplus)$.

DEFINITION 8.20

Given groups $(G, *)$ and $(H, \#)$, a function $f: G \to H$ is called an *isomorphism* provided the following conditions hold:

1. f is one-to-one and onto.
2. $f(a * b) = f(a)\#f(b)$ for all $a, b \in G$.

We say that $(G, *)$ *is isomorphic to* $(H, \#)$ and write $(G, *) \simeq (H, \#)$.

Example 8.33   Consider the function $f: (\mathbb{R}, +) \to (\mathbb{R}^+, \cdot)$, defined by $f(x) = 2^x$. (Recall that $\mathbb{R}^+$ denotes the set of positive reals.) Show that f is an isomorphism.

Solution   The function f is one-to-one since $f(x_1) = f(x_2)$ implies that $2^{x_1} = 2^{x_2}$, which implies that $x_1 = x_2$. Also f is onto, since, for any $y \in \mathbb{R}^+$, there is an $x \in \mathbb{R}$ such that $2^x = y$, namely, $\log_2 y$. Finally, for $a, b \in \mathbb{R}$, we see that $f(a + b) = 2^{a+b} = 2^a \cdot 2^b = f(a) \cdot f(b)$. Therefore f is an isomorphism, and the groups $(\mathbb{R}, +)$ and $(\mathbb{R}^+, \cdot)$ are isomorphic. ∎

Example 8.34   Determine which of the following functions are isomorphisms.

(a) $f_1: (\mathbb{Z}, +) \to (\mathbb{Z}, +)$; $f_1(m) = m + 1$
(b) $f_2: (\mathbb{Z}, +) \to (\mathbb{Z}_7, \oplus)$; $f_2(m) = [m]$
(c) $f_3: (\mathbb{Z}_6, \oplus) \to (\mathbb{Z}_6, \oplus)$; $f_3([a]) = [5a]$
(d) $f_4: (\mathbb{Z}, +) \to (\mathbb{Q}, +)$; $f_4(m) = m$

Solution (a) The function f_1 is clearly one-to-one and onto. However, for m, $n \in \mathbb{Z}$, we see that $f_1(m + n) = m + n + 1$, while $f_1(m) + f_1(n) = (m + 1) + (n + 1) = m + n + 2$. Thus $f_1(m + n) \neq f_1(m) + f_1(n)$, and so f_1 is not an isomorphism. (Clearly, however, every group is isomorphic to itself, the identity function being an isomorphism. This points out the distinction between determining whether two given groups are isomorphic and verifying whether a given function is an isomorphism.)

(b) The function f_2 is not one-to-one since, for instance, $f_2(0) = f_2(7)$. Hence f_2 is not an isomorphism. Also since $\mathbb{Z}_7$ has only seven elements, any function from $\mathbb{Z}$ to $\mathbb{Z}_7$ fails to be one-to-one. So we can, in fact, conclude that the groups $(\mathbb{Z}, +)$ and $(\mathbb{Z}_7, \oplus)$ are not isomorphic.

(c) We first show that f_3 is one-to-one. Suppose that $f_3([a]) = f_3([b])$ for some $[a], [b] \in \mathbb{Z}_6$. Then we have

$$[5a] = [5b] \rightarrow 6|(5a - 5b)$$
$$\rightarrow 6|5(a - b)$$
$$\rightarrow 6|(a - b) \qquad \text{since } \gcd(5,6) = 1$$

so $[a] = [b]$. Thus f_3 is one-to-one. Since $\mathbb{Z}_6$ is finite, Theorem 5.1 tells us that f_3 is also onto. In addition, $f_3([a] \oplus [b]) = f_3([a + b]) = [5(a + b)] = [5a + 5b] = [5a] \oplus [5b] = f_3([a]) \oplus f_3([b])$. Hence f_3 is an isomorphism.

(d) Here im $f = \mathbb{Z}$, so f_4 is clearly not onto. (In the Chapter Problems the student is asked to show that the groups $(\mathbb{Z}, +)$ and $(\mathbb{Q}, +)$ are not isomorphic.) ∎

We have seen that the groups $(\langle \rho \rangle, \circ)$ and $(\mathbb{Z}_4, \oplus)$ are isomorphic. It is sometimes said that these groups "are the same, up to isomorphism." A central problem in group theory is to determine, for a given positive integer n, all nonisomorphic groups of order n. For example, the student is asked in Exercises 14 and 15 to show that there are exactly two nonisomorphic groups of order 4 (namely, the group $(\mathbb{Z}_4, \oplus)$ and the Klein four-group of Exercise 12 in Exercises 8.2).

We can compare two algebraic structures of other types, semigroups or rings, for example, and in each case we can define the concept of isomorphism in an appropriate manner. In each setting, the notion of isomorphism is used to determine when two structures are essentially the same.

Consider the function $f_2: \mathbb{Z} \rightarrow \mathbb{Z}_7$ of Example 8.34(b), where $f(m) = [m]$. As was mentioned, f_2 is not one-to-one; however, we do have

$$f_2(m + n) = [m + n]$$
$$= [m] \oplus [n]$$
$$= f_2(m) \oplus f_2(n)$$

So f_2 does satisfy property 2 of Definition 8.20 of isomorphism for groups. Thus, even though $(\mathbb{Z}, +)$ is not isomorphic to $(\mathbb{Z}_7, \oplus)$, the two groups are related in some sense. We shall call f_2 a *homomorphism*.

DEFINITION 8.21

Given groups $(G, *)$ and $(H, \#)$, a function $h\colon G \to H$ is called a *homomorphism* provided $h(a * b) = h(a) \# h(b)$ for all $a, b \in G$.

Example 8.35   In this example we wish to exhibit a homomorphism from the group $(\mathbb{Z}_{12}, \oplus)$ to the group $(\mathbb{Z}_4, \oplus)$. To avoid confusion, we use the notation $[x]_m$ to denote the element $[x]$ in the group $\mathbb{Z}_m$.

Define $h\colon \mathbb{Z}_{12} \to \mathbb{Z}_4$ by $h([x]_{12}) = [x]_4$. Then

$$h([0]_{12}) = h([4]_{12}) = h([8]_{12}) = [0]_4$$
$$h([1]_{12}) = h([5]_{12}) = h([9]_{12}) = [1]_4$$
$$h([2]_{12}) = h([6]_{12}) = h([10]_{12}) = [2]_4$$
$$h([3]_{12}) = h([7]_{12}) = h([11]_{12}) = [3]_4$$

The fact that h is a homomorphism can be seen from the following:

$$h([a]_{12} \oplus [b]_{12}) = h([a + b]_{12})$$
$$= [a + b]_4$$
$$= [a]_4 \oplus [b]_4$$
$$= h([a]_{12}) \oplus h([b]_{12}) \qquad \blacksquare$$

We do not attempt to develop any further the theory of isomorphisms or homomorphisms, leaving this for a course in abstract algebra. It suffices for our purposes to address some of the elementary properties in the exercises and Chapter Problems.

Before leaving this topic, however, we would like to consider the notion of isomorphism for Boolean algebras. Using this idea, we can present a very important and fundamental isomorphism theorem for Boolean algebras.

DEFINITION 8.22

Let $(B_1, \vee, \wedge)$ and $(B_2, \vee', \wedge')$ be Boolean algebras. A function $\theta\colon B_1 \to B_2$ is called an *isomorphism* provided the following conditions hold:

1. θ is one-to-one and onto.
2. $\theta(a \vee b) = \theta(a) \vee' \theta(b)$
3. $\theta(a \wedge b) = \theta(a) \wedge' \theta(b)$

for all $a, b \in B_1$. We write $(B_1, \vee, \wedge) \simeq (B_2, \vee', \wedge')$.

Example 8.36   Let $(L, \vee, \wedge)$ be the Boolean algebra of positive divisors of 30 and let $(\mathscr{P}(A), \cup, \cap)$ be the Boolean algebra of all subsets of the set $A = \{a, b, c\}$. Prove that $(L, \vee, \wedge) \simeq (\mathscr{P}(A), \cup, \cap)$.

Solution   We must define a mapping $\theta: L \to \mathscr{P}(A)$ that satisfies the conditions of Definition 8.22. It would seem quite natural to expect that, if t is an atom of L, then $\theta(t)$ should be an atom of $\mathscr{P}(A)$. The atoms of L are 2, 3, and 5, while those of $\mathscr{P}(A)$ are $\{a\}$, $\{b\}$, and $\{c\}$. Thus we begin by defining $\theta(2) = \{a\}$, $\theta(3) = \{b\}$, and $\theta(5) = \{c\}$. Conditions 2 and 3 of the definition require that

$$\text{(i)} \quad \theta(x \vee y) = \theta(x) \cup \theta(y)$$
$$\text{(ii)} \quad \theta(x \wedge y) = \theta(x) \cap \theta(y)$$

for all $x, y \in L$. These conditions require that

$$\theta(6) = \theta(2 \vee 3) = \theta(2) \cup \theta(3) = \{a\} \cup \{b\} = \{a, b\}$$
$$\theta(10) = \theta(2 \vee 5) = \{a, c\}$$
$$\theta(15) = \theta(3 \vee 5) = \{b, c\}$$
$$\theta(30) = \theta(2 \vee 15) = \{a\} \cup \{b, c\} = \{a, b, c\}$$
$$\theta(1) = \theta(2 \wedge 3) = \theta(2) \cap \theta(3) = \emptyset$$

The mapping θ is clearly one-to-one and onto. We must still prove that θ satisfies conditions (i) and (ii) above. To show this, it suffices to consider expressions of the following types:

$1 \vee x$	$x \wedge 1$	$x \vee x$	$x \wedge x$	$x \vee 30$	$x \wedge 30$
$p \vee q$	$p \wedge q$	$p \vee qr$	$p \wedge qr$	$pq \vee qr$	$pq \wedge qr$

where p, q, and r are distinct prime factors of 30 and x is any element of L. We have already established that $\theta(p \vee q) = \theta(p) \cup \theta(q)$. Consider $x \wedge 30$ and $p \vee qr$. For $x \wedge 30$ we see that

$$\theta(x \wedge 30) = \theta(x) = \theta(x) \cap A = \theta(x) \cap \theta(30)$$

and for $p \vee qr$ we have

$$\begin{aligned}
\theta(p \vee qr) = \theta(30) = A &= \{a, b, c\} = \{a\} \cup \{b\} \cup \{c\} \\
&= \theta(p) \cup [\theta(q) \cup \theta(r)] \\
&= \theta(p) \cup \theta(q \vee r) \\
&= \theta(p) \cup \theta(qr)
\end{aligned}$$

The remaining cases are handled in a similar fashion and are left for the student. Thus it follows that $(L, \vee, \wedge) \simeq (\mathscr{P}(A), \cup, \cap)$.   ∎

Example 8.36 shows that the finite Boolean algebra $(L, \vee, \wedge)$ is isomorphic to $(\mathscr{P}(A), \cup, \cap)$. Is every finite Boolean algebra isomorphic to the Boolean algebra of subsets of some finite set A? The answer to this question is yes, and the general result is known as the *fundamental isomorphism theorem for Boolean algebras*.

THEOREM 8.27   If $(B, \vee, \wedge)$ is a finite Boolean algebra, then $(B, \vee, \wedge)$ is isomorphic to the Boolean algebra $(\mathscr{P}(A), \cup, \cap)$ of subsets of some finite set A.

In other words, Theorem 8.27 states that if $(B, \vee, \wedge)$ is a finite Boolean algebra, then, up to isomorphism, $(B, \vee, \wedge)$ is nothing more than the Boolean algebra of all subsets of a set with m elements, for some nonnegative integer m. Recall that, if A is a set with m elements, then the power set $\mathscr{P}(A)$ consists of 2^m elements. So we have the following consequence of Theorem 8.27 (see Exercise 22 of Exercises 8.6).

COROLLARY 8.28   If $(B, \vee, \wedge)$ is a finite Boolean algebra, then $|B| = 2^m$ for some nonnegative integer m.

Exercises 8.9

1. Let $(G, *)$ and $(H, \#)$ be groups with respective identities e_G and e_H, and let $f: G \to H$ be a homomorphism. For any $n \in \mathbb{N}$ and any $a \in G$, prove each of the following properties:
   (a) $f(e_G) = e_H$ (b) $f(a^{-1}) = f(a)^{-1}$
   (c) $f(a^n) = [f(a)]^n$ (d) $f(a^{-n}) = [f(a)]^{-n}$
2. Show that any function $f: (\mathbb{Z}_6, \oplus) \to (\mathbb{Z}_4, \oplus)$ with $f([1]) = [1]$ cannot be a homomorphism.
3. Let $(G, *)$ be a group.
   (a) If $f: (\mathbb{Z}_m, \oplus) \to (G, *)$ is a homomorphism, then show that f is completely determined by $f([1])$.
   (b) If $g: (\mathbb{Z}, +) \to (G, *)$ is a homomorphism, then show that g is completely determined by $g(1)$.
4. Verify that $f: (\mathbb{Z}_4, \oplus) \to (\mathbb{Z}_5^\#, \odot)$, determined by $f([1]) = [2]$, is an isomorphism.
5. In each of the following cases, find the homomorphism h determined by the given information.
   (a) $h: (\mathbb{Z}_{12}, \oplus) \to (\mathbb{Z}_8, \oplus)$, with $h([1]) = [2]$
   (b) $h: (\mathbb{Z}_{12}, \oplus) \to (\mathbb{Z}_8, \oplus)$, with $h([1]) = [4]$
   (c) $h: (\mathbb{Z}, +) \to (\mathbb{Z}, +)$, with $h(1) = 2$
   (d) $h: (\mathbb{Z}, +) \to (\mathbb{Z}_8, \oplus)$, with $h(1) = [2]$
6. Let H be the set of rational numbers of the form 2^m, where $m \in \mathbb{Z}$; that is,

$$h = \left\{ \ldots, \tfrac{1}{4}, \tfrac{1}{2}, 1, 2, 4, \ldots \right\}$$

   (a) Show that $(H, \cdot)$ is a subgroup of $(\mathbb{Q}^+, \cdot)$.
   (b) Show that $(H, \cdot)$ is isomorphic to $(\mathbb{Z}, +)$.

7. Recall that the elements δ and γ of the symmetric group S_3 are given by

$$\begin{array}{ll} \delta: & 1 \to 2 \qquad \gamma: \quad 1 \to 2 \\ & 2 \to 3 \qquad \qquad \;\; 2 \to 1 \\ & 3 \to 1 \qquad \qquad \;\; 3 \to 3 \end{array}$$

   (a) Find the homomorphism $f: \mathbb{Z}_6 \to S_3$ such that $f([1]) = \delta$.
   (b) Is there a homomorphism $g: S_3 \to \mathbb{Z}_6$ with $g(\delta) = [2]$ and $g(\gamma) = [0]$?
   (c) Is there a homomorphism $h: S_3 \to \mathbb{Z}_6$ with $h(\delta) = [0]$ and $h(\gamma) = [3]$?

8. Let M_k denote the set of integer multiples of $k \in \mathbb{N}$. Show that the groups $(M_k, +)$ and $(\mathbb{Z}, +)$ are isomorphic.

9. Show that the groups $(\mathbb{Z}_6, \oplus)$ and $(\mathbb{Z}_7^\#, \odot)$ are isomorphic.

10. Consider the dihedral group $(D_4, \circ)$ that was introduced in Exercise 19 of Exercises 8.2.

   (a) Let $H = \{\iota, \rho^2, \sigma, \rho^2 \circ \sigma\}$. Show that $(H, \circ)$ is a subgroup of $(D_4, \circ)$.
   (b) Show that $(H, \circ)$ is isomorphic to the Klein four-group (see Exercise 12 of Exercises 8.2).

11. Let $(G, *)$ and $(H, \#)$ be groups and let $f: G \to H$ be a homomorphism. Show that, if $x \in G$ has finite order, then the order of $f(x)$ is finite and divides the order of x.

12. Let $(G, *)$ and $(H, \#)$ be groups and let $f: G \to H$ be an isomorphism. Prove that $|f(x)| = |x|$ for all $x \in G$. (In other words, if $|x| = m$ in $(G, *)$, then $|f(x)| = m$ in $(H, \#)$; if $|x| = \infty$, then $|f(x)| = \infty$.) What does this say about the number of elements of a given order in both G and H?

13. Let $(G, *)$ be a group of order 3, say $G = \{e, a, b\}$. Show that $(G, *)$ is isomorphic to $(\mathbb{Z}_3, \oplus)$.

14. Show that $(\mathbb{Z}_4, \oplus)$ and the Klein four-group are nonisomorphic groups.

15. Let $(G, *)$ be a group of order 4, say $G = \{e, a, b, c\}$. Show that $(G, *)$ is isomorphic either to $(\mathbb{Z}_4, \oplus)$ or to the Klein four-group. (Note: This shows that, up to isomorphism, there are exactly two groups of order 4.)

16. Show that $(\mathbb{Z}_6, \oplus)$ and $(S_3, \circ)$ are nonisomorphic groups. (In fact, these are the only groups of order 6, up to isomorphism.)

17. Let $(G, *)$ and $(H, \#)$ be groups and let $f: G \to H$ be a homomorphism.

   (a) Show that $(\text{im } f, \#)$ is a subgroup of $(H, \#)$.
   (b) Show that $(\text{im } f, \#)$ is abelian if $(G, *)$ is abelian.
   (c) If $(G, *)$ is abelian, does it follow necessarily that $(H, \#)$ is abelian?

18. Let $g: (\mathbb{Z}_m, \oplus) \to (\mathbb{Z}_n, \oplus)$ be a mapping that satisfies $g([1]) = [a]$, where $0 \le a \le n - 1$. If g is a homomorphism, what can be said about a? (Hint: Use the result of Exercise 11.)

19. Use the result of Exercise 18 to determine all homomorphisms for each of the following pairs of groups:
    (a) from $(\mathbb{Z}_6, \oplus)$ to $(\mathbb{Z}_4, \oplus)$ **(b)** from $(\mathbb{Z}_6, \oplus)$ to $(\mathbb{Z}_9, \oplus)$
    (c) from $(\mathbb{Z}_6, \oplus)$ to $(\mathbb{Z}_7, \oplus)$
    (d) from $(\mathbb{Z}_d, \oplus)$ to $(\mathbb{Z}_c, \oplus)$ where $\gcd(c, d) = 1$
20. Find all isomorphisms for each of the following pairs of groups:
    (a) from $(\mathbb{Z}_6, \oplus)$ to $(\mathbb{Z}_7^{\#}, \odot)$ **(b)** from $(\mathbb{Z}_8, \oplus)$ to $(\mathbb{Z}_8, \oplus)$
    (c) from $(\mathbb{Z}, +)$ to $(E, +)$, where E is the set of even integers
    (d) from $(\mathbb{Z}, +)$ to $(\mathbb{Z}, +)$
21. Complete the verification in Example 8.36 that $(L, \vee, \wedge)$ and $(\mathcal{P}(A), \cup, \cap)$ are isomorphic Boolean algebras.
22. The purpose of this exercise is to prove Theorem 8.27. Let $(B, \preccurlyeq)$ be a finite Boolean algebra. You should apply Theorem 8.23 and the results of Exercise 22 of Exercises 8.6. Thus A denotes the set of atoms of B and $A_b = \{a \in A \mid a \preccurlyeq b\}$.
    (a) Let $C = \{c_1, c_2, \ldots, c_r\}$ and $D = \{d_1, d_2, \ldots, d_t\}$ be subsets of A. Show that $c_1 \vee c_2 \vee \cdots \vee c_r = d_1 \vee d_2 \vee \cdots \vee d_t$ if and only if $C = D$.
    (b) Define $\theta\colon B \to \mathcal{P}(A)$ by $\theta(b) = A_b$ and show that θ is a Boolean algebra isomorphism.
23. Let $(G, *)$ and $(H, \#)$ be groups and let $\alpha\colon G \to H$ be a homomorphism. Define the *kernel of* α, denoted $\ker \alpha$, to be the set

$$\ker \alpha = \{x \in G \mid \alpha(x) = e_H\}$$

where e_H is the identity element of $(H, \#)$.
    (a) Show that $(\ker \alpha, *)$ is a subgroup of $(G, *)$.
    (b) Show that, if $a \in \ker \alpha$, then $x^{-1} * a * x \in \ker \alpha$ for all $x \in G$.
24. Define $\alpha\colon GL_2(\mathbb{R}) \to \mathbb{R}^{\#}$ by $\alpha(A) = \det(A)$.
    (a) Show that α is a homomorphism.
    (b) Find im α.
    (c) Find ker α.

CHAPTER PROBLEMS

1. For each of the following definitions of $*$, verify that $*$ is a binary operation on $\mathbb{Q}$. Then determine whether $*$: (i) is associative, (ii) is commutative, and (iii) has an identity. In case $*$ has an identity, determine which rational numbers have inverses.
    (a) $x * y = (x + y)/2$ **(b)** $x * y = x + y + 2$
    (c) $x * y = 3xy$ **(d)** $x * y = xy/2$
    (e) $x * y = x(y + 1)$ **(f)** $x * y = x + y + xy$
2. The operations we have studied in this chapter have been *binary operations*, so-called because such an operation requires *two operands*.

In general, operations on a set may have one, two, or more operands. The case of operations requiring a *single operand* is particularly important; such operations are called *unary operations*. For example, consider the operations of negation (minus) on the set $\mathbb{R}$; given any $r \in \mathbb{R}$, this operation returns as the result the number $-r$.

(a) Give an example of a unary operation on the set $\mathbb{R}^{\#}$ of nonzero reals.

(b) Give an example of a unary operation on the power set $\mathscr{P}(A)$ of a given set A.

(c) Give as many examples as you can of unary operations in Pascal or Ada.

(d) What is another name for a unary operation on a set A?

3. In the usual notation for binary operations, the operator is written between the two operands; this is referred to as *infix notation*. An alternative notation is the *postfix* (or *reverse Polish*) notation; here each operator is written immediately after its rightmost operand. For example, in postfix notation $x + y$ is written as $xy +$. Postfix notation has several advantages over infix notation. For one, it is suitable for both unary and binary operations, whereas true infix notation assumes binary operations are being used. Perhaps more important, postfix notation is completely unambiguous. The problem of ambiguity for infix notation is illustrated by the expression $x + y * z$. Does it mean $(x + y) * z$ or $x + (y * z)$? Note that with infix notation this problem must be resolved by using parentheses, or by agreeing on certain precedence rules for the operations. With postfix notation, however, $(x + y) * z$ is written as $xy + z *$, whereas $x + (y * z)$ is written as $xyz * +$, so there is no problem! Find the value of each of the following postfix expressions. (Here $-$ denotes unary negation and $*$ means multiplication.)

(a) $3\ 5 + 7 *$ **(b)** $3\ -\ 8 +$ **(c)** $5\ 7 * 4 - 6 * +$

Write each of the following expressions using postfix notation.

(d) $-\ 3 * (2 + 6)$ **(e)** $-((3 * 8) + (2 * 7))$

4. Another common notation for expressions is known as *prefix* (or *Polish*) notation, in which the operator symbol is written immediately before its leftmost operand; for example, $x + y$ is written $+xy$. A special version of prefix notation is called *Cambridge Polish* notation and it is used in the programming language LISP. It encloses each (sub)expression within parentheses, so, for instance, $x + y$ is written as $(+xy)$ and $-x * y$ is written as $(*(-x)y)$. (Here again $*$ denotes multiplication and $-$ denotes unary negation.) Write each of the following expressions using Cambridge Polish notation.

(a) $(3 + 5) * 7$ **(b)** $3 + (5 * 7)$

(c) $-(4 * 9)$ **(d)** $(5 * 7) + (-4 * 6)$

Find the value of each of the following expressions.

(e) $(*(-3)(+2\ 6))$ **(f)** $(-(+(*3\ 8)(*2\ 7)))$

5. Let S be a finite set consisting of m elements. Determine the number of each:

   (a) binary operations on S

   (b) commutative binary operations on S

   (c) binary operations on S having an identity

6. For each of the following subsets of $\mathscr{F}(\mathbb{R})$, determine whether composition is a binary operation on the subset. If so, discuss the properties of the resulting semigroup.

   (a) the set $\mathscr{C}$ of continuous functions

   (b) the set $\mathscr{D}$ of differentiable functions

   (c) th set $\mathscr{P}$ of polynomial functions

   (d) the set $\mathscr{P}_n$ of polynomials of degree less than or equal to n

7. Let $(G, *)$ be a group and let H be a subset of G. Prove that $(H, *)$ is a subgroup of $(G, *)$ if and only if (i) H is nonempty and (ii) $a * b^{-1} \in H$ for all $a, b \in H$.

8. Let $\mathscr{F}(\mathbb{R})^*$ denote the set of all $f: \mathbb{R} \to \mathbb{R}$ such that $f(a) \neq 0$ for all $a \in \mathbb{R}$. For $f, g \in \mathscr{F}(\mathbb{R})^*$, define $f \cdot g$ by $(f \cdot g)(x) = f(x)g(x)$. Show that $\cdot$ is a binary operation on $\mathscr{F}(\mathbb{R})^*$ and $(\mathscr{F}(\mathbb{R})^*, \cdot)$ is a group.

9. Let $(G, *)$ be a semigroup and suppose that, for any two elements $a, b \in G$, both of the equations $a * x = b$ and $y * a = b$ have solutions for x and y in G. Prove that $(G, *)$ is a group.

10. Let $\mathscr{L}_G$ denote the collection of all subgroups of the group $(G, *)$.

    (a) Show that $\mathscr{L}_G$ is partially ordered by inclusion.

    In fact, it can be shown that the poset $(\mathscr{L}_G, \subseteq)$ is a lattice, called the *lattice of subgroups* of the group $(G, *)$. Draw the Hasse diagram for the lattice of subgroups of the following groups.

    (b) $(S_3, \circ)$ **(c)** $(\mathbb{Z}_{12}, \oplus)$ **(d)** $(D_4, \circ)$

11. Find two elements $f, g \in S(\mathbb{N})$ (the symmetric group on the set of positive integers) such that each of f and g has finite order in $S(\mathbb{N})$, but $f \circ g$ has infinite order.

12. Let $(G, *)$ be an abelian group and let $H = \{a \in G \mid a^n = e\}$, where n is a fixed positive integer (as usual, e is the identity element of $(G, *)$).

    (a) Show that $(H, *)$ is a subgroup of $(G, *)$.

    (b) Give an example to show that the result of part (a) may not hold if $(G, *)$ is nonabelian.

13. Let A be a nonempty subset of $\{1, 2, \ldots, n\}$ and let

    $$H_A = \{\alpha \in S_n \mid \alpha(a) = a \text{ for all } a \in A\}$$

    Show that $(H_A, \circ)$ is a subgroup of $(S_n, \circ)$.

14. An element s of a ring $(S, +, \cdot)$ is called an *idempotent* if $s \cdot s = s$. Find the idempotent elements of each of the following rings.

    (a) $(\mathbb{Z}, +, \cdot)$ **(b)** $(\mathbb{Z}[\sqrt{2}], +, \cdot)$ **(c)** $(\mathscr{F}(\mathbb{R}), +, \cdot)$

    (d) $(\mathbb{Z}_6, \oplus, \odot)$ **(e)** $(M_2(\mathbb{R}), +, \cdot)$

15. Let $(S, +, \cdot)$ be a ring and let T be a subset of S. We say that T is a *subring* of S provided $(T, +, \cdot)$ is itself a ring. If L is a nonempty subset of S, then show that L is a subring of S if and only if the following conditions hold:

   (i) $(L, +)$ is a subgroup of $(S, +)$.

   (ii) If $t_1, t_2 \in L$, then $t_1 \cdot t_2 \in L$.

16. Let $(S, +, \cdot)$ be a commutative ring with identity. An element $a \in S$ is called *invertible* provided a has a multiplicative inverse in $(S, \cdot)$.
   (a) Show that the invertible elements of S form a group under the ring multiplication.
   (b) Find the group of invertible elements of $(\mathbb{Z}_m, \oplus, \odot)$. (Hint: You have seen this group before.)

17. Consider the ring $(\mathscr{F}(\mathbb{R}), +, \cdot)$. Determine which of the subsets $\mathscr{C}$, $\mathscr{D}$, $\mathscr{P}$, and $\mathscr{P}_n$, defined in Problem 6, are subrings of the ring $(\mathscr{F}(\mathbb{R}), +, \cdot)$.

18. Let $(F, +, \cdot)$ be a field. For $a, b \in F$, $b \neq z$, define the operation $/$ (called *division*) on F by $a/b = a \cdot b^{-1}$. Show the following:
   (a) $(a/b) \cdot (c/d) = (a \cdot c)/(b \cdot d)$
   (b) $(a/b) + (c/d) = [(a \cdot d) + (b \cdot c)]/(b \cdot d)$

19. Let $(S, +, \cdot)$ be a ring with identity. We call $(S, +, \cdot)$ a *Boolean ring* provided every element of S is idempotent (see Problem 14).
   (a) Prove that, if $(S, +, \cdot)$ is Boolean, then $(S, +, \cdot)$ is commutative and $2s = z$ for all $s \in S$.
   (b) If $*$ denotes symmetric difference, then show that the ring $(\mathscr{P}(A), *, \cap)$ is Boolean.

20. Let $(S, +, \cdot)$ be an integral domain. If there exists an $n \in \mathbb{N}$ such that $ne = z$, then the least such n is called the *characteristic* of S. If no such n exists, then S is said to have *characteristic zero*.
   (a) Show that, if S has characteristic n, then $ns = z$ for all $s \in S$.
   (b) Show that, if S has characteristic $n > 0$, then n is a prime.
   (c) For each prime p, give an example of an integral domain with characteristic p. Also give an example of an integral domain with characteristic zero.

21. Let $(G, *)$ be an abelian group and define $\theta: G \to G$ by $\theta(x) = x^2$. Show that θ is a homomorphism and describe the kernel of θ.

22. Let $(B, \vee, \wedge)$ be a Boolean algebra and let $a, b, c \in B$.
   (a) Prove: If $a \vee c = b \vee c$ and $a \vee c' = b \vee c'$, then $a = b$.
   (b) Give an example to show that the condition $a \vee c = b \vee c$ alone is not sufficient for $a = b$.

23. If $\theta: (A, \vee, \wedge) \to (B, \vee', \wedge')$ is a Boolean algebra isomorphism, prove the following:
   (a) θ maps the unity element (zero element) of A to the unity element (zero element) of B.
   (b) $\theta(a') = \theta(a)'$ for all $a \in A$.

24. For each of the following Boolean forms, draw the associated lg-circuit.

(a) $(a \wedge b \wedge c) \vee d$ (b) $(a \vee b) \wedge (c \vee d)$

(c) $a \vee (b \wedge c) \vee d$ (d) $((a \vee b) \wedge c) \vee d$

(e) $(a \wedge (b \vee (c \wedge d))) \vee e$

25. Determine the normal form of each of the following switching functions in $(\mathscr{F}_4, \vee, \wedge)$.

(a) $f_1(x, y, z, w) = (x \wedge y \wedge z) \vee w$

(b) $f_2(x, y, z, w) = (x' \vee y) \wedge (z \wedge w)$

(c) $f_3(x, y, z, w) = ((x \vee y) \wedge z) \vee w$

(d) $f_4(x, y, z, w) = x \wedge (y \vee (z \wedge w))$

26. The *XOR operator* $\oplus$ on $\mathscr{B}_n$ is defined as follows: For Boolean variables x and y, $x \oplus y = (x \vee y) \wedge (x \wedge y)'$.

(a) Complete the table of values for $x \oplus y$.

(b) Design an lg-circuit for $x \oplus y$. One uses the diagram given in Figure 8.31 to denote this circuit and calls it an *XOR gate*.

(c) Prove that $x \oplus (y \oplus z) = (x \oplus y) \oplus z$.

(d) Prove that $x \vee y = x \oplus y \oplus (x \wedge y)$.

Figure 8.31 The XOR gate

27. Determine switching functions for each of the lg-circuits in Figure 8.32. Simplify the switching functions obtained so as to decrease the number of gates with two inputs in the associated lg-circuit. Exhibit the lg-circuits for these simplified forms.

28. Let $(G, *)$ be a group. An isomorphism $f: G \to G$ is called an *automorphism of* $(G, *)$. Let a be a fixed element of G.

(a) Define $f_a: G \to G$ by $f_a(x) = a * x * a^{-1}$. Show that f_a is an automorphism of $(G, *)$. What is f_a if $(G, *)$ is abelian?

(b) Show that $g_a: G \to G$, defined by $g_a(x) = a * x$, is a permutation of G. Is g_a an automorphism of $(G, *)$?

(c) Define $h: G \to G$ by $h(x) = x^{-1}$. Show that h is an automorphism of $(G, *)$ if and only if $(G, *)$ is abelian.

29. Prove that the group $(\mathbb{Q}, +)$ is not cyclic.

30. Let a and b be real numbers with $a \neq 0$. Define $f: \mathbb{R} \to \mathbb{R}$ by $f(x) = ax + b$.

(a) Show that f is one-to-one and onto.

(b) Show that the set of all such functions forms a group under composition.

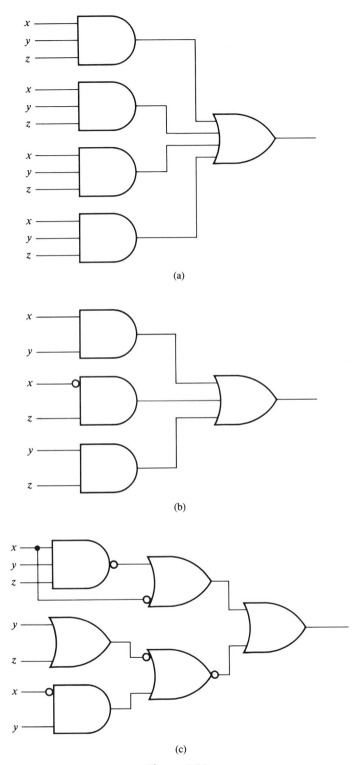

Figure 8.32

31. Define a binary operation $*$ on $\mathbb{R}$ by $a * b = \max(a, b)$. Show that $(\mathbb{R}, *)$ is a semigroup. Is $*$ commutative? Does $(\mathbb{R}, *)$ have an identity?

32. Let $(L, \preccurlyeq)$ be a totally ordered set. Show that $(L, \preccurlyeq)$ is a distributive lattice.

33. Show that any finite lattice contains a zero element and unit element. (Hint: See Chapter 4 Problem 31.)

34. Define $\alpha \colon C^{\#} \to GL_2(\mathbb{R})$ by

$$\alpha(a + ib) = \begin{bmatrix} a & b \\ -b & a \end{bmatrix}$$

(a) Show that α is a homomorphism of the groups $(C^{\#}, \cdot)$ and $(GL_2(\mathbb{R}), \cdot)$,

(b) Show that α is one-to-one.

Appendix A: Matrices

Recall from Section 4.1 that an $m \times n$ matrix is a rectangular array of real numbers of the form

$$A = \begin{bmatrix} a_{11} & a_{12} & \cdots & a_{1n} \\ a_{21} & a_{22} & \cdots & a_{2n} \\ & \vdots & & \\ a_{m1} & a_{m2} & \cdots & a_{mn} \end{bmatrix}$$

The ith *row* of A is $[a_{i1}, a_{i2}, \ldots, a_{in}]$, where $1 \le i \le m$, and the jth *column* of A is

$$\begin{bmatrix} a_{1j} \\ a_{2j} \\ \vdots \\ a_{mj} \end{bmatrix}$$

We say that A has *size* $m \times n$ and use the customary shorthand notation $[a_{ij}]$ to denote A. The element a_{ij}, located in the ith row and jth column of A, is called the (i, j)-*entry* of A. For example,

$$C = \begin{bmatrix} 2 & \frac{2}{3} & 3 & 4 \\ 3 & 1 & 0 & -6 \\ \frac{1}{2} & 0 & \sqrt{2} & 0 \end{bmatrix}$$

is a 3×4 matrix and, for instance, its $(2, 3)$-entry is 0.

In this section we discuss some properties satisfied by the basic matrix operations. In order to do this properly, we must first decide when two matrices are equal. In what follows, we assume that all matrices have real-number entries.

DEFINITION A.1

Let $A = [a_{ij}]$ and $B = [b_{ij}]$ be $m \times n$ matrices. We say that A and B are *equal*, denoted $A = B$, provided $a_{ij} = b_{ij}$ for $1 \leq i \leq m$ and $1 \leq j \leq n$.

Example A.1   Under what conditions are the following matrices equal?

$$A = \begin{bmatrix} 1 & 1 & 2 \\ 1 & z & -1 \end{bmatrix} \quad \text{and} \quad B = \begin{bmatrix} 1 & y & 2 \\ 1 & 4 & -1 \end{bmatrix}$$

Solution   According to the definition of equality for matrices, we see that $A = B$ if and only if $y = 1$ and $z = 4$.   ■

It should be noted that, in order for two matrices to be equal, they must be of the same size.

We now introduce the operations of addition and scalar multiplication.

DEFINITION A.2

Let $A = [a_{ij}]$ and $B = [b_{ij}]$ be $m \times n$ matrices.

1. The *sum* of A and B, denoted $A + B$, is the $m \times n$ matrix $C = [c_{ij}]$, where

$$c_{ij} = a_{ij} + b_{ij}$$

   for $1 \leq i \leq m$ and $1 \leq j \leq n$.

2. For $r \in \mathbb{R}$, the *scalar product* of r and A, denoted rA, is the $m \times n$ matrix whose (i, j)-entry is ra_{ij} for $1 \leq i \leq m$ and $1 \leq j \leq n$.

Example A.2   Let

$$A = \begin{bmatrix} 1 & 4 & -3 & 5 \\ 0 & -2 & 8 & 0 \\ 2 & 1 & -4 & 6 \\ 4 & -2 & 0 & -3 \end{bmatrix} \quad \text{and} \quad B = \begin{bmatrix} 0 & 2 & 0 & -2 \\ 1 & 1 & 4 & -6 \\ 0 & 2 & -3 & 8 \\ 1 & 1 & -1 & 0 \end{bmatrix}$$

Evaluate the following matrices: (a) $A + B$, (b) $(-2)B$, and (c) $3A + 4B$.

Solution   Following the definitions given, we see that

(a)

$$A + B = \begin{bmatrix} 1 & 6 & -3 & 3 \\ 1 & -1 & 12 & -6 \\ 2 & 3 & -7 & 14 \\ 5 & -1 & -1 & -3 \end{bmatrix}$$

(b)

$$(-2)B = \begin{bmatrix} 0 & -4 & 0 & 4 \\ -2 & -2 & -8 & 12 \\ 0 & -4 & 6 & -16 \\ -2 & -2 & 2 & 0 \end{bmatrix}$$

(c)

$$3A + 4B = \begin{bmatrix} 3 & 20 & -9 & 7 \\ 4 & -2 & 40 & -24 \\ 6 & 11 & -24 & 50 \\ 16 & -2 & -4 & -9 \end{bmatrix} \qquad ■$$

DEFINITION A.3

Let $A = [a_{ij}]$ be an $m \times n$ matrix and let $B = [b_{ij}]$ be an $n \times p$ matrix. The *product* of A with B, denoted AB, is the $m \times p$ matrix $C = [c_{ij}]$, where

$$c_{ij} = a_{i1}b_{1j} + a_{i2}b_{2j} + \cdots + a_{in}b_{nj}$$

for $1 \le i \le m$ and $1 \le j \le p$.

Notice that the (i, j)-entry of the matrix AB in the above definition is obtained from the ith row of A and the jth column of B. This can be more appropriately focused by viewing AB as follows:

$$AB = \begin{bmatrix} a_{11} & \cdots & a_{1t} & \cdots & a_{1n} \\ & & \vdots & & \\ a_{i1} & \cdots & a_{it} & \cdots & a_{in} \\ & & \vdots & & \\ a_{m1} & \cdots & a_{mt} & \cdots & a_{mn} \end{bmatrix} \begin{bmatrix} b_{11} & \cdots & b_{1j} & \cdots & b_{1p} \\ & & \vdots & & \\ b_{l1} & \cdots & b_{lj} & \cdots & b_{lp} \\ & & \vdots & & \\ b_{n1} & \cdots & b_{nj} & \cdots & b_{np} \end{bmatrix}$$

and the (i, j)-entry is $a_{i1}b_{1j} + a_{i2}b_{2j} + \cdots + a_{in}b_{nj}$. We call this expression the *inner product* of the ith row of A with the jth column of B.

Example A.3   Let

$$A = \begin{bmatrix} 1 & -1 & 2 & 0 \\ 0 & 4 & -2 & 6 \end{bmatrix} \quad \text{and} \quad B = \begin{bmatrix} 1 & -5 \\ 0 & 3 \\ 6 & -6 \\ 0 & 9 \end{bmatrix}$$

Evaluate the product AB.

Solution   Let $C = AB$. Note that C is a 2×2 matrix. Using the definition and the idea of inner product, we obtain

$$c_{11} = (1)(1) + (-1)(0) + (2)(6) + (0)(0) = 13$$
$$c_{12} = (1)(-5) + (-1)(3) + (2)(-6) + (0)(9) = -20$$
$$c_{21} = (0)(1) + (4)(0) + (-2)(6) + (6)(0) = -12$$
$$c_{22} = (0)(-5) + (4)(3) + (-2)(-6) + (6)(9) = 78$$

Thus

$$C = \begin{bmatrix} 13 & -20 \\ -12 & 78 \end{bmatrix}$$ ∎

Notice, in the preceding example, that the product BA is also defined and that $AB \neq BA$; in fact, the two matrices AB and BA have different sizes (AB is a 2×2 matrix and BA is a 4×4 matrix). In general, it may happen that, for two matrices A and B, the product AB is defined but the product BA is not; for example, if A is a 3×4 matrix and B is a 4×6 matrix, then AB is defined but BA is not. We now proceed to some basic properties of the matrix operations.

THEOREM A.1   For $m \times n$ matrices A and B and $a, b \in \mathbb{R}$, the following properties hold:

1. $A + B = B + A$
2. $a(A + B) = aA + aB$
3. $(a + b)A = aA + bA$
4. $(ab)A = a(bA)$

Proof   We prove parts 1 and 4 and leave the remaining parts as an exercise for the student.

Set $A = [a_{ij}]$ and $B = [b_{ij}]$, and for $1 \leq i \leq m$ and $1 \leq j \leq n$, let c_{ij} and d_{ij} denote the (i, j)-entries of $A + B$ and $B + A$, respectively. Then it follows that $c_{ij} = a_{ij} + b_{ij}$ and $d_{ij} = b_{ij} + a_{ij}$ and, since a_{ij} and b_{ij} are real numbers, we have $c_{ij} = d_{ij}$ for $1 \leq i \leq m$ and $1 \leq j \leq n$. This proves part 1.

To see part 4 we compare the (i, j)-entries of $(ab)A$ and $a(bA)$ for $1 \leq i \leq m$ and $1 \leq j \leq n$. The (i, j)-entry of $(ab)A$ is $(ab)a_{ij}$ and that of $a(bA)$ is $a(ba_{ij})$. Again, since we are dealing with real numbers, we have that $(ab)a_{ij} = a(ba_{ij})$. Hence we obtain $(ab)A = a(bA)$. □

The next theorem addresses properties satisfied by matrix multiplication.

THEOREM A.2 Let A be an $m \times n$ matrix, let B be an $n \times p$ matrix, and let $a \in \mathbb{R}$.

1. For any $p \times r$ matrix C, we have $A(BC) = (AB)C$.
2. $a(AB) = (aA)B = A(aB)$
3. For any $n \times p$ matrix C and any $p \times r$ matrix D, we have

$$A(B + C) = AB + AC \quad \text{and} \quad (B + C)D = BD + CD$$

Proof We prove only part 1 and leave the remaining parts as exercises.

Set $A = [a_{ij}]$, $B = [b_{ij}]$, and $C = [c_{ij}]$. We compare the (i, j)-entries of $(AB)C$ and $A(BC)$ for $1 \le i \le m$ and $1 \le j \le r$. The (i, j)-entry of $(AB)C$ is obtained from the ith row of AB and the jth column of C. In order to see this product better, let $AB = D = [d_{ij}]$ and $BC = E = [e_{ij}]$. Also let $(AB)C = T = [t_{ij}]$ and $A(BC) = S = [s_{ij}]$. Then

$$t_{ij} = \sum_{k=1}^{p} d_{ik}c_{kj} = \sum_{k=1}^{p} \left(\sum_{h=1}^{n} a_{ih}b_{hk} \right) c_{kj} = \sum_{k=1}^{p} \sum_{h=1}^{n} a_{ih}b_{hk}c_{kj}$$

and

$$s_{ij} = \sum_{h=1}^{n} a_{ih}s_{hj} = \sum_{h=1}^{n} a_{ih} \left(\sum_{k=1}^{p} b_{hk}c_{kj} \right)$$

$$= \sum_{h=1}^{n} \sum_{k=1}^{p} a_{ih}b_{hk}c_{kj} = \sum_{k=1}^{p} \sum_{h=1}^{n} a_{ih}b_{hk}c_{kj}$$

Thus we see that $t_{ij} = s_{ij}$ for $1 \le i \le m$ and $1 \le j \le r$, and the result follows. $\square$

The $m \times n$ matrix, all of whose entries are 0, is called the $m \times n$ *zero matrix* and is denoted by O_{mn}. If $A = [a_{ij}]$ is an $n \times n$ matrix, then the entries $a_{11}, a_{22}, \ldots, a_{nn}$ are called the *main diagonal entries* of A. The $n \times n$ matrix whose main diagonal entries are all 1 and all of whose other entries are 0 is called the $n \times n$ *identity matrix*. It is denoted by I_n. For example, notice that

$$O_{32} = \begin{bmatrix} 0 & 0 \\ 0 & 0 \\ 0 & 0 \end{bmatrix} \qquad I_2 = \begin{bmatrix} 1 & 0 \\ 0 & 1 \end{bmatrix}$$

The following result is very basic; its proof is left to Exercise 6.

THEOREM A.3
1. If A is any $m \times n$ matrix, then $A + O_{mn} = O_{mn} + A = A$.
2. If B is any $n \times n$ matrix, then $BI_n = I_nB = B$.

Exercises A

1. Prove parts 2 and 3 of Theorem A.1.
2. Prove parts 2 and 3 of Theorem A.2.

The following matrices are to be used in Exercises 3–5.

$$A = \begin{bmatrix} 1 & -2 \\ 4 & 12 \\ -1 & 3 \end{bmatrix} \qquad B = \begin{bmatrix} 2 & 1 & 4 \\ 0 & -3 & 4 \end{bmatrix}$$

$$C = \begin{bmatrix} 1 & -1 \\ 6 & 10 \end{bmatrix} \qquad D = \begin{bmatrix} 0 & 1 \\ -1 & 4 \\ 6 & 3 \end{bmatrix}$$

3. Compute the following matrices:
   (a) $3A$ **(b)** $A - D$ **(c)** $3A - 6D$ **(d)** AB
   (e) $O_2 C$ **(f)** BA **(g)** AC **(h)** $(3C)(2B)$
4. Find a matrix E such that $3A + 7D + E$ is the 3×2 zero matrix.
5. Find a matrix E such that $EC = CE = I_2$.
6. Prove Theorem A.3.
7. An $n \times n$ matrix A is called *invertible* if there is an $n \times n$ matrix B such that $AB = BA = I_n$. We call B an *inverse* of A.
   (a) Prove: If A is an invertible $n \times n$ matrix, then A has a unique inverse. It is denoted by A^{-1}.
   (b) If

$$A = \begin{bmatrix} a & b \\ c & d \end{bmatrix}$$

   then prove that A is invertible if and only if $ad - bc \neq 0$. The expression $ad - bc$ is called the *determinant* of A and is denoted by $\det(A)$.
   (c) Given that $ad - bc \neq 0$, find the inverse of the matrix A of part (b).
8. If A is an $n \times n$ matrix and $k \in \mathbb{N}$, then the power A^k is defined inductively as follows:

$$\textbf{(i)} \quad A^1 = A$$

$$\textbf{(ii)} \quad A^k = A^{k-1}A \text{ for } k \geq 2$$

   (a) If

$$A = \begin{bmatrix} 0 & 1 & 0 & 0 \\ 0 & 0 & 1 & 0 \\ 0 & 0 & 0 & 1 \\ 0 & 0 & 0 & 0 \end{bmatrix}$$

   then find A^2, A^3, A^4, and A^5.

(b) If

$$A = \begin{bmatrix} 1 & 1 \\ 0 & 1 \end{bmatrix}$$

then find a general formula for A^m for all $m \in \mathbb{N}$. Use induction to prove the formula is correct.

9. Prove: If A and B are 2×2 matrices, then $\det(AB) = \det(A)\det(B)$.

10. If $A = [a_{ij}]$ is an $m \times n$ matrix, then the *transpose* of A, denoted A^t, is the $n \times m$ matrix whose ith row is the ith column of A for $1 \le i \le n$.

   (a) Find each of the following matrices from the matrices A, B, C, and D given at the beginning of this exercise set.

   (i) A^t **(ii)** B^t **(iii)** C^t

   (iv) $(A + D)^t$ **(v)** $A^t + D^t$ **(vi)** $(AB)^t$ **(vii)** $B^t A^t$

   (b) If $A = [a_{ij}]$ and $B = [b_{ij}]$ are $m \times n$ matrices, then prove that $(A + B)^t = A^t + B^t$.

   (c) If $A = [a_{ij}]$ is an $m \times n$ matrix and $B = [b_{jk}]$ is an $n \times p$ matrix, then prove that $(AB)^t = B^t A^t$.

Appendix B:
Answers to Selected
Exercises and Problems

Exercises 1.1

1. (a), (c), (d), (f), (g), and **(h)** are declarative sentences; **(a), (c), (f),** and **(h)** are statements.
2. (a) $\{-3, -2, -1, 0, 1, 2, 3, 4\}$ **(b)** $\{-1, 0, 2\}$
3. (a) $\{n \in \mathbb{Z} \mid n < 0\}$
  (b) $\{n \in \mathbb{Z} \mid n$ is a perfect square$\}$
  (c) $\{n \in \mathbb{Z} \mid n$ is a perfect cube$\}$
  (d) $\{n \in \mathbb{Z} \mid n$ is a multiple of 4$\}$
  (e) $\{n \in \mathbb{Z} \mid n$ is an odd multiple of 3$\}$

Exercises 1.2

1. (a) $p \wedge q$ **(b)** $p \vee r$ **(c)** $p \rightarrow \sim q$
  (b) $p \leftrightarrow r$ **(e)** $\sim(r \rightarrow p)$ **(f)** $(q \vee r) \wedge \sim(q \wedge r)$
2. (a) Ralph reads the *New York Times* and jogs 3 miles.
  (b) Ralph watches the *MacNeil, Lehrer Newshour* or jogs 3 miles.
  (c) Either Ralph reads the *New York Times* and watches the *MacNeil, Lehrer Newshour* or he jogs 3 miles.
  (d) Ralph doesn't read the *New York Times* or he doesn't watch the *MacNeil, Lehrer Newshour*.
  (e) If Ralph reads the *New York Times*, then he watches the *MacNeil, Lehrer Newshour*.
  (f) Ralph watches the *MacNeil, Lehrer Newshour* if and only if he jogs 3 miles.
3. (a) $p \vee q$ is true. **(b)** $p \wedge q$ is false. **(c)** $p \rightarrow q$ is true.
  (d) $\sim q \rightarrow p$ is true. **(e)** $p \leftrightarrow q$ is false. **(f)** $q \leftrightarrow \sim p$ is true.
4. (a) p: $\sqrt{2709} \in \mathbb{Z}$, q: $\sqrt{2709}$ is even, r: $\sqrt{2709}$ is odd; $p \rightarrow (q \vee r)$
  (b) p: 53 is prime, q: $53 > 2$, r: 53 is odd; $(p \wedge q) \rightarrow r$
  (c) p: $\sqrt[3]{7} < 0$, q: $\left(\sqrt[3]{7}\right)^2 < 4$, r: $\sqrt[3]{7} = 0$, s: $\sqrt[3]{7} > 0$, t: $\sqrt[3]{7} < 2$;
    $(\sim p \wedge q) \rightarrow [r \vee (s \wedge t)]$
  (d) p: 26 is even, q: $26 > 2$, r: 26 is prime; $(p \wedge q) \rightarrow \sim r$
5. (a) triangle ABC is equilateral $\rightarrow$ triangle ABC is isosceles
  (b) $x = 3 \leftrightarrow 2x - 3 = 3$

   (c) π^{π} is a real number $\rightarrow$ (π^{π} is rational $\vee$ π^{π} is irrational)

   (d) $xy = 0 \leftrightarrow (x = 0 \vee y = 0)$

   (e) $\sqrt{47,089} > 200 \rightarrow (\sqrt{47,089}$ is prime $\rightarrow \sqrt{47,089} > 210)$

   (f) (line k is perpendicular to line m $\wedge$ line m is parallel to line n) $\rightarrow$ line k is perpendicular to line n

   (g) $n \in \mathbb{Z} \rightarrow (n > 0 \vee n < 0 \vee n = 0)$

   (h) $(a, b \in \mathbb{Z} \wedge b \neq 0) \rightarrow a/b$ is a rational number

8. (a) NOT has highest precedence; $*$, $/$, DIV, MOD, and AND have the next highest, followed next by $+$, $-$, and OR; the relational operators $<$, $<=$, $=$, $<>$, $>=$, and $>$ have lowest precedence.

   (b) The expression "A > 14 AND A $-$ B > 1" is not valid. It is interpreted as
"A > ((14 AND A) $-$ B) > 1."

   (c) The expression "NOT FOUND AND (CURRENT <= MAX)" is valid. The parentheses cause the Boolean expression CURRENT <= MAX to be evaluated first; then NOT FOUND is evaluated; then AND is applied.

   (d) The expressions in both parts **(b)** and **(c)** are valid in Ada. A major difference in the precedence rules for Ada operators from those of Pascal is that the Boolean operators have lower precedence than the relational operators.

10. (a) ASERVES AND AWINSVOLLEY

   (b) (NOT ASERVES) AND (NOT AWINSVOLLEY)

   (c) (ASERVES AND (NOT AWINSVOLLEY)) OR ((NOT ASERVES) AND AWINSVOLLEY)
Alternate solution: ASERVES <> AWINSVOLLEY

   (d) ASERVES := (NOT ASERVES)

Exercises 1.3

3. $q \rightarrow p \equiv \sim q \vee p \equiv p \vee \sim q \equiv \sim(\sim p) \vee \sim q \equiv \sim p \rightarrow \sim q$

4. (a) If $x \neq 2$, then $x^4 \neq 16$. If $x^4 = 16$, then $x = 2$. If $x^4 \neq 16$, then $x \neq 2$.

   (b) If $y \leq 0$, then $y = -3$. If $y \neq -3$, then $y > 0$. If $y = -3$, then $y \leq 0$.

   (c) If x is even or y is even, then xy is even. If xy is odd, then x is odd and y is odd. If xy is even, then x is even or y is even.

   (d) If $x^2 \neq x$, then both $x \neq 0$ and $x \neq 1$. If $x = 0$ or $x = 1$, then $x^2 = x$. If $x \neq 0$ and $x \neq 1$, then $x^2 \neq x$.

   (e) If $x \neq 17$ and $x^3 \neq 8$, then x is not prime. If x is prime, then $x = 17$ or $x^3 = 8$. If x is not prime, then both $x \neq 17$ and $x^3 \neq 8$.

   (f) If $xy = 0$, then either $x = 0$ or $y = 0$. If $x \neq 0$ and $y \neq 0$, then $xy \neq 0$. If $x = 0$ or $y = 0$, then $xy = 0$.

5. (a) If quadrilateral $ABCD$ is not a rectangle, then $ABCD$ is not a parallelogram.
If quadrilateral $ABCD$ is a parallelogram, then $ABCD$ is a rectangle.
If quadrilateral $ABCD$ is not a parallelogram, then $ABCD$ is not a rectangle.

   (c) If quadrilateral $ABCD$ is not a square, then either $ABCD$ is not a rectangle or $ABCD$ is not a rhombus.
If quadrilateral $ABCD$ is both a rectangle and a rhombus, then $ABCD$ is a square.
If quadrilateral $ABCD$ is not a rectangle or is not a rhombus, then $ABCD$ is not a square.

   (e) If polygon P has the property that either it is both equiangular and not equilateral or it is both equilateral and not equiangular, then P is not a triangle.
If polygon P is a triangle, then it has the property that it is equiangular if and only if it is equilateral.
If polygon P is not a triangle, then it has the property that either it is both equiangular and not equilateral or it is both equilateral and not equiangular.

12. Both p and r are true.

13. (a) If p, then q. $x = -2$ implies $x^3 = -8$. $x = -2$ only if $x^3 = -8$. $x = -2$ is sufficient for $x^3 = -8$. $x^3 = -8$ is necessary for $x = -2$.

(b) q is necessary for p. If Randy passes this course, then he is intelligent. Randy's passing this course implies that he is intelligent. Randy passes this course only if he is intelligent. Randy's passing this course is sufficient for his being intelligent.

(c) p is sufficient for q. If Susan works hard, then Susan passes this course. Working hard implies that Susan passes this course. Susan's passing this course is necessary for Susan working hard.

(f) p implies q. If Susan is a good student, then Susan studies hard. Susan is a good student only if Susan studies hard. Being a good student is sufficient for the fact that Susan studies hard. Studying hard is necessary for the fact that Susan is a good student.

(i) p only if q. If Randy passes this course, then Randy passes the final exam. Randy's passing this course implies his passing the final exam. Passing this course is sufficient for Randy's passing the final exam. Passing the final exam is necessary for Randy's passing this course.

14. (a) $x^3 - x^2 + x - 1 = 0$ if and only if $x = 1$.

(b) Randy's passing this course is necessary and sufficient for Susan's helping him study.

Exercises 1.4

1. (a) There exists an x such that x is prime.

(b) For every x, x is greater than 2.

(c) There exists an x such that x is both prime and even.

(d) For every x, if x is greater than 2, then either x is prime or x is even.

(e) There exists an x such that both x is prime and either x is even or x is greater than 2.

(f) For all x, if x is both prime and even, then x is not greater than 2.

2. (a) $\forall n \in \mathbb{Z}(n$ is even $\rightarrow \exists m \in \mathbb{Z}(n = 2m))$

(b) $\forall n \in \mathbb{Z}(n \in \mathbb{Q})$ **(c)** $\exists T(T$ is a right triangle $\land$ T is an isosceles triangle$)$

(d) $\forall Q((Q$ is a parallelogram $\land$ Q has two adjacent sides that are perpendicular$) \rightarrow Q$ is a rectangle$)$

(e) $\exists n \in \mathbb{Z}(n$ is both even and prime$)$

(f) $\exists s, t \in \mathbb{Z}(1 < s < t < 187 \land st = 187)$

(g) $\exists x \in \mathbb{Z}(x/2 \in \mathbb{Z} \land \forall y \in \mathbb{Z}(x/(2y) \notin \mathbb{Z}))$ **(h)** $\forall x, y \in \mathbb{R}(2x^2 - xy + 5 > 0)$

3. (a) $\exists x[\sim p(x) \land \sim q(x)]$ **(b)** $\exists x \, \exists y[p(x, y) \land \sim q(x, y)]$

(c) $\exists x[\forall y(p(x, y) \land \sim q(x, y))]$ **(d)** $\forall x[(\exists y p(x, y) \land \sim q(x, y)) \lor \forall z \sim r(x, z)]$

6. (a) $\exists x \in A[p(x) \lor q(x)]$ is true $\leftrightarrow p(x) \lor q(x)$ is true for some $x \in A \leftrightarrow (p(x)$ is true for some $x \in A) \lor (q(x)$ is true for some $x \in A) \leftrightarrow [\exists x \in Ap(x)] \lor [\exists x \in Aq(x)]$ is true

(b) $\forall x \in A[p(x) \land q(x)]$ is true $\leftrightarrow p(x) \land q(x)$ is true for all $x \in A \leftrightarrow (p(x)$ is true for all $x \in A) \land (q(x)$ is true for all $x \in A) \leftrightarrow [\forall x \in Ap(x)] \land [\forall x \in A \, q(x)]$ is true

(c) $\exists x \in A[p(x) \rightarrow q(x)] \equiv \exists x \in A[\sim p(x) \lor q(x)] \equiv [\exists x \in A \sim p(x)] \lor [\exists x \in Aq(x)]$ (by part **(a)**) $\equiv \sim[\forall x \in Ap(x)] \lor [\exists x \in Aq(x)] \equiv [\forall x \in Ap(x)] \rightarrow [\exists x \in Aq(x)]$

7. (a) $\exists I V[I] = 0$ **(b)** $\forall I V[I] = 0$ **(c)** $\forall I \, \forall J(I \neq J \rightarrow V[I] \neq V[J])$

(d) $\exists I \, \exists J(V[I] + V[J] = 0)$ **(e)** $\forall I[\exists J(|V[I] - V[J]| = 3)]$

(f) $\exists I[\forall J(I \neq J \rightarrow V[I] > V[J])]$ **(g)** $\forall I \, \forall J(I < J \rightarrow V[I] < V[J])$

(h) $\forall I[I$ even $\rightarrow \forall J(J$ odd $\rightarrow V[I] > V[J])]$

Exercises 1.5

1. (a) $[\sim q \land (p \rightarrow q)] \rightarrow \sim p$; valid   **(b)** $[(p \rightarrow q) \land \sim p] \rightarrow \sim q$; invalid

(c) $[(p \lor q) \land \sim q] \rightarrow p$; valid   **(d)** $[(p \rightarrow q) \land (r \lor \sim q) \land \sim r] \rightarrow \sim p$; valid

(e) $[(p \rightarrow q) \land (p \rightarrow r) \land \sim r] \rightarrow \sim q$; invalid

3. (a) $x = y \rightarrow x \cdot x = y \cdot y \rightarrow x^2 = y^2$

(b) $x^2 \neq y^2 \rightarrow x^2 - y^2 \neq 0 \rightarrow (x + y)(x - y) \neq 0 \rightarrow x - y \neq 0 \rightarrow x \neq y$

7. **(a)** If m is even, then $m = 2k$ for some $k \in \mathbb{Z}$. Hence $m^2 = 4k^2 = 2(2k^2)$. This shows that m^2 is even.

(b) If 3 is a factor of m, then $m = 3k$ for some $k \in \mathbb{Z}$. Hence $m^2 = 9k^2 = 3(3k^2)$. This shows that 3 is a factor of m^2.

(c) If m is not even, then $m = 2k + 1$ for some $k \in \mathbb{Z}$. Hence $m^2 = (2k + 1)^2 = 4k^2 + 4k + 1 = 2(2k^2 + 2k) + 1$. Therefore m^2 is not even.

(d) If m is not even, then $m = 2k + 1$ for some $k \in \mathbb{Z}$. Hence $m^3 = (2k + 1)^3 = 8k^3 + 4k^2 + 4k + 1 = 2(4k^3 + 2k^2 + 2k) + 1$. Therefore m^3 is not even.

(e) If 3 is not a factor of m, then $m = 3k + r$ for some $k \in \mathbb{Z}$ and $r = 1$ or 2. Hence $m^2 = (3k + r)^2 = (9k^2 + 6kr + r^2) = 3(3k^2 + 2kr) + r^2$. Note that $r^2 = 1$ or 4, so that 3 is not a factor of m^2.

9. Suppose $x^3 + 4x = 0$ and $x \neq 0$. Since $x^3 + 4x = x(x^2 + 4) = 0$ and $x \neq 0$, it must be that $x^2 + 4 = 0$. But this contradicts the fact that $x^2 + 4 > 0$ for any real number x.

10. Let $p = 11$, for example.

Chapter 1 Problems

1. **(b)** Either Ralph reads the *New York Times* or he watches the *MacNeil, Lehrer Newshour*, but not both.

(f) If Ralph jogs 3 miles, then he reads the *New York Times* but does not watch the *MacNeil, Lehrer Newshour*.

(h) If Ralph does not read the *New York Times* or he watches the *MacNeil, Lehrer Newshour*, then he jogs 3 miles.

3. **(a)** Fred has taken CS 260 if and only if either Fred knows BASIC or Fred knows Pascal.

(d) If Fred knows Pascal, then both Fred has taken CS 260 and Fred does not know BASIC.

(e) Either Fred knows both BASIC and Pascal or Fred has not taken CS 260.

(f) Fred knows BASIC and, if Fred has taken CS 260, then Fred knows Pascal.

4. **(a)** $p \wedge q$; $\sim p \vee \sim q$; The function $f(x) = x^3 - x$ is not one-to-one or is not onto.

(b) $p \vee q$; $\sim p \wedge \sim q$; $2^{1234} - 1$ is not prime and is not even.

(c) $p \to q$; $p \wedge \sim q$; G is hamiltonian and G is not connected.

(d) $(p \wedge q) \to r$; $p \wedge q \wedge \sim r$;
6 is related to 18 and 18 is related to 72, but 6 is not related to 72.

(e) $p \leftrightarrow q$; $(p \wedge \sim q) \vee (\sim p \wedge q)$; Either the function $f(x) = 2^x$ is one-to-one but not onto or it is onto but not one-to-one.

5. **(a)** $\sim p \vee \sim q \vee \sim r$ **(b)** $(p \wedge \sim q) \vee (\sim p \wedge q)$

(c) $p \wedge q \wedge \sim r$ **(d)** $\sim p \vee (\sim q \wedge \sim r)$

(e) $\sim p \vee (p \wedge \sim q) \vee (q \wedge \sim r) \equiv (\sim p \vee \sim q \vee \sim r)$

(f) $p \vee q$ **(g)** $\sim p \vee (q \wedge \sim r)$ **(h)** $p \wedge \sim q \wedge \sim r$

6. **(a)** $(p \wedge q) \to r \equiv \sim(p \wedge q) \vee r \equiv (\sim p \vee \sim q) \vee r \equiv \sim p \vee (\sim q \vee r) \equiv p \to (\sim q \vee r) \equiv p \to (q \to r)$. To prove $(p \wedge q) \to r$, we may prove $q \to r$ under the assumption that p is true.

(b) $(p \vee q) \to r \equiv \sim(p \vee q) \vee r \equiv (\sim p \wedge \sim q) \vee r \equiv (\sim p \vee r) \wedge (\sim q \vee r) \equiv (p \to r) \wedge (q \to r)$. To prove $(p \vee q) \to r$, we may give two separate proofs of $p \to r$ and of $q \to r$.

(c) $p \to (q \wedge r) \equiv \sim p \vee (q \wedge r) \equiv (\sim p \vee q) \wedge (\sim p \vee r) \equiv (p \to q) \wedge (p \to r)$. To prove $p \to (q \wedge r)$, we may prove separately the two implications $p \to q$ and $p \to r$.

(d) $p \to (q \vee r) \equiv \sim p \vee (q \vee r) \equiv (\sim p \vee q) \vee r \equiv (\sim p \vee r) \vee q \equiv \sim(p \wedge \sim r) \vee q \equiv (p \wedge \sim r) \to q \equiv (\sim r \wedge p) \to q \equiv \sim r \to (p \to q)$, by part (a). To prove $p \to (q \vee r)$, we may prove $p \to q$ under the assumption that r is false.

7. Using the basic syllogism, $[(p \to q) \wedge (q \to r) \wedge (r \to p)] \to [(p \to r) \wedge (r \to p)] \equiv p \leftrightarrow r$. Similarly, $[(p \to q) \wedge (q \to r) \wedge (r \to p)] \to [(p \to q) \wedge (q \to p)] \equiv p \leftrightarrow q$. Finally, $[(p \to q) \wedge (q \to r) \wedge (r \to p)] \equiv [(r \to p) \wedge (p \to q) \wedge (q \to r)] \to (r \to q) \wedge (q \to r) \equiv r \leftrightarrow q$.

8. **(a)** $\sim(\sim p \wedge \sim q)$ **(b)** $\sim(p \wedge \sim q)$ **(c)** $\sim(p \wedge \sim q) \wedge \sim(\sim p \wedge q)$

(d) $\sim(\sim p \wedge q \wedge r)$

11. **(a)** If x is not odd, then x^2 is not odd.
   If x^2 is odd, then x is odd.
   If x^2 is not odd, then x is not odd.
   (c) If f is not differentiable, then f is not continuous.
   If f is continuous, then f is differentiable.
   If f is not continuous, then f is not differentiable.
   (e) If $F(x) \le F(y)$, then $x \le y$.
   If $x > y$, then $F(x) > F(y)$.
   If $x \le y$, then $F(x) \le F(y)$.
   (g) If f is not defined at 0, then f is not continuous at 0.
   If f is continuous at 0, then f is defined at 0.
   If f is not continuous at 0, then f is not defined at 0.

13. **(a)** $(p \wedge q) \to r \equiv \, \sim (p \wedge q) \vee r \equiv (\sim p \vee \sim q) \vee r \equiv \, \sim p \vee (\sim q \vee r) \equiv \, \sim p \vee (r \vee \sim q) \equiv$
   $(\sim p \vee r) \vee \sim q \equiv \, \sim (p \wedge \sim r) \vee \sim q \equiv (p \wedge \sim r) \to \sim q$
   (c) If n is prime and n is even, then $n \le 2$.
   If $n > 2$ and n is even, then n is not prime.
   (d) If $f'(x) = 2x + 1$ and $f(x) \ne x^2 + x + 3$, then $f(0) \ne 3$.
   If $f(0) = 3$ and $f(x) \ne x^2 + x + 3$, then $f'(x) \ne 2x + 1$.

14. **(a)** $\forall x, y \in \mathbb{R} \ (2^x = 2^y \leftrightarrow x = y)$
   (b) (7 is the greatest common divisor of 119 and 154) $\leftrightarrow \exists s, t \in \mathbb{Z} \ (7 = 119s + 154t)$
   (c) $\forall n \in \mathbb{N} \ (n = 1 \vee n$ is prime $\vee \exists s, t \in \mathbb{Z}(1 < s \le t < n \wedge n = st))$
   (d) $\exists f(f$ is a function $\wedge f$ is one-to-one $\wedge \forall x \in \mathbb{R} \ (f(x) \ne x))$

17. **(a)** $\forall y[\exists x(xy < 3)]$ **(b)** $\exists x[\forall y(\exists z(x + y \ne z))]$
   (c) $\exists x(x > 0 \wedge x^2 < x)$ **(d)** $\exists x[\forall y(x < y \to 2^x < 2^y)]$
   (e) $\forall x[x \in \mathbb{Q} \to \exists y \ (y \notin \mathbb{Q} \wedge x + y \in Q)]$
   (f) $\exists x \ \exists y[(x > 0 \wedge y > 0) \wedge \forall n \ (nx \le y)]$ **(g)** $\exists y[y > 0 \wedge \forall x \ (\log x \le y)]$

19. **(a)** If the relation R is not symmetric, then there exist $x, y \in A$ such that x is related to y but y is not related to x.
   (b) If, for all $s, t \in \mathbb{Z}$, $7 \ne 119s + 154t$, then 7 is not the greatest common divisor of 119 and 154.
   (c) If, for some $b \in \mathbb{R}$, $\log a \ne b$ for every $a \in \mathbb{R}$, $a > 0$, then the function $\log x$ is not onto.
   (d) If, for some $a, b, c \in A$, a is related to b and b is related to c but a is not related to c, then the relation R is not transitive.

21. **(a)** valid   **(b)** valid   **(c)** invalid   **(d)** invalid   **(e)** valid

23. **(a)** $x + \dfrac{1}{x} < 2 \to x + \dfrac{1}{x} - 2 < 0 \to \dfrac{x^2 - 2x + 1}{x} < 0 \to \dfrac{(x - 1)^2}{x} < 0 \to x < 0$ (since $(x - 1)^2$ is nonnegative)
   (b) $(x > 0 \wedge x + \dfrac{1}{x} < 2) \to x^2 + 1 < 2x \to x^2 - 2x + 1 < 0 \to (x - 1)^2 < 0$, a contradiction

25. If he wears white kid gloves, then he is not an opium eater.

Exercises 2.1

1. **(a)** false   **(b)** true   **(c)** true   **(d)** false   **(e)** true   **(f)** false   **(g)** true   **(h)** true
2. $A = B = D, C = E$
3. **(a)** $\{\emptyset\}$ **(b)** $\{\emptyset, \{1\}\}$ **(c)** $\{\emptyset, \{1\}, \{2\}, \{1, 2\}\}$
   (d) $\{\emptyset, \{1\}, \{2\}, \{3\}, \{1, 2\}, \{1, 3\}, \{2, 3\}, \{1, 2, 3\}\}$
   (e) $\{\emptyset, \{\emptyset\}, \{\{1\}\}, \{\emptyset, \{1\}\}\}$ **(f)** $\mathscr{P}(\mathscr{P}(\emptyset)) = \{\emptyset, \{\emptyset\}\}$
5. 2^n

9. **(a)** $A = \{1,2\}$, $B = \{1\}$, $C = \emptyset$ **(b)** $A = \{1,2\}$, $B = \{1\}$, $C = \{2\}$
   (c) $A = \{1\}$, $B = \{2\}$, $C = \{3\}$ **(d)** $A = \{1,2\}$, $B = \{1\}$, $C = \{2\}$, $D = \emptyset$
   (e) $A = \{1,2,3\}$, $B = \{1\}$, $C = \{2\}$, $D = \{3\}$, $E = \emptyset$
   (f) $A = \{1,2,3\}$, $B = \{1,2\}$, $C = \{1,3\}$, $D = \{2\}$, $E = \{1\}$, $F = \emptyset$
10. **(a)** $A = \emptyset$, $B = \{\emptyset\}$, $C = \{\{\emptyset\}\}$ **(b)** $A = \emptyset$, $B = \{\emptyset\}$, $C = \{\emptyset, \{\emptyset\}\}$

Exercises 2.2

3. **(a)** $\{\ldots, -12, -6, 0, 6, 12, \ldots\}$ **(b)** $\{\ldots, -15, -9, -3, 3, 9, 15, \ldots\}$
   (c) $\{\ldots, -10, -6, -2, 2, 6, 10, \ldots\}$ **(d)** $\{\ldots, -10, -6, -2, 2, 6, 10, \ldots\}$
   (e) $\emptyset$ **(f)** $\{m \mid m$ is a multiple of 3 or m is a multiple of 4$\}$
   (g) C **(h)** $\{m \mid m$ is both even or a multiple of 3 and not a multiple of 4$\}$
7. **(a)** A **(b)** B **(c)** $\emptyset$ **(d)** $A \subset B \leftrightarrow (A - B = \emptyset \wedge B - A \neq \emptyset)$
   (e) $A \subset B \leftrightarrow (A \cap B = A \wedge A \cap B \neq B)$ **(f)** $A = B$
9. **(a)** $\{(x,0), (x,1), (y,0), (y,1)\}$ **(b)** $\{(0,-1), (0,0), (0,1), (1,-1), (1,0), (1,1)\}$
   (c) $\{(x,0,-1), (x,0,0), (x,0,1), (x,1,-1), (x,1,0), (x,1,1), (y,0,-1), (y,0,0), (y,0,1), (y,1,-1),$
   $(y,1,0), (y,1,1)\}$
   (e) $\{(x,(0,-1)), (x,(0,0)), (x,(0,1)), (x,(1,-1)), (x,(1,0)), (x,(1,1)), (y,(0,-1)), (y,(0,0)),$
   $(y,(0,1)), (y,(1,-1)), (y,(1,0)), (y,(1,1))\}$
   (f) $\{(0,0,0,0), (0,0,0,1), (0,0,1,0), (0,0,1,1), (0,1,0,0), (0,1,0,1), (0,1,1,0), (0,1,1,1), (1,0,0,0),$
   $(1,0,0,1), (1,0,1,0), (1,0,1,1), (1,1,0,0), (1,1,0,1), (1,1,1,0), (1,1,1,1)\}$
   (g) $\{\emptyset, \{(0,0)\}, \{(0,1)\}, \{(1,0)\}, \{(1,1)\}, \{(0,0),(0,1)\}, \{(0,0),(1,0)\}, \{(0,0),(1,1)\}, \{(0,1),(1,0)\},$
   $\{(0,1),(1,1)\}, \{(1,0),(1,1)\}, \{(0,0),(0,1),(1,0)\}, \{(0,0),(0,1),(1,1)\}, \{(0,0),(1,0),(1,1)\},$
   $\{(0,1),(1,0),(1,1)\}, \{(0,0),(0,1),(1,0),(1,1)\}\}$
   (h) $\mathscr{P}(B) \times \mathscr{P}(B) = \{(\emptyset,\emptyset), (\emptyset,\{0\}), (\emptyset,\{1\}), (\emptyset,B), (\{0\},\emptyset), (\{0\},\{0\}), (\{0\},\{1\}), (\{0\},B), (\{1\},\emptyset),$
   $(\{1\},\{0\}), (\{1\},\{1\}), (\{1\},B), (B,\emptyset), (B,\{0\}), (B,\{1\}), (B,B)\}$
12. **(a)** $A \times B = B \times A \leftrightarrow A = B$ **(b)** $(A \times B) \cap (B \times A) = \emptyset \leftrightarrow A \cap B = \emptyset$
14. **(a)** C := $[2,3,5,7,11,13,17,19,23]$ **(b)** C := $[3..17]$
   (c) C := $[1..25] - [7,14,21]$ **(d)** C := $(A - B) + (B - A)$
   (alternately, C := $(A + B) - (A*B))$ **(e)** C := $C - (A + B)$
18. **(a)** A[J] or B[J]   **(b)** A[J] AND B[J]   **(c)** NOT A[J]

Exercises 2.3

1. The union is $(-1, 100)$; the intersection is $(-\frac{1}{50}, 2)$.
3. **(a)** The union is $\mathbb{Z}$; the intersection is $A_1 = \{-1, 0, 1, 2\}$.
   (b) The union is $(-1, 2)$; the intersection is $\{0\}$.
   (c) The union is $\mathbb{N}$; the intersection is $\emptyset$.
   (d) The union is $\mathbb{Z}$; the intersection is $\{0\}$.
5. **(a)** The union is $(0, \infty)$; the intersection is $\{1\}$.
   (b) The union is $\mathbb{R} \times \mathbb{R}$; the intersection is $\emptyset$.
   (c) The union is $U - \{(0,0,\ldots,0)\}$; the intersection is $\{(1,1,\ldots,1)\}$.
   (d) The union is U; the intersection is the set of "global" identifiers.
   (e) The union is $U - \{\emptyset\}$; the intersection is B.
   (f) The union is $\{(0,0)\} \cup \{(x,y) \mid x > 0\}$ (given (a,b) with $a > 0$, $(a,b) \in A_i$ for $2ai = (a^2 + b^2)$);
   the intersection is $\{(0,0)\}$.

Exercises 2.4

1. 3
2. **(a)** 18   **(b)** 27

3. (a) 10^9 **(b)** 9^9 **(c)** $9^2 10^7$ **(d)** $10 \cdot 9 \cdot 8 \cdot 7 \cdot 6 \cdot 5 \cdot 4 \cdot 3 \cdot 2$
5. (a) 34,632   **(b)** 1,280,448   **(c)** 47,309,184   **(d)** 33,696   **(e)** 47,342,880
7. (a) 52^4 **(b)** 39^4 **(c)** 13^4 **(d)** $4 \cdot 13^4$ **(e)** $13 \cdot 48 \cdot 52^2$
   (f) $13 \cdot 52^3 + 48 \cdot 52^3 - 13 \cdot 48 \cdot 52^2$
8. (a) 184   **(b)** 41
9. (a) 27   **(b)** 12   **(c)** 7
10. (a) 4   **(b)** 10   **(c)** 35   **(d)** 70
11. (a) 3125   **(b)** 120   **(c)** 1200   **(d)** 240
12. (a) 1728   **(b)** 60   **(c)** 525   **(d)** 27   **(e)** 216   **(f)** 972   **(g)** 999
13. (a) $4 \cdot 25^3$ **(b)** $25^4 + 4 \cdot 25^3$ **(c)** $26^4 - 25^4$ **(d)** $5 \cdot 4 \cdot 21^3$ **(e)** $26^4 - 21^4$
14. (a) 1296   **(b)** 360
15. 7

Chapter 2 Problems

2. (a) R' **(b)** $R \cap W$ **(c)** $D \cap R \cap W'$ **(d)** $W \cap (D' \cup R)$ **(e)** $R \cap D' \cap W'$
   (f) $D \cup W \cup R'$
3. (a) $D \cap E'$ **(b)** $B \cup E$ **(c)** $A \cap D \cap E'$ **(d)** $A' \cap (C \cup D) \cap E$
13. 19 students
14. The union is A_{100}; the intersection is A_1.
15. The union is $\mathbb{N}$; the intersection is $\emptyset$.
16. The union is $\mathbb{Z}$; the intersection is A_1.
17. The union is $(0, \infty)$; the intersection is $\{1\}$.
18. The union is $\mathbb{Q}$; the intersection is $\mathbb{Z}$.
19. The union is $(-3, 1)$; the intersection is $\emptyset$.
20. The union is $[0, \infty)$; the intersection is $\emptyset$.
21. The union is $(\frac{1}{2}, 3)$; the intersection is $(1, 2)$.
22. (a) 40,320   **(b)** 1,373,568
23. The union is $\{(0,0)\} \cup \{(x, y) \mid x \neq 0 \land y > 0\}$. (Hint: Given (a, b), $a \neq 0$, $b > 0$, we have $(a, b) \in$
   A_i for $a^2 i = b$.) The intersection is $\{(0,0)\}$.
24. (a) 2^{10} **(b)** 4^{10}
25. (a) $9 \cdot 10^6$ **(b)** 10^5 **(c)** $9^2 \cdot 10^8$

Exercises 3.1

1. (a) yes   **(b)** no   **(c)** no   **(d)** yes
3. (a) 3   **(b)** 2   **(c)** 2   **(d)** 1

Exercises 3.2

1. (a) 297 DIV 11 = 27, 297 MOD 11 = 0
   (b) -63 DIV 9 = -7, -63 MOD 9 = 0
   (c) 77 DIV 8 = 9, 77 MOD 8 = 5
   (d) -71 DIV 6 = -12, -71 MOD 6 = 1
   (e) 35 DIV $-5 = -7$, 35 MOD $-5 = 0$
   (f) 39 DIV 6 = 6, 39 MOD 6 = 3
4. Since $ac \mid bc$, there exists $q \in \mathbb{Z}$ such that $(ac)q = bc$. Since $c \neq 0$, we may divide both sides of this last
   identity by c to obtain $aq = b$. Thus $a \mid b$.
5. This is false; let $a = 6$, $b = 8$, and $c = 9$.
9. (a) 0   **(b)** 4   **(c)** 1   **(d)** 1   **(e)** 1   **(f)** 5

11. Let n, $n + 1$, and $n + 2$ be the three integers. If n MOD $3 = 0$, then $3 | n$.
If n MOD $3 = 1$, then there is an integer q such that $n = 3q + 1$. Thus $n + 2 = 3q + 3 = 3(q + 1)$, showing that $3 | (n + 2)$. Similarly, if n MOD $3 = 2$, then it can be shown that $3 | (n + 1)$.

12. (c) By repeated squaring, 3^2 MOD $7 = 2$, 3^4 MOD $7 = 4$, 3^8 MOD $7 = 16$ MOD $7 = 2$, 3^{16} MOD $7 = 4$, and 3^{32} MOD $7 = 2$. Then 3^{40} MOD $7 = (3^{32} \cdot 3^8)$ MOD $7 = (3^{32}$ MOD $7)(3^8$ MOD $7) = (2)(2) = 4$.

    (d) $(1^3 + 2^3 + 3^3 + 4^3 + \cdots + 100^3)$ MOD $4 = [(1^3$ MOD $4) + (2^3$ MOD $4) + (3^3$ MOD $4) + (4^3$ MOD $4) + \cdots + (97^3$ MOD $4) + (98^3$ MOD $4) + (99^3$ MOD $4) + (100^3$ MOD $4)]$ MOD $4 = (1 + 0 + 3 + 0 + \cdots + 1 + 0 + 3 + 0)$ MOD $4 = 100$ MOD $4 = 0$

13. The hint is that x^2 MOD $5 = (x$ MOD $5)^2$ MOD 5, and x MOD $5 \in \{0, 1, 2, 3, 4\}$. So 0^2 MOD $5 = 0$, 1^2 MOD $5 = 1$, 2^2 MOD $5 = 4$, 3^2 MOD $5 = 4$, 4^2 MOD $5 = 1$. Thus, for any $x \in \mathbb{Z}$, x^2 MOD $5 \in \{0, 1, 4\}$.

Exercises 3.3

1. (a) $d = 4$, $s = 25$, $t = -11$ **(b)** $d = 14$, $s = -7$, $t = 5$

3. This implies that $\gcd(a, b) = 1$.

5. (a) Let m and $m + 1$ be two consecutive integers. Then $1 = (-1)(m) + (1)(m + 1)$, showing that m and $m + 1$ are relatively prime.

    (b) This is false; for example, let $m = 3$.

    (c) This is true, since $1 = (2m + 1)(-3) + (3m + 2)(2)$.

7. (a) Let r_{i+1} be the remainder at the next stage. Then $\gcd(a, b) = \gcd(r_{i+1}, r_i) = 1$ or r_i.

    (b) Check if $r_i | a$ and $r_i | b$.

    (c) 371 MOD $40 = 11$, which is prime. Since 11 does not divide 40, $\gcd(40, 371) = 1$.

    (d) 325 MOD $52 = 13$, which is prime. Since $13 | 52$ and $13 | 325$, $\gcd(52, 325) = 13$.

13. $d = 1$, $s = -11$, $t = 4$

Exercises 3.4

1. (a) $4725 = 3^3 \cdot 5^2 \cdot 7$ **(b)** $9702 = 2 \cdot 3^2 \cdot 7^2 \cdot 11$ **(c)** $25,625 = 5^4 \cdot 41$

2. (a) $\gcd(a, b) = p_1^{\delta_1} \cdot p_2^{\delta_2} \cdot \cdots \cdot p_n^{\delta_n}$, where $\delta_i = \min\{\alpha_i, \beta_i\}$, $1 \le i \le n$.

    (b) $4725 = 2^0 \cdot 3^3 \cdot 5^2 \cdot 7^1 \cdot 11^0$ and $9702 = 2^1 \cdot 3^2 \cdot 5^0 \cdot 7^2 \cdot 11^1$, so $\gcd(4725, 9702) = 2^0 \cdot 3^2 \cdot 5^0 \cdot 7^1 \cdot 11^0 = 63$.

4. (a) Since p is an integer, p MOD $4 \in \{0, 1, 2, 3\}$, and since p is odd, p MOD 4 is either 1 or 3.

    (b) Since p is an odd integer, we have p MOD $6 \in \{1, 3, 5\}$. Since p is prime, p MOD $6 \ne 3$, since 3 is a factor of $6n + 3$ for any $n \in \mathbb{Z}$.

Exercises 3.5

1. Let $S = \{n \in \mathbb{N} \mid 1 + 3 + \cdots + (2n - 1) = n^2\}$. It is easy to see that $1 \in S$. Assume $k \in S$ for some arbitrary integer $k \ge 1$; that is, assume $1 + 3 + \cdots + (2k - 1) = k^2$. Adding $2k + 1$ to both sides of this identity yields $1 + 3 + \cdots + (2k - 1) + (2k + 1) = k^2 + (2k + 1) = (k + 1)^2$. This shows that $k + 1 \in S$. Therefore, by PMI, $S = \mathbb{N}$.

3. Let $S = \{n \in \mathbb{N} \mid 1^2 + 2^2 + \cdots + n^2 = n(n + 1)(2n + 1)/6\}$. It is easy to check that $1 \in S$. Assume $k \in S$ for some arbitrary integer $k \ge 1$; that is, assume $1^2 + 2^2 + \cdots + k^2 = k(k + 1)(2k + 1)/6$. Adding $(k + 1)^2$ to both sides of this identity yields $1^2 + 2^2 + \cdots + k^2 + (k + 1)^2 = k(k + 1)(2k + 1)/6 + (k + 1)^2 = (k + 1)[k(2k + 1) + 6(k + 1)]/6 = (k + 1)(2k^2 + 7k + 6)/6 = (k + 1)(k + 2)(2k + 3)/6$. This shows that $k + 1 \in S$. Therefore, by PMI, $S = \mathbb{N}$.

5. Let $S = \{n \in \mathbb{N}$ such that $6 | (n^3 + 5n)\}$. Since $6 | (1^3 + 5 \cdot 1)$, we have $1 \in S$. Assume $k \in S$ for some arbitrary integer $k \ge 1$; that is, assume $6 | (k^3 + 5k)$. Also it can be argued that $6 | [3k(k + 1)]$, since either

k or $k + 1$ must be even. It then follows by Theorem 3.2, part 2, that $6|[(k^3 + 5k) + 3k(k + 1) + 6]$—namely, that $6|[(k + 1)^3 + 5(k + 1)]$. This shows that $k + 1 \in S$. Therefore, by PMI, $S = \mathbb{N}$.

6. (a) No; in fact, both $P_1(10)$ and $P_1(11)$ are false.
   (b) No; in fact, $P_2(n)$ is false for every $n \in \mathbb{N}$.

Exercises 3.6

1. Let $M = \{2, 3, 4, \ldots\}$ and let $S = \{n \in M \mid n$ can be factored as a product of primes$\}$. We note that $2 \in S$. Let n be an arbitrary integer, $n > 2$, and assume that $k \in S$ for each integer k, $2 \le k < n$. We wish to show that $n \in S$. This is certainly the case if n is prime, so we may assume that n is composite, say, $n = s \cdot t$, where $s, t \in \mathbb{N}$ and $1 < s \le t < n$. Then, by the induction hypothesis, both $s \in S$ and $t \in S$. Thus both s and t can be factored as a product of primes and, since $n = s \cdot t$, this means that n can be so factored also. So $n \in S$. Therefore, by PMI, $S = M$.

Chapter 3 Problems

1. (a) 100 DIV 13 = 7, 100 MOD 13 = 9,   **(b)** -100 DIV 13 $= -8$, -100 MOD 13 = 4
   (c) 100 DIV $-13 = -7$, 100 MOD $-13 = 9$
   (d) -100 DIV $-13 = 8$, -100 MOD $-13 = 4$
2. (a) $\mathbb{N} \cup \{0\}$ **(b)** $\{\ldots, -4, -2, 0, 2, 4, \ldots\}$
5. (a) $24{,}635{,}975 = 5^2 \cdot 7^3 \cdot 13^2 \cdot 17^1$ **(b)** $1{,}376{,}375 = 5^3 \cdot 7^1 \cdot 11^2 \cdot 13^1$
   (c) $\gcd(24{,}635{,}975, 1{,}376{,}375) = 5^2 \cdot 7^1 \cdot 13^1$
7. $[m|(35n + 26) \wedge m|(7n + 3)] \to m|[(35n + 26) + (7n + 3)(-5)] \to m|11$. Now, since $m|11$ and $m > 1$, we have that $m = 11$.
9. (a) $[m \text{ MOD } 6 = 5 \wedge m \text{ DIV } 6 = q] \to m = 6q + 5 \to m = 3(2q + 1) + 2 \to m \text{ MOD } 3 = 2$
   (b) This is false; let $m = 8$, for example.
11. (a) $17 = (357)(-7) + (629)(4)$ **(b)** $1 = (1109)(-2353) + (4999)(522)$
12. $2n$
13. Assume $n = m^2 = t^3$, where $m \text{ MOD } 7 = r$ and $t \text{ MOD } 7 = s$, $0 \le r, s < 7$. Then $m^2 \text{ MOD } 7 = r^2 \text{ MOD } 7 \in \{0, 1, 2, 4\}$ and $t^3 \text{ MOD } 7 = s^3 \text{ MOD } 7 \in \{0, 1, 6\}$. Thus $n \text{ MOD } 7 \in \{0, 1\}$.
21. (a) 3, 5   **(b)** 2, 6, 7, 8
25. (a) Each α_i is even.   **(b)** Each β_i is a multiple of 3.   **(c)** $\alpha_i \le \beta_i$ for each i.
27. (a) Let $S = \{n \in \mathbb{N} \mid 1 \cdot 1 + 2 \cdot 2 \cdot 1 + \cdots + n \cdot n \cdot (n - 1) \cdot \cdots \cdot 1 = ((n + 1) \cdot n \cdot \cdots \cdot 1) - 1\}$.
   Note that $1 \in S$. Assume $k \in S$ for some arbitrary integer $k \ge 1$; that is, assume $1 \cdot 1 + 2 \cdot 2 \cdot 1 + \cdots + k \cdot k \cdot (k - 1) \cdot \cdots \cdot 1 = ((k + 1) \cdot k \cdot \cdots \cdot 1) - 1$. Adding $(k + 1) \cdot (k + 1) \cdot k \cdots \cdots 1$ to both sides of this identity yields $1 \cdot 1 + 2 \cdot 2 \cdot 1 + \cdots + k \cdot k \cdot (k - 1) \cdot \cdots \cdot 1 + (k + 1) \cdot (k + 1) \cdot k \cdot \cdots \cdot 1 = [((k + 1) \cdot k \cdot \cdots \cdot 1) - 1] + (k + 1) \cdot (k + 1) \cdot k \cdot \cdots \cdot 1 = [((k + 1) \cdot k \cdot \cdots \cdot 1)((k + 1) + 1)] - 1 = [(k + 2) \cdot (k + 1) \cdot \cdots \cdot 1] - 1$, showing that $k + 1 \in S$. Therefore, by PMI, $S = \mathbb{N}$.
   (c) Let $S = \{n \in \mathbb{N} \mid 1^2 - 2^2 + \cdots + (-1)^{n+1}n^2 = (-1)^{n+1}n(n + 1)/2\}$. Note that $1 \in S$. Assume $k \in S$ for some arbitrary integer $k \ge 1$; that is, assume $1^2 - 2^2 + \cdots + (-1)^{k+1}k^2 = (-1)^{k+1}k(k + 1)/2$. Adding $(-1)^{k+2}(k + 1)^2$ to both sides of this identity yields $1^2 - 2^2 + \cdots + (-1)^{k+1}k^2 + (-1)^{k+2}(k + 1)^2 = (-1)^{k+1}k(k + 1)/2 + (-1)^{k+2}(k + 1)^2 = (-1)^{k+2}(k + 1)(-k + (2k + 2))/2 = (-1)^{k+2}(k + 1)(k + 2)/2$, showing that $k + 1 \in S$. Therefore, by PMI, $S = \mathbb{N}$.
   (e) Let $S = \{n \in \mathbb{N} \mid 1 \cdot 2 + 2 \cdot 3 + \cdots + n(n + 1) = n(n + 1)(n + 2)/3\}$. Note that $1 \in S$. Assume $k \in S$ for some arbitrary integer $k \ge 1$; that is, assume $1 \cdot 2 + 2 \cdot 3 + \cdots + k(k + 1) = k(k + 1)(k + 2)/3$. Adding $(k + 1)(k + 2)$ to both sides of this identity yields

$1 \cdot 2 + 2 \cdot 3 + \cdots + k(k + 1) + (k + 1)(k + 2) = k(k + 1)(k + 2)/3 + (k + 1)(k + 2) = (k + 1)(k + 2)(k + 3)/3$, showing that $k + 1 \in S$. Therefore, by PMI, $S = \mathbb{N}$.

33. The flaw in the argument is in step 3; if $k = 1$, then it is not possible to choose subsets A and B of S such that $|A| = |B| = k$, $A \cap B \neq \emptyset$, and $A \cup B = S$.

Exercises 4.1

3. We give the adjacency matrices.

(a) $\begin{bmatrix} 1 & 1 & 1 & 1 \\ 0 & 1 & 0 & 1 \\ 0 & 0 & 1 & 1 \\ 0 & 0 & 0 & 1 \end{bmatrix}$
(b) $\begin{bmatrix} 0 & 1 & 1 & 1 \\ 0 & 0 & 1 & 1 \\ 0 & 0 & 0 & 1 \\ 0 & 0 & 0 & 0 \end{bmatrix}$

(c) $\begin{bmatrix} 1 & 0 & 0 & 0 \\ 0 & 1 & 0 & 0 \\ 0 & 0 & 1 & 0 \\ 0 & 0 & 0 & 1 \end{bmatrix}$
(d) $\begin{bmatrix} 0 & 1 & 1 & 1 \\ 1 & 0 & 1 & 1 \\ 1 & 1 & 0 & 1 \\ 1 & 1 & 1 & 0 \end{bmatrix}$

(e) $\begin{bmatrix} 1 & 1 & 1 & 1 \\ 0 & 0 & 0 & 1 \\ 0 & 0 & 0 & 0 \\ 0 & 0 & 0 & 0 \end{bmatrix}$
(f) $\begin{bmatrix} 0 & 1 & 0 & 1 \\ 1 & 0 & 1 & 0 \\ 0 & 1 & 0 & 1 \\ 1 & 0 & 1 & 0 \end{bmatrix}$

5. **(a), (c), (d), (f)** The domain and image are both $\{1, 2, 3, 6\}$.
   (b) The domain is $\{1, 2, 3\}$ and the image is $\{2, 3, 6\}$.
   (e) The domain is $\{1, 2\}$; the image is $\{1, 2, 3, 6\}$.

6. The adjacency matrix is $\begin{bmatrix} 0 & 1 & 1 & 1 & 1 & 1 \\ 0 & 0 & 1 & 0 & 1 & 1 \\ 0 & 0 & 0 & 0 & 0 & 1 \\ 0 & 0 & 0 & 0 & 1 & 1 \\ 0 & 0 & 0 & 0 & 0 & 1 \\ 0 & 0 & 0 & 0 & 0 & 0 \end{bmatrix}$

Exercises 4.2

1. (a) reflexive, symmetric   **(b)** irreflexive, antisymmetric, transitive
   (c) irreflexive, symmetric   **(d)** reflexive, symmetric, transitive   **(e)** none
3. (a) reflexive, symmetric, transitive   **(b)** irreflexive, antisymmetric
   (c) reflexive, transitive   **(d)** reflexive, symmetric, transitive
5. (a) irreflexive, symmetric   **(b)** reflexive, symmetric
   (c) reflexive, antisymmetric, transitive   **(d)** reflexive, symmetric, transitive
   (e) irreflexive, symmetric   **(f)** symmetric, transitive
7. (a) reflexive, antisymmetric, transitive   **(b)** reflexive, symmetric, transitive
   (c) reflexive, transitive   **(d)** irreflexive, symmetric
   (e) reflexive, symmetric, transitive   **(f)** reflexive, symmetric, transitive
8. (a) R is not reflexive, since $(X, X) \notin R$; R is not irreflexive, since $(\emptyset, \emptyset) \in R$.
   (b) R is not transitive, since $(X, \emptyset) \in R$ and $(\emptyset, X) \in R$ but $(X, X) \notin R$.
9. (a) reflexive, transitive   **(b)** irreflexive, symmetric
   (c) reflexive, symmetric, transitive   **(d)** irreflexive, antisymmetric, transitive
   (e) reflexive, antisymmetric, transitive   **(f)** none
10. Only **(k)**, **(n)** and **(o)** are false.

Exercises 4.3

1. The relation R_4 is an equivalence relation; the distinct equivalence classes are $[1] = \{1, 3, 5\}$ and $[2] = \{2, 4\}$.

3. The relation R_1, being the complete relation on $\{-2, -1, 0, 1, 2\}$, is an equivalence relation; the only equivalence class is $[0] = \{-2, -1, 0, 1, 2\}$. The relation R_4 is also an equivalence relation; the distinct equivalence classes are $[0] = \{0\}$, $[1] = \{-1, 1\}$, and $[2] = \{-2, 2\}$.

5. The relation R_4 is an equivalence relation; the distinct equivalence classes are $[2] = \{2, 4, 6, 8, \ldots\}$ and $[3] = \{3, 5, 7, 9, \ldots\}$.

6. (a) $\{\{\text{triangles}\}, \{\text{quadrilaterals}\}, \{\text{pentagons}\}, \{\text{hexagons}\}, \ldots\}$
   (b) $\{\{2\}, \{1, 3\}, \{0, 4\}, \{-1, 5\}, \{-2, 6\}, \ldots\}$

7. The relations R_2, R_5, and R_6 are equivalence relations. For R_2, the distinct equivalence classes are $[1] = \{\ldots, -3, -1, 1, 3, \ldots\}$ and $[2] = \{\ldots, -4, -2, 2, 4, \ldots\}$. For R_5, the distinct equivalence classes are $[1] = \{\ldots, -5, -2, 1, 4, 7, \ldots\}$, $[2] = \{\ldots, -4, -1, 2, 5, 8, \ldots\}$, and $[3] = \{\ldots, -6, -3, 3, 6, \ldots\}$. For R_6, the distinct equivalence classes are $[1] = \{1, 2, 3, \ldots\}$ and $[-1] = \{\ldots, -3, -2, -1\}$.

8. (a) $[-22] = [-12] = [-6] = [-2] = [6] = [8] = [44]$, $[-3] = [3] = [39]$
   (b) $[-22] = [8] = [44]$, $[-12] = [-6] = [-3] = [3] = [6] = [39]$
   (c) $[-22] = [-12] = [-2] = [3] = [8]$, $[-6] = [39] = [44]$
   (d) $[-12] = [-3] = [6]$, $[-6] = [3] = [39]$, $[8] = [44]$

9. The relation of part (c) is the only equivalence relation; $[1] = \{1\}$, $[2] = \{2, 3, 4\}$.

10. (a) true   (b) false   (c) false

12. m

13. (a) $R_1 = \{(1, 1), (2, 2), (3, 3), (4, 4), (5, 5), (6, 6)\}$
    (b) $R_2 = \{(1, 1), (2, 2), (2, 3), (3, 2), (3, 3), (4, 4), (4, 5), (4, 6), (5, 4), (5, 5), (5, 6), (6, 4), (6, 5), (6, 6)\}$
    (c) $R_3 = \{(1, 1), (1, 3), (1, 5), (3, 1), (3, 3), (3, 5), (5, 1), (5, 3), (5, 5), (2, 2), (2, 4), (2, 6), (4, 2), (4, 4),$
       $(4, 6), (6, 2), (6, 4), (6, 6)\}$
    (d) $R_4 = \{1, 2, 3, 4, 5, 6\} \times \{1, 2, 3, 4, 5, 6\}$

15. (a) 1   (b) 2   (c) 5   (d) 18

Exercises 4.4

1. (a) yes   (b) yes   (c) yes   (d) yes   (e) no, not reflexive   (f) yes   (g) yes
   (h) no, not transitive   (i) no, not transitive   (j) no, not antisymmetric

3. The relation of part (e) is a partial-order relation.

7. (b) $\text{glb}(\{(-1, 4), (3, 2)\}) = (-1, 2)$; $\text{lub}(\{(-1, 4), (3, 2)\}) = (3, 4)$
   (c) $(a, b) \wedge (c, d) = (\min\{a, c\}, \min\{b, d\})$; $(a, b) \vee (c, d) = (\max\{a, c\}, \max\{b, d\})$

9. (b) $\text{glb}(\{(-1, 4), (3, 2)\}) = (-1, 4)$; $\text{lub}(\{(-1, 4), (3, 2)\}) = (3, 2)$
   (c) $\text{glb}(\{(-1, 4), (-1, 7)\}) = (-1, 4)$; $\text{lub}(\{(-1, 4), (-1, 7)\}) = (-1, 7)$

12. Note that $x \wedge x = x = x \vee x$ for all $x \in \{a, b, c, d, e, f\}$. Also $a \wedge x = x$ and $a \vee x = a$ for all $x \in \{b, c, d, e, f\}$, and $f \vee x = x$ and $f \wedge x = f$ for all $x \in \{a, b, c, d, e\}$. Moreover, $x \wedge y = y \wedge x$ and $x \vee y = y \vee x$ for all $x, y \in \{a, b, c, d, e, f\}$. Now $b \wedge c = d$, $b \wedge d = d$, $b \wedge e = f$, $c \wedge d = d$, $c \wedge e = e$, and $d \wedge e = f$. Also $b \vee c = a$, $b \vee d = b$, $b \vee e = a$, $c \vee d = c$, $c \vee e = c$, and $d \vee e = c$.

14. $(S_{n_0}, \mid)$ is a totally ordered set if and only if $n_0 = p^k$, where p is prime and k is a positive integer.

15. (b) $\begin{bmatrix} 1 & 0 \\ 0 & 1 \end{bmatrix} \wedge \begin{bmatrix} 0 & 1 \\ 0 & 1 \end{bmatrix} = \begin{bmatrix} 0 & 0 \\ 0 & 1 \end{bmatrix}$

$\begin{bmatrix} 1 & 0 \\ 0 & 1 \end{bmatrix} \vee \begin{bmatrix} 0 & 1 \\ 0 & 1 \end{bmatrix} = \begin{bmatrix} 1 & 1 \\ 0 & 1 \end{bmatrix}$

(c) $\begin{bmatrix} 1 & 0 \\ 0 & 1 \end{bmatrix} \wedge \begin{bmatrix} 0 & 1 \\ 1 & 1 \end{bmatrix} = \begin{bmatrix} 0 & 0 \\ 0 & 1 \end{bmatrix}$

$\begin{bmatrix} 1 & 0 \\ 0 & 1 \end{bmatrix} \vee \begin{bmatrix} 0 & 1 \\ 1 & 1 \end{bmatrix} = \begin{bmatrix} 1 & 1 \\ 1 & 1 \end{bmatrix}$

17. (a) i (b) a (c) does not exist   (d) does not exist   (e) does not exist   (f) b

Chapter 4 Problems

1. (a) R is not reflexive provided $(a, a) \notin R$ for some $a \in A$.
  (b) R is not irreflexive provided $(a, a) \in R$ for some $a \in A$.
  (c) R is not symmetric provided, for some $a, b \in R$, $(a, b) \in R$ and $(b, a) \notin R$.
  (d) R is not antisymmetric provided, for some $a, b \in R$, $a \neq b$, $(a, b) \in R$ and $(b, a) \in R$.
  (e) R is not transitive provided, for some $a, b, c \in R$, $(a, b) \in R$ and $(b, c) \in R$ and $(a, c) \notin R$.
3. Only (g), (h), (i), (m), and (n) are false.
5. (a) Yes; $[(a, b)]$ is the line $y = x + (b - a)$ with slope 1 and y-intercept $b - a$.
  (b) Yes; $[(1, 0)] = \{(1, 0)\}$; for $(a, b) \neq (1, 0)$, $[(a, b)]$ is the circle centered at $(1, 0)$ with radius $(a - 1)^2 + b^2$.
  (c) No
  (d) Yes; $[(0, 0)] = \{(0, 0)\}$; for $(a, b) \neq (0, 0)$, $[(a, b)]$ is the square with vertices at $(r, 0)$, $(0, r)$, $(-r, 0)$, and $(0, -r)$, where $r = |a| + |b|$.
  (e) Yes; $[(0, 0)] = \{(x, y) \mid x = 0 \vee y = 0\}$; for $ab \neq 0$, $[(a, b)]$ is the hyperbola $y = ab/x$.
6. (a) For $m \in \mathbb{Z}$, $[m] = [m, m + 1)$.
  (b) For $m \in \mathbb{Z}$, $[m] = [(2m - 1)/2, (2m + 1)/2)$.
  (c) $[0] = \{0\}$; for $x > 0$, $[x] = \{-x, x\}$.
  (d) $[0] = \mathbb{Z}$; for $x \in \mathbb{R} - \mathbb{Z}$, $[x] = \{\ldots, x - 2, x - 1, x, x + 1, x + 2, \ldots\}$.
  (e) $[0] = \mathbb{Q}$; for $x \in \mathbb{R} - \mathbb{Q}$, $[x] = \{x + r \mid r \in \mathbb{Q}\}$.
16. (a) reflexive, transitive   (b) irreflexive, symmetric   (c) irreflexive, symmetric
  (d) reflexive, symmetric   (e) symmetric
17. (a) i (b) b (c) j (d) b (e) does not exist   (f) does not exist
18. (a) $\neq$ (b) $<$ (c) Define R by $(m, n) \in R \leftrightarrow |m - n| \leq 1$.   (d) $\leq$
19. (a) a (b) i (c) b (d) h (e) m (f) a (g) j (h) m
22. Only (f) is false.
29. All the statements are true.

Exercises 5.1

1. f_1, f_2, and f_4 are functions.
2. g_1, g_2, and g_4 are well-defined.
3. (a) f is a function; im $f = [0, \infty)$.
  (b) g is not a function, since, for example, $(0, 2) \in g$ and $(0, -2) \in g$.
  (c) h is a function; im $h = P$.
  (d) This f is not in general a function, since a given x may have several uncles.
  (e) This g is a function; im $g = \{y \mid y$ is an eldest son and y is a parent or uncle$\}$.
4. f_3, f_4, f_5, and f_7 are well-defined.

Exercises 5.2

1. The functions g, p, and c are one-to-one; n is one-to-one iff $|A| = 1$.
3. The functions g, h, n, and c are onto.
4. Hint: There are 12 such functions.

5. The functions f_2, f_4, f_5, f_6, and f_8 are one-to-one.

7. The functions f_1, f_4, f_5, f_7, and f_8 are onto.

8. **(a)** If $r < m$, then f is not onto.   **(b)** If $m < r$, then f is not one-to-one.

9. The functions f_1, f_4, f_5, f_6, and f_9 are one-to-one.

10. **(a)** f_1 is not a permutation; $f_1(C) = \{[2], [10]\}$, $f_1^{-1}(D) = \{[2], [4], [8], [10]\}$
    (b) f_2 is not a permutation; $f_2(C) = \{[4], [8]\}$, $f_2^{-1}(D) = \{[1], [2], [4], [5], [7], [8], [10], [11]\}$
    (c) f_3 is a permutation; $f_3(C) = \{[1], [5], [7], [11]\}$, $f_3^{-1}(D) = D$

11. The functions f_1, f_4, f_5, and f_6 are onto.

12.**(a)** $f_1(n) = 1$ **(b)** $f_2(n) = 2n$ **(c)** the function f_7 of Exercise 5   **(d)** $f_3(n) = n$

13.**(a)** $f_1(C) = \{2, 3\}$, $f_1^{-1}(D) = \{1, 3, 4\}$ **(b)** $f_2(C) = \{1, 3\}$, $f_2^{-1}(D) = \{2\}$
    (c) $f_3(C) = \{2\}$, $f_3^{-1}(D) = \emptyset$ **(d)** $f_4(C) = \{4\} = f_4^{-1}(D)$
    (e) $f_5(C) = C$, $f_5^{-1}(D) = \{\dots, -3, -2, -1\}$
    (f) $f_6(C) = \{2, 4, 6, \dots\}$, $f_6^{-1}(D) = \{\dots, -6, -4, -2, 0\} \cup \mathbb{N}$
    (g) $f_7(C) = \mathbb{N}$, $f_7^{-1}(D) = \{3, 4, 7, 8, 11, 12, \dots\}$
    (h) $f_8(C) = \{1, 3, 5, \dots\} = f_8^{-1}(D)$

Exercises 5.3

1. **(a)** $f_1^{-1} : \mathbb{Q} \to \mathbb{Q}$; $f_1^{-1}(r) = (r - 2)/4$
   (b) $f_2^{-1} : \mathbb{R} - \{1\} \to \mathbb{R} - \{1\}$; $f_2^{-1}(x) = x/(x - 1)$
   (c) $f_3^{-1} : \mathbb{Z}_{12} \to \mathbb{Z}_{12}$; $f_3^{-1}([x]) = [5x]$
   (d) $f_4^{-1} : \mathbb{Z}_{39} \to \mathbb{Z}_{39}$; $f_4^{-1}([x]) = [8(x - 2)]$
   (e) $f_5^{-1} : \mathbb{Z} \to \mathbb{Z}$; $f_5^{-1}(m) = m - 1$

   (f) $f_6^{-1} : \mathbb{N} \to \mathbb{Z}$; $f_6^{-1}(m) = \begin{cases} (m - 1)/2, & \text{if } m \text{ is odd} \\ -m/2, & \text{if } m \text{ is even} \end{cases}$

   (g) $f_7^{-1} : \{1, 2, 3, 4\} \to \{1, 2, 3, 4\}$; $f_7^{-1}(1) = 2$, $f_7^{-1}(2) = 3$, $f_7^{-1}(3) = 4$, $f_7^{-1}(4) = 1$
   (h) $f_8^{-1} : \{1, 2, 3, 4\} \to \{1, 2, 3, 4\}$; $f_8^{-1}(1) = 3$, $f_8^{-1}(2) = 4$, $f_8^{-1}(3) = 1$, $f_8^{-1}(4) = 2$

3. **(a)** $g \circ f : \mathbb{Z} \to (0, \infty)$; $(g \circ f)(m) = 1/(|m| + 1)$
   (b) $g \circ f : \mathbb{R} \to (0, 1)$; $(g \circ f)(x) = x^2/(x^2 + 1)$
   (c) $g \circ f : \mathbb{R} - \{2\} \to \mathbb{R} - \{0\}$; $(g \circ f)(x) = x - 2$
   (d) $g \circ f : \mathbb{R} \to [0, \infty)$; $(g \circ f)(x) = |x|$

   (e) $g \circ f : \mathbb{Q} - \{10/3\} \to \mathbb{Q} - \{2\}$; $(g \circ f)(r) = \dfrac{6r - 14}{3r - 10}$

   (f) $g \circ f : \mathbb{Z} \to \mathbb{Z}_5$; $(g \circ f)(m) = [m + 1]$
   (g) $g \circ f : \mathbb{Z}_8 \to \mathbb{Z}_6$; $(g \circ f)([m]) = [0]$
   (h) $g \circ f : \{1, 2, 3, 4\} \to \{1, 2, 3, 4\}$; $(g \circ f)(1) = 2$, $(g \circ f)(2) = 3$, $(g \circ f)(3) = 4$, $(g \circ f)(4) = 1$

4. Define $f : \{0, 1\} \to \{0, 1, 2\}$ by $f(0) = 0$, $f(1) = 1$, and $g : \{0, 1, 2\} \to \{0, 1\}$ by $g(0) = 0$, $g(1) = 1$, $g(2) = 1$.

5. **(a)** All are functions on $\mathbb{Z}$: $f^{-1}(m) = m - 1$, $g^{-1}(m) = 2 - m$, $(f \circ g)(m) = 3 - m$, $(f \circ g)^{-1}(m) = 3 - m = (g^{-1} \circ f^{-1})(m)$.
   (b) All are functions on $\mathbb{Z}_7$: $f^{-1}([m]) = [m - 3] = [m + 4]$, $g^{-1}([m]) = [4m]$, $(f \circ g)([m]) = [2m + 3]$, $(f \circ g)^{-1}([m]) = [4m + 2] = (g^{-1} \circ f^{-1})([m])$. Again, $g^{-1} \circ f^{-1} = (f \circ g)^{-1}$.
   (c) All are functions on $\{1, 2, 3, 4\}$: $f^{-1}(1) = 2$, $f^{-1}(2) = 3$, $f^{-1}(3) = 4$, $f^{-1}(4) = 1$, $g^{-1}(1) = 3$, $g^{-1}(2) = 4$, $g^{-1}(3) = 1$, $g^{-1}(4) = 2$, $(f \circ g)(1) = 2$, $(f \circ g)(2) = 3$, $(f \circ g)(3) = 4$, $(f \circ g)(4) = 1$, $(f \circ g)^{-1}(1) = 4 = (g^{-1} \circ f^{-1})(1)$, $(f \circ g)^{-1}(2) = 1 = (g^{-1} \circ f^{-1})(2)$, $(f \circ g)^{-1}(3) = 2 = (g^{-1} \circ f^{-1})(3)$, $(f \circ g)^{-1}(4) = 3 = (g^{-1} \circ f^{-1})(4)$.
   (d) All are functions on $\{1, 2, 3, 4\}$: $f^{-1}(1) = 4$, $f^{-1}(2) = 1$, $f^{-1}(3) = 3$, $f^{-1}(4) = 2$, $g^{-1}(1) = 1$, $g^{-1}(2) = 4$, $g^{-1}(3) = 2$, $g^{-1}(4) = 3$, $(f \circ g)(1) = 2$, $(f \circ g)(2) = 3$, $(f \circ g)(3) = 1$, $(f \circ g)(4) = 4$, $(f \circ g)^{-1}(1) = 3$, $(f \circ g)^{-1}(2) = 1$, $(f \circ g)^{-1}(3) = 2$, $(f \circ g)^{-1}(4) = 4$. Again, $(f \circ g)^{-1} = g^{-1} \circ f^{-1}$.

(e) All are functions on $\mathbf{Q}$: $f^{-1}(r) = r/4$, $g^{-1}(r) = 2r + 3$, $(f \circ g)(r) = 2(r - 3)$, $(f \circ g)^{-1}(r) = (r + 6)/2 = (g^{-1} \circ f^{-1})(r)$.

(f) All are functions on $\mathbf{Q} - \{1\}$: $f^{-1}(r) = (r + 1)/2$, $g^{-1}(r) = r/(r - 1)$, $(f \circ g)(r) = (r + 1)/(r - 1)$, $(f \circ g)^{-1}(r) = (r + 1)/(r - 1) = (g^{-1} \circ f^{-1})(r)$.

6. See the answer to Exercise 4.

7. (a) $(g \circ f)(m) = 1 - m = (f^{-1} \circ g^{-1})(m)$

(b) $(g \circ f)([m]) = [2m + 6]$, $(f^{-1} \circ g^{-1})(m) = [4m - 3]$

(c) $(g \circ f)(1) = 2$, $(g \circ f)(2) = 3$, $(g \circ f)(3) = 4$, $(g \circ f)(4) = 1$, $(f^{-1} \circ g^{-1})(1) = 4$, $(f^{-1} \circ g^{-1})(2) = 1$, $(f^{-1} \circ g^{-1})(3) = 2$, $(f^{-1} \circ g^{-1})(4) = 3$

(d) $(g \circ f)(1) = 3$, $(g \circ f)(2) = 2$, $(g \circ f)(3) = 4$, $(g \circ f)(4) = 1$, $(f^{-1} \circ g^{-1})(1) = 4$, $(f^{-1} \circ g^{-1})(2) = 2$, $(f^{-1} \circ g^{-1})(3) = 1$, $(f^{-1} \circ g^{-1})(4) = 3$

(e) $(g \circ f)(r) = (4r - 3)/2$, $(f^{-1} \circ g^{-1})(r) = (2r + 3)/4$

(f) $(g \circ f)(r) = (2r - 1)/(2r - 2) = (f^{-1} \circ g^{-1})(r)$

8. (a) $(f \circ f)(x) =$ the paternal grandfather of x (b) $(f \circ g)(x) = f(x)$

(c) $(g \circ f)(x) =$ the eldest sibling of x (d) $(g \circ g)(x) = g(x)$

Exercises 5.4

1. (a) $\log_2 n$ (b) 1   (c) n^2 (d) 2^n

6. n^4

7. $n \log_2 n$

Exercises 5.5

1. Hint: Define $f: (0, 1) \to (-1, 1)$ by $f(x) = 2x - 1$; show that f is a bijection.

5. Hint: Define $f: [0, 1) \to [0, \infty)$ by $f(x) = x/(1 - x)$; show that f is a bijection.

13. Hint: Use the facts that $\mathbf{Q}$ is denumerable, $\mathbf{R}$ is nondenumerable, and $\mathbf{R} = \mathbf{Q} \cup (\mathbf{R} - \mathbf{Q})$; see Theorem 5.14.

Chapter 5 Problems

1. (a) $f(x) = 1 - x$ (b) $f(x) = x/2$ (c) $f(x) = 4(x - 0.5)^2$ (d) $f(x) = 1/2$

2. (a) $f(x) = 1 - x$ if $x \le 0$; $f(x) = x$ if $x > 0$

(b) $f(x) = 1 - x$ if $x < 0$; $f(x) = x + 2$ if $x \ge 0$

(c) $f(x) = 1 - x$ if $x \le 0$; $f(x) = x/2$ if $x > 0$

(d) $f(x) = 1$

3. (a) one-to-one, not onto   (b) one-to-one and onto   (c) onto, not one-to-one

(d) neither one-to-one nor onto   (c) one-to-one and onto   (f) one-to-one, not onto

5. (a) Both are functions on $\mathbf{R}$: $(f \circ g)(x) = 9x^2 + 27x + 20$, $(g \circ f)(x) = 3(x^2 + x) + 4$

(b) Both are functions on $(0, \infty)$: $(f \circ g)(x) = f(x)$, $(g \circ f)(x) = (x^2 + 1)/x$

9. (a) $(f \circ g)(m) = 2m + 1$ (b) $(g \circ f)(m) = 2(m + 1)$

(c) $(f \circ h)(m) = \begin{cases} 1 & \text{if } m \text{ is even} \\ 2 & \text{if } m \text{ is odd} \end{cases}$

(d) $(h \circ f)(m) = \begin{cases} 1 & \text{if } m \text{ is even} \\ 0 & \text{if } m \text{ is odd} \end{cases}$ (e) $(g \circ h)(m) = \begin{cases} 0 & \text{if } m \text{ is even} \\ 2 & \text{if } m \text{ is odd} \end{cases}$

(f) $(h \circ g)(m) = 0$ (g) $(g \circ g)(m) = 4m$ (h) $(h \circ f \circ g)(m) = 1$

10. (a) Both are functions on $\mathbf{N}$: $(f + g)(x) = x^2 + 2x + 1$, $(f \cdot g)(x) = x^2(2x + 1)$.

(b) Both are functions on $\mathbf{Q}$: $(f + g)(x) = 2(5x + 3)/3$, $(f \cdot g)(x) = x(3x + 2)/3$.

(c) Both are functions on $\mathbf{R} - \{0\}$: $(f + g)(x) = (x^4 + 3x^2 + 1)/[x(x^2 + 1)]$, $(f \cdot g)(x) = 1$.

(d) Both are functions on $\mathbf{R}$: $(f + g)(x) = 0$, $(f \cdot g)(x) = -x^4 + 4x^3 - 10x^2 + 12x - 9$.

15. (b) $g^{-1}: \mathbb{Q} - \{3/2\} \to \mathbb{Q} - \{1/2\}; \; g^{-1}(x) = x/(2x - 3)$
   (e) $t^{-1}: \mathbb{Z}_{149} \to \mathbb{Z}_{149}; \; t^{-1}([n]) = [25n]$

19. We show that f is one-to-one: $f([x_1]) = f([x_2]) \to [15x_1] = [15x_2] \to 119|(15x_1 - 15x_2) \to$
   $119|15(x_1 - x_2) \to 119|(x_1 - x_2) \to [x_1] = [x_2]$. It then follows from Theorem 5.1 that f is a
   permutation. To find f^{-1}, note that $[y] = f([x]) \to [y] = [15x] \to 119|(y - 15x) \to 119q = y - 15x$
   (for some $q \in \mathbb{Z}) \to 8(119q) = 8y - 120x \to 119(8q + x) = 8y - x \to 119|(8y - x) \to [x] = [8y]$.
   Hence $f^{-1}([y]) = [8y]$.

20. Note that $g(1) = 2 = g(4)$, so g is not one-to-one; thus g is also not onto by Theorem 5.1. So g is not a
   permutation of $\{1, 2, \ldots, 100\}$.

21. $g \circ f: \mathbb{Q} - \{1/4\} \to \mathbb{Q} - \{3/2\}; \; (g \circ f)(x) = (1 - 6x)/(1 - 4x),$
   $(g \circ f)^{-1}: \mathbb{Q} - \{3/2\} \to \mathbb{Q} - \{1/4\}; \; (g \circ f)^{-1}(x) = (1 - x)/(6 - 4x),$
   $f^{-1}: \mathbb{Q} - \{0\} \to \mathbb{Q} - \{1/4\}; \; f^{-1}(x) = (1 - x)/4, \; g^{-1}: \mathbb{Q} - \{3/2\} \to \mathbb{Q} - \{0\};$
   $g^{-1}(x) = 1/(3 - 2x), \; f^{-1} \circ g^{-1} = (g \circ f)^{-1}$

27. (a) $f(n) = \Theta(g(n))$ **(b)** $f(n) = \Theta(g(n))$
   (c) $f(n) = O(g(n))$ **(d)** $f(n) = \Theta(g(n))$
   (e) $f(n) = O(g(n))$ **(f)** $f(n) = O(g(n))$
   (g) $f(n) = \Theta(g(n))$ **(h)** $f(n) = O(g(n))$
   (i) $f(n) = \Theta(g(n))$

Exercises 6.1

1. (a) (1)   **(c)** $(1, 2, 3), (1, 3, 2), (2, 1, 3), (2, 3, 1), (3, 1, 2), (3, 2, 1)$
2. (b) $(1, 2, 3, 4), (1, 2, 4, 3), (1, 3, 2, 4), (1, 3, 4, 2), (1, 4, 2, 3), (1, 4, 3, 2), (2, 1, 3, 4), (2, 1, 4, 3), (2, 3, 1, 4),$
   $(2, 3, 4, 1), (2, 4, 1, 3), (2, 4, 3, 1), (3, 1, 2, 4), (3, 1, 4, 2), (3, 2, 1, 4), (3, 2, 4, 1), (3, 4, 1, 2), (3, 4, 2, 1),$
   $(4, 1, 2, 3), (4, 1, 3, 2), (4, 2, 1, 3), (4, 2, 3, 1), (4, 3, 1, 2), (4, 3, 2, 1)$
   (c) $(1, 2, 3, 4, 7, 5, 6)$ **(d)** $(1, 6, 3, 2, 4, 5, 7)$
3. (a) $\{1, 2, 3\}, \{1, 2, 4\}, \{1, 2, 5\}, \{1, 3, 4\}, \{1, 3, 5\}, \{1, 4, 5\}, \{2, 3, 4\}, \{2, 3, 5\}, \{2, 4, 5\}, \{3, 4, 5\}$
4. (b) $\{1, 2, 3\}, \{1, 2, 4\}, \{1, 2, 5\}, \{1, 2, 6\}, \{1, 3, 4\}, \{1, 3, 5\}, \{1, 3, 6\}, \{1, 4, 5\}, \{1, 4, 6\}, \{1, 5, 6\},$
   $\{2, 3, 4\}, \{2, 3, 5\}, \{2, 3, 6\}, \{2, 4, 5\}, \{2, 4, 6\}, \{2, 5, 6\}, \{3, 4, 5\}, \{3, 4, 6\}, \{3, 5, 6\}, \{4, 5, 6\}$
   (c) $\{1, 3, 4, 6\}$ **(d)** $\{2, 5, 6, 7\}$
5. (b) $\emptyset, \{1\}, \{1, 2\}, \{1, 2, 3\}, \{1, 2, 3, 4\}, \{1, 2, 4\}, \{1, 3\}, \{1, 3, 4\}, \{1, 4\}, \{2\}, \{2, 3\}, \{2, 3, 4\}, \{2, 4\}, \{3\},$
   $\{3, 4\}, \{4\}$
   (c) $\{1, 3, 4, 5, 6\}$ **(d)** $\{1, 2, 3, 4, 5, 7\}$ **(e)** $\{2\}$

Exercises 6.2

1. 30
2. 5
3. (a) $8^2 \cdot 10^5$ **(b)** $2 \cdot 8^2 \cdot 10^7$
4. $2^{(2^n)}$
5. $26^3 + 1$
6. (a) 1080   **(b)** $(a_1 + 1)(a_2 + 1)(a_3 + 1)(a_4 + 1)$
9. 8
17. $50^3, 50 \cdot 49^2$

Exercises 6.3

1. 210
2. $C(12, 8) \cdot C(15, 8)$
3. (a) 6   **(b)** 210   **(c)** 720   **(d)** 360

5. $n \geq 10$

6. $1296, 360$

9. (a) 1035 (b) 576 (c) 912

10. $C(20, 12)$

11. (a) 720 (b) 120 (c) 20 (d) 64

12. (a) 24 (b) 6 (c) 42 (d) 78 (e) 114

13. $C(16, 6)$

14. 219

15. $C(52, 5) - 4C(39, 5) + 6C(26, 5) - 4C(13, 5)$

16. (a) 60 (b) 48 (c) 42 (d) 78

17. 720

18. (a) $P(9)$ (b) $P(4) \cdot P(5)$ (c) $P(4) \cdot P(2)^5$ (d) $P(5)^2$

19. (a) $C(13, 1) \cdot C(4, 2) \cdot C(12, 3) \cdot C(4, 1)^3$ (b) $C(13, 1) \cdot C(4, 3) \cdot C(12, 2) \cdot C(4, 1)^2$
   (c) $C(13, 1) \cdot C(4, 3) \cdot C(12, 1) \cdot C(4, 2)$ (d) $C(4, 1) \cdot C(9, 1)$
   (e) $10 \cdot 4^5 - 10 \cdot 4$ (f) $C(4, 1)C(13, 5) - C(4, 1)C(10, 1)$

20. $C(52, 6)C(48, 4)$; $C(52, 6)C(48, 4)C(6, 1)C(4, 1)$

21. (a) 256 (b) 35 (c) 32

22. 2^{33}

23. 286

24. $C(32, 8)$

25. (a) $C(14, 8)$ (b) $C(19, 15)C(14, 10)$

26. (a) 1287 (b) 165 (c) 1056 (d) 657

Exercises 6.4

1. (a) $C(4, 2) = C(3, 2) + C(3, 1) = C(2, 2) + C(2, 1) + C(2, 1) + C(2, 0) = 1 + C(1, 1) + C(1, 0) +$
   $C(1, 1) + C(1, 0) + 1 = 6$

3. (a) $(x + y)^5 = x^5 + 5x^4y + 10x^3y^2 + 10x^2y^3 + 5xy^4 + y^5$
   (b) $(x + y)^6 = x^6 + 6x^5y + 15x^4y^2 + 20x^3y^3 + 15x^2y^4 + 6xy^5 + y^6$
   (c) $(x + 3y)^7 = x^7 + 21x^6y + 189x^5y^2 + 945x^4y^3 + 2835x^3y^4 + 5103x^2y^5 + 5103xy^6 + 2187y^7$

7. (a) $C(15, 4)$ (b) $2^6C(10, 4)$ (c) $-3^3 \cdot 2^5 \cdot C(8, 3)$ (d) $C(6, 5) + C(6, 4) + C(6, 3)$

9. (a) $C(13, 5)$ (b) $C(12, 6)$ (c) 2^{10} (d) $9 \cdot 2^8$

13. (b) $C(6, 3) = 6C(5, 3)/3 = 2C(5, 3) = 2 \cdot 5 \cdot C(4, 3)/2 = 5C(4, 3) = 5 \cdot 4 \cdot C(3, 3) = 20$

Exercises 6.5

1. 1918

2. 120

3. $49, 974, 120$

4. $C(52, 5) - 3C(48, 5) + 3C(44, 5) - C(40, 5)$

5. 64800

6. 366

7. 2830

9. $2(9!) + 2(8!) - 8! - 4(7!) - 6! + 2(6!) + 2(5!) - 4!$

11. (a) 2 (b) 32 (c) 1488 (d) $\sum_{k=0}^{n}(-2)^k(2n - k - 1)!C(n, k)$

12. $10^{30} - 10^{15} - 10^{10} - 10^6 + 10^5 + 10^3 + 10^2 - 10$

13. $P(C(8, 3), 5) - C(8, 1)P(C(7, 3), 5) + C(8, 2)P(C(6, 3), 5) - C(8, 3)P(C(5, 3), 5)$

15. 315

16. $C(n, k)D_{n-k}$

Exercises 6.6

1. $k(m) = (n - m + 1)k(m - 1)/m$
3. (a) $h_1(0) = 1$, $h_1(1) = 3$, $h_1(n) = 2(h_1(n - 1) + h_1(n - 2))$, $n \geq 2$
   (b) $h_2(0) = 1$, $h_2(1) = 3$, $h_2(n) = 2h_2(n - 1) + h_2(n - 2)$, $n \geq 2$
   (c) $h_3(0) = 1$, $h_3(1) = 3$, $h_3(2) = 9$,
       $h_3(n) = 2(h_3(n - 1) + h_3(n - 2) + h_3(n - 3))$, $n \geq 3$
   (d) $h_4(0) = 1$, $h_4(1) = 3$, and, for $n \geq 2$, $h_4(n) = 1 + 2h_4(n - 1) + \sum_{k=0}^{n-2}h_4(k)$
5. $t(1) = 1$, $t(n) = (2n - 1)t(n - 1)$, $n \geq 2$
7. $w(n) = w(n - 1) + w(n - 2)$
9. For convenience, define $c(2) = 1$; then $c(n) = \sum_{k=3}^{n}c(k - 1)c(n - k + 2)$.
11. $b(n) = b(n - 1) + b(n - 2)$
13. $p(n) = (100)2^{n-1} + 2p(n - 1)$
15. $g(n) = P(2n - 2, 2)g(n - 1)$

Exercises 6.8

1. (a) $h_1(n) = (4^n - (-1)^n)/5$ (b) $h_2(n) = 2$
   (c) $h_3(n) = (1 + (-1)^n)/2$ (d) $h_4(n) = [(-1)^n + (2 - \sqrt{3})^n + (2 + \sqrt{3})^n]/3$
   (e) $h_5(n) = (1 - (-1)^n)2^{n-2}$ (f) $h_6(n) = (n + 1)!$
   (g) $h_7(n) = [(4)3^n - 3 - (-3)^n]/12$ (h) $h_8(n) = (n - 2)4^{n-1}$
   (i) $h_9(n) = [(-2)^n + 8 - 6n]/9$ (j) $h_0(n) = (-1)^n + (n^2 - n - 1)2^n$
3. (b) $g(n) = 3^n - 2^{n+1}$
5. (b) $h(n) = [i^n - (-i)^n]/(2i)$
7. (a) $a(n) = (-1)^n + (3n - 1)2^n$ (b) $a_2(n) = 2(-1)^n + (4n - 1)3^n$

Chapter 6 Problems

1. (a) 120   (b) 84
3. 11,550
4. (a) $3! = 6$ (b) $5!/2! = 60$ (c) $5!/(2! \cdot 2!) = 30$ (d) $9!/(3! \cdot 2!) = 20,160$
   (e) $n!/(n_1!)(n_2!) \cdots (n_m!)$
5. (a) $C(8, 3)25^5$ (b) $C(8, 3)5^3 21^5 + C(8, 4)5^4 21^4$ (c) $P(26, 8)$
   (d) $25^8 + C(8, 2)25^6 + C(8, 4)25^4 + C(8, 6)25^2 + 1$
6. 90%
7. (a) 8568   (b) 4806   (c) 7098   (d) 2184   (e) 6150
9. (a) 2^{20} (b) 1024   (c) 0   (d) 240   (e) 448
11. (a) 288   (b) 360   (c) 3039   (d) 2925
13. 99
15. (a) 1 7 21 35 35 21 7 1   (b) 1 8 28 56 70 56 28 8 1
23. (a) $P(26)$ (b) $P(21)P(22, 5)$ (c) $P(25)$ (d) $P(20)P(21, 5)$
25. (a) 495   (b) 105
27. (a) 55   (b) 52
31. (a) 10^{20} (b) $\prod_{k=1}^{10}k(2k - 1)$
32. (a) $P(9)$ (b) 22,680   (c) 1680   (d) 36
33. (a) $C(23, 8)$ (b) $C(16, 8)$ (c) 2520   (d) 35
   (e) If order matters, the answer is 4096; if order does not matter, 816.
34. (a) 120   (b) 56
35. (a) 30   (b) 720   (c) 180   (d) 474   (e) 66
37. (a) $P(10)$ (b) 8640
38. 72
41. 940

47. (a) $w(1) = 1 = w(2)$, $w(3) = 2$ **(b)** $w(n) = w(n-1) + w(n-3)$

48. $g_3(1) = 1$, $g_3(n) = C(3n-1,2)g_3(n-1)$, $n \geq 2$

50. (a) $h_1(n) = (5)2^n - (4)3^n$ **(b)** $h_2(n) = [3^{n-1} + (-3)^{n-1}]/2$

  (c) $h_3(n) = 4^n + 3(-1)^n - 4$ **(d)** $h_4(n) = (n-1)5^n$

  (e) $h_5(n) = (3-n)(-2^n) + 4^n$ **(f)** $h_6(n) = 4n - 1 + 2(-3)^n$

  (g) $h_7(n) = 2^n n!$ **(h)** $h_8(n) = (2n^2 - 3)2^n$

51. (a) $f_1(n) = an + b$ **(b)** $f_2(n) = an^2 + c$ **(c)** $f_3(n) = an^2 + bn + c$

  (d) $f_4(n) = a^n$ **(e)** $f_5(n) = a^n + b$ **(f)** $f_6(n) = a$

53. (a) $f_7(n) = 3^n + 4$ **(b)** $f_8(n) = n + 2 - 2^n$

  (c) $f_9(n) = (-2)^n + 9n^2 + 21n + 8$ **(d)** $f_0(n) = [10(-3)^n + 4^{n+1}]/7$

60. (a) 60   **(b)** $-P(12,5)3^6$

62. (a) 35   **(b)** $-6P(15,10)$

64. 50,000

Exercises 7.1

1. $G_1 = (\{u, v, w, x, y, z\}, \{uv, vw, vz, wx, xy, xz, yz\})$; 7

2. $\Delta(G) \leq 1$

3 (b) Suppose $\deg(x) = \deg(y) = n-1$ for some x, $y \in V(G)$. Then $vx, vy \in E(G)$ for every vertex $v \in V(G) - \{x, y\}$, which implies that $\delta(G) \geq 2$.

5. (a) K_n **(b)** Let $G_n = (\{v_1, v_2, \ldots, v_{2n}\}, \{v_1v_2, v_3v_4, \ldots, v_{2n-1}v_{2n}\})$

  (c) C_n **(d)** Let $H_n = (\{v_1, v_2, \ldots, v_{2n}\}, \{v_1v_2, v_2v_3, \ldots, v_{2n-1}v_{2n}, v_{2n}v_1\} \cup \{v_1v_{n+1}, v_2v_{n+2}, \ldots, v_nv_{2n}\})$

7. (a) Define $\phi: V(G_1) \to V(G_2)$ by $\phi(u_1) = v_4$, $\phi(u_2) = v_2$, $\phi(u_3) = v_5$, $\phi(u_4) = v_1$, $\phi(u_5) = v_6$, $\phi(u_6) = v_3$; then ϕ is an isomorphism. The graph G_3 is not isomorphic to G_1, since G_3 has size 6 whereas G_1 has size 7.

  (b) Define $\phi: V(G_4) \to V(G_6)$ by $\phi(u_1) = x_1$, $\phi(u_2) = x_2$, $\phi(u_3) = x_4$, $\phi(u_4) = x_3$, $\phi(u_5) = x_5$; then ϕ is an isomorphism. The graph G_5 is not isomorphic to G_4, since G_5 has a vertex of degree 2 and G_4 does not.

  (c) Define $\phi: V(G_7) \to V(G_9)$ by $\phi(u_1) = x_2$, $\phi(u_2) = x_3$, $\phi(u_3) = x_4$, $\phi(u_4) = x_6$, $\phi(u_5) = x_1$, $\phi(u_6) = x_5$; then ϕ is an isomorphism. The graph G_8 contains a cycle of length 4 whereas G_7 does not, so G_8 is not isomorphic to G_7.

8. (a) Let $G_1 = (\{u, v, w, x, y\}, \{uv, vw, vy, wx, xy\})$.

  (b) Such a graph would have three vertices of odd degree, contradicting Corollary 7.2.

  (c) Let G be a graph of order 5 and let $x \in V(G)$. If $\deg(x) = 4$, then $vx \in E(G)$ for all $v \in V(G) - \{x\}$. Thus $\delta(G) \geq 1$.

  (d) Let $G_2 = (\{u, v, w, x, y, z\}, \{uv, uz, vw, wx, wy, wz, xy, xz, yz\})$.

  (e) Let $G_3 = (\{u, v, w, x, y, z\}, \{uz, vy, vz, wx, wy, wz, xy, xz, yz\})$.

  (f) Let $G_4 = (\{t, u, v, w, x, y, z\}, \{ty, tz, ux, uy, uz, vw, vx, vy, vz, wx, wy, wz, xy, xz\})$.

9. (a) Let $G_1 = (\{u, v\}, \{uv\})$, $G_2 = (\{u, v\}, \emptyset)$.

  (b) Let $G_1 = (\{u, v, x, y\}, \{uv, xy\})$, $G_2 = (\{u, v, x, y\}, \{uv, ux\})$.

  (c) Let $G_1 = (\{u, v, w, x, y\}, \{uv, vw, vy, wx, xy\})$, $G_2 = (\{u, v, w, x, y\}, \{uv, vw, wx, wy, xy\})$.

11. (b) Note that $\{v_1, v_3, v_5, v_7\}$ and $\{v_2, v_4, v_6\}$ are partite sets of G_1.

  (d) Suppose V_1 and V_2 are partite sets of G_2 and $v_1 \in V_1$. Since v_1 and v_2 are adjacent, $v_2 \in V_2$. But now consider v_6. Since v_1 is adjacent to v_6, we have $v_6 \notin V_1$, and since v_2 is adjacent to v_6, we have $v_6 \notin V_2$. This gives a contradiction.

16. (d) Let $G_r = (\{u_1, u_2, \ldots, u_{r+1}\} \cup \{v_1, v_2, \ldots, v_{r+1}\}, \{u_iv_j \mid i \neq j\})$.

17. (a) Either G is isomorphic to $K_{3,3}$ or G is isomorphic to the graph $(\{u_1, u_2, u_3, v_1, v_2, v_3\}, \{u_1u_2, u_2u_3, u_3u_1, v_1v_2, v_2v_3, v_3v_1, u_1v_1, u_2v_2, u_3v_3\})$.

  (b) G is isomorphic to K_5.

19. Hint: Let n_q be the number of graphs of order 5 and size q, up to isomorphism. Then $n_q = n_{10-q}$, $0 \le q \le 4$ (why?), and $n_0 = n_1 = 1$, $n_2 = 2$, $n_3 = 4$, $n_4 = n_5 = 6$.

22. (a) $C(p,2) - q$ **(b)** $p - 1 - d$ **(c)** $\delta(\overline{G}) = p - 1 - \Delta(G)$ **(d)** $\Delta(\overline{G}) = p - 1 - \delta(G)$

23. (a) $u, x, w, z, y, x, w, t, v$ **(b)** u, t, w, z, y, x, w, v **(c)** u, t, v

    (d) u, t, w, x, y, z, v **(e)** $u, t, v, z, y, v, w, x, u$ **(f)** u, t, w, x, u

    (g) u, t, v, w, z, y, x, u

Exercises 7.2

1. The graphs G_3 and G_6 are eulerian.

    (c) $u_1, u_2, u_3, u_1, u_4, u_3, u_6, u_5, u_1$ is an eulerian circuit of G_3.

    (f) $v_1, v_2, v_3, v_4, v_5, v_6, v_3, v_1, v_4, v_2, v_6, v_1$ is an eulerian circuit of G_6.

3. The graphs G_2 and G_4 contain eulerian trails.

    (b) $v_3, v_2, v_1, v_5, v_2, v_4, v_3, v_5, v_4$ is an eulerian trail of G_2.

    (d) $v_2, v_1, v_6, v_2, v_3, v_4, v_5, v_3$ is an eulerian trail of G_4.

4. (a) Note that vertices a, b, c, and d all have odd degree.

    (b) Cross from region a to region b using one bridge and then cross back to region a using the other bridge; next cross from a to c and then back to a using the two bridges joining regions a and c; then cross to region d; then cross from d to b and back using the two bridges joining these regions; finally, cross from d to c and then from c to region b.

Exercises 7.3

1. (a) u, x, w, v, u **(b)** u, z, w, y, v, x **(c)** $z, u, z, w, x, w, v, u, x$

    (d) If we proceed "algorithmically," we may find the walk

      $u, x, w, x, w, v, u, z, u, x, w, v, y, x, w, v, u, z, w, v, u, z, y$, which has length 22. However, this is not the shortest walk that includes each arc.

2. The distance matrix is as follows, where the rows and columns correspond to u, v, x, y, and z, respectively.

$$\begin{bmatrix} 0 & 1 & 4 & 3 & 2 \\ 2 & 0 & 3 & 2 & 1 \\ 3 & 1 & 0 & 3 & 2 \\ 4 & 2 & 1 & 0 & 3 \\ 1 & 2 & 2 & 1 & 0 \end{bmatrix}$$

3. The distance matrix is as follows, where the rows and columns correspond to u, v, w, x, y, and z, respectively.

$$\begin{bmatrix} 0 & 3 & 2 & 1 & 2 & 1 \\ 1 & 0 & 3 & 2 & 1 & 2 \\ 2 & 1 & 0 & 1 & 2 & 3 \\ 3 & 2 & 1 & 0 & 3 & 4 \\ 4 & 3 & 2 & 1 & 0 & 5 \\ 1 & 2 & 1 & 2 & 1 & 0 \end{bmatrix}$$

4. $G = (\{u, v, x, y\}, \{(u, x), (v, u), (x, v), (x, y), (y, u)\})$

5. $\text{id}(u) = 2$, $\text{id}(v) = 1$, $\text{id}(w) = 2$, $\text{id}(x) = 3$, $\text{id}(y) = 2$, $\text{id}(z) = 1$; $\text{od}(u) = 3$, $\text{od}(v) = 2$, $\text{od}(w) = 2$, $\text{od}(x) = 1$, $\text{od}(y) = 1$, $\text{od}(z) = 3$

6. This is false; see the digraph G in the answer to Exercise 4.

7. Let $A = \{(v_i, v_j) \mid j < i\}$.

13. (a) Define $\phi: V(G_1) \rightarrow V(G_2)$ by $\phi(a) = y$, $\phi(b) = v$, $\phi(c) = x$, $\phi(d) = z$, $\phi(e) = u$ and verify that ϕ is an isomorphism.

   (b) Suppose that $\phi: V(G_3) \rightarrow V(G)$ is an isomorphism. Since G_3 has exactly one loop at a and G has a loop at z, it must be that $\phi(a) = z$. Then, since $(a, e) \in A(G_3)$ and $(z, u) \in A(G)$, we have that $\phi(e) = u$. But then $(e, b) \in A(G_3)$ implies $(u, \phi(b)) \in A(G)$, a contradiction since u has outdegree 0 in G.

15. (a) $G_1 = (\{u\}, \emptyset)$, $G_2 = (\{u\}, \{(u, u)\})$

   (b) $G_1 = (\{u, v\}, \{(u, v)\})$, $G_2 = (\{u, v\}, \{(u, u)\})$

   (c) $G_1 = (\{u, v\}, \{(u, v), (v, u)\})$, $G_2 = (\{u, v\}, \{(u, u), (v, v)\})$; or if loops are not allowed, let $G_1 = (\{u, v, x, y\}, \{(u, v), (v, x), (x, y), (y, u)\})$ and $G_2 = (\{u, v, x, y\}, \{(u, v), (v, u), (x, y), (y, x)\})$.

21. Hint: There are 16 of them.

22. $t, z, y, x, y, v, z, v, u, x, w, v, t, z, u, t$

Exercises 7.4

1. The shortest c_0-c_i paths, $2 \le i \le 6$ are as follows:

$$c_0, c_2 \qquad c_0, c_3 \qquad c_0, c_3, c_4 \qquad c_0, c_3, c_4, c_5 \qquad c_0, c_3, c_6$$

3. Note that each digraph in Figure 7.20 is strong, so the reachability matrix is a matrix of all 1's in each case.

4. (a) The final computed values are as follows: $d(u_1) = 11$, $d(u_2) = 16$, $d(u_3) = 6$, $d(u_4) = 1$, $d(u_5) = 4$, $p(u_1) = u_3$, $p(u_2) = u_3$, $p(u_3) = u_0$, $p(u_4) = u_0$, $p(u_5) = u_4$.

   (b) The shortest u_0-u_i paths, $1 \le i \le 5$, are as follows:

$$u_0, u_3, u_1 \qquad u_0, u_3, u_2 \qquad u_0, u_3 \qquad u_0, u_4 \qquad u_0, u_4, u_5$$

5. We give the final computed values in each part.

   (a) $d(u_1) = 2$, $d(u_2) = 3$, $d(u_3) = 5$, $d(u_4) = 4$, $p(u_1) = u_0$, $p(u_2) = u_0$, $p(u_3) = u_0$, $p(u_4) = u_2$

   (b) $d(u_1) = 14$, $d(u_2) = 5$, $d(u_3) = 10$, $d(u_4) = 8$, $d(u_5) = 1$, $p(u_1) = u_2$, $p(u_2) = u_5$, $p(u_3) = u_4$, $p(u_4) = u_5$, $p(u_5) = u_0$

   (c) $d(u_1) = 5$, $d(u_2) = 2$, $d(u_3) = 7$, $d(u_4) = 7$, $d(u_5) = 17$, $p(u_1) = u_2$, $p(u_2) = u_0$, $p(u_3) = u_2$, $p(u_4) = u_0$, $p(u_5) = u_3$

   (d) $d(u_1) = 16$, $d(u_2) = 17$, $d(u_3) = 11$, $d(u_4) = 16$, $d(u_5) = 18$, $d(u_6) = 9$, $p(u_1) = u_3$, $p(u_2) = u_1$, $p(u_3) = u_0$, $p(u_4) = u_6$, $p(u_5) = u_6$, $p(u_6) = u_0$

6. The sum of the elements in the ith row of M is $od(v_i)$; the sum of the elements in the jth column is $id(v_j)$.

8. (a) True; for $U \subseteq V(G) = V$, the subgraph $\langle U \rangle$ can be obtained by deleting the vertices of $V - U$.

   (b) False; note that a subgraph obtained by deleting arcs must be a spanning subgraph, whereas an arc-induced subgraph need not be spanning.

9. (a) Let $G = (\{u_0, u_1, u_2\}, \{(u_0, u_1), (u_0, u_2), (u_1, u_2)\})$ and weight the arcs so that $w((u_0, u_2)) < w((u_0, u_1)) + w((u_1, u_2))$.

   (b) A sufficient condition is that the arc of smallest weight is (u_0, u_i) for some i, $1 \le i \le p - 1$.

   (c) No; in part (a), for example, let $w((u_0, u_2)) = w((u_0, u_1)) + w((u_1, u_2))$.

Exercises 7.5

1. (a) $(\{b, d, e, s, x, y, z\}, \{(b, d), (d, t), (d, x), (d, y), (e, s), (s, z), (x, e)\})$

   (b) $(\{e, s, z\}, \{(e, s), (s, z)\})$ (c) $(\{e, s, x, z\}, \{(e, s), (s, z), (x, e)\})$

   (d) $(\{b, d, e, s, x, y, z\}, \{(b, d), (d, t), (d, x), (e, s), (s, z), (x, e), (y, b)\})$

2. (a) $(\{u_1, u_2, \ldots, u_8\}, \{(u_1, u_2), (u_2, u_5), (u_4, u_7), (u_5, u_4), (u_5, u_6), (u_7, u_3), (u_7, u_8)\})$
  (b) $(\{u_1, u_2, \ldots, u_9\}, \{(u_1, u_2), (u_1, u_7), (u_2, u_3), (u_3, u_4), (u_3, u_8), (u_4, u_6), (u_6, u_5), (u_8, u_9)\})$
5. (a) We proceed by induction on p. Clearly the result holds when $p = 1$. Let k be a positive integer and assume the result holds when $p = k$. Let $T = (V_0, A_0)$ be a rooted tree of order $k + 1$ and size q; we must show that $q = k$. Let y be a leaf of T, let x be the parent of y, and consider the rooted tree $T_1 = (V_0 - \{y\}, A_0 - \{(x, y)\})$. This tree has order k and size $q - 1$ so, by the induction hypothesis, $q - 1 = k - 1$. Thus $q = k$ and the proof is complete.
  (b) This is false; let $H = (\{u, v, w\}, \{(u, w), (v, w)\})$.
6. (a) k **(b)** $(m^{h-k+1} - 1)/(m - 1)$, $m > 1$
7. (a) Let $T = (\{1, 2, \ldots, 15\}, \{(m, n) \mid n = 2m \vee n = 2m + 1\})$.
  (b) Let $T = (\{1, 2, \ldots, 10\}, \{(m, n) \mid m \bmod 3 = 1 \wedge (n = m + 1 \vee n = m + 2 \vee n = m + 3)\})$.
  (c) $mh + 1$
8. $(m^{h+1} - 1)/(m - 1)$, $m > 1$

Exercises 7.6

1. There is only one, isomorphic to $(\{u, v, w, x, y, z\}, \{uv, uw, ux, uy, yz\})$.
3. There are five of them, isomorphic to the spanning trees T_1, T_2, T_3, T_4, and T_5, where $E(T_1) = \{uv, ux, uy, uz, wx\}$, $E(T_2) = \{uv, ux, uz, wz, yz\}$, $E(T_3) = \{uv, ux, uy, wx, yz\}$, $E(T_4) = \{uv, ux, uy, wz, yz\}$, and $E(T_5) = \{uv, vy, wx, wz, xy\}$.
5. There are two minimum spanning trees with weight 25; one has edge set $\{rx, ry, sx, tx, yz\}$ and the other has edge set $\{ry, sx, sy, tx, yz\}$.
6. This is true since, in Step 2 of Algorithm 7.5, there is a unique choice for the list L.
7. One possibility is the tree with edge set $\{ad, au, az, bv, cx, uv, vx, vy\}$.
9. It is isomorphic to $(\{u_1, u_2, \ldots, u_n\}, \{u_1u_2, u_1u_3, \ldots, u_1u_{n-1}, u_{n-1}u_n\})$.
10. They are $T_1 = (V, E_1)$, $T_2 = (V, E_2)$, and $T_3 = (V, E_3)$, where $V = \{u_1, u_2, \ldots, u_n\}$, $E_1 = \{u_1u_2, u_1u_3, \ldots, u_1u_{n-2}, u_{n-2}u_{n-1}, u_{n-2}u_n\}$, $E_2 = \{u_1u_2, u_1u_3, \ldots, u_1u_{n-2}, u_{n-3}u_{n-1}, u_{n-2}u_n\}$, and $E_3 = \{u_1u_2, u_1u_3, \ldots, u_1u_{n-2}, u_{n-2}u_{n-1}, u_{n-1}u_n\}$.
17. (b) Consider the trees T_2 and T_3 in the answer to Exercise 10.
20. $q = p - m$

Exercises 7.7

3. (a) Proceed by contrapositive; suppose G does not satisfy Ore's condition. Then there exist distinct, nonadjacent vertices x and y of G such that $\deg(x) + \deg(y) < n$. Partition the edge set of G as $E_1 \cup E_2$ where E_1 includes those edges incident with x or y. Then $|E_1| < n$ and $|E_2| \le C(n - 2, 2)$. Thus $q = |E_1| + |E_2| < n + C(n - 2, 2) = (n^2 - 3n + 6)/2$.
  (b) For n odd, $n \ge 5$, let $G = (\{v_1, v_2, \ldots, v_n\}, E)$ such that $\deg(v_i) = n - 1$, $2 \le 2i \le n - 1$, $\deg(v_i) = n - 2$, $n - 1 < 2i \le 2n - 2$, and $\deg(v_n) = (n - 1)/2$. Then $2q = (n - 1)^2 \ge n^2 - 3n + 6$ and G does not satisfy Dirac's condition.
4. (a) $E(Cl(G)) = E(G) \cup \{sw, sy, tz, uy\}$ **(b)** $Cl(G) = G$ **(c)** $Cl(G) \simeq K_8$
5. The closure is isomorphic to a complete graph in each part.
6. (a) $Cl(G_1) = (V_1, E_1 \cup \{ty\})$; G_1 is not hamiltonian, since $G_1 - y$ has two components.
  (b) $Cl(G_2) = G_2$; the cycle t, w, z, v, y, u, x, t is a hamiltonian cycle of G_2.
  (c) $Cl(G_3) \simeq K_7$; thus G_3 is hamiltonian.
9. (a) Let G_c have order 6 and degree sequence $(2, 2, 3, 4, 4, 5)$.
  (b) Let G_d have order 5 and degree sequence $(2, 2, 3, 3, 4)$.
  (c) Let G_e have order 6 and degree sequence $(2, 3, 3, 3, 3, 4)$.
  (d) Let G_f have order 5 and degree sequence $(2, 3, 3, 4, 4)$.

10. This digraph is not strong and hence not hamiltonian. It is also not traceable.
11. (a) hamiltonian   **(b)** traceable, not hamiltonian
    (c) hamiltonian   **(d)** traceable, not hamiltonian

Exercises 7.8

1.

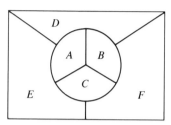

3. (c) Suppose the map can be colored using red, blue, and green. We may assume, without loss of generality, that J is colored red, K blue, and L green. Then D must be green, E blue, and F red. But then G is adjacent to a country of each color, a contradiction.
4. (b)

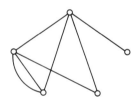

5. (a) 5   **(b)** 4   **(c)** 4
7. 3
9. Since T is nonempty and bipartite, $\chi(T) = 2$.
10. (b) $G_1 + G_2 = (\{u, v, w, x, y, z\}, \{uv, uw, ux, uy, uz, vw, vx, vy, vz, wx, xy, xz\})$
    (c) Note that G_1 and G_2 are subgraphs of $G_1 + G_2$. Let $k_1 = \chi(G_1)$ and $k_2 = \chi(G_2)$. In $G_1 + G_2$, color the subgraph G_1 using colors $1, 2, \ldots, k_1$ and the subgraph G_2 using colors $k_1 + 1, k_1 + 2, \ldots, k_1 + k_2$ and observe that this gives a $(k_1 + k_2)$-coloring of $G_1 + G_2$. Thus $\chi(G_1 + G_2) \le k_1 + k_2$. Furthermore let an n-coloring of $G_1 + G_2$ be given. Let C_i be the set of colors used to color the subgraph G_i, $i = 1, 2$. Then $|C_1| \ge k_1$, $|C_2| \ge k_2$, and, since every vertex of G_1 is adjacent to every vertex of G_2 in $G_1 + G_2$, we have that $C_1 \cap C_2 = \emptyset$. Thus $n \ge k_1 + k_2$, showing that $\chi(G_1 + G_2) = k_1 + k_2$.
11. We use the result of Exercise 10(c). Thus $\chi(W_{2n}) = \chi(K_1 + C_{2n-1}) = \chi(K_1) + \chi(C_{2n-1}) = 1 + 3 = 4$ and $\chi(W_{2n+1}) = \chi(K_1 + C_{2n}) = \chi(K_1) + \chi(C_{2n}) = 1 + 2 = 3$.

12. (a) Let G be a 3-critical graph and let C: $u_1, u_2, \ldots, u_{2n+1}, u_1$ be an odd cycle of G. Since $\chi(C) = 3$ and G is 3-critical, C is a spanning cycle of G. Suppose $u_i u_j$ is an edge of G, where $1 \leq i < j \leq n$ and $i + 1 < j$. Then one of the cycles $u_1, \ldots, u_i, u_j, \ldots, u_{2n+1}, u_1$ or $u_i, u_{i+1}, \ldots, u_j, u_i$ is odd, contradicting the fact that G is 3-critical. Thus the edges of C are the only edges of G, so that G is an odd cycle.

(b) Use the graph G of Figure 7.45(c).

(c) Use the result of Exercise 14; thus, for the graph G of Figure 7.45(c), $K_1 + G$ is 5-critical.

13. (a) 3

(b) Each color class is a pairwise disjoint collection of subsets of U.

(c) We require a maximal independent set S of H such that the union of the subsets in S is U. Since each of the subsets $A, B, \ldots, G$ has at most three elements, S is required to have cardinality at least 4. However no such S exists.

15. Let $\chi(G) = k$ and let $V_1, V_2, \ldots, V_k$ be the color classes of a k-coloring of G. Then, for each i, the set $V(H) \cap V_i$ either is empty or is an independent set in G and the union of the sets $V(H) \cap V_i$ is $V(H)$. Hence the nonempty sets $V(H) \cap V_i$ provide an m-coloring of H for some $m \leq k$. Therefore $\chi(H) \leq \chi(G)$.

25. An example is the graph of the icosahedron shown in Figure 7.33.

27. (a) planar   **(b)** nonplanar (see Chapter Problem 52)   **(c)** nonplanar   **(d)** planar

28. The maximal independent sets are $\{t, u\}$, $\{t, v\}$, $\{u, w, y\}$, $\{u, z\}$, and $\{x, z\}$.

Exercises 7.9

1. (a) $\{011, 1, 10, 11\}$ **(b)** $\{1\}$ **(c)** $\{10011, 101, 1011, 11, 111\}$

(d) $\{0111, 01110, 11, 110, 111, 1110\}$

(e) $\{10101, 101010, 1011, 10110, 1101, 11010, 111, 1110\}$

(f) $\{011011, 0111, 01111, 1011, 11, 11011, 111, 1111\}$

(g) $X^* = \{\varepsilon\} \cup \{\alpha \mid \alpha$ begins with a 1 and has no two 0's consecutive$\}$

(h) $Y^* = \{\beta \mid \beta$ has at least two 1's between any two 0's and between the rightmost 0 and the end of $\beta\}$

2. We follow the convention of Example 7.46; also we simplify the notation $t((s, x))$ to $t(s, x)$ and note that, in each part, $t(-1, 0) = t(-1, 1) = -1$.

(a) $S = \{-1, 0\}$; $A = \emptyset$; $t(0, 0) = t(0, 1) = -1$

(b) $S = \{-1, 0\}$; $A = \{0\}$; $t(0, 0) = t(0, 1) = -1$

(c) $S = \{-1, 0, 1\}$; $A = \{1\}$; $t(0, 0) = 1$, $t(0, 1) = t(1, 0) = t(1, 1) = -1$

(d) $S = \{-1, 0, 1\}$; $A = \{1\}$; $t(0, 1) = 1$, $t(0, 0) = t(1, 0) = t(1, 1) = -1$

(e) $S = \{-1, 0, 1, 2\}$; $A = \{2\}$; $t(0, 0) = t(1, 1) = 1$, $t(1, 0) = 2$, $t(0, 1) = t(1, 0) = t(2, 0) = t(2, 1) = -1$

(f) $S = \{-1, 0, 1, 2\}$; $A = \{0, 1, 2\}$; $t(0, 0) = t(2, 0) = 2$, $t(0, 1) = t(1, 1) = 1$, $t(1, 0) = t(2, 1) = -1$

(g) $S = \{-1, 0, 1, \ldots, 6\}$; $A = \{0, 5, 6\}$; $t(0, 0) = t(5, 0) = t(6, 0) = 2$,
$t(0, 1) = t(5, 1) = t(6, 1) = 1$, $t(1, 0) = 3$, $t(2, 0) = 4$, $t(3, 1) = 5$, $t(4, 0) = 6$,
$t(1, 1) = t(2, 1) = t(3, 0) = t(4, 1) = -1$

(h) $S = \{-1, 0, 1, 2, 3, 4\}$; $A = \{0, 3, 4\}$; $t(0, 0) = t(3, 0) = t(4, 0) = 2$,
$t(0, 1) = t(3, 1) = t(4, 1) = 1$, $t(1, 1) = 3$, $t(2, 0) = 4$, $t(1, 0) = t(2, 1) = -1$

3. We follow the convention and notation of the answers to Exercise 2.

(a) $S = \{-1, 0, 1, \ldots, 5\}$; $A = \{5\}$; $t(0, 0) = t(1, 0) = t(2, 0) = t(3, 0) = t(4, 0) = 2$,
$t(0, 1) = t(3, 1) = 1$, $t(1, 1) = t(4, 1) = 3$, $t(2, 1) = 4$, $t(4, 2) = 5$, $t(0, 2) =$
$t(1, 2) = t(2, 2) = t(3, 2) = t(5, 0) = t(5, 1) = t(5, 2) = -1$

(b) $S = \{-1, 0, 1, 2, 3, 4\}$; $A = \{1, 3, 4\}$; $t(0, 2) = 1$, $t(1, 0) = t(3, 0) = t(4, 0) = 2$,
$t(1, 1) = t(3, 1) = t(4, 1) = 3$, $t(2, 0) = 4$, $t(0, 0) = t(0, 1) = t(1, 2) = t(2, 1) =$
$t(2, 2) = t(3, 2) = t(4, 2) = -1$

(c) $S = \{-1, 0, 1, 2\}$; $A = \{0, 1\}$; $t(0, 0) = t(1, 0) = t(1, 1) = t(2, 0) = 1$,
$t(0, 2) = t(1, 2) = 2$, $t(0, 1) = t(2, 1) = t(2, 2) = -1$

(d) $S = \{-1, 0, 1, 2, 3, 4\}$; $A = \{4\}$; $t(0,0) = t(0,1) = 1$, $t(1,1) = t(2,1) = 3$,
$t(1,2) = t(2,2) = 2$, $t(2,0) = t(3,2) = 4$, $t(0,2) = t(1,0) = t(3,0) = t(3,1) = t(4,0) = t(4,1) = t(4,2) = -1$

4. We follow the convention and notation of the answers to Exercise 2.
(a) $S = \{-1, 0, 1, 2\}$; $A = \{2\}$; $t(0,0) = t(1,0) = t(2,0) = 1$, $t(1,1) = t(2,1) = 2$, $t(0,1) = -1$
(b) $S = \{-1, 0, 1, 2, 3, 4\}$; $A = \{0, 3, 4\}$; $t(0,0) = t(2,1) = t(4,1) = 2$, $t(0,0) = t(1,0) = t(3,0) = 1$, $t(1,1) = 3$, $t(2,0) = 4$, $t(3,1) = t(4,0) = -1$
(c) $S = \{0, 1\}$; $A = \{0\}$; $t(0,0) = t(1,1) = 0$, $t(0,1) = t(1,0) = 1$
(d) $S = \{-1, 0, 1, \ldots, 7\}$; $A = \{1, 3, 5, 7\}$; $t(0,0) = t(6,0) = 3$, $t(0,1) = t(2,1) = t(4,1) = 1$, $t(1,0) = 2$, $t(1,1) = t(3,1) = t(7,1) = 4$, $t(2,0) = 5$, $t(3,0) = t(5,0) = 0$, $t(4,0) = 7$, $t(7,0) = 6$, $t(5,1) = t(6,1) = -1$

6. Let $G = (\{0, 1, 2, 3, 4\}, \{0, 1\}, t, 0, \{3, 4\})$, where $t(0,0) = t(1,0) = t(2,0) = t(4,0) = 2$, $t(0,1) = t(1,1) = t(3,1) = 1$, $t(2,1) = t(4,1) = 3$, and $t(3,0) = 4$.

Chapter 7 Problems

2. (a) These are not isomorphic because $G_1 \simeq K_{3,3}$, whereas G_2 is not bipartite.
(b) An isomorphism $\phi: V(G_1) \to V(G_2)$ is defined by $\phi(u_1) = v_1$, $\phi(u_2) = v_6$, $\phi(u_3) = v_2$, $\phi(u_4) = v_3$, $\phi(u_5) = v_5$, $\phi(u_6) = v_4$, $\phi(u_7) = v_7$.
3. (a) Let G have size q. Then $\overline{G}$ has size q also. Thus $2q = p(p-1)/2$ or $4q = p(p-1)$. Thus $p \equiv 0$ or $1 \pmod 4$.
(b) We give the construction for $p \equiv 0 \pmod 4$. Start with two copies of $K_{p/4}$ and a copy of $K_{p/4, p/4}$. Add all edges joining the first $K_{p/4}$ with one partite set of $K_{p/4, p/4}$ and all edges joining the second $K_{p/4}$ with the other partite set. The resulting graph is self-complementary.
5. Hint for part (b):   Let $\deg(v_i) = d_i$, $1 \le i \le p$, and let $H = \langle\langle v_1, v_2, \ldots, v_n \rangle\rangle$. Then $d_1 + d_2 + \cdots + d_n \le x + y$, where x is the sum of the degrees of the v_i in H and y is the number of edges of G that join $V(H)$ with $V(G) - V(H)$.
6. (b) $s, w, v, z, x, u, y, w, t, v, x, t, z, s$
7. (a) $q \ge p - 1$ (b) $q \le C(p-1, 2)$
(c) Any tree of order p shows that the bound in part (a) is sharp. For part (b), let $G = (\{v_1, v_2, \ldots, v_p\}, E)$ such that $\deg(v_1) = \cdots = \deg(v_{p-1}) = p - 2$ and $\deg(v_p) = 0$.
9. (a) 3-partite   (b) 2-partite   (c) 4-partite
11. The order is $p = p_1 + p_2 + \cdots + p_n$ and the size is
$[p_1(p - p_1) + p_2(p - p_2) + \cdots + p_n(p - p_n)]/2 = [p^2 - p_1^2 - p_2^2 - \cdots - p_n^2]/2$.
12. For the general case, let $p = ns + r$, $0 \le r < n$. Let $p_i = s$, $1 \le i \le n - r$, and $p_i = s + 1$, $n - r < i \le n$. Show that q is at most the size of $K_{p_1, p_2, \ldots, p_n}$.
13. (a) $H_1 = (\{v, w, x, y, z\}, \{vw, vy, wx, wz, xy, yz\})$
(b) $H_2 = \langle\langle u, w, x, y \rangle\rangle$
(c) $H_3 = (V, \{tu, tw, ux, vz, vw, xy, yz\})$, $H_4 = (V, \{tu, tw, ux, vy, vz, wx, yz\})$
(d) $K_{1,1,3}$ has two nonadjacent vertices of degree 4, whereas G does not.
14. They are H, $\langle\langle uv, ux, uz, vx, yz \rangle\rangle$, $\langle\langle uv, ux, uz, vz, yz \rangle\rangle$, $\langle\langle ux, uz, vx, vz, yz \rangle\rangle$, $\langle\langle uv, ux, uz, yz \rangle\rangle$, and $\langle\langle uv, uz, vx, yz \rangle\rangle$.
15. A graph G of order p has the property that every induced subgraph is connected if and only if $G \simeq K_p$.
17. Let $m \ge 2$. A graph G of order p has the property that every induced subgraph on m or more vertices is connected if and only if $G \simeq K_{p_1, p_2, \ldots, p_n}$ for some $n \ge 2$, where $p_1 \le p_2 \le \cdots \le p_n < m$ and $p_1 + p_2 + \cdots + p_n = p$.
19. (a) $d(a) = 2$, $d(b) = 5$, $d(c) = 6$, $d(t) = 10$, $d(u) = 3$, $d(v) = 6$, $d(w) = 1$, $d(y) = 8$, $d(z) = 3$, $p(a) = p(u) = p(w) = x$, $p(b) = p(t) = p(z) = w$, $p(c) = u$, $p(v) = z$, $p(y) = b$

(b) $d(s) = 4$, $d(t) = 23$, $d(u) = 8$, $d(v) = 6$, $d(w) = 12$, $d(y) = 1$, $d(z) = 4$,
$p(s) = p(u) = p(v) = p(w) = p(z) = y$, $p(t) = z$, $p(y) = x$

20. (a) A possible edge set is $\{ ax, bw, by, cu, tv, ux, vz, wx, wz \}$.
   (b) A possible edge set is $\{ sy, tw, uy, vy, wz, xy, yz \}$.

22. (a) yes   **(b)** yes   **(c)** no; c must occur before y.

23. A possibility is the spanning rooted tree $(V, \{(a, b), (a, c), (b, d), (b, x), (d, e), (x, y), (y, z)\})$.

31. A graph G is isomorphic to the graph of an equivalence relation if and only if each component of G is complete.

32. (a) If G is a $(2, n)$-cage, then G contains an n-cycle, and, since C_n is a 2-regular graph with girth n, it must be that $G \simeq C_n$.
   (b) Let G be an $(r, 3)$-cage. Then G has order at least $r + 1$. Furthermore K_{r+1} is an r-regular graph with girth 3. Thus $G \simeq K_{r+1}$.
   (c) The unique $(r, 4)$-cage is $K_{r, r}$.
   (d) Petersen's graph (Figure 7.40) is the unique $(3, 5)$-cage.
   (e) Let $H = (\{ a, b, c, d, e, f, s, t, u, v, w, x, y, z \}$,
   $\{ ab, af, az, bc, bw, cd, ct, de, dy, ef, ev, fs, st, sx, uv, uz, vw, wx, xy, yz \})$. The graph H is called the *Heawood graph* and, up to isomorphism, is the unique $(3, 6)$-cage.

34. (a) $G^2 = (V, A_2)$, where $A_2 = A \cup \{(s, u), (t, v), (u, w), (v, t), (v, x), (w, u), (w, y), (x, w), (x, z),$
   $(y, s), (y, t), (y, x), (z, t), (z, y)\}$;
   $G^3 = (V, A_3)$, where $A_3 = A_2 \cup \{(s, v), (t, w), (u, t), (u, x), (v, u), (v, y), (w, v), (w, z), (x, s),$
   $(x, t), (y, u), (z, u), (z, w)\}$;
   $G^4 = (V, A_4)$, where $A_4 = A_3 \cup \{(s, w), (t, x), (u, y), (v, z), (w, s), (x, u), (y, v), (z, v)\}$;
   $G^5 = (V, A_5)$, where $A_5 = A_4 \cup \{(s, x), (t, y), (u, z), (v, s), (x, v)\}$;
   $G^6 = (V, A_6)$, where $A_6 = A_5 \cup \{(s, y), (t, z), (u, s)\}$;
   $G^7 = (V, A_7)$, where $A_7 = A_6 \cup \{(s, z), (t, s)\}$.
   (c) Let $T = (\{ s, u, v, w, x, y, z \}, \{ su, sv, sw, ux, vy, wz \})$.

37. $\tilde{A} = \{(a, b), (b, c), (b, e), (c, d), (c, g), (d, h), (e, f), (e, k), (g, m), (h, i), (i, j), (m, n)\}$;
   $p(b) = $ left, $p(c) = $ right, $p(d) = $ right, $p(e) = $ left, $p(f) = $ right, $p(g) = $ left,
   $p(h) = $ left, $p(i) = $ right, $p(j) = $ right, $p(k) = $ left, $p(m) = $ left, $p(n) = $ right

38. (a) A vertex v of T is a leaf of $\tilde{T}$ if and only if v is a leaf of T and v is the youngest child of its parent in T.
   (b) mh
   (c) We can recover T completely.

39. (a) $a, b, e, h, f, c, g, m, n, d, h, i, j$

40. Let $G = (\{ v_1, v_2, \ldots, v_p \}, E)$, where $\langle \{ v_1, v_2, \ldots, v_k \} \rangle \simeq K_k$ and $\deg(v_i) = 0$, $k < i \leq p$.

46. (a) $G_1 \times G_2 = (\{(u, x), (u, y), (u, z), (v, x), (v, y), (v, z)\}, \{(u, x)(u, y), (u, x)(u, z), (u, x)(v, x),$
   $(u, y)(u, z), (u, y)(v, y), (u, z)(v, z), (v, x)(v, y), (v, x)(v, z), (v, y)(v, z)\})$
   (b) $G_1 \times G_2 = (\{(t, w), (t, x), (t, y), (t, z), (u, w), (u, x), (u, y), (u, z), (v, w), (v, x), (v, y), (v, z)\},$
   $\{(t, w)(t, x), (t, w)(u, w), (t, x)(t, y), (t, x)(u, x), (t, y)(t, z), (t, y)(u, y), (t, z)(u, z), (u, w)(u, x),$
   $(u, w)(v, w), (u, x)(u, y), (u, x)(v, x), (u, y)(u, z), (u, y)(v, y), (u, z)(v, z), (v, w)(v, x),$
   $(v, x)(v, y), (v, y)(v, z)\})$

47. (b) Q_n has order 2^n and size $n2^{n-1}$.
   (c) We proceed by induction on n. The result is clear if $n = 1$.
   Let k be a positive integer and assume that Q_k is bipartite, with partite sets U_0 and U_1. Now consider Q_{k+1}. The subgraph induced by $\{0\} \times V(Q_k)$ is isomorphic to Q_k, and hence is bipartite with partite sets $\{0\} \times U_0$ and $\{0\} \times U_1$. Similarly, the subgraph induced by $\{1\} \times V(Q_k)$ is bipartite with partite sets $\{1\} \times U_0$ and $\{1\} \times U_1$. Let $V_0 = (\{0\} \times U_0) \cup (\{1\} \times U_1)$ and $V_1 = (\{0\} \times U_1) \cup (\{1\} \times U_0)$. Then it can be shown that V_0 and V_1 are partite sets of Q_{k+1}.
   (d) We proceed by induction on n. The result is clear for $n = 2$. Let k be a positive integer, $k \geq 2$, and assume that Q_k is hamiltonian. Then, in particular, Q_k is traceable; let $m = 2^k$ and let $v_1, v_2, \ldots, v_m$

be a hamiltonian path in Q_k. Then

$$(0, v_1), (0, v_2), \ldots, (0, v_m), (1, v_m), \ldots, (1, v_2), (1, v_1), (0, v_1)$$

is a hamiltonian cycle of Q_{k+1}.

(e) n **(f)** Q_n is planar if and only if $n \le 3$.

48. Use the graph T^2 of Problem 34, part **(c)**.

49. Assume $m \le n \le p$. Then $K_{m, n, p}$ is hamiltonian if and only if $m + n \ge p$.

50. Assume, as usual, that $p_1 \le p_2 \le \cdots \le p_n$. This graph is planar if and only if either $n = 2$ and $p_1 \le 2$, or $n = 3$ and $p_2 = 1$ or $p_2 = p_3 = 2$, or $n = 4$, $p_3 = 1$, and $p_4 \le 2$.

51. (a) We prove the contrapositive. Assume G is not planar. If G contains a subgraph H that is a subdivision of $K_{3,3}$, then H has order $n + 6$ and size $n + 9$ for some nonnegative integer n. Since G is connected, $q - p \ge (n + 9) - (n + 6) = 3$. A similar argument can be made if G contains a subgraph H that is a subdivision of K_5.

(b) Let G be a subdivision of $K_{3,3}$ such that G has order p. Then the size of G is $p + 3$ and G is nonplanar.

53. Step 0: Initially, choose any vertex u; $W := \{u\}$, $U := V - \{u\}$. Step 1: Repeat Step 2 until U is empty. Step 2: Choose $v \in U$; if v is not adjacent to any vertex of W then $W := W \cup \{v\}$; $U := U - \{v\}$.

54. Let $G = (\{u, v, x, y\}, \{uv, vx, xy\}) \cong P_4$. If the first W selected in Step 3 is $W = \{u, y\}$, then the "algorithm" will compute $\chi(G) = 3$, whereas the correct value is $\chi(G) = 2$. However, this method does produce an upper bound—namely, n—on $\chi(G)$.

Exercises 8.1

1. (a) For each $b_1, b_2 \in B$, $b_1 * b_2$ is a uniquely determined element of B.

(b) Let $b_1, b_2, b_3 \in B$. Since $b_1, b_2, b_3 \in A$ and $*$ is associative on A, $(b_1 * b_2) * b_3 = b_1 * (b_2 * b_3)$.

(d) $(\mathbb{Z}, \cdot)$, has identity 1. If E is the set of even integers, then $(E, \cdot)$ has no identity.

(e) Let $\mathbb{Q}^\# = \mathbb{Q} - \{0\}$. $(\mathbb{Q}^\#, \cdot)$ has identity 1 and each element of $\mathbb{Q}^\#$ has an inverse. $\mathbb{Z}^\# = \mathbb{Z} - \{0\}$; then $(\mathbb{Z}^\#, \cdot)$ has an identity, but not all of its elements have inverses.

2. Consider $f: A \to A$, with $f(a) = b$, $f(b) = a$, $f(c) = c$, and $g: A \to A$, with $g(a) = b$, $g(b) = c$, $g(c) = a$.

5. (a) Let $g, h \in G_a$. Then $(h \circ g)(a) = h(g(a)) = h(a) = a$, so $g \circ h \in G_a$.

(b) because $\circ$ is an associative binary operation on $S(A)$ and is a binary operation on G_a

(c) $i_A(a) = a$, so $i_A \in G_a$. Let $g \in G_a$ and consider g^{-1}; since $a = i_A(a) = (g^{-1} \circ g)(a) = g^{-1}(g(a)) = g^{-1}(a)$, we have $g^{-1} \in G_a$.

6. Define $a * x = x * a = x$ for all $x \in A$ and define $b * b = b * c = a$.

7. (a) $*$ is a commutative binary operation, but is not associative.

(b) $*$ is a binary operation; however, it is neither commutative nor associative.

(c) $*$ is not a binary operation.

9. (a) There are elements $a, b, c \in A$ such that $a * (b * c) \ne (a * b) * c$.

(b) There are elements $a, b \in A$ such that $a * b \ne b * a$.

(c) There is an element $b \in A$ such that $a * b \ne b$.

(d) $b * c \ne e$ for all $c \in A$.

11. $a * b = b, a * d = d, b * a = b, b * b = c, b * c = d, c * a = c, c * d = b, d * a = d, d * b = a, d * d = c$

13.

$\circ$	f_1	f_2	f_3	f_4	f_5	f_6
f_1	f_1	f_2	f_3	f_4	f_5	f_6
f_2	f_2	f_1	f_4	f_3	f_6	f_5
f_3	f_3	f_5	f_1	f_6	f_2	f_4
f_4	f_4	f_6	f_2	f_5	f_1	f_3
f_5	f_5	f_3	f_6	f_1	f_4	f_2
f_6	f_6	f_4	f_5	f_2	f_3	f_1

15. (a) No; $(\frac{1}{2} \div \frac{1}{3}) \div \frac{1}{4} = 6$ and $\frac{1}{2} \div (\frac{1}{3} \div \frac{1}{4}) = \frac{3}{8}$.
　　(b) No; $\frac{1}{2} \div \frac{1}{3} = \frac{3}{2}$ and $\frac{1}{3} \div \frac{1}{2} = \frac{2}{3}$.
　　(c) No; $\frac{1}{2} \div 1 = \frac{1}{2}$, but $1 \div \frac{1}{2} = 2$ (for example).
18. $C = \{2^r 5^s \mid r, s \in \mathbb{N}\}$

Exercises 8.2

4. [1] is the identity. Associativity is inherited. If $[a] \in U_m$, then $\gcd(a, m) = 1$, so there exist $s, t \in U_m$ such that $as + mt = 1$. Thus $[a] \odot [s] = [1]$.
6. Note that $(a * b)^{-1} = a^{-1} * b^{-1}$ if and only if $((a * b)^{-1})^{-1} = (a^{-1} * b^{-1})^{-1}$; that is, $a * b = b * a$.
8. (a) $a * b = a * c \rightarrow a^{-1} * (a * b) = a^{-1} * (a * c) \rightarrow (a^{-1} * a) * b =$
$(a^{-1} * a) * c \rightarrow e * b = e * c \rightarrow b = c$
9. $x * y = y \rightarrow x * y = e * y \rightarrow x = e$ (by cancellation)
11.

*	x	y	z
x	x	y	z
y	y	z	x
z	z	x	y

12.

∘	i	r	h	v
i	i	r	h	v
r	r	i	v	h
h	h	v	i	r
v	v	h	r	i

14. For $\alpha, \beta \in G_S$, $(\alpha \circ \beta)(S) = \alpha(\beta(S)) = \alpha(S) = S$, so $\alpha \circ \beta \in G_S$. Associativity is inherited. Since $i(S) = S$, $i \in G_S$. For $\alpha \in G_S$, $S = i(S) = (\alpha^{-1} \circ \alpha)(S) = \alpha^{-1}(\alpha(S)) = \alpha^{-1}(S)$, so $\alpha^{-1} \in G_S$. Hence G_S is a group.
16. For $A, B \in GL_2(\mathbb{R})$, $(B^{-1}A^{-1})(AB) = A^{-1}(B^{-1}B)A = A^{-1}(I_2 A) = A^{-1}A = I_2$, so $AB \in GL_2(\mathbb{R})$. Associativity is inherited. $I_2 \in GL_2(\mathbb{R})$, so there is an identity element. For each $A \in GL_2(\mathbb{R})$, $A^{-1} \in GL_2(\mathbb{R})$, so $GL_2(\mathbb{R})$ is a group. If

$$A = \begin{bmatrix} 1 & 1 \\ 0 & 1 \end{bmatrix} \qquad B = \begin{bmatrix} 1 & 0 \\ 1 & 1 \end{bmatrix}$$

then $A, B \in GL_2(\mathbb{R})$ and $AB \neq BA$, so $GL_2(\mathbb{R})$ is a nonabelian group.
20. We show only the inductive step. Assuming $k \in S$, let $a_1, a_2, \ldots, a_k, a_{k+1}$ be elements of G. Then, using Theorem 8.6 and the IHOP,

$$(a_1 * a_2 * \cdots * a_k * a_{k+1})^{-1} = ((a_1 * a_2 * \cdots * a_k) * a_{k+1})^{-1}$$
$$= a_{k+1}^{-1} * (a_1 * a_2 * \cdots * a_k)^{-1}$$
$$= a_{k+1}^{-1} * (a_k^{-1} * \cdots * a_2^{-1} * a_1^{-1})$$
$$= a_{k+1}^{-1} * a_k^{-1} * \cdots * a_2^{-1} * a_1^{-1}$$

so $k + 1 \in S$. Hence $S = \mathbb{N}$.

Exercises 8.3

2. (a) $|[2]| = 11$ in $(\mathbb{Z}_{11}, \oplus\,)$,   **(b)** $|[2]| = 10$ in $(\mathbb{Z}_{11}^{\#}, \odot)$
   (c) $|\cos(2\pi/5) + i\sin(2\pi/5)| = 5$ **(d)** $|A| = 2$, $|B| = \infty$ **(e)** $|[3]| = 4$
   (f) $|\alpha| = 4$ **(g)** $|c^2| = 15$, $|c^3| = 10$, $|c^5| = 6$, $|c^{10}| = 3$, $|c^{12}| = 5$
4. Hint: Show that each element of S_3 has order 1, 2, or 3.
5. Hint: Use the fact that $(x^{-1})^n = (x^n)^{-1}$, so that $x^n = e \leftrightarrow (x^{-1})^n = e$.
8. Any element of the form c^r, where $\gcd(r, 24) = 1$, is a generator of $\langle c \rangle$.
9. Show that each element of U_8 has order 1 or 2. Show that $[2]$ is a generator for U_9.
12. (a) Case 1: $m \geq 0$. Use induction on m, with S as usual. Then $(a * b)^0 = e = e * e = a^0 * b^0$, so $0 \in S$.
   Assume $k \in S$ for an arbitrary $k \geq 0$.
   So $(a * b)^k = a^k * b^k$. To show that $k + 1 \in S$, we use the fact G is abelian:
   $(a * b)^{k+1} = (a * b)^k * (a * b) = (a^k * b^k) * (a * b) = (a^k * a) * (b^k * b) =$
   $a^{k+1} * b^{k+1}$, so $k + 1 \in S$. Hence $S = \mathbb{N}$. Case 2: $m < 0$. Say $m = -n$ (so $n \in \mathbb{N}$). Then
   $(a * b)^m = (a * b)^{-n} = ((a * b)^n)^{-1} = (a^n * b^n)^{-1} = (a^n)^{-1} * (b^n)^{-1} = a^m * b^m$.
   (b) Let $|a * b| = t$. Then $(a * b)^{mn} = a^{mn} * b^{mn} = (a^m)^n * (a^n)^m = e^n * e^m = e$. Therefore $t \mid mn$.
13. (a) Note that $a = b^2 * a * b^{-2} = b * (b * a * b^{-1}) * b^{-1} = b * a^2 * b^{-1} =$
   $b * a * a * b^{-1} = b * a * (b^{-1} * b) * a * b^{-1} = (b * a * b^{-1}) * (b * a * b^{-1}) =$
   $a^2 * a^2 = a^4$. So $a = a^4$ and hence $a^3 = e$. From this it can be shown that $|a| = 3$.
16. We have $(a * b)^2 = a^2 * b^2 \leftrightarrow (a * b) * (a * b) = (a * a) * (b * b) \leftrightarrow a * (b * (a * b)) = a * (a * (b * b)) \leftrightarrow$
   $b * (a * b) = a * (b * b) \leftrightarrow (b * a) * b = (a * b) * b \leftrightarrow b * a = a * b$.

Exercises 8.4

1. (b) Note that $H \neq \emptyset$ since $0 = 0 + 0\sqrt{5}$; for $a + b\sqrt{5}$, $c + d\sqrt{5} \in \mathbb{Z}[\sqrt{5}\,]$, see that $(a + b\sqrt{5}) + (c + d\sqrt{5}) = (a + c) + (b + d)\sqrt{5} \in \mathbb{Z}[\sqrt{5}\,]$; also $-(a + b\sqrt{5}) = (-a) + (-b)\sqrt{5} \in \mathbb{Z}[\sqrt{5}\,]$.
   (e) Clearly $H \neq \emptyset$. For

$$A = \begin{bmatrix} 1 & a \\ 0 & 1 \end{bmatrix} \quad \text{and} \quad B = \begin{bmatrix} 1 & b \\ 0 & 1 \end{bmatrix}$$

   see that

$$AB = \begin{bmatrix} 1 & a + b \\ 0 & 1 \end{bmatrix} \quad \text{and} \quad A^{-1} = \begin{bmatrix} 1 & -a \\ 0 & 1 \end{bmatrix}$$

   so $AB \in H$ and $A^{-1} \in H$. Thus H is a subgroup.
   (h) Note that $H \neq \emptyset$ since $f_0 \in H$. If $f, g \in H$, then $(fg)(m) = f(m)g(m) = 0$ for all $m \in \mathbb{Z}$, so $fg \in H$. If $f \in H$, then $(-f)(m) = -f(m) = 0$ for all $m \in \mathbb{Z}$, so $-f \in H$.
3. (a) We have $H \neq \emptyset$, since $e = e^2 \in H$. For a^2, $b^2 \in H$, note that $a^2 * b^2 = (a * b)^2 \in H$. Here we used Exercise 12 of Exercises 8.3. For $a^2 \in H$, $(a^2)^{-1} = (a^{-1})^2 \in H$. Therefore $H \leq G$.
5. (b) Since $x^{-1} * a * x \in H$, $H \neq \emptyset$. For $x^{-1} * a^m * x$, $x^{-1} * a^n * x \in H$, $(x^{-1} * a^m * x) * (x^{-1} * a^n * x) = (x^{-1} * a^m) * (x * x^{-1}) * (a^n * x) = (x^{-1} * a^m) * e * (a^n * x) = x^{-1} * (a^m * a^n) * x = x^{-1} * a^{m+n} * x \in H$. And $(x^{-1} * a^m * x)^{-1} = x^{-1} * a^{-m} * x \in H$. Thus $H \leq G$.
7. (b) $|\iota| = 1$, $\iota^{-1} = \iota$; $|r_1| = |r_3| = 4$, $r_1^{-1} = r_3$; $|r_2| = |h| = |v| = |d_1| = |d_2| = 2$ and each of these elements is its own inverse.
   (c) Use part **(b)** to see that D_4 is not cyclic. The subgroup $V = \{i, r_2, h, v\}$ is not cyclic.
   (d) $i \to \varepsilon$, $r_1 \to \rho$, $r_2 \to \rho^2$, $r_3 \to \rho^3$, $v \to \sigma$, $d_1 \to \rho\sigma$, $h \to \rho^2\sigma$, $d_3 \to \rho^3\sigma$
13. (a) Expressed in terms of α and β,

$$A_4 = \{i, \alpha, \alpha^2, \beta, \beta^2, \alpha\beta, \alpha^2\beta, \alpha\beta^2, \alpha^2\beta^2, \beta^2\alpha, \beta\alpha^2, \beta\alpha^2\beta\}$$

(b) The unique subgroup of order 4 is $V = \{i, \alpha\beta, \alpha^2\beta^2, \beta\alpha^2\beta\}$.

(c) The subgroups of order 3 are $\langle\alpha\rangle$, $\langle\beta\rangle$, $\langle\alpha^2\beta\rangle$, $\langle\alpha\beta^2\rangle$.

14. (a) Let $a \in G$, $a \neq e$, and consider $H = \langle a\rangle$; H is a proper subgroup of G and $H \neq \langle e\rangle$. Hence $H = G$.

(b) $\langle[0]\rangle$, $\langle[2]\rangle$, $\langle[3]\rangle$, $\langle[4]\rangle$, $\langle[6]\rangle$

(c) If m is a prime, then choose $[a] \in \mathbb{Z}_m$ such that $[a] \neq [0]$ and $m \nmid a$. So $\gcd(a, m) = 1$, and there exist integers s and t such that $as + mt = 1$. Then $[m] \odot [s] \oplus [a] \odot [t] = [1]$. Since $[m] = [0]$, we have $[a] \odot [t] = [1]$. Hence $[1] \in \langle[a]\rangle$. So $\langle[a]\rangle = \mathbb{Z}_m$ and, by part **(a)**, $\mathbb{Z}_m$ has no proper subgroups. If m is not prime, say $m = bc$, where $b > 1$ and $c > 1$, then show that $\langle[b]\rangle$ is a proper subgroup of $\mathbb{Z}_m$.

Exercises 8.5

1. (a) This is not a ring since the distributive laws fail to hold: $a \cdot' (b +' c) = a(b + c - 1) = ab + ac - a$, while $a \cdot' b +' a \cdot' c = ab + ac - 1$.

(b) Under $+'$, we have the following: $a +' b = a + b - 1 = b +' a$; $a +' (b +' c) = a + (b + c - 1) - 1 = (a + b - 1) + c - 1 = (a +' b) +' c$; $a +' 1 = a + 1 - 1 = a$, so 1 is a zero element; $a +' (2 - a) = a + (2 - a) - 1 = 1$, so $2 - a$ is an additive inverse for a. Under $\cdot'$, $a \cdot' (b \cdot' c) = a + b \cdot' c - a(b \cdot' c) = a + (b + c - bc) - a(b + c - bc) = a + b + c - ab - bc - ac + abc$, and $(a \cdot' b) \cdot' c = (a + b - ab) + c - (a + b - ab)c = a + b + c - ab - bc - ac + abc$, so $\cdot'$ is associative. For the distributive laws, note that $a \cdot' (b +' c) = a + (b + c - 1) - a(b + c - 1) = 2a + b + c - ab - ac - 1$, and $a \cdot' b +' a \cdot' c = (a + b - ab) + (a + c - ac) - 1 = 2a + b + c - ab - ac - 1$. So $(S, +', \cdot')$ is a ring.

(c) The distributive laws fail; for example, $2 \cdot' (3 +' 4) = 2 + 3 \cdot 4 = 14$, while $2 \cdot' 3 +' 2 \cdot' 4 = (2 + 3)(2 + 4) = 30$.

2. (a) 1   **(b)** 0   **(c)** yes   **(d)** yes   **(e)** No; for example, there is no inverse for 3.

5. Case 1: Use induction on m to show that $(mn)(ab) = (ma)(nb)$ for all $m \geq 0$ and all $n \in \mathbb{Z}$. Let S be as usual. For $m = 0$, $(mn)(ab) = (0 \cdot n)(ab) = 0 \cdot (ab) = z = z \cdot (nb) = (0 \cdot a)(nb)$ for all $n \in \mathbb{Z}$, so $0 \in S$. Assume $k \in S$ for some arbitrary $k \geq 0$. The induction hypothesis is then

$$\text{IHOP:} \quad (kn)(ab) = (ka)(nb) \quad \text{for all } n \in \mathbb{Z}$$

To show that $k + 1 \in S$, we have, using IHOP and part 1 of Theorem 8.15, $((k + 1)n)(ab) = (kn + n)(ab) = (kn)(ab) + n(ab) = (ka)(nb) + a(nb) = (ka + a)(nb) = ((k + 1)a)(nb)$. Hence $k + 1 \in S$. So S is the set of all nonnegative integers.

Case 2: For $m < 0$, say $m = -r$, we have $(mn)(ab) = (-rn)(ab) = (r(-n))(ab) = (ra)((-n)b) = (ra)(-nb) = (-ra)(nb) = ((-r)a)(nb) = (ma)(nb)$.

6. For example, $\sqrt{2} \in \mathbb{Z}[\sqrt{2}]$ and $\sqrt{2}$ has no multiplicative inverse in $\mathbb{Z}[\sqrt{2}]$.

9. Under $*$, we use the relation $B - C = B \cap C'$ to show the following properties: $B * (C * D) = (B \cap (C * D)') \cup ((C * D) \cap B') = [B \cap ((C \cap D') \cup (D \cap C'))'] \cup [((C \cap D') \cup (D \cap C')) \cap B'] = [B \cap ((C' \cup D) \cap (C \cup D'))] \cup [(B' \cap C \cap D') \cup (B' \cap C' \cap D)] = (B \cap C' \cap D') \cup (B \cap C \cap D) \cup (B' \cap C \cap D') \cup (B' \cap C' \cap D)$ and $(B * C) * D = ((B * C) \cap D') \cup (D \cap (B * C)') = [((B \cap C') \cup (C \cap B')) \cap D'] \cup [D \cap ((B \cap C') \cup (C \cap B'))'] = (B \cap C' \cap D') \cup (B' \cap C \cap D') \cup [D \cap ((B' \cup C) \cap (B \cup C'))] = (B \cap C' \cap D') \cup (B' \cap C \cap D') \cup (B' \cap C' \cap D) \cup (B \cap C \cap D)$, so $B * (C * D) = (B * C) * D$ and $*$ is associative; the zero element is $\emptyset$; the additive inverse of B is B. Under $\cap$, associativity was verified in Chapter 2.

For the distributive laws, note that $B \cap (C * D) = B \cap ((C \cap D') \cup (D \cap C')) = (B \cap C \cap D') \cup (B \cap C' \cap D)$ and $(B \cap C) * (B \cap D) = ((B \cap C) \cap (B \cap D)') \cup ((B \cap D) \cap (B \cap C)') = ((B \cap C) \cap (B' \cup D')) \cup ((B \cap D) \cap (B' \cup C')) = (B \cap C \cap D') \cap (B \cap C' \cap D)$, so that $B \cap (C * D) = (B \cap C) * (B \cap D)$.

10. No; $f \circ (g + h) \neq f \circ g + f \circ h$, in general.

12. **(a)** If $\gcd(a, m) = d > 1$, then $m = m_1 d$ and $a = a_1 d$ for some $m_1, a_1 \in \mathbb{Z}^{\#}$. Then $[m_1] \neq [0]$, so $[a] \odot [m_1] = [a_1 d m_1] = [a_1 m] = [0]$.

   (b) By part **(a)**, if $\gcd(a, m) \neq 1$, then $[a]$ has no multiplicative inverse. So assume $\gcd(a, m) = 1$. Then there exist integers s and t such that $as + mt = 1$. Thus $[as + mt] = [1]$, or $[as] \oplus [mt] = [1]$. But $[mt] = [0]$ and so $[as] = [a] \odot [s] = [1]$, so $[a]$ is invertible.

   (c) Use part **(a)** to get zero divisors.

13. Consider $(E, *)$. For commutativity, we have $a * b = ab/2 = ba/2 = b * a$. For associativity, note that $a * (b * c) = a * (bc/2) = (a(bc/2))/2 = (abc)/4$ and $(a * b) * c = (ab/2) * c = ((ab/2)c)/2 = (abc)/4 = a * (b * c)$. The identity is 2, since $a * 2 = (a \cdot 2)/2 = a$. Also $*$ distributes over $+$, as shown by $a * (b + c) = a(b + c)/2 = (ab + ac)/2 = ab/2 + ac/2 = a * b + a * c$.

Exercises 8.6

1. **(a)** For complements, note that $a' = b$, $b' = a$, $c' = d$, $d' = c$, $0' = 1$, and $1' = 0$. This lattice is not distributive; for example, $a \wedge (b \vee c) = a$ and $(a \wedge b) \vee (a \wedge c) = c$. Therefore it is not a Boolean algebra.

   (b) This lattice is a Boolean algebra; for complements, we have $1' = 0$, $a' = f$, $b' = e$, $c' = d$, $d' = c$, $e' = b$, $f' = a$.

   (c) This is not a Boolean algebra; for instance, b has no complement.

2. (a)   **(b)**

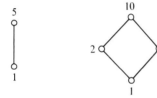

   (c) See Exercise 4(c), with $p = 2$, $q = 3$, and $r = 5$.

4. (c)

$\vee$	n	pq	pr	qr	p	q	r	1
n	n	n	n	n	n	n	n	n
pq	n	pq	n	n	pq	pq	n	pq
pr	n	n	pr	n	pr	n	pr	pr
qr	n	n	n	qr	n	qr	qr	qr
p	n	pq	pr	n	p	pq	pr	p
q	n	pq	n	qr	pq	q	qr	q
r	n	n	pr	qr	pr	qr	r	r
1	n	pq	pr	qr	p	q	r	1

$\wedge$	n	pq	pr	qr	p	q	r	1
n	n	pq	pr	qr	p	q	r	1
pq	pq	pq	p	q	p	q	1	1
pr	pr	p	pr	r	p	1	r	1
qr	qr	q	r	qr	1	q	r	1
p	p	p	p	1	p	1	1	1
q	q	q	1	q	1	q	1	1
r	r	1	r	r	1	1	r	1
1	1	1	1	1	1	1	1	1

5. (a) First, if n is a product of distinct primes, then for each positive divisor a of n, $\gcd(a, n/a) = 1$ and $\mathrm{lcm}(a, n/a) = n$. Therefore, $a' = n/a$. Suppose now that $p^2 \mid n$ for some prime p. Then for any positive divisor b of n, if $p \mid b$, then $\gcd(p, b) = p$, while if $p \nmid b$, then $\mathrm{lcm}(p, b) < n$. This shows that p has no complement.

(b) We show that the lattice $(L, \preccurlyeq)$ of positive divisors of n is distributive. Let $a, b, c \in L$, and consider the distributive law $a \wedge (b \vee c) = (a \wedge b) \vee (a \wedge c)$. To prove that this law holds, we must verify that $\gcd(a, \mathrm{lcm}(b, c)) = \mathrm{lcm}(\gcd(a, b), \gcd(a, c))$. Let $p_1, p_2, \ldots, p_m$ be the distinct prime factors of n; then we may write $a = p_1^{\alpha_1} \cdots p_m^{\alpha_m}$, $b = p_1^{\beta_1} \cdots p_m^{\beta_m}$, and $c = p_1^{\gamma_1} \cdots p_m^{\gamma_m}$. Hence $\gcd(a, \mathrm{lcm}(b, c)) = p_1^{\delta_1} \cdots p_m^{\delta_m}$, where $\delta_i = \min(\alpha_i, \max(\beta_i, \gamma_i))$, and $\mathrm{lcm}(\gcd(a, b), \gcd(a, c)) = p_1^{\varepsilon_1} \cdots p_m^{\varepsilon_m}$, where $\varepsilon_i = \max(\min(\alpha_i, \beta_i), \min(\alpha_i, \gamma_i))$. To complete the proof, one must show that $\delta_i = \varepsilon_i$, $1 \leq i \leq m$. The other distributive law is verified in a similar manner.

6. 2^m

8. (a) The commutative and associative laws hold; notice that $(x_i + y_i - x_i y_i) + z_i - (x_i + y_i - x_i y_i)z_i = x_i + (y_i + z_i - y_i z_i) - x_i(y_i + z_i - y_i z_i)$ for $i = 1, 2, \ldots, n$. The distributive laws hold; notice that $x_i(y_i + z_i - y_i z_i) = x_i y_i + x_i z_i - (x_i y_i)(x_i z_i)$ for each $i = 1, 2, \ldots, n$. The zero and unit elements are $(0, 0, \ldots, 0)$ and $(1, 1, \ldots, 1)$, respectively. The complement of $(x_1, x_2, \ldots, x_n)$ is $(y_1, y_2, \ldots, y_n)$, where $y_i = 1$ if $x_i = 0$ and $y_i = 0$ if $x_i = 1$.

(b) $\preccurlyeq$ is a partial ordering. For instance, consider transitivity; if
$(x_1, x_2, \ldots, x_n) \preccurlyeq (y_1, y_2, \ldots, y_n)$ and $(y_1, y_2, \ldots, y_n) \preccurlyeq (z_1, z_2, \ldots, z_n)$, then
$(x_1, x_2, \ldots, x_n) \vee (z_1, z_2, \ldots, z_n) = (x_1, x_2, \ldots, x_n) \vee ((y_1, y_2, \ldots, y_n) \vee (z_1, z_2, \ldots, z_n)) =$
$((x_1, x_2, \ldots, x_n) \vee (y_1, y_2, \ldots, y_n)) \vee (z_1, z_2, \ldots, z_n) = (y_1, y_2, \ldots, y_n) \vee (z_1, z_2, \ldots, z_n) =$
$(z_1, z_2, \ldots, z_n)$. So $(x_1, x_2, \ldots, x_n) \preccurlyeq (z_1, z_2, \ldots, z_n)$.

9. (a) p **(c)** r **(e)** r **(g)** qr

15. $a \wedge (b \vee c) = a \neq b = (a \wedge b) \vee (a \wedge c)$

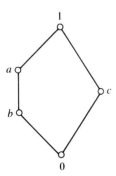

17. Suppose a_1 and a_2 are complements for a. Then $a_1 \wedge a = 0$, $a_1 \vee a = 1$, $a_2 \wedge a = 0$, and $a_2 \vee a = 1$. Then $a_1 = a_1 \wedge 1 = a_1 \wedge (a \vee a_2) = (a_1 \wedge a) \vee (a_1 \wedge a_2) = 0 \vee (a_1 \wedge a_2) = a_1 \wedge a_2$ and $a_2 = a_2 \wedge 1 = a_2 \wedge (a \vee a_1) = (a_2 \wedge a) \vee (a_2 \wedge a_1) = 0 \vee (a_2 \wedge a_1) = a_1 \wedge a_2$. So $a_1 = a_2$.

19. We verify the partial-order properties. For the reflexive property, $a = a \vee 0 = a \vee (a \wedge a') = (a \vee a) \wedge (a \vee a') = (a \vee a) \wedge 1 = a \vee a$. So $a \vee a = a$ and $a \preccurlyeq a$. For antisymmetry, if $a \vee b = b$ and $b \vee a = a$, then $a = b \vee a = a \vee b = b$. So $a \preccurlyeq b$ and $b \preccurlyeq a$ implies $a = b$. For transitivity, if $a \vee b = b$ and $b \vee c = c$, then $a \vee c = a \vee (b \vee c) = (a \vee b) \vee c = b \vee c = c$. Hence $a \leqslant b$ and $b \leqslant c$ imply $a \leqslant c$.

Exercises 8.7

1. (a) $x_1' \wedge x_2 \wedge x_3 \wedge x_4'$ **(b)** $x_1 \wedge x_2' \wedge x_3' \wedge x_4$

2. (b)

x_1	x_2	x_3	$f(x_1\, x_2, x_3)$
0	0	0	0
0	0	1	1
0	1	0	0
0	1	1	1
1	0	0	1
1	0	1	1
1	1	0	0
1	1	1	1

$f(x_1, x_2, x_3) = (x_1' \wedge x_2' \wedge x_3) \vee (x_1' \wedge x_2 \wedge x_3) \vee (x_1 \wedge x_2' \wedge x_3') \vee (x_1 \wedge x_2' \wedge x_3) \vee (x_1 \wedge x_2 \wedge x_3)$

4. (a) $(x_1 \wedge x_2) \vee (x_1' \wedge x_2 \wedge x_3') = ((x_1 \wedge x_2) \wedge (x_3 \vee x_3')) \vee (x_1' \wedge x_2 \wedge x_3') = (x_1 \wedge x_2 \wedge x_3) \vee (x_1 \wedge x_2 \wedge x_3') \vee (x_1' \wedge x_2 \wedge x_3')$

(c) $(x_1 \vee x_3) \wedge x_2 = (x_1 \wedge x_2) \vee (x_2 \wedge x_3) = [(x_1 \wedge x_2) \wedge (x_3 \vee x_3')] \vee [(x_2 \wedge x_3) \wedge (x_1 \vee x_1')] = (x_1 \wedge x_2 \wedge x_3) \vee (x_1 \wedge x_2 \wedge x_3') \vee (x_1 \wedge x_2 \wedge x_3) \vee (x_1' \wedge x_2 \wedge x_3) = (x_1 \wedge x_2 \wedge x_3) \vee (x_1 \wedge x_2 \wedge x_3') \vee (x_1' \wedge x_2 \wedge x_3)$

5. For g, we have $g(x_1, x_2, x_2) = (x_1' \wedge x_2 \wedge x_3) \vee (x_1 \wedge x_2' \wedge x_3) \vee (x_1 \wedge x_2 \wedge x_3') \vee (x_1 \wedge x_2 \wedge x_3) = [(x_1' \wedge x_2 \wedge x_3) \vee (x_1 \wedge x_2 \wedge x_3)] \vee (x_1 \wedge x_2' \wedge x_3) \vee (x_1 \wedge x_2 \wedge x_3') = (x_2 \wedge x_3) \vee (x_1 \wedge x_2' \wedge x_3) \vee (x_1 \wedge x_2 \wedge x_3') = [(x_2 \vee (x_1 \wedge x_2')) \wedge x_3] \vee (x_1 \wedge x_2 \wedge x_3') = [((x_2 \vee x_1) \wedge (x_2 \vee x_2')) \wedge x_3] \vee (x_1 \wedge x_2 \wedge x_3') = ((x_2 \vee x_1) \wedge x_3) \vee (x_1 \wedge x_2 \wedge x_3') = (x_2 \wedge x_3) \vee (x_1 \wedge x_3) \vee (x_1 \wedge x_2 \wedge x_3') = (x_2 \wedge x_3) \vee [x_1 \wedge (x_3 \vee (x_2 \wedge x_3'))] = (x_2 \wedge x_3) \vee [x_1 \wedge ((x_2 \vee x_3) \wedge (x_3 \vee x_3'))] = (x_2 \wedge x_3) \vee [(x_1 \wedge x_2) \vee (x_1 \wedge x_3)] = (x_1 \wedge x_2) \vee (x_1 \wedge x_3) \vee (x_2 \wedge x_3)$.

8. Since $p(x_1, x_2, \ldots, x_n)$ and $q(x_1, x_2, \ldots, x_n)$ are different, there must be some index i, $1 \leq i \leq n$, such that x_i appears in one of the minterms and x_i' in the other. Hence, using commutativity and associativity, one has $p(x_1, x_2, \ldots, x_n) \wedge q(x_1, x_2, \ldots, x_n) = (x_i \wedge x_i') \wedge p(x_1, x_2, \ldots, x_n) \wedge q(x_1, x_2, \ldots, x_n) = 0$.

Exercises 8.8

1. (a)   **(c)**

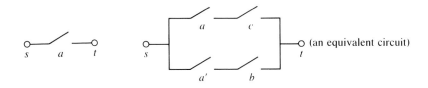

(an equivalent circuit)

(e)

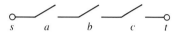

4. (a)

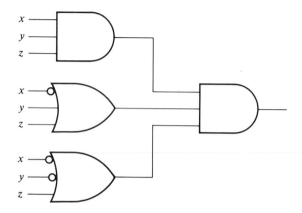

5. (b) $(x \lor y \lor z') \lor (x' \land y' \land z) \lor (x \land y' \land z')$ **(c)** $[((x \land y) \land z)' \lor (y' \lor z)]'$

6. (a) the lg-circuit corresponding to $(x \land y') \lor (x' \land y)$
   (c) the lg-circuit corresponding to $(x' \land y \land z) \lor (x \land y' \land z) \lor (x \land y \land z')$

9. If B, G, and H represent the switches on the barn, garage, and house, respectively, then the lg-circuit that represents the situation corresponds to the expression $(B \land G \land H) \lor (B' \land G' \land H) \lor (B' \land G \land H') \lor (B \land G' \land H')$.

Exercises 8.9

1. (a) $f(e_G) = f(e_G * e_G) = f(e_G) \# f(e_G)$, so by Exercise 9 of Exercises 8.2, $f(e_G) = e_H$.
   (b) $e_H = f(e_G) = f(a * a^{-1}) = f(a) \# f(a^{-1})$, so $f(a^{-1}) = (f(a))^{-1}$.
   (c) Use induction on n, with S as usual. Note that (i) $f(a^1) = f(a) = (f(a))^1$, so $1 \in S$; (ii) IHOP: $f(a^k) = (f(a))^k$ for some arbitrary $k \geq 1$; (iii) $f(a^{k+1}) = f(a^k * a) = f(a^k) \# f(a) = (f(a))^k \# f(a) = (f(a))^{k+1}$, so $k + 1 \in S$.
   (d) $f(a^{-n}) = f((a^n)^{-1}) = (f(a^n))^{-1} = ((f(a))^n)^{-1} = (f(a))^{-n}$.

2. f is not well-defined; $[0] = [6]$ in $\mathbb{Z}_6$, while $f([0]) = [0]$ and $f([6]) = [2]$.

3. (a) If $f([1]) = h$, then $f([t]) = f(t[1]) = (f([1]))^t = h^t$.

4. Notice that $f([t]) = f(t[1]) = (f([1]))^t = [2]^t$, so $f([s] \oplus [t]) = f([s + t]) = [2]^{s+t} = [2]^s \odot [2]^t = f([s]) \odot f([t])$ and im $f = \{[2], [2]^2, [2]^3, [2]^4\} = \{[2], [4], [3], [1]\}$. Thus f must be one-to-one. So f is an isomorphism.

7. (a) $f([k]) = \delta^k$ for all $k \in \mathbb{Z}$.
   (b) No; $g(\gamma\delta\gamma) = g(\delta^2) = [4]$ and $g(\gamma) \oplus g(\delta) \oplus g(\gamma) = [0] \oplus [2] \oplus [0] = [2]$.

(c) Yes; $h(\gamma) = h(\delta\gamma) = h(\delta^2\gamma) = [3]$. An arbitrary element of S_3 has the form $\delta^k\gamma^r$, where $0 \le k \le 3$ and $r = 0$ or 1. We must check that $g(xy) = g(x) \oplus g(y)$ for all $x, y \in S_3$. Say, $x = \delta^k\gamma^r$ and $y = \delta^s\gamma^t$. We consider four cases, according to whether $r = 0$ or 1 and $t = 0$ or 1.

   Case 1: $r = 0$ and $t = 0$. Then $g(xy) = g((\delta^k\gamma^0)(\delta^s\gamma^0)) = g(\delta^k\delta^s) = g(\delta^{k+s}) = [0] = g(\delta^k) \oplus g(\delta^s)$.

   Case 2: $r = 0$ and $t = 1$. Then $g(xy) = g((\delta^k\gamma^0)(\delta^r\gamma^1)) = g((\delta^k\delta^r)\gamma) = g(\delta^{k+r}\gamma) = [3] = g(\delta^k) \oplus g(\delta^r\gamma)$.

   The remaining cases are handled similarly.

11. Say, $|x| = m$. Then $(f(x))^m = f(x^m) = f(e_G) = e_H$; hence $|f(x)|$ divides m.

12. Case 1: $|x| = m$. Then we have $|f(x)|$ divides m. If $|f(x)| = n$, then $e_H = (f(x))^n = f(x^n)$, so $f(x^n) = f(e_G)$. Since f is one-to-one, we get $x^n = e_G$. So $m \mid n$ and hence $m = n$.
   Case 2: $|x| = \infty$. If $e_H = (f(x))^n$, then $f(x^n) = f(e_G)$, so that $x^n = e_G$. Since $|x| = \infty$, we have $n = 0$. Thus $|f(x)| = \infty$.

17. (a) im $f = \{f(g) \mid g \in G\}$. Note that (i) im $f \ne \emptyset$, since $e_H = f(e_G) \in$ im f; (ii) for $f(a), f(b) \in$ im f, $f(a) \,\#\, f(b) = f(a * b) \in$ im f; (iii) for $f(a) \in$ im f, $(f(a))^{-1} = f(a^{-1}) \in$ im f. So im f is a subgroup of H.

   (b) Notice that $f(a) \,\#\, f(b) = f(a * b) = f(b * a) = f(b) \,\#\, f(a)$.

   (c) No; see Exercise 7.

18. Applying Exercise 11 directly, we see that $|\,[a]\,|$ divides m.

19. (a) Use the maps g_1 and g_2 determined by $g_1([1]) = [0]$ and $g_2([1]) = [2]$.

   (b) Use the maps g_1, g_2, g_3 determined by $g_1([1]) = [0]$, $g_2([1]) = [3]$, $g_3([1]) = [6]$.

   (c) Use any mapping of the form $g: \mathbb{Z} \to E$, where $g(1) = 2t$ for some $t \in \mathbb{Z}$.

   (d) Use any mapping $g: \mathbb{Z} \to \mathbb{Z}$, where $g(1) = t$ for some $t \in \mathbb{Z}$.

23. (a) First $e_G \in$ ker α, so ker $\alpha \ne \emptyset$. Next, if $a, b \in$ ker α, then $\alpha(a * b) = \alpha(a) \,\#\, \alpha(b) = e_H \,\#\, e_H = e_H$, so $a * b \in$ ker α. Finally, if $a \in$ ker α, then $e_H = \alpha(e_G) = \alpha(a * a^{-1}) = \alpha(a) \,\#\, \alpha(a^{-1}) = e_H \,\#\, \alpha(a^{-1}) = \alpha(a^{-1})$; thus $a^{-1} \in$ ker α. So ker α is a subgroup of G.

   (b) For $x \in G$ and $a \in$ ker α, we have $\alpha(x^{-1} * a * x) = \alpha(x^{-1}) \,\#\, \alpha(a) \,\#\, \alpha(x) = \alpha(x^{-1}) \,\#\, e_H \,\#\, \alpha(x)$
   $= \alpha(x^{-1}) \,\#\, \alpha(x) = \alpha(x^{-1} * x) = \alpha(e_G) = e_H$. So $x^{-1} * a * x \in$ ker α.

Chapter 8 Problems

3. (a) 56   (c) 11   (e) $3\,8*27* \,+\, -$

5. (a) m^{m^2} (b) $m^{m(m+1)/2}$ (c) m^{m^2-2m+2}

7. Let e denote the identity of $(G, *)$. If $(H, *)$ is a subgroup of $(G, *)$, then $e \in H$. Therefore (i) $H \ne \emptyset$. Also $a, b \in H \to a, b^{-1} \in H \to a * b^{-1} \in H$. Thus (ii) $a, b \in H \to a * b^{-1} \in H$. Conversely, assume conditions (i) and (ii) hold. Since $H \ne \emptyset$, there is some $x \in G$ such that $x \in H$. Hence (iii) $x \in H \to x * x^{-1} = e \in H$, (iv) $e \in H \wedge x \in H \to e * x^{-1} = x^{-1} \in H$, and (v) $x, y \in H \to x, y^{-1} \in H \to x * (y^{-1})^{-1} = x * y \in H$. Also $*$ is associative on H, since $*$ is associative on G and $H \subseteq G$. Therefore $(H, *)$ is a subgroup of $(G, *)$.

8. Hint: Let $e: \mathbb{R} \to \mathbb{R}$ be defined by $e(x) = 1$. Show that e is the identity of $(\mathcal{F}(\mathbb{R})^{\#}, \cdot)$. Also, given $f \in \mathcal{F}(\mathbb{R})^{\#}$, define $g: \mathbb{R} \to \mathbb{R}$ by $g(x) = 1/f(x)$. Show that $g \in \mathcal{F}(\mathbb{R})^{\#}$ and that g is the inverse of f.

9. Let $a \in G$ be fixed. Then the equations $a * x = a$ and $y * a = a$ have solutions e' and e'', respectively. Also there exist $a', a'' \in G$ such that $a * a' = e'$ and $a'' * a = e''$. So $e' = a * a' = (e'' * a) * a' = e'' * (a * a') = e'' * e' = (a'' * a) * e' = a'' * (a * e') = a'' * a = e''$. Let $e = e' = e''$. Notice that e is idempotent; that is, $e = e * e$. Next let d be an arbitrary element of G, and let $b, c \in G$ be the solutions of $e * x = d$ and $y * e = d$, respectively. Then $e * d = e * (e * b) = (e * e) * b = e * b = d$, and similarly $d * e = (c * e) * e = c * (e * e) = c * e = d$. This shows that $(G, *)$ has identity e. Recall that for any $a \in G$ there exist $a', a'' \in G$ such that $a * a' = e = a'' * a$. Thus $a' = e * a' = (a'' * a) * a' = a'' * (a * a') = a'' * e = a''$. Therefore $a^{-1} = a' = a''$. This shows that $(G, *)$ is a group.

10. (b) **(c)**

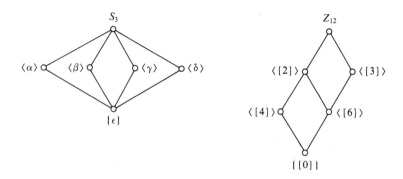

(d)

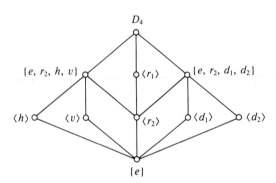

20. (a) $ns = n(e \cdot s) = (ne) \cdot s = z \cdot s = z$

(b) Suppose, to the contrary, that n is composite, say $n = ab$, $1 < a \le b < n$. Then $z = ne = (ab)e = (ab)(e \cdot e) = ae \cdot be$. This says that ae is a zero divisor, a contradiction.

22. (a) $a = a \vee (c \wedge c') = (a \vee c) \wedge (a \vee c') = (b \vee c) \wedge (b \vee c') = b \vee (c \wedge c') = b$

23. (a) Let $a \in A$ and $b \in B$ with $\theta(a) = b$. Then $\theta(1) \wedge b = \theta(1) \wedge \theta(a) = \theta(1 \wedge a) = \theta(a) = b$, and so $\theta(1)$ is the unity for B. Also $\theta(0) \vee b = \theta(0) \vee \theta(a) = \theta(0 \vee a) = \theta(a) = b$, so that $\theta(0)$ is the zero for B.

(b) Moreover $\theta(a) \vee \theta(a') = \theta(a \vee a') = \theta(1)$, and $\theta(a) \wedge \theta(a') = \theta(a \wedge a') = \theta(0)$, which shows that $\theta(a') = (\theta(a))'$.

26. (a)

x	y	$x \oplus y$
0	0	0
0	1	1
1	0	1
1	1	0

(b)

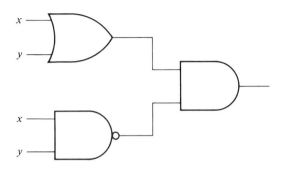

(c) $x \oplus (y \oplus z) = (x \vee (y \oplus z)) \wedge (x \wedge (y \oplus z))'$
$= [x \vee ((y \vee z) \wedge (y \wedge z)')] \wedge [x \wedge ((y \vee z) \wedge (y \wedge z)')]'$
$= [x \vee ((y \vee z) \wedge (y' \vee z'))] \wedge [x' \vee (y' \wedge z') \vee (y \wedge z)]$
$= (x \vee (y' \wedge z) \vee (y \wedge z)) \wedge (x' \vee (y' \wedge z') \vee (y \wedge z))$
$= (x \wedge y' \wedge z') \vee (x \wedge y \wedge z) \vee (x' \wedge y \wedge z') \vee (x' \wedge y' \wedge z)$
Use a similar technique to show that $(x \oplus y) \oplus z$ yields the same expression.

(d) $x \oplus y \oplus (x \wedge y) = [(x \vee y) \wedge (x \wedge y)' \vee (x \wedge y)] \wedge$
$[((x \vee y) \wedge (x \wedge y)' \wedge (x \wedge y)]' = [((x \vee y) \wedge (x' \vee y') \vee (x \wedge y)] \wedge$
$[(x' \wedge y') \vee (x \wedge y) \vee (x' \vee y')] = [(x \wedge y') \vee (x' \wedge y) \vee (x \wedge y)] \wedge$
$[(x' \wedge y') \vee (x \wedge y) \vee (x' \vee y')] = (x \wedge y') \vee (x' \wedge y) \vee (x \wedge y) =$
$x \vee (x' \wedge y) = (x \vee x') \wedge (x \vee y) = x \vee y$

28. Let $x, y \in G$ and let e_G denote the identity of G.

(a) First $f_a(x) = f_a(y) \rightarrow a * x * a^{-1} = a * y * a^{-1} \rightarrow x = y$, so f_a is one-to-one. Next $f_a(a^{-1} * x * a)$
$= x$, so f_a is onto. Last $f_a(x * y) = a * (x * y) * a^{-1} = (a * x * a^{-1}) * (a * y * a^{-1}) = f_a(x) * f_a(y)$.
Therefore f_a is an automorphism.

(b) $g_a(x) = g_a(y) \rightarrow a * x = a * y \rightarrow a^{-1} * a * x = a^{-1} * a * y \rightarrow e_G * x = e_G * y \rightarrow x = y$, so g_a is
one-to-one. Also $g_a(a^{-1} * x) = x$, which shows that g_a is onto. If $a \neq e_G$, then $g_a(e_G) = a * e_G = a$;
thus g_a is not an automorphism.

29. Suppose, to the contrary, that $Q = \langle r \rangle$, where $r = m/n$ and $\gcd(m, n) = 1$. Let p be any prime such
that $p \nmid n$. Since $1/p \in Q = \langle r \rangle$, it must be that $1/p = (m/n) \cdot t$ for some integer t. But then
$mtp = n$, so that $p \mid n$. This yields a contradiction. Therefore $(Q, +)$ is not cyclic.

32. Let $(L, \preccurlyeq)$ be a totally ordered set. Then it is easy to see that $(L, \preccurlyeq)$ is a lattice. We must show that the
distributive laws are satisfied. Let $a, b, c \in L$. There are six cases to consider, depending on the order of
$a, b,$ and c. For example, if $a \preccurlyeq b \preccurlyeq c$, then $a \wedge (b \vee c) = a \wedge c = a = a \vee a = (a \wedge b) \vee (a \wedge c)$
and $a \vee (b \wedge c) = a \vee b = b = b \wedge c = (a \vee b) \wedge (a \vee c)$. The other five cases are similar.

Index